Nonlinear Phenomena at Phase Transitions and Instabilities

NATO ADVANCED STUDY INSTITUTES SERIES

A series of edited volumes comprising multifaceted studies of contemporary scientific issues by some of the best scientific minds in the world, assembled in cooperation with NATO Scientific Affairs Division.

Series B: Physics

Recent Volumes in this Series

Volume 71 – Atomic and Molecular Collision Theory
edited by Franco A. Gianturco

Volume 72 – Phase Transitions: *Cargèse 1980*
edited by Maurice Lévy, Jean-Claude Le Guillou, and Jean Zinn-Justin

Volume 73 – Scattering Techniques Applied to Supramolecular and Nonequilibrium Systems
edited by Sow-Hsin Chen, Benjamin Chu, and Ralph Nossal

Volume 74 – Rigorous Atomic and Molecular Physics
edited by G. Velo and A. S. Wightman

Volume 75 – Nonlinear Phenomena in Physics and Biology
edited by Richard H. Enns, Billy L. Jones, Robert M. Miura, and Sadanand S. Rangnekar

Volume 76 – Metal Hydrides
edited by Gust Bambakidis

Volume 77 – Nonlinear Phenomena at Phase Transitions and Instabilities
edited by T. Riste

Volume 78 – Excitations in Disordered Systems
edited by M. F. Thorpe

Volume 79 – Artificial Particle Beams in Space Plasma Studies
edited by Bjørn Grandal

This series is published by an international board of publishers in conjunction with NATO Scientific Affairs Division

A Life Sciences B Physics	Plenum Publishing Corporation London and New York
C Mathematical and Physical Sciences	D. Reidel Publishing Company Dordrecht, The Netherlands and Hingham, Massachusetts, USA
D Behavioral and Social Sciences E Applied Sciences	Martinus Nijhoff Publishers The Hague, The Netherlands

Nonlinear Phenomena at Phase Transitions and Instabilities

Edited by
T. Riste
Institute for Energy Technology
Kjeller, Norway

SPRINGER SCIENCE+BUSINESS MEDIA, LLC

Library of Congress Cataloging in Publication Data

NATO Advanced Study Institute (1981 : Geilo, Norway)
Nonlinear phenomena at phase transitions and instabilities.

(NATO advanced study institutes series. Series B, Physics ; v. 77)
"Proceedings of a NATO Advanced Study Institute held at Geilo Norway, March 29-April 9, 1981"—Verso of t.p.
Includes bibliographies and index.
1. Nonlinear theories—Congresses. 2. Phase transitions (Statistical physics)—Congresses. 3. Hydrodynamics—Congresses. 4. Solitons—Congresses. I. Riste, Tormod, 1925- . II. North Atlantic Treaty Organization. Division of Scientific Affairs. III. Series. IV. Title: Instabilities.
QC20.7.N6N28 1981 530.4 81-17780
AACR2
ISBN 978-1-4684-4129-1 ISBN 978-1-4684-4127-7 (eBook)
DOI 10.1007/978-1-4684-4127-7

Proceedings of a NATO Advanced Study Institute held at Geilo, Norway, March 29–April 9, 1981

Originally published by Plenum Press, New York 1982

PREFACE

This NATO Advanced Study Institute, held in Geilo between March 29th and April 9th 1981, was the sixth in a series devoted to the subject of phase transitions and instabilities. The present institute was intended to provide a forum for discussion of the importance of nonlinear phenomena associated with instabilities in systems as seemingly disparate as ferroelectrics and rotating buckets of oil.

Ten years ago, at the first Geilo school, the report of a central peak in the fluctuation spectrum of $SrTiO_3$ close to its 106 K structural phase transition demonstrated that the simple soft-mode theory of such transitions was incomplete. The missing ingredient was the essential nonlinearity of the system. Participants at this year's Geilo school heard assessments of a decade of experimental and theoretical effort which has been expended to elucidate the nature of this nonlinearity. The importance of ordered clusters and the walls which bound them was stressed in this context.

A specific type of wall, the soliton, was discussed by a number of speakers. New experimental results which purport to demonstrate the existence of solitons in a one-dimensional ferromagnet were presented. A detailed discussion was given of the role of solitons in transport phenomena in driven multistable systems, typified by a sine-Gordon chain.

In the lectures much attention was paid to hydrodynamic instabilities for which nonlinearities and boundary constraints act to select and stabilize certain structures and patterns. Several variants of the Rayleigh-Bénard problem were described, and it was shown that the Lorenz model could be applied not only to this problem but also to the single-mode laser and to the proliferation of insect colonies! The model was used to demonstrate the possibility of achieving chaos (turbulence) by a sequence of period-doubling transitions governed by a universal constant first introduced by Feigenbaum. Recent experimental evidence for the occurrence of such period-doubling bifurcations in Rayleigh-Bénard cells containing either helium or water was presented. The fact that the bifurcation theory is relevant both to hydrodynamic instabilities and to transitions between modulated crystal structures opens intriguing possibilities for future studies in both fields.

While the significance of spatial patterns is only beginning to become clear in studies of phase transitions, it is evident in hydrodynamically unstable systems and even more striking in the growth of dendrites (on snow flakes for example). This subject and the related field of eutectic solidification were presented in detail. Participants of the school were much stimulated by the first report of a periodic fluctuation in the facet of a solidification front.

The final lectures dealt with recent progress in the study of two-dimensional melting. In this context experiments which confirm or deny the existence of continuous melting or of the hexatic liquid phase are eagerly awaited (for the next Geilo meeting?). Nevertheless, participants were able to watch a film of one example of two-dimensional "melting": the breakdown of two-dimensional convective forms in an elliptic shear instability sequence in a thin liquid crystal layer.

The material contained in this book consists of longer, tutorial papers and of shorter research reports. The former were presented as invited talks, as indicated in the list of contents. The short papers were all presented orally during the study institute, and concern problem areas introduced in the invited lectures.

The committee joins the other participants in expressing their sincere thanks to Eigil Andersen and Gerd Jarrett and other members of the staff of the Institute for Energy Technology, Kjeller, Norway, for their careful planning and assistance, upon which largely rested the success of this NATO Advanced Study Institute.

May 1981

H.Z. Cummins
E.H. Hauge
J.G. Feder
R. Pynn
T. Riste
H. Thomas

CONTENTS

ANHARMONIC PROPERTIES NEAR STRUCTURAL PHASE TRANSITIONS

K. Alex Müller

IBM Zurich Research Laboratory
8803 Rüschlikon-ZH Switzerland

INTENTION AND SUMMARY:

In these three lectures an attempt is made to describe, in a certain context, experimental evidence for anharmonic behavior near structural phase transitions (SPT's): Theoretically, it is expected that systems with lower effective lattice dimensionality and shorter ranges of forces are, near T_c, more dominated by correlated fluctuations (in space and time). Thus, intrinsic anharmonic behavior due to such *correlated* fluctuations is expected to become more pronounced, the lower the effective lattice dimensionality and the shorter the range of forces. The more important experiments on SPT's carried out in the past decade are discussed in the above context. The manuscript is subdivided into three sections and progresses on the average, from higher to lower dimensionality and from earlier to more recent results.

The first section tries to summarize, in an introductory way, the most important findings known up to three years ago: The first experiments in $SrTiO_3$ pointing towards collective anharmonic effects are recalled. Then the role of impurities is discussed. The important notion of an intrinsic displacive-to-precursor order crossover for short-range systems is presented.

In the second section, experiments carried out in $SrTiO_3$-type and ferroelectric systems over the past three years are reviewed. Evidence for the displacive-to-order disorder crossover by EPR in $SrTiO_3$ is presented first. Then it is further substantiated in infrared absorption, scattering, magnetic resonance and ultrasound experiments in other SPT's. The evidence is more qualitative for higher dimensionality, and more quantitative for anisotropic systems, equivalent to lower effective lattice dimensionality. In the latter, slow critical dynamics above T_c different from normal soft-mode behavior, the so-called dynamic central peak, has been observed. Two such time-scale responses are expected for displacive (soft-mode) to precursor order (central-peak) crossover.

The third section, presented verbally only due to space restrictions, is devoted to incommensurate phases. The incommensurate amplitude modulation can occur in two or one dimensions. In the latter case, the system is quasi-one-dimensional for the particular excitation, and shows the most pronounced anharmonic structural behavior: the whole phase is critical, a low-frequency phase-on mode is present, and upon cooling through the incommensurate phase, the amplitude modulation crosses over to a strongly anharmonic soliton regime. Evidence from inelastic neturon-scattering, ultrasound absorption and nuclear magnetic resonance is summarized.

I. FROM EARLY EXPERIMENTS TO THE NEW VIEW

These Lecture Notes have been kept as short as possible to cope with the necessary space limitations. Those who feel the first Lecture is not a sufficient introduction into the field are referred to the previous volumes of the regular NATO Spring Schools.[1] From these, especially the first volume published ten years ago has become a classic in the field of SPT's.[2] More complete texts on SPT's have appeared recently in the form of two books, one by Bruce and Cowley with Taylor and Francis which reviews the whole field,[3] the other, volume 1 of a series with Springer edited by Müller and Thomas.[4] It summarizes optical studies, inelastic neutron scattering and ultrasound experiments. Part of the present Lecture is taken from an introductory note in the Springer book[4] and quite a large part from a review on "Intrinsic and Extrinsic Central-Peak Properties" written by the author two years ago.[5] The present Lecture can actually be regarded as a summary of the most important parts of that rather complete review.

a) Classical and Nonclassical Behavior

The change of structure in a solid can occur by reconstruction or by slightly distorting the lattice without disrupting the linkage of the net. Under SPT, solid-state physicists understand only the latter type of transformation. Depending on the microscopic interaction, one distinguishes five main categories: ferroelectric, ferroelastic, Jahn-Teller, charge density wave SPT's, and those induced by short-range forces. The ferroelectric transitions are those to which most of the earliest investigations were devoted. It was then recognized that a distortion can occur as a result of small *displacements* in the lattice position of single atoms or molecular units on the one hand, or the *ordering* of atoms or molecules among various equivalent positions on the other.

A distinction between displacive and order-disorder SPT is possible on the basis of atomic single-cell potentials: Consider an anharmonic local potential as a function of one spatial coordinate $Q: V(Q) = \bar{A}\, Q^2 + B\, Q^4$ with constants $\bar{A} < 0$ and $B > 0$. This double well potential has two minima and a maximum, their difference ΔE being $\Delta E = \bar{A}^2/4B$. If the depth of the two energy wells is $\Delta E \gg kT_c$, i.e., the parameter $g = \Delta E/kT \gg 1$, the transition occurs due to ordering between fixed displacements $Q_{0\pm} = \pm\sqrt{-\bar{A}/2B}$. On the other hand, if $\Delta E \ll kT_c$, $g \ll 1$, a continuous cooperative displacement $\langle Q(T) \rangle$ of atoms along Q below T_c is found as a function of temperature, at least in mean-field theory.[2,4]

For pure order-disorder SPT's, it is then easy to define the order parameter of the transition as $\langle Q(T) \rangle = V^{-1} \sum_i (Q^i_{0+} - Q^i_{0-})$, the sum extending over the i cells of the lattice. The transitions themselves result from coupling different cells, i and j in the lattice; there is an interaction energy $\bar{V} = \sum_{i>j} v_{ij}\, Q_i\, Q_j$. Using mean-field theory, one derives the free energy of the system and if the latter does not contain a third-order invariant, a second-order transition at T_c results with $\langle Q \rangle \propto (T_c - T)^{1/2}$ below T_c.[2] The same result is also found for displacive transitions. Note that a nonzero single-cell anharmonicity is, however, also needed in displacive systems. A parabolic single-cell potential does not yield a phase transition despite $\bar{V} \neq 0$.[2]

The dynamic behavior of the two types of distortive SPT is quite different. Above T_c, the order-disorder systems behave like magnetic ones, in which the transitions come about by the ordering of magnetic moments. Their excitation spectrum shows relaxational character and is centered around $\omega = 0$. The inverse relaxation time behaves like $1/\tau \propto (T - T_c)$ for $T \geq T_c$. On the other hand, in displacive systems, a mode of finite frequency exists above T_c, and tends to freeze-out on approaching T_c from above. It is due to the nonzero mass of the atoms in motion. This soft mode $\omega_s(t)$ was a concept introduced two decades ago by Cochran[6] and was extremely fruitful for experimental and theoretical research in displacive SPT. With temperature near T_c, ω_s varies like $(T - T_c)^{1/2}$. However, if the soft mode is overdamped, its temperature dependence cannot be distinguished from relaxational response.[4] The soft-mode concept has thus been extended to order-disorder relaxational systems by some researchers.

The static and dynamic behavior of SPT's so far depicted is known as classical. In ferroelectrics, coupling via long-range dipolar forces is present. At the same time, such coupling was assumed always to lead to mean-field behavior. Thus, one believed this to occur for all SPT's. In the perovskite lattice, with general formula ABX_3 an SPT can occur which is controlled by short-range interactions in addition to the well-known ferroelectric one in which the B sublattice is displaced. The lattice consists of BX_6 octahedra interconnected at the corners formed by X ions and dodecahedrally coordinated A ions. Due to this interconnection, the near-rigid BX_6 octahedra can rotate in an alternative manner by angles $\pm \vec{\varphi}(T)$, the order parameter of the transition. The soft mode $\omega_s(T)$ of the transition occurs at the R-corner of the Brillouin zone due to this alternate rotation. The near-rigidity rotating building blocks of the lattice imply short-range coupling of neighboring cells.[2]

In $SrTiO_3$, critical ultrasound absorption was observed[7] indicating fluctuations above the second-order SPT occurring at $T_c \pm 105$ K. In this crystal, the TiO_6 octahedra rotate below T_c around [100] axes, whereas in $LaAlO_3$ around [111] axes.[2] It was then found by EPR that the order parameter close to T_c would vary rather like $|T-T_c|^{1/3}$ in $SrTiO_3$ and $LaAlO_3$, a dependence common to short-range critical behavior for a three-dimensional lattice.[8]

Riste et al. then undertook a neutron-scattering experiment in $SrTiO_3$.[9] The intensity I(T) of the (1/2, 1/2, 3/2) Bragg peak allowed below $T_c \simeq 105$ K on symmetry grounds, and is classically expected to behave like $I(T) \propto \overline{\varphi}(T)^2 \propto (T - T_c)$, followed rather a $|T-T_c|^{2/3}$ dependence for $T < (T_c - 4\ K)$ but showed finite quasielastic scattering above T_c (see Fig. 1). To investigate the quasielastic behavior, an inelastic neutron-scattering experiment revealed that in the van Hove function, the R-corner soft mode did not freeze-out. It remained finite and lost intensity on approaching T_c at the expense of a *central peak* (c.p.) in the structure factor $S(\vec{q},\omega)$. Its width was beyond resolution. A subsequent study by Shapiro et al.[10] on $SrTiO_3$ and $KMnF_3$, and by Kjems et al. on $LaAlO_3$,[11] confirmed the existence of this peak critically rising for $T \to T_c^+$ but its width could not be resolved although the experimental resolution was improved over an order of magnitude to 0.02 meV. The soft mode itself is underdamped in $SrTiO_3$ for $T > T_c + 5$ K and overdamped in $KMnF_3$ and $LaAlO_3$. The discovery of the c.p. in $SrTiO_3$ up to about 50 K above T_c, and the observation of EPR linewidth broadening over that temperature span[12] started large-scale research into the possible existence of a new fundamental excitation of solids.

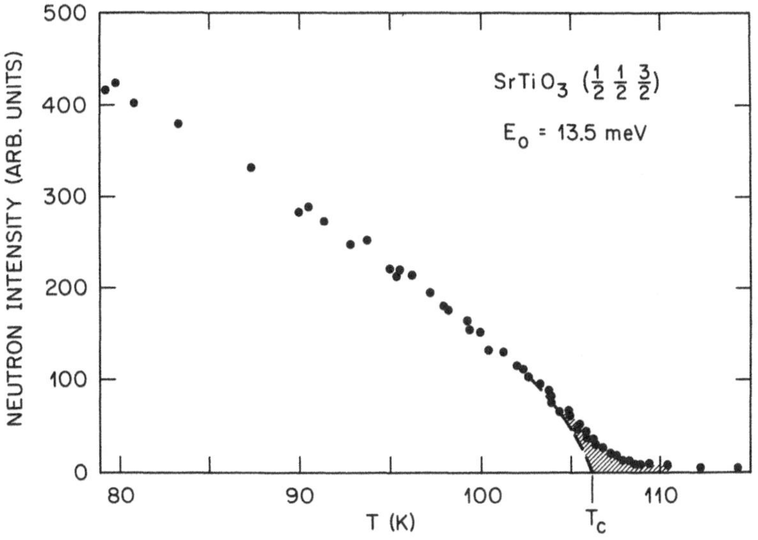

Fig. 1. The temperature dependence of the superlattice reflection $K = 2\pi/\alpha$ (1/2, 1/2, 3/2) in $SrTiO_3$. The dotted line indicates the $<\phi^2>$ dependence obtained from EPR. (After Riste et al., Ref. 9.)

b) Central-Peak Research

Analysis: Due originally to Hohenberg,[13] the form factor

$$S(\vec{q},\omega) = \frac{1}{2\pi}\int_{-\infty}^{+\infty} < \varphi_q(t)\ \varphi_q(0) > e^{i\omega t}\ dt, \tag{1}$$

with $\varphi_q(t)$ the Fourier transform of the R-corner rotational order parameter $\varphi(\vec{r})$, has been cast in the following phenomenological form:

$$S(\vec{q},\omega) \simeq \frac{n(\omega)+1}{\pi\omega}\ \mathrm{Im}\left\{\frac{1}{\omega_{s0}^2(\vec{q},T)-\omega^2-i\omega[\Gamma_0+\Gamma(\omega,T)]}\right\}, \tag{2}$$

where $n(\omega)$ is the phonon occupation number, and

$$\Gamma(\omega,T) = \delta^2(T)/(\gamma - i\omega)\ . \tag{3}$$

For $\Gamma_0 << \delta^2/\gamma$ and $\omega >> \gamma$, one obtains the *soft mode* with $\omega_{s\infty}^2 = \omega_{s0}^2 + \delta^2(T)$ and width Γ_0 in the so-called collision-less regime,[14] and a central peak for small ω in the collision-dominated regime

$$S_{cp}(\vec{q},\omega) = \frac{kT}{\pi} \frac{\delta^2(T)}{\omega_{s\infty}^2 \omega_{s0}^2} \left(\frac{\gamma'}{\omega^2 + \gamma'^2} \right) , \tag{4}$$

where $\gamma' = \gamma\omega_s^2/[\omega_{s0}^2 + \delta^2(T)]$ is the central-peak width, and $\omega_{s0}^2(\vec{q},T)$ is the critical quantity proportional to $\chi^{-1}(\vec{q},T)$; χ being the static susceptibility of the system.

Theoretically and experimentally, attention was focused on getting the two unknown quantities in Eq. (2): (i) $\delta^2(T,\vec{q})$, responsible for the intensity of the c.p. as well as for the finite soft-mode frequency $\omega_{s\infty}(T = T_c) = \delta(T = T_c)$, and (ii) $\gamma(\vec{q})$, responsible for the width of the c.p. Theoretically, in view of the form $\Gamma = \delta^2/(\gamma - i\omega)$, one looked for some *coupling of the soft phonon to slowly-relaxing intrinsic degrees of freedom* which, because of the symmetry of the $SrTiO_3$ system, had to be nonlinear. These efforts included coupling to acoustic modes, phonon-density fluctuations, thermal-diffusion modes, etc. None of them appeared satisfactory at the time and provoked criticism. However, Feder in a review:[15] expressed the opinion that *the central mode is intimately related with the critical fluctuations near the structural phase transitions*, barring the possibility of being due purely to impurity effects, as proposed by Axe.[16]

Central-peak widths. Central peaks with critically enhanced intensity in displacive SPT's have been reported in addition to the already-mentioned $SrTiO_3$ and $KMnF_3$ in the following displacive SPT: $NaNbO_3$, $RbCaF_3$ and $Pb_5Ge_3O_{11}$, SbSI. The former two undergo antiferrodistortive SPT's like the first three, the latter undergo ferroelectric SPT's.[5] To exclude static defects as a cause for the c.p.'s, for $T > T_c$, one tried to resolve an experimental width $\gamma' \equiv \Gamma'(\vec{q},T)$, i.e., to determine the first of the two parameters characterizing the phenomenon, see Eq. (4). This effort turned out to be as unsatisfactory as the theoretical ones: With inelastic neutron scattering, no finite width could be resolved in any case. This, despite Töpler et al.[17] increased the resolution from the early 0.02 meV to 0.08 μeV, which corresponds to 20 MHz. This was achieved with a refined backscattering technique and applied to $SrTiO_3$.

γ-ray scattering provides a method with still higher resolution, of the order of 2.5 MHz. Experiments in $SrTiO_3$, and $KMnF_3$ and $NaNbO_3$ did not resolve any finite width, either.[18] However, a detailed analysis of the depth dependence of critical γ- and X-ray scattering in $SrTiO_3$ showed that the penetration depth at 14.4 keV is of the order of 1 to 10 μm.[18] The thickness is about two orders of magnitude thinner than the known region of surface strain, i.e., highly perturbed. Thus, there will be a sufficient number of static defects of the proper symmetry to couple linearly with the soft mode and produce a narrow static central peak.

Ultrasound absorption, NMR and EPR are experimental techniques much better suited to probe relaxation phenomena in the KHz to MHz frequency range, furthermore, they probe bulk and not surface properties. In $KMnF_3$, Hatta et al.[19] arrived at a critical Γ' width of 20 MHz at $T = T_c + 1$ K and 170 MHz at $T = T_c + 4$ K, from ultrasound absorption. Because his results were in conflict with the γ-ray experiments, insufficient attention was paid to them. We shall revert to $KMnF_3$ at the end of the second section because of its large phonon anisotropy. Owing to the very flat soft-phonon branch between the cubic R and M points in $KMnF_3$,[10] this results in near two-dimensional correlated octahedral sheets above T_c. This is also the case in $NaNbO_3$. Earlier than the γ-ray experiments, Rigamonti's[20] group accounted for their ^{23}Na relaxation measurements by using a constant and large cubic-phonon anisotropy of $\Delta = 1/50$ down to $T_c + 4$ K and a local relaxation rate of $\Gamma_\ell = 300$ MHz at that temperature. In $NaNbO_3$, a c.p. has been found by Dénoyer et al.[21]

The highest resolution attained was an extended analysis of EPR linewidth data in $SrTiO_3$.[22] It yielded a local relaxation rate of Γ_ℓ of less than 6 MHz between T_c and $T_c + 11$ K. Under certain assumptions, one calculates a collective central peak of width $\Gamma' \sim (1/10)\Gamma_\ell$, i.e., Γ' was narrower than 0.6 MHz. We shall describe this technique in the first paragraph of the second lecture. *Here we repeat that dynamic c.p.'s were observed in those two crystals, $NaNbO_3$ and $KMnF_3$, which* exhibit the flattest phonon dispersion between the R and M points of the Brillouin zone, i.e., *allow quasi-two-dimensional correlations, of the octahedra, in (100) planes of the cubic crystal.*

Coupling to impurities. Due to the narrowness of the c.p.'s observed, scattering from static strain fields was discussed quite early by Axe et al.[16] But his formula was not obeyed experimentally. Furthermore, neither could the temperature dependence of the EPR linewidth close to T_c, $\Delta H(T) \propto (T - T_c)^{-0.65}$,[12] be accounted for by inhomogeneous broadening by Folk and Schwabl.[23]

The possibility of defects or impurities being relevant for c.p.'s near structural phase transitions (SPT) received a new and strong impetus from a paper by Halperin and Varma (HV).[24] HV used mean-field theory for a displacive system containing random static or slowly-relaxing defect cells coupled linearly to the order parameter. The defects breaking the symmetry make transitions between two wells at $\pm x_d$ with a transition rate $\nu = \nu_0 \exp\Delta E/kT$, where ΔE is an activation energy of the order of 0.1 to 0.2 eV, $\nu_0 \approx 10^{-12}$

sec, their concentration is c. With these assumptions, they found a c.p. and calculated the two quantities characterizing its dynamic structure factor [Eqs. (2)-(4)],

$$\delta^2(T) = c\left(\frac{m\omega_{an}^{2}(T)v_{q0}x_d^{2}}{kT}\right)\gamma \qquad \nu = \nu_0 \exp\Delta E/kT \ . \qquad (5)$$

Here, m is the mass of the intrinsic atom, $\omega_{an}(T)$ the *uncoupled* intrinsic single-particle anharmonic frequency, and v_{q0} the Fourier transform of the bilinear interaction $v_{\ell\ell'}$ of the particles in cells ℓ and ℓ'. An important condition for a defect to be operative in producing a c.p. is that $dT_c/dc > 0$. From Eq. (5), one expects $\delta^2(T) \propto 1/T$ and $\propto c$. Close to T_c, where the correlation length ξ becomes larger than the distance between particles, Eq. (5) is not expected to hold due to cluster interactions.

Halperin and Varma compared their formula quantitatively to the known neutron and EPR data in $SrTiO_3$. They obtained agreement for a concentration of $c \approx 10^{-5}$ a ratio of defect-to-intrinsic rotation of the octahedra of $x_d/x_i \approx 5$, i.e., a very reasonable result, although the 1/T dependence was missing. Their findings prompted further theoretical mean-field investigations. Petschek[25] could account quantitatively for the Lorentzian EPR line shape in $SrTiO_3$ between 5 and 12 K above the phase transition. He calculated the nonlinear interaction of the defect cell with nearby cells and estimated the defect-cell concentration from neutron data. This substantiated considerably the defect-induced c.p. picture.

The impurities considered by HV break the translational symmetry of the crystal and may give rise to local modes which can be investigated by resonance experiments. Due to coupling to the soft mode, these local modes can become strongly temperature dependent. Höck and Thomas [26(a)] calculated their behavior self-consistently with mean-field theory as well. Quadratic coupling to the order parameter is another possibility.[26(b)] Here, the impurity is at a symmetry-conserving site in contrast to that which couples linearly.

Central-peak intensities. The theories of defect-induced c.p.'s redirected the experimental efforts from the c.p. widths Γ' towards their intensity I. Most recently, Hastings et al.[27] observed a systematic enhancement of the central-peak intensity upon reduction of $SrTiO_3$ with hydrogen. The results provide direct experimental evidence for the involvement of a defect mechanism for the central-peak formation in $SrTiO_3$. Increasing the amount of Ti^{3+} upon reduction from 6×10^{17} to 20^{20} cm^{-3}, they found an enhancement of only a few times the least-reduced sample. That is, the $I \propto c$ relation from Eq. (5) does not seem to hold.

Cubic $KTaO_3$ doped with several percent of Nb and other ions was investigated by Yacoby.[28] From the first-order Raman spectra, he concluded that Nb^{5+} on Ta^{5+} sites lacks inversion symmetry on the time scale of the Raman experiment, and relaxes with a time constant slower than 10^{-10} sec. The first-order TO_2 Raman intensity $I_N(T)$ increases critically on approaching T_c^+ from above as $I_N(T)^{-1}\ T \propto (T - T_c)$, and is compatible with the Halperin-Varma model. It thus appears to be the first scattering investigation confirming the H-V model for a particular *known* impurity.

From the calculation of Petschek, the Raman scattering in $KTaO_3$:Nb, Brillouin scattering in $Pb_5Ge_3O_{11}$, and EPR in doped KH_2PO_4 — discussion of which is beyond the scope of this Lecture — the occurrence of c.p.'s due to linear coupling of the soft mode to

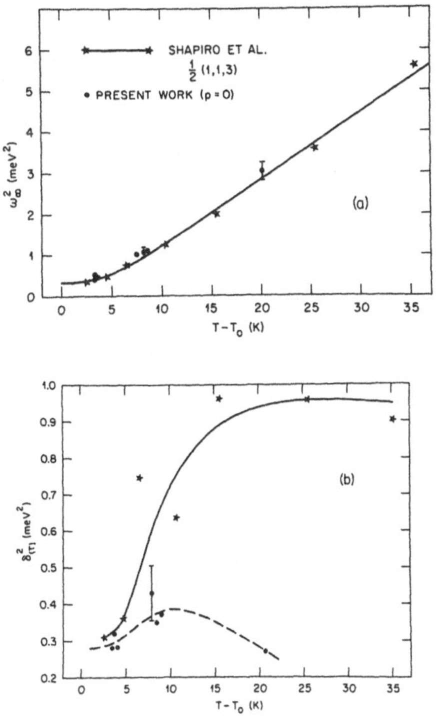

Fig. 2. R_{25} soft mode in $SrTiO_3$. (a) Comparison between two determinations of the quasiharmonic frequency $\omega_\infty{}^2(T)$. (b) Same comparison in terms of $\delta^2(T)$, the parameter which characterizes the central-peak strength. (After Currat et al., Ref. 29.)

static or very slowly relaxing defects is well documented. All theories involving dynamic defects so far use the mean-field approximation. Thus, they cannot predict the c.p. behavior on approaching T_c. Here, an inelastic neutron experiment by Currat et al.[29] aimed at understanding the soft-mode behavior of [111] uniaxially stressed $SrTiO_3$, inadvertently brought new insight.

From Fig. 2, one can see that away from T_c, the parameter $\delta^2(T)$, which characterizes the central-peak strength, differs for the crystal used in Grenoble and the one of Shapiro et al.[10] in Brookhaven. This points to the fact that the central peak in the bulk is extrinsic in neutron scattering away from T_c. *However, on approaching T_c, the magnitude of these quantities merged for the crystals used. Actually, the intensity at T_c is the same as the one obtained by Lyons and Fleury*[30] *from light scattering approaching T_c from below for a slow relaxing c.p.:* Their experiments indicate two characteristic times of the relaxing self-energy τ_1 and τ_2: a shorter τ_1 with $\tau_1{}^{-1} \sim 15$ GHz is observable direct in the depolarized scattering for $T < T_c$. From the Brillouin spectrum, a slower process τ_2 with $\tau_2{}^{-1} < 0.3$ GHz was inferred for $T \to T_c^-$. Its existence is what Courtens predicted from an analysis of neutron-scattering, specific-heat and birefringence data for $T > T_c$.[31] Thus, the finite frequency of the soft mode $\omega_{s\infty}(T - T_c)$ in $SrTiO_3$, independent of the sample at T_c pointed to a possible intrinsic observable quantity, i.e., $\delta^2(T = T_c)$.

c) Computer Simulation, and Anharmonic Theory

The difficulties with conventional anharmonic lattice perturbation theory led to the suspicion that close to T_c, the c.p. phenomenon might result from strongly anharmonic effects, not tractable by perturbation expansions of the harmonic lattice. To see what additional excitations, if any, may be present, large-scale computer simulations were undertaken. The first paper of a series pointing towards the heart of the problem was that by Schneider and Stoll.[32] They simulated a two-dimensional d = 2 lattice with a model Hamiltonian of the form

$$\mathcal{H} = \sum_{\ell} \frac{M\dot{x}_{\ell}^{2}}{2} + \frac{\bar{A}}{2}\sum_{\ell} x_{\ell}^{2} + \frac{B}{4}\sum_{\ell} x_{\ell}^{4} + \frac{C}{2}\sum_{<\ell\ell'>} (x_{\ell} - x_{\ell'})^{2} , \qquad (6)$$

where ℓ labels the particle with mass M in the ℓ-th unit cell. $M\dot{x}_{\ell}$ and x_{ℓ} are the momentum and displacement with respect to the square rigid lattice with constant a. $\bar{A}$, B and C are model parameters chosen to represent antiferrodistortive motions of the particles with short-range interactions only, as symbolized by

$<\ell\ell'>$. The second and third terms in Eq. (6) represent the harmonic and anharmonic single-particle potentials, and the fourth their interaction.

At T = 0, the order parameter is given by $x = (-\bar{A}/B)^{1/2}$, thus for $\bar{A} < 0$, a finite T_c is guaranteed. The larger the $|\bar{A}|$ value, the more order-disorder character the system will have, with a deep double-well single-particle potential. The measure of the order-disorder versus displacive character is the ratio g defined in section a), i.e., $g = \bar{A}^2/4B\,kT_c$. For $|\bar{A}| \to 0$, $g \to 0$, one approaches the classical displacive limit which has $T_c = 0$. Schneider and Stoll employed the molecular-dynamics technique, and using a lattice of 40 × 40 points, with periodic boundary conditions, they computed the displacement-displacement correlation function $<x_q(t)x_{-q}(0)>$, and from it the corresponding dynamic structure factor $S(\vec{q},\omega)$. They simulated an order-disorder and a displacive system; the result for the latter with g = 0.2 is shown in Fig. 3 as a function of frequency at $(T - T_c)/T_c = 0.06$ for various wave vectors close to the reciprocal lattice point {3.1} which becomes a Bragg spot below T_c.

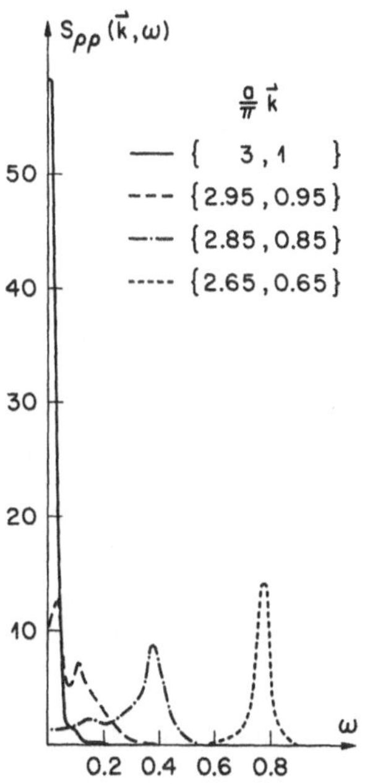

Fig. 3. The frequency- and wave-vector dependence of the spectral response of a d = 2 displacive model with g = 0.19 at t ≈ 0.06, as determined by molecular dynamics. (After Schneider and Stoll, Ref. 32.)

One sees *in addition* to the soft mode, a c.p. of *comparable width.* The soft mode stopped at a finite $\omega_{s\infty}^2$ and its intensity became weaker relative to the c.p. on approaching T_c. Simulations of the order-disorder system with $g > 1$ yielded only the well-known diffusive $\omega = 0$ centered excitation and no soft mode as expected.

To understand the physical meaning of their result, they displayed the dynamical-rotational motion of the particles graphically. These figures revealed near T_c the existence of short-range clusters separated by walls where the local displacements change sign. The existence and dynamics of these clusters give rise to the c.p. and inhibit the soft mode from freezing-out because the symmetry is broken locally.

Results similar to those reviewed here were published later by Aubry[33] and Koehler et al.[34] for a linear array. In such a linear chain, no phase transition occurs at finite temperature but, classically, $T = 0$ can be viewed as a critical temperature. Consequently, the system exhibits all features of interest for $T \geq T_c$. Krumhansl and Schrieffer[35] then showed analyticallly, for such an array, the existence of an important class of non-linear solutions which could not be represented by usual phonon perturbation expansions: the motion of "domain walls" or kink and anti-kink *solitons*. The reader is referred to a recent conference[36] held on the subject which shows the significance these highly-nonlinear mathematical solutions have attained.

On the molecular-dynamics side, Schneider and Stoll have considerably substantiated the picture by simulations for $d = 3$ and $d = 4$ dimensions and have also considered other systems. To discuss all these results here would lead us too far, therefore, we refer to a review of theirs.[36] *The essential point is that for $T > T_c$, the picture we described for $d = 2$ remains qualitatively unaltered, but the cluster-induced c.p. feature occurs closer to T_c^+, on lowering T, for larger lattice dimensionalities.* A further step in understanding the role of impurities was achieved by a subsequent computer simulation by Schneider and Stoll.[37] In their $d = 2$ system, they introduced random slowly-relaxing defects coupled to the displacive system. The result was a c.p. considerably narrower than that found in their intrinsic simulations, plus the essentially unchanged broad soft phonon at finite frequency $\omega_\infty(T)$. The critical dynamics of the narrowed c.p. $\propto (T - T_c)^{-\gamma}$ remained unchanged. This simulation which avoids the shortcomings of the mean-field approximation, and includes impurity interactions near T_c, indicated where further progress could be expected experimentally in real systems.

d) The New View

This picture emanating from computer simulations in d = 1 to 4 dimensions[32,38] as well as with the analytical analysis in d = 1 dimensions[35] was then reviewed by Bruce.[39] Starting with a comparison of the two extreme idealized cases of SPT's, the deep-well order-disorder system on the one hand, and the displacive system on the other, he pointed out the shortcomings of both approaches. In the former, only the dynamical hopping across the deep local double well is taken into account, leading to the existence of cluster walls and their diffusion, but neglecting the phonons in the quasi-statically disordered lattice. In the other extreme, we have the classical soft-phonon picture. From the two extreme cases of order-disorder and near-harmonic displacive models in which only domain-wall motion or soft modes result, respectively, Bruce summarized the contemporary theoretical view as follows: ".....*With the onset of criticality, the growth in short-range order promotes a crossover from a regime in which collective behavior has the classical displacive form, to a regime in which the collective behavior displays features better described in the language traditionally reserved for order-disorder systems*".

This view — which is more detailed in Dr. Bruce's Lectures — constitutes an appealing graphic expression of universality, in that it offers a vivid description of the underlying character of the spectrum of excitations and of why deviations from classical static exponents have been observed in SPT. In other words, it correlates the occurrence of non-classical static behavior with the formation of order-disorder-type precursor clusters. Bruce amplified his statements by considering the local potential V_s* of cluster coordinates after application of renormalization-group transformations: $V_b = |R_b V$. In d = 3, $g^* = V_s^*/kT_c = 0.55$ is of the order of unity. Referring to our former discussion, this suggests that a near-displacive system with $g \approx 1/10$ before renormalization becomes more order-disorder-type after, or equivalently, when the correlation length increases on approaching T_c. For d = 2 and 1, this behavior is more pronounced than for d = 3. Renormalization-group calculations in d = 2 and 3 dimensions were then used to calculate critical distribution functions confirming the picture.[40] For $d > d_c$ when the Gaussian fixed point is stable, the harmonic approximation should hold, and $g^* \approx 0$. It appears that for d = 1 and 2, clusters are "mutually exclusive", i.e., at any one time a given region of the crystal can only be occupied by a single cluster. In d = 4,[41] clusters of ordering and anti-ordering can interpenetrate, and the idea of a cluster wall seems less important.

However, the appealing picture reviewed above lacked experimental verification. Since the local potential V_s is of importance, it was natural to look for experiments which can give information, if not on the V_s of a cluster, then on a closely-related property: the local probability distribution $P(\varphi)$ of the order parameter φ. In the following section, it is shown that EPR achieved this goal.

II. RECENT CONSOLIDATING RESULTS

In this section, we review recent experimental evidence for the occurrence of intrinsic anharmonic behavior due to *correlated* fluctuations above T_c, which may also be called precursor order. Strongly correlated non-linear systems have been analyzed during recent years by the renormalization-group (RG) approach. Depending on the symmetry of a system and its range of forces, a critical lattice dimensionality d_c is calculated.[42] Systems with $d > d_c$ have so-called stable Gaussian fixed point, they follow strictly static mean-field or Landau behavior, and the Fourier coordinates of their fluctuations are *uncorrelated*. For $d < d_c$, correlated fluctuations near T_c are relevant and determine the behavior of the system. All the experiments discussed below have $d = 3$, (in contrast to Lecture III, where $d \simeq 1$ for the incommensurate phases reviewed), but d_c varies. The lattice dimensionality parameter in RG, $\varepsilon = d_c - d$, indicates whether or nor clusters occur on a local scale. For ε, the more positive, the cluster picture is dominant near T_c, whereas for $\varepsilon < 0$ it is not relevant. If the dimensionality of the system is marginal, i.e., $d = d_c$, logarithmic corrections to the Gaussian fixed-point behavior are predicted[43] and observed;[44] thus, we also expect an intrinsic c.p. to exist. In computer simulations for a short-range system with $d = d_c = 4$, this was found.[36]

The antiferrodistortive $SrTiO_3$ and $LaAlO_3$ SPT's with their short-range character have $d_c = 4$, $\varepsilon = 1$. The same may be true for the cubic ferro-electric crystals $BaTiO_3$ and $KNbO_3$, whereas the $RbCaF_3$ and $KMnF_3$ antiferrodistortive SPT's have $d_c = 4.5$, i.e., $\varepsilon = 1.5$[45] due to the strong R-phonon anisotropy: Thus, one expects a more pronounced anharmonic behavior to be observable. From the above, it is inferred that for $d > d_c$, $\varepsilon < 0$, no intrinsic cluster-induced c.p. behavior should occur. In a series of impressive experiments on KH_2PO_4, Courtens has shown this to be the case. He was actually the first to suggest the possible importance of the critical dimensionality d_c.[46] In the uniaxial hydrogen-bonded ferroelectric KH_2PO_4, $d_c = 2.5$, $\varepsilon = -0.5$, due to strong-coupling of the soft optic to the acoustic modes.[47] The transition is in fact slightly first order. Application of an electric ordering field to KH_2PO_4 in the paraelectric phase produces linear coupling between energy and polarization.[46]

A heat-diffusion central peak appears on the polarization autocorrelation function, the statics and dynamics of which are *quantitatively* mean field[46(c)] and the c.p. does not narrow down on approaching T_c. Other less quantitative heat-diffusion c.p's for $T < T_c$ have been discussed in the previous review[5] as well as other extrinsic c.p.'s disappearing upon specific treatments or appearing on doping (so-called Halperin-Varma centers) and are outside the scope of these Lectures.

a) Anharmonicity at T_c in $SrTiO_3$

EPR as a probe for the local probability distribution. The accurate values $\langle\varphi(T)\rangle$ which lead to the discovery of static critical phenomena in $SrTiO_3$ were obtained by measuring Fe^{3+} EPR line positions. The EPR of Fe^{3+} ions substitutional on B sites reflects the local orientation of BO_6 octahedra.[2] For a fixed direction θ of the external magnetic field $\vec{H}$ in a (001) crystallographic plane, the secular resonance field H_r is given by[48]

$$H_r = H_0(\theta,\nu) + A_1(\theta,\nu)\phi\,[001] + A_2(\theta,\nu)\phi^2\,[001]\,, \qquad (7)$$

where φ [001] is the octahedral rotation around the [001] axis, and H_0 is the resonance field in the absence of rotation. The sensitivity parameter $A(\theta) = \partial H_r/\partial\theta$ measures the shift of the resonance field upon rotation by θ. Both H_0 and A_1 also depend on the applied microwave frequency. It could be shown that the second-order term, proportional to φ^2, becomes negligibly small for certain values of θ, $A_2 \simeq 0$. The constant A_1 is largest for the $Fe^{3+} - V_0$ pair center due to the large anisotropy of H_0 along the pair axis, and the rotation φ [001] it measures is proportional to that measured by the Fe^{3+}.[2] The $Fe^{3+} - V_0$ center consists of a trivalent iron impurity on a B-site with a nearest-neighbor oxygen vacancy. For H || [110] and K-band, A = 26 Gauss/degree for an [001] octahedral rotation. The local random fluctuations φ_ℓ are of the order of a degree, whereas the background linewidth is about 3 Gauss or 0.10 degrees. Thus, owing to the high sensitivity of this center, the stochastic variation $\delta\varphi_\ell(t) = \varphi_\ell(t) - \varphi$ could be observed. The possibly time-dependent departure $\delta H(t)$ from $H_0 + A_1\langle\phi\rangle$ is a random function proportional to $\delta\varphi_\ell(t)$ [from Eq. (7)].

In magnetic resonance, one can distinguish two extreme cases with respect to the fluctuation time τ of the order parameter. If τ is much faster than the characteristic measuring time of the experiment τ_m, it is called a "fast-motion regime", if τ is much slower than τ_m, a "slow-motion regime". τ_m is essentially given by the inverse homogeneous EPR linewidth $\tau_m = 1/\Delta\nu = \hbar/g\beta\Delta H$.[49] In the slow-motion regime, the main critical part of the fluctuation time τ of $\varphi(t)$ becomes quasi-static compared to the EPR measuring time τ_m. This means that at each site ℓ of an Fe^{3+} - V_0 center, the local rotation $\varphi(\vec{R}_\ell,t)$ is seen at rest by the EPR experiment. Thus, the shape of the EPR line $L(H) = L(\varphi)$ [from Eq. (7)] is proportional to the probability distribution $P(\varphi)$ of the ensemble. In EPR, the derivatives $dL(H)/dH$ of the absorption lines $L(H)$ are recorded using Zeeman modulation. Thus for *slow motion*, the curves are proportional to $dP/d\varphi_\ell$. From this proportionality, it also follows from Eq. (7) that the full EPR linewidth $\Delta H(T)$ in the slow-motion regime is proportional to the mean-square fluctuation amplitude[48]

$$\Delta H_s(T) = 2\, A_1 < \delta\, \varphi^2 >^{1/2} \quad . \tag{8}$$

Thus, at T_c the linewidth is always finite as is the local fluctuation amplitude.

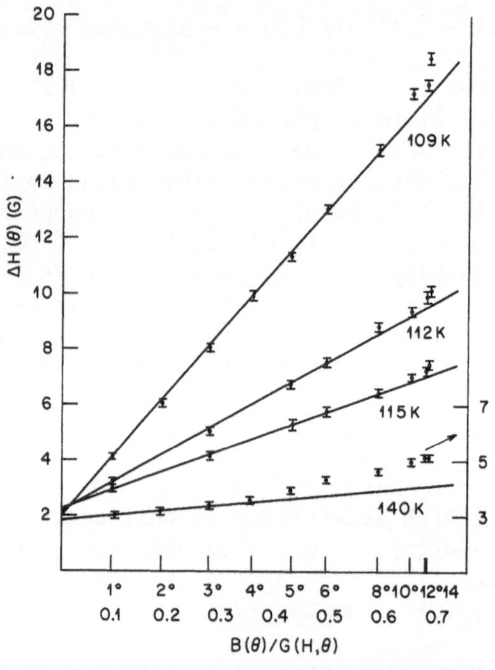

Fig. 4. Variation of the EPR linewidth $\Delta H(\theta)$ as a function of the sensitivity parameter $A(\theta)$ for various temperatures between 109 and 140 K of a Verneuil-grown $SrTiO_3$ crystal. (After Reiter et al., Ref. 22.)

On increasing the temperature above T_c, the extra EPR linewidth $\Delta H_s(T)$ narrows down to the background width ΔH_b (see Fig. 4). In a recent detailed study, the sensitivity parameter $A_1(\theta)$ was varied at a set of fixed temperatures. It showed unambiguously that $\Delta H_s(T)$ is strictly proportional to $A_1(\theta)$[22] for Verneuil-grown crystals (see Fig. 4). In the fast-motion regime, it would be proportional to $A_1(\theta)^2$ $H(\theta)$. Thus, in these crystals, one sees static local fluctuations throughout the critical regime for the secular part of the line, at least up to $T_c + 11$ K = 116 K. The static Lorentzian shape away from T_c is caused by impurity-induced anharmonic interactions, the theory of which was developed by Petschek.[25]

The background with $\Delta H_b(\theta)$ also varies with θ but in an entirely different manner. Therefore, $\Delta H_s(\theta)$ and $\Delta H_b(\theta)$ could be separated analytically, since for both variations, analytic expressions could be computed. $\Delta H_b(T)$ is determined by the spectral density

$$J(\omega) = \sum_{\vec{q}} S(\vec{q},\omega)$$

at $\omega_1 = 5.2$ cm^{-1} and $\omega_2 = 0.4$ cm^{-1} of the quasi-harmonic fluctuations of the octahedra. It is essentially a Raman nonsecular process between the $\pm 1/2$ ground state of the $Fe^{3+}(S = 5/2)$ ion and its two excited Kramer's doublets $\pm 3/2$ and $\pm 5/2$[22] located at energies $\hbar\omega_{1,2}$. Because the fluctuations are much faster than the reciprocal EPR linewidth, this background contribution is in the fast-motion regime which explains its narrow 2-3 Gauss width. The latter is the limit of resolution of the local relaxation rate and corresponds to 6 MHz, as mentioned under I.b).

From analysis of the line shapes of the $Fe^{3+} - V_0$ center in the slow-motion regime, it was found that the probability $P(\varphi_\ell)$ to observe φ_ℓ deviated from the expected Gaussian, above and below T_c. Above T_c, the shoulders of the lines fell more rapidly to zero than a Gaussian, and $P(\varphi_\ell)$ was better described by adding a term $e^{-\varphi_\ell^4}$. Close to T_c, up to 20% of a normalized function of this form had to be used. This deviation from a Gaussian will be the basis for the new analysis described in the next subsection.

Evidence for short-range order at T_c in $SrTiO_3$. Motivated by the cluster picture, it was recently supposed that the time-dependent local scalar coordinate $\varphi(t)$, whose ensemble average is the order parameter for the displacive SPT, can be written in the form.[50]

$$\varphi(t) = \sigma(t) + y(t) \; . \tag{9}$$

Here, the coordinate y(t) is taken to be a Gaussian random variable, with correlation time τ_y, and mean-square amplitude $(\langle y^2 \rangle)^{1/2}$: this variable describes the quasi-harmonic fluctuations about the instantaneous quasi-equilibrium position set by the value of the coordinate $\sigma(t)$, which reflects the influence of the clusters. y may also include the distribution resulting from impurities and strains always present in a real crystal. The simplest variant (sv) of the cluster picture suggests that the coordinate σ be taken to undergo Markovian hopping between *two* values $\pm\,\sigma_0$, with a transition probability per unit time $1/\tau_\sigma$. This is clearly an oversimplification: a more refined picture would allow for a continuous *distribution*, $P(\sigma)$.

From a comparison of the temperature dependencies of the order parameter $\langle\varphi(t)\rangle$ and that of the soft mode $\omega_s(T)$, one can estimate σ_0 as follows: for $T < T_c$, the long-range order $\langle\varphi\rangle_{\ell r}$ in monodomain single crystals as measured by EPR, was found by Steigmeier and Auderset[51] to be proportional to

$$\omega_s(T) = 0.69\ \langle\varphi(T)\rangle_{\ell r}\ , \tag{10}$$

where ω_s is measured in THz, and φ in angular degrees.[52] At T_c, the soft mode is not frozen-out due to the existence of clusters which prefigurate the low-temperature phase. $\omega_{s\infty}(T = T_c) = 0.13$ THz[10,29] due to the existence of the short-range order $\langle\varphi(T)\rangle_{sr}$. One estimates from Eq. (10)

$$\langle\varphi(T)\rangle_{sr} = 0.19^\circ\ . \tag{11}$$

High-resolution structural X-ray studies attempted to measure this quantity.[53] However, X-rays and neutron scattering probe the crystal much faster than the harmonic thermal fluctuations of the lattice. The latter have amplitudes for the rotational order parameter of $(\langle y^2\rangle)^{1/2} = 2.1^\circ$.[53] Thus $\langle y^2\rangle / \langle\varphi(T)^2\rangle_{sr} = 122$, and X-ray and neutron-scattering structural investigations which determine the probability distribution function through its Fourier transform (the Debye-Waller factor), in general, cannot be expected to detect the cluster-induced nonlinear local behavior. The quasi-harmonic fluctuations y mask the nonlinear σ_0 character.

Consider now EPR as a probe to determine $P(\varphi)$. We have seen in the preceding sub-section, that the quasi-harmonic fluctuations with spectral density $J(\omega)$ are only contributing to the non-critical background EPR linewidth $\Delta H_b(\varphi) \approx 2$ Gauss. In other words, for the discrimination against the harmonic fluctuations it is sufficient that $P[\sigma(t)]$ can be probed in the present case.

There are, however, limitations to this undertaking for the following reasons: (i) The ratio $r = (\sigma_s{}^2\tau_\sigma / \langle y^2 \rangle \tau_y) \gg 1$ has to be fulfilled. This is easily the case because the cluster fluctuations τ_σ have been slowed down by impurities to a frequency slower than $\Delta H_b(\varphi)\ g\beta/h \simeq 6$ MHz as compared to τ_y which should be, at most, 10^{-11} sec from the soft-phonon width.[10] (ii) The distribution of $P(y)$ also contains contributions from static strains y_{st}, thus $\langle y_{st}{}^2 \rangle \lesssim \sigma_0^2$. (iii) If the sensitivity parameter $B = \partial\omega/\partial\varphi = A\ H_r\ \partial g/\partial\varphi$ gets smaller than $1/\sigma_0\tau_\sigma$, i.e., B (or A) is too small and τ_σ too fast, the double-peaked form of $P(\varphi)$ is erased. Now, $B = 1.6 \times 10^8$ sec/deg[22] and, expecting $\sigma_0 = 0.2^\circ$ with $\tau_\sigma > 10^{-7}$ sec, $1/(\sigma_0 \cdot 10^{-7}) < 5 \cdot 10^7$ sec/deg $< B$ is also fulfilled for Verneuil-grown crystals.

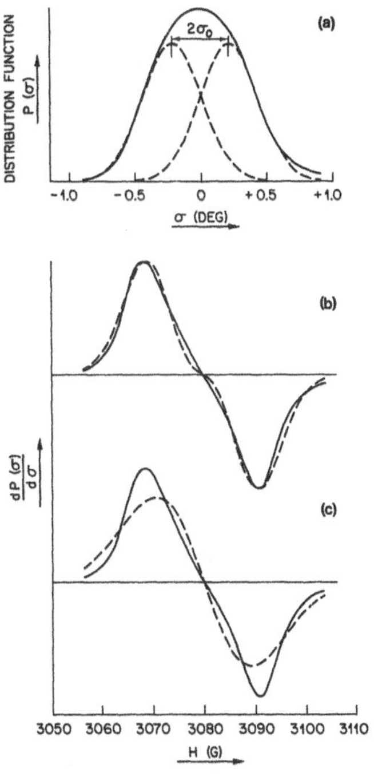

Fig. 5. (a) The experimental distribution function $P(\sigma)$ at $T = 105.5$ K (solid line) together with two displaced Gaussians with whose superposition (not shown) the spectrum is modeled. (b) The derivative of the experimental $P(\sigma)$ (solid line), and of its double-Gaussian representation. (c) The derivative of $P(\sigma)$ and its single-Gaussian representation. (After Bruce et al., Ref. 50.)

We now return to discuss the distribution function $P(\sigma)$ itself in EPR, as recently published. A comparison of the *derivative* of this function at T_c with that of a Gaussian of the same peak height, clearly reveals the non-Gaussian nature of $P(\sigma)$ [Fig. 5(c)]. The extent of its anharmonic character is characterized by its representation as a superposition of two symmetrically displaced Gaussians for y [Fig. 5(a)], whose separation provides one adjustable parameter σ_0. The Ising variable σ_0 is chosen to optimize the corresponding representation of the derivative spectrum [Fig. 5(b)]. Note that $\langle y_{st}^2 \rangle \approx \sigma_0^2$ and condition b), the strain-induced broadening, limit the resolution of the experiment.

These results established, unambiguously and for the first time, the existence of local precursor order persisting for times long in comparison with 10^{-7} sec. More explicitly, Fig. 5 suggests that, near but above T_c = 105.3 K, the oxygen octahedra in $SrTiO_3$ oscillate typically about quasi-equilibrium positions displaced by $\sigma_0 \approx 0.22^0$ from the high-symmetry position. The striking accord with the value in Eq. (11) derived from $\omega_{s\infty}$ constitutes strong evidence of the overall coherence for the displacive order-disorder crossover as an intrinsic property of displacive phase transitions on approaching T_c. Calibrating the c.p. intensity measured by neutron scattering[54] by the Bragg intensity below T_c gives an estimate of $\langle \varphi(T_c) \rangle \approx 0.28^0$ in rough agreement with Eq. (11).

b) Displacive-Order-Disorder Crossover in Ferroelectric Oxides

Uniaxial ferroelectrics like TGS with their long-range dipolar interactions have a critical dimensionality $d_c = 3$. In this case, Larkin and Khmelnitzki[43] showed in their pioneering work that critical behavior is given by logarithmic corrections to the classical Landau results. Thus, fluctuation effects are already present but not dominating. In $Pb_5Ge_3O_{11}$, the dynamic portion of the c.p. spectrum has been reported to diverge logarithmically.[55] Cubic dipolar systems have been investigated by Aharony, Fisher and Bruce[56] using R.G. theory. The static critical behavior calculated is nearly indistinguishable from that in cubic short-range systems. In real cubic ferroelectrics, long-range dipolar and short-range interactions are present. Their fixed-point behavior is not yet known, but it is unlikely to be more mean-field than the pure dipolar one. Indeed, Kind and Müller[57] measured the exponent of the dielectric susceptibility ε above the second-order phase transition in a $KTa_{0.9}Nb_{0.1}O_3$ crystal to be $\gamma = 1.7$, and if appropriate corrections are taken into account, $\gamma \geq 1.4$, well outside the mean-field behavior of $\gamma = 1$. Thus, one might expect that for cubic ferroelectrics like $BaTiO_3$ and $KNbO_3$, a crossover from displacive to order-disorder behavior may exist on approaching T_c^+. Most recent infrared reflectivity measurements show this occurs beyond doubt as described below. This is also quite relevant in the context of critical behavior as these systems have traditionally been analyzed by Landau theory. Because they undergo first-order transitions, this is always possible, but such procedures are relativized by the new findings.

The new soft-mode data have been obtained by Gervais and his group[58] using a Fourier transform scanning interferometer equipped to work up to high temperatures. These modern i.r. spectrometers are extremely well-suited to measure accurately Raman active modes in the frequency range from 10 to 4000 cm^{-1}. The motivation for this work was the long-standing controversy between the soft mode versus order-disorder description in $BaTiO_3$ and $KNbO_3$. The former was supported by early infrared reflectivity measurement and later inelastic neutron-scattering data of the Brookhaven group.[59] The order-disorder picture stemmed from the observation of anomalous strong diffuse X-ray scattering by Comes, Lambert and Guinier[60] as well as Raman activity *above* the tetragonal-cubic transition, and more recent Raman investigations in the tetragonal phase.[61]

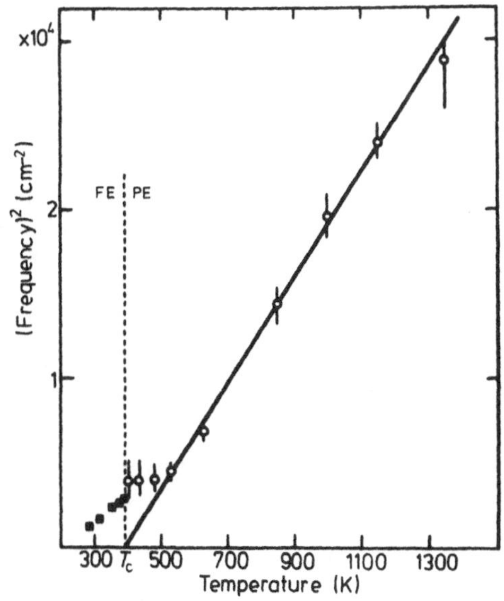

Fig. 6. Temperature dependence of the frequency squared of the lowest transverse mode (0) for BaTiO. (After Y. Luspin et al., Ref. 58). (■) are Raman data.

The infrared reflectivity data were analyzed on the basis of a factorized form of the dielectric functions,[58]

$$\varepsilon(\omega) = \varepsilon_\infty \pi_j \frac{\Omega_{iLO}^2 - \omega^2 + i\omega\gamma_{iLO}}{\Omega_{jTO}^2 - \omega^2 + i\omega\gamma_{iTO}} .$$

The Ω_{jTO} and Ω_{j1O} are frequencies, and the γ_{iTO} and γ_{iLO} the corresponding damping of the transverse and longitudinal optical modes, respectively. This function fitted the measured reflectivity scans with high accuracy. Results for the soft modes in $BaTiO_3$ are shown in Fig. 6. The curve shows a clear-cut soft-mode behavior over a wide range of temperatures in the cubic P.E. phase whose extrapolation intersects the temperature axis near T_c. *However, on approaching T_c, ω_{TO} starts saturating some 120 K above the phase transition.* At T_c, an A_1-type component is abruptly shifted to $\approx$ 280 cm^{-1} consistent with the first-order transition, while an E component continues to soften to the next tetragonal-to-orthorhombic transition *thus, the E-mode shows no significant discontinuity at T_c.* Similar behavior has been observed in the isomorphous phase of $KNbO_3$. Furthermore, in both crystals the damping is heavy, over ten times that of the next modes, and also persists into the tetragonal phase.

The ratio of the dielectric constants $\varepsilon_{cap}/\varepsilon_{lat}$, deduced from piezoelectric resonances and the Lyddane-Sachs-Teller relation, shows the following: The soft mode roughly explains ε_{cap} in $BaTiO_3$ at sufficiently high temperatures, but this becomes less and less true on approaching T_c from above. The theory of "dirty" displacive ferroelectrics is found to be inappropriate since, while the dielectric-constant ratio "diverges" at T_c, the soft-mode frequency is nearly constant. Thus, on approaching the cubic-tetragonal transitions from above, a dynamical disorder affects the displacive behavior and an additional relaxation can be related to disorder, as has been invoked by Comes et al.[60] We arrive, therefore, at the conclusion that a displacive-to-order-disorder crossover occurs as shown to be the case in the antiferrodistortive SPT of $SrTiO_3$. These new findings also settle the controversy whether the cubic-to-tetragonal transition in $BaTiO_3$ and $KNBO_3$ are displacive or order-disorder: they are both, and it depends at which temperature one probes the dynamics.

The uniaxial ferroelectric $LiTaO_3$ has been investigated by the same group with the same method. ω_{soft} again does not freeze-out completely, and $\varepsilon_{cap}/\varepsilon_{lat}$ increases on approaching T_c but is less "divergent", as ε_{lat} shows itself a peak at T_c. Thus, the behavior is all in all more "classical". This is precisely expected, in view of the fact that for this uniaxial crystal $d = d_c = 3$, i.e., the dimensionality is marginal, and correlated fluctuations are present but less dominating.

c) Evidence for Precursor Order in $RbCaF_3$ and $KMnF_3$

$RbCaF_3$ and $KMnF_3$ crystallize also in the perovskite structure, and undergo the same kind of antiferrodistortive SPT as $SrTiO_3$. However, their soft R-corner mode dispersion is practically flat between the R and M points of the cubic Brillouin zone.[62] This dispersion has been approximated by $\omega_R(\vec{q},T) = \omega_s(T) + \lambda_{RM} q_{RM}^2$, where $\vec{q}_{RM}$ is the wave vector pointing from the R towards the M point of the zone. The ratio of the constant λ_{RM} to the corresponding one characterizing the dispersion towards the Γ point of the zone, $\lambda_{R\Gamma}$, $a = \lambda_{RM}/\lambda_{R\Gamma} = 0.013 \pm 0.07$ in both crystals, as compared to $SrTiO_3$ with $a = 0.036 \pm 0.012$. This means that octahedral rotational fluctuations are highly correlated in (100), (010) and (001) sheets, and almost uncorrelated between one sheet and the next. These mutually perpendicular correlated planar sheets drive the transition first-order in $RbCaF_3$ and $KMnF_3$[63] in contrast to $SrTiO_3$ in which the correlations are more three-dimensional, i.e., pancake-like. The correctness of a fluctuation-driven first-order transition has been proven by combined hydrostatic and uniaxial stress EPR experiments on $RbCaF_3$.[64] The hydrostatic pressure does not break the symmetry, and thus does not alter the amount of first-order discontinuity at T_c.[64(b)] On the other hand, a very modest uniaxial stress causes the transition to become continuous, i.e., only one order-parameter component, say [100] is allowed to become critical.[64(a)] The octahedral fluctuations occur then in (100) sheets around the particular [100] direction chosen. The critical dimension for this so-called "Lifshitz tricritical point" is $d_c = 4.5$, i.e., $\varepsilon = 1.5$ larger than for $SrTiO_3$ ($\varepsilon = 1$). The correlated fluctuations are "almost" two dimensional and thus —— within our context —— one expects anharmonic effects to be observed.

NMR study on ^{87}Rb. The Rb^+ ion is at a dodecahedral or A site of the ABX_3 structure. In the cubic phase, its point symmetry is also cubic. The nuclear spin of ^{87}Rb is $I = 3/2$, and its ground state is four-fold degenerate. Upon application of an external magnetic field, the four Zeeman levels split equidistantly, and one NMR line $H_r = H_0(\nu)$ is observed in terms of Eq. (7). In the tetragonal phase, the ground state is split by quadrapolar interaction into two doublets: $\pm 3/2$ and $\pm 1/2$. There are then three transitions allowed $\pm 3/2 \leftrightarrow \pm 1/2$ and $+1/2 \leftrightarrow -1/2$. The latter is nearly undisplaced, whereas for the former, on symmetry grounds $A_1 = 0$ and they are thus shifted by $A_2\varphi^2$, referring to Eq. (7).

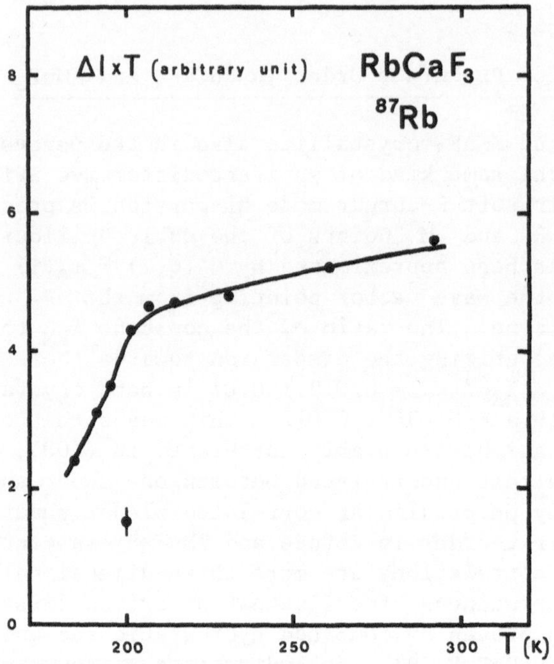

Fig. 7. Temperature dependence of $\Delta I \times T$ showing the failure of Curie law for the NMR signal in the cubic phase of $RbCaF_3$. (After Bulou et al., Ref. 65.)

Bulou et al.[65] have studied the ^{87}Rb NMR above and below, cubic-to-tetragonal phase transition occurring at 198 K. Normally, the intensities I of NMR lines follow a Curie law $I \propto 1/T$ down to the mK range. Thus, $I \times T$ should be approximately constant. Figure 7 shows a plot of this product in the cubic phase. It appears that the NMR signal fails to follow the Curie law below 210 K. This is the temperature below which the c.p. in $RbCaF_3$ gains in intensity.[66] The ^{87}Rb cubic-line intensity decreases below 121 K because it goes into quadrupole split satellite lines arising from axial interactions, although these lines could not be observed because of their frequency spread. The origin of these quadrupolar interactions might be attributed to strains or defects. However, the *same* results were observed in powders as in single crystals prepared in a different manner. This points quite strongly toward an intrinsic cause for the occurrence of the NMR intensity drop, i.e., quadrupole splittings resulting from tetragonal clusters above T_c, which also cause the c.p.

The authors of Ref. 65 also determined by NMR the temperature variation of the order parameter $\varphi(T)$ below T_c. From their data, one finds that $\varphi(T)$ is almost the same as that deduced from neutron scattering, and proportional to the T-dependence obtained by EPR. The authors compared their $\varphi(T)$ dependence to the one they calculated from a model by Rousseau,[67] who assumes a *fixed* order-disorder component in the potential. A strong discrepancy is found which rules out Rousseau's simple model.

Hard-mode Raman activity. Modes not directly involved in the phase transitions — so-called hard modes — are of interest if, on symmetry grounds, they are "Raman inactive" well above T_c, and "Raman active" below T_c. If precursor order exists above the SPT, some intensity will still be present above T_c. Such experiments have been mentioned in Section I for $KTa_{1-x}Nb_xO_3$, and II for $BaTiO_3$.[61] In $RbCaF_3$ and $KMnF_3$, the intensity and width of the hard Eg (and B_{2g}) modes at 203 cm^{-1} and 230 cm^{-1}, respectively, have been measured by a group in Edinburgh.[68] Eg and B_{2g} are the symmetry designation in the tetragonal D_{4h} phase; their parent mode in O_h^1 is R_{25}'. Figure 8 reproduces these measurements in $KMnF_3$ above and below T_c. There is

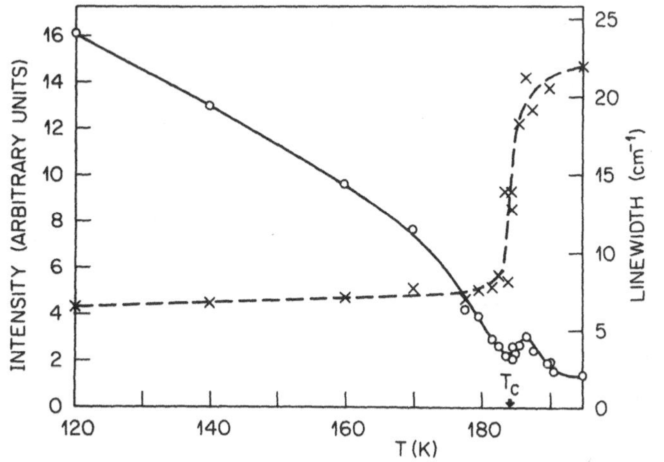

Fig. 8. The temperature dependence of the intensity and the *effective* line width of the *hard* 227 cm^{-1} E_g-mode feature in the light-scattering spectrum of $KMnF_3$. (After Bruce et al., Ref. 68.)

a peak in intensity above T_c as found by birefringence in $SrTiO_3$ by Courtens[69] and assigned to homogeneous strains.

Bruce[68] analyzed these data in terms of precursor order σ_0, see Eq. (7) for $T \to T_c^+$, i.e., "characterizing the degree of quasi-static long but finite-ranged order which implies the existence of a central peak." The Raman intensity J could be written as the sum of three terms,

$$J = J^{LR} + J^{cp} + J^{pH} . \qquad (12)$$

The first term represents the first-order scattering present in the low-temperature phase by macroscopic order, and is proportional to $<\phi>^2$ well below T_c. J^{cp} results from first-order hard-mode scattering induced by the short-range order σ_0, and J^{pH} represents the conventional second-order Raman scattering involving hard and soft-phonon quasi-harmonic interactions. $J^{LR} = 0$ for $T \geq T_c$, but dominates the spectrum well below T_c. Because for J^{LR} and J^{cp} the same coupling constant appears, the former could be used to estimate σ_0 from J^{cp} above T_c. Bruce presents arguments that J^{cp} dominates for T_c^+, but J^{pH} becomes larger on heating and accounts for the broadening of the scattered line. It should also be noted that from the theory, J^{cp} is an integral over-all frequency and wave-vector components contained in the van Howe function of the c.p.

Making use of the estimated saturating values of the soft mode for $T \to T_c$, $\omega_{s\infty}(T = T_c) = \delta(T = T_c)$ [Eq. (7)] and assuming $J(T = T_c) = J^{cp}$, Bruce obtained the upper bounds for $\sigma_0(T = T_c)$ as 3.6 and 3.3 angular degrees in $KMnF_3$ and $RbCaF_3$, respectively. These values are comparable to the measured first-order discontinuities of 3.5°($KMnF_3$) and 2°($RbCaF_3$), respectively. This becomes intelligible if one views the discontinuous process occurring at the first-order instability temperature as an alignment of pre-ordered regions. The authors compared the above σ_0 values with that found in $SrTiO_3$. To do so, they defined a ratio $C_I = \sigma_0/<\varphi(T_0)>$ for $(T_c - T_0)/T_0 = 0.5$. The parameter C_I gives a measure of Ising or order-disorder "content" of critical behavior. For a pure Ising system: $C_I \simeq 1$. The values deduced are 0.3 and 0.4 for $KMnF_3$ and $RbCaF_3$, respectively, as compared to 0.1 in $SrTiO_3$.[50] The markedly larger C_I values than $SrTiO_3$ are consistent with their generally stronger anharmonic character.

d) Dynamic Components in $SrTiO_3$

Up to last year, *all* the c.p. width determinations in $SrTiO_3$ had been carried out on Verneuil-grown crystals. But the dynamic c.p.'s reported earlier in $KMnF_3$ by Hatta[19] or the one in $NaNbO_3$ by Rigamonti's group[20] were observed in flux-grown single crystals or powders produced by chemical reactions. Such crystals and powders contain appreciably fewer defects than those grown by the "brutal" Verneuil method. $SrTiO_3$ crystals synthetized by the latter method contain 10^6 dislocations/cm^2. Recently, $SrTiO_3$ single crystals became available which Scheel had grown in our Laboratory from a borate flux.[70] Their dislocation density is of the order of 1/cm^2, i.e., essentially dislocation-free. However, other defects may, of course, be present.

It seemed worthwhile therefore to carry out EPR experiments in such crystals and analyze them in the manner described in the first section of this Lecture for the Verneuil ones, i.e., to measure for a given set of temperatures $T > T_c$ the EPR linewidth of the $Fe^{3+} - V_0$ center as a function of the sensitivity parameter A_1, i.e., magnetic-field directions, and plot them as a function of A_1 and $A_1{}^2g$.[71] We

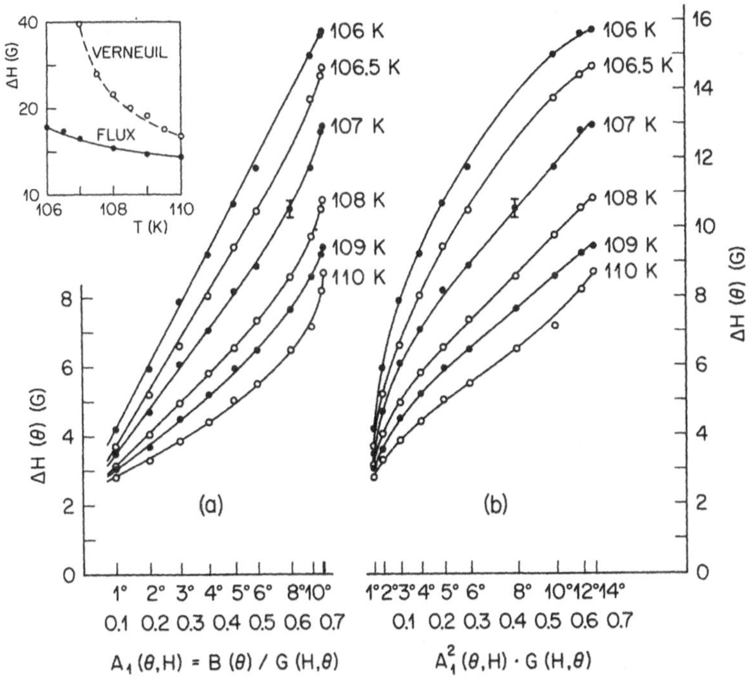

Fig. 9. EPR linewidths $\Delta H(\theta)$ in flux-grown $SrTiO_3$ as a function of sensitivity parameters: (a) A_1 for slow motion; (b) $A_1{}^2 \cdot 6$ for fast motion. Inset compares $\Delta H(\theta = 14^{\circ})$ for Verneuil and flux-grown crystals.

recall that a linear dependence on A_1 implies static, and on $A_1{}^2G$, dynamic behavior. This is done in Fig. 9. Comparing it to Fig. 4, one sees that the dependence is between linear and quadratic at $T - T_c = 3$ K. Thus, there are indeed dynamic components in the c.p. close to T_c! This is the first time that dynamic c.p. components have been observed in $SrTiO_3$! The background width for $A_1 \to 0$ is 3.3 Gauss. Using the argument exposed in the opening paragraph, this means that the local relaxation rate is of the order of 10 MHz. At T = 109 K, the behavior is nearly dynamic, i.e., follows the $A_1{}^2g$ dependence. Thus, the collective relaxation rate is then estimated to be

$$\Gamma' \approx 1/10\ \Gamma_\ell = 1\ \text{MHz}. \qquad (13)$$

The inset of the figure compares the EPR linewidths of the $Fe^{3+} - V_0$ center measured in a Verneuil-grown and a flux-grown crystal above T_c.[71] The magnetic field was parallel to a [110] direction in both experiments, and the rotational order parameter $<\varphi(T)>$ measured below T_c was within experimental accuracy *identical* for the two crystals for given $(T_c - T)$'s. The interpretation follows from the discussion at the beginning of this section, and the one given immediately above: In the Verneuil-grown crystal, the fluctuations in φ, $\delta\varphi(t)$, are static on the experimental time scale, i.e., pinned, and according to Eq. (8), $<\delta\varphi^2> = \Delta H_s^V(T)/2A_1$. In the flux-grown crystals, there are dynamic components averaged out by the EPR experiment with a time scale of 10 MHz, and $\Delta H_s^f < \Delta H_s^V$, because Eq. (8) is invalid.

e) Critical Dynamics in $KMnF_3$

The attenuation of ultrasound at SPT's indicated the existence of critical fluctuations in $SrTiO_3$[7] before other experiments proved it,[8] and the existence of slow dynamics near T_c was then revealed with it[19] in $KMnF_3$. The technique normally employed is to bond a transducer to a very plane face of the crystal, and generate an ultrasound pulse train of frequency ω which travels along the sample. It gets reflected at the other end of the sample, and travels back where it is detected by the transducer. This reflected pulse train is often badly distorted by interference, rendering data inadequate for determination of critical exponents. A new technique has been invented by Shiren and Melcher[72] which avoids these difficulties. The forward acoustic pulse emitted from the transducer is passed through the sample into the echo-active crystal bonded to it, which is located in a microwave cavity. Here a homogeneous electric field $E = E_0\exp(i2\omega t)$ is applied to the forward wave. By a mixing process, a backward acoustic wave is generated. The outcome of the mixing is to *reverse* the wave vector, i.e., $\vec{k} \rightarrow \vec{k}$, at all points on the wave front. The echo wave front now reconstructs continuously, replicating the original form during backward propagation, and finally reaches the transducer as a plane wave. At this stage a *correct* measurement of acoustic amplitude is made.

Recently, Fossheim and Holt[73] used the phonon-echo technique to remeasure very accurately the critical ultrasound absorption α in $KMnF_3$ from 95 to 470 MHz well above T_c. They could fit their data by a law $\alpha \propto \omega^2(T - T_c)^{-\rho}$ with $\rho = 1.87 \pm 0.05$ in good agreement with R.G. calculations of Murata[74] for the n = d = 3 Heisenberg model, $\rho = 1.34 + 2(\phi - 1)$ where ϕ is the crossover exponent $\phi = 1.26$. This appears appropriate, as away from T_c, the static critical exponent is $\beta \approx 0.3$ .. 0.4 from EPR and other measurements[75] in agreement with the Heisenberg model.

On approaching T_c, the attenuation did *not* continue to diverge as $(T - T_c)^{-\rho}$, rather, a rollover was found. To study this behavior, the data were analyzed with respect to a dynamic scaling function $G(\omega\tau)$[76], i.e.,

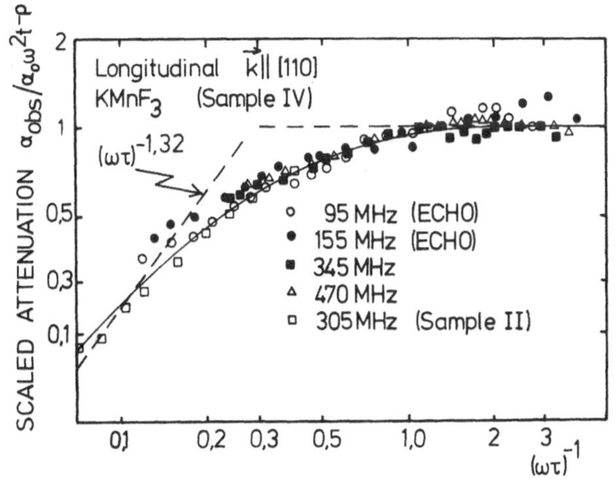

Fig. 10. Dynamic scaling function for ultrasonic attenuation in $KMnF_3$. The fully drawn curve represents the average of all data from a large number of frequencies with use of four different modes. Data points refer to one particular mode: L[100]. The dashed lines show the expected asymptotic behavior as discussed in the text. (After Modine et al., Ref. 75.)

$$\alpha \sim \omega^2 t^{-1.87} G(\omega\tau).$$

$\tau = \xi^z = \xi_0 t^{-\nu z}$ is the relaxation time of correlated regions, and ξ is the correlation length. $G(\omega\tau)$ may be determined from a plot of $\alpha_{obs}/\omega^2 t^{-1.87}$ vs $\omega\tau$. This plot is shown in Fig. 10. Clearly for $\omega\tau \ll 1$, i.e., far from T_c, where τ is short, $G(\omega\tau) = 1$. For $T \to T_c^+$, the ultrasonic attenuation is not expected to exhibit singular behavior. This means that $G(\omega\tau)$ must have the following asymptotic form in the limit $\omega\tau \gg 1$:

$$G(\omega\tau) \sim (\omega\tau)^{-\rho/\nu z}.$$

The authors now assumed that ρ and νz are still those calculated from Heisenberg fixed-point behavior, $\rho = 1.87$, $\nu = 0.65$, $z = 2$, i.e., $G(\omega\tau) \sim (\omega\tau)^{-1.32}$. This asymptote also shown is not well tangential to their data. This is not unexpected because the Heisenberg fixed point is *not* accessible in $KMnF_3$[63] as in $RbCaF_3$ the cubic fluctuations drive the system to a first-order transition.[64]

To obtain the τ-dependence for the argument in $G(\omega\tau)$, or the abscissa $(\omega\tau)^{-1}$ in the figure shown, Fossheim and Holt wrote: $\omega\tau = (\omega/\omega_0)t^{-\nu z}$ with $t = (T - T_c)/T_c$. The condition $\omega\tau = 1$ corresponds to a crossover temperature $t_\Delta = (\omega/\omega_0)^{1/\nu z}$. Plotting t_Δ versus ω gave the exponent $1/\nu z = 0.7$ in agreement with the Heisenberg model. But one can take it as an experimentally determined quantity. The characteristic frequency ω_0 was also found: $\omega_0/2\pi \approx 1.8 \times 10^{11}\ s^{-1}$. The relaxation time in units of seconds is then

$$\tau \approx 9 \times 10^{-13}\ t^{-\nu z} = 9 \times 10^{-13}\ t^{-1.4}\ .$$

The time constant τ, determined in this manner, may be interpreted as the characteristic time of the cluster dynamics near T_c. Also, it may be seen as a direct manifestation of the inverse width of the central peak which is too narrow to be determined from neutron data.

III. INCOMMENSURATE SYSTEMS

Due to space limitations, this lecture has not been put into writing. Well-written introductions to these systems can be found in Chapter I of Ref. 3 and Chapters II and III of Ref. 4. This lecture is in part adapted from them and especially from a review by R. Blinc, entitled "Structurally Incommensurate Systems — Magnetic Resonance Detects Phonons and Solitons" to appear in the J. of Magnetic Resonance.

REFERENCES

1. From 1973 on, these volumes were edited by T. Riste with Noordhoff, Leiden, and Plenum, New York.
2. "Structural Phase Transitions and Soft Modes," E. J. Samuelsen, E. Andersen, and J. Feder, eds., Universitetsforlaget, Oslo (1971); J. F. Scott, Rev. Mod. Phys. 46:83 (1974).
3. A. D. Bruce and R. A. Cowley, "Structural Phase Transitions," Taylor and Francis Ltd., London, (1981).
4. "Structural Phase Transitions," K. A. Müller and H. Thomas, eds., Current Topics in Physics 23, Springer, Berlin (1981).
5. K. A. Müller, in: "Dynamical Critical Phenomena and Related Topics," p. 210, C. P. Enz, ed., Lecture Notes in Physics 104, Springer, Berlin (1979).
6. W. Cochran, Adv. Phys. 9:387 (1960).
7. B. Berre, K. Fossheim, and K. A. Müller, Phys. Rev. Lett. 23:589 (1969).
8. K. A. Müller and W. Berlinger, Phys. Rev. Lett. 26:13 (1971).
9. T. Riste, E. J. Samuelsen, K. Otnes, and J. Feder, Solid State Commun. 9:1455 (1971).
10. S. M. Shapiro, J. D. Axe, G. Shirane, and T. Riste, Phys. Rev. B 6:4332 (1972).
11. J. K. Kjems, G. Shirane, K. A. Müller, and H. J. Scheel, Phys. Rev. B 8:1119 (1973).
12. Th. von Waldkirch, K. A. Müller, W. Berlinger, and H. Thomas, Phys. Rev. Lett. 28:503 (1972).
13. P. Hohenberg, (1971) unpublished.
14. R. A. Cowley, Ferroelectrics 6:163 (1974).
15. J. Feder, in: "Local Properties at Structural Phase Transitions," K. A. Müller and A. Rigamonti, eds., North-Holland, Amsterdam (1976) p. 312.
16. J. D. Axe, S. M. Shapiro, G. Shirane, and T. Riste, in: "Anharmonic Lattices, Structural Transitions and Melting," T. Riste, ed., Noordhoff-Leiden, (1974) p. 23.
17. J. Töpler, B. Alefeld, and A. Heidemann, J. Phys. C 10:635 (1977).
18. C. N. W. Darlington and D. A. O'Connor, J. Phys. C 9:3561 (1976) and "Lattice Dynamics," Proc. Int. Conf., H. Balkanski, ed., Flammarion, Paris (1977) p. 750.
19. I. Hatta, M. Matsudo, and S. Sawada, J. Phys. C 7:L299 (1974).
20. A. Avogadro, G. Bonera, F. Borsa, and A. Rigamonti, Phys. Rev. B 9:3905 (1974).
21. F. Denoyer and R. Currat, "Neutron Inelastic Scattering," Proc. Symp. IAEA, Vienna (1977) p. 273.
22. G. F. Reiter, W. Berlinger, K. A. Müller, and P. Heller, Phys. Rev. B 21:1 (1980).
23. R. Folk and F. Schwabl, Solid State Commun. 15:937 (1974).

24. B. I. Halperin and C. M. Varma, Phys. Rev. B 14:4030 (1976).
25. R. G. Petschek, Phys. Rev. B 22:1409 (1980).
26. (a) K. H. Höck and H. Thomas, Z. Phys. B 27:267 (1977).
(b) H. Schmidt and F. Schwabl, Phys. Lett. 61A:476 (1977).
27. J. B. Hastings, S. M. Shapiro, and B. C. Frazer, Phys. Rev. Lett. 40:237 (1978).
28. Y. Yacoby, Z. Phys. B 31:275 (1978) and references therein.
29. R. Currat, K. A. Müller, W. Berlinger, and F. Denoyer, Phys. Rev. B 17:2937 (1978).
30. K. B. Lyons and P. A. Fleury, Solid State Commun. 23:477 (1977).
31. E. Courtens, Phys. Rev. Lett. 37:1584 (1976).
32. T. Schneider and E. Stoll, Phys. Rev. Lett. 31:1254 (1973); Phys. Rev. B 13:1216 (1976).
33. S. Aubry, J. Chem. Phys. 62:3217 (1975); 64:3392 (1976).
34. T. R. Koehler, A. R. Bishop, J. A. Krumhansl, and J. R. Schrieffer, Solid. State Commun. 17:1515 (1975).
35. J. A. Krumhansl and J. R. Schrieffer, Phys. Rev. B 11:3535 (1975).
36. "Solitons and Condensed Matter Physics," A.R. Bishop and T. Schneider, eds., Springer Series in Solid State Sciences, Springer, Heidelberg (1978).
37. T. Schneider and E. Stoll, Phys. Rev. B 16:2220 (1977).
38. T. Schneider and E. Stoll, Phys. Rev. B 10:2004 (1974).
39. A. D. Bruce, in: "Solitons and Condensed Matter-Physics," A. R. Bishop and T. Schneider, eds., Springer, Heidelberg (1978) p. 116.
40. A. D. Bruce, T. Schneider, and E. Stoll, Phys. Rev. Lett. 43:1284 (1979).
41. E. Stoll, Phys. Lett. 58A:121 (1976).
42. M. E. Fisher, Rev. Mod. Phys. 46:597 (1974).
43. A. I. Larkin and D. E. Khmel'nitskii, Sov. Phys.—JETP 29:1123 (1969).
44. B. A. Strukov, Ferroelectrics 12:97 (1976); K. H. Ehses and H. E. Müser, Ferroelectrics 12:247 (1976).
45. A. Aharony and A. D. Bruce, Phys. Rev. Lett. 42:462 (1979).
46. (a) E. Courtens, Phys. Rev. Lett. 41:1171 (1978);
(b) E. Courtens, R. Gammon, and S. Alexander, Phys. Rev. Lett. 43:1026 (1979), and (c) to be published.
47. R. Folk, H. Iro, and F. Schwabl, Z. Phys. B 25:69 (1976).
48. Th. von Waldkirch, K. A. Müller, and W. Berlinger, Phys. Rev. B 5:4324 (1972) and B 7:1052 (1973).
49. A. Abragam, in: "The Principles of Nuclear Magnetism," Clarendon, Oxford (1961) Chap. 10, p. 424.
50. A. D. Bruce, K. A. Müller, and W. Berlinger, Phys. Rev. Lett. 42:185 (1979).
51. E. F. Steigmeier and H. Auderset, Solid State Commun. 12:565 (1973).

52. A theoretical rationale for Eq. (10) to hold is given by G. Meissner, N. Menyhàrd, and P. Szépfalusy in this issue.
53. G. M. Meyer, R. J. Nelmes, and I. Hutton, "Lattice Dynamics", Proc. Int. Conf., H. Balkanski, ed., Flammarion, Paris (1977) p. 652.
54. A. D. Bruce and R. A. Cowley, Adv. Phys. 29:219 (1980).
55. K. B. Lyons and P. A. Fleury, Phys. Rev. B 17:2403 (1978).
56. A. Aharony and M. E. Fisher, Phys. Rev. B 8:3323 (1973); A.D. Bruce and A. Aharony, Phys. Rev. B 10: 2078 (1974).
57. R. Kind and K. A. Müller, Commun. Phys. 1:223 (1976) and new data analysis (unpublished).
58. Y. Luspin, J. L. Servoin, and F. Gervais, J. Phys. C 13:3761 (1980); J. L. Servoin, F. Gervais, and K. A. Müller, to be published.
59. W. G. Spitzer, R. C. Miller, D. A. Kleinman, and L. E. Howarth, Phys. Rev. 126:1710 (1962); J. Harada, J. D. Axe, and G. Shirane, Phys. Rev. B 4:155 (1971).
60. R. Comes, M. Lambert, and A. Guinier, Solid State Commun. 6:715 (1968). M. Lambert and R. Comes, Solid State Commun. 7:305 (1969).
61. A. M. Quittet and M. Lambert, Solid State Commun. 12:1053 (1973); D. Heiman and S. Ushioda, Phys. Rev. 9:2122 (1974).
62. M. Rousseau, J. Nouet, and R. Almairac, J. Phys. (Paris) 38:1423 (1977), see Table II.
63. T. Nattermann, J. Phys. C 9:3337 (1976)
64. (a) J.Y. Buzaré, J. C. Fayet, W. Berlinger, and K. A. Müller, Phys. Rev. Lett. 42:465 (1979). (b) K. A. Müller, W. Berlinger, J. Y. Buzaré, and J. C. Fayet, Phys. Rev. B 21:1763 (1980).
65. A. Bulou, H. Théveneau, A. Trokiner, and P. Papon, J. Phys. (Paris) 40:L277 (1979).
66. R. Almairac, M. Rousseau, J. Y. Gesland, J. Nouet, and B. Hennion, J. Phys. (Paris) 38:1429 (1977).
67. M. Rousseau, J. Phys. (Paris) 40:L439 (1979).
68. A. D. Bruce, W. Taylor, and A. F. Murray, J. Phys. C 13:483 (1980).
69. E. Courtens, Phys. Rev. Lett. 29:1380 (1972).
70. H. J. Scheel, J. G. Bednorz, and P. Dill, Ferroelectrics 13:507 (1976).
71. K. A. Müller, W. Berlinger, and H. J. Scheel, to be published.
72. N. S. Shiren and R. L. Melcher, in: "Phonon Scattering in Solids", L. J. Challis, V. W. Rampton, and A. F. G. Wyatt, eds., Plenum, New York (1976) p. 405.
73. K. Fossheim and R. M. Holt, Phys. Rev. Lett. 45:730 (1980).
74. K. K. Murata, Phys. Rev. B 13:4015 (1976).

75. F. A. Modine, E. Sonder, W. P. Unruh, C. B. Finch, and R. D. Westbrook, Phys. Rev. B 10:1623 (1974).
76. P. C. Hohenberg and B. I. Halperin, Rev. Mod. Phys. 49:435 (1977).

THE THEORY OF STRUCTURAL PHASE TRANSITIONS:

UNIVERSALITY AND QUASI-ELASTIC SCATTERING PHENOMENA

Alastair D. Bruce

Department of Physics
University of Edinburgh
Scotland, U.K.

PREAMBLE

The theoretical study of structural phase transitions over the last decade has been dominated by two continuing concerns. Firstly, much effort has been devoted to the determination of the universal quantities — principally critical exponents — characterising critical point singularities. The surge of activity in this area is mainly attributable to the development of renormalisation group (R.G.) methods[1], making it possible to explore a great wealth of critical phenomena largely inaccessible to the techniques previously available. The second dominant theme has been an ever-growing interest in the patterns of short range order (s.r.o.) underlying critical point behaviour, and the qualitative characteristics which these patterns imply for the collective excitation spectrum near criticality. The motivation for much of the work in this field has been the evidence assembled from, successively, experiments[2], computer-simulation hybrids of theory and experiment[3] and theory itself[4], suggesting that both the excitation spectrum and the underlying configurations of s.r.o. display important features, associated with cluster walls, which are not captured by classical phonon-based theories of structural phase transitions (s.p.t.'s).

It is not my intention here to survey these two territories in detail: recent reviews are to be found elsewhere[5,6]. I wish, rather, to present the results of recent studies in each area. In the process I shall endeavour to give some impression of the current state and prospects of each one, and to emphasize that although research in each area has largely been pursued independently there is, in fact, much common ground.

More specifically I will begin (Section 1) with a brief review of the elements of s.p.t. theory, introducing the important observables, and using a R.G.-based argument to establish the network of links between their critical singularities which forms the heart of universality.

Section 2 is designed to illustrate the fact that the spectrum of universal quantities emerging from contemporary theory includes quantities considerably more exotic than familiar exponents and amplitude ratios. In particular we will consider two relatively novel universal quantities, linked by their common relevance to quasi-elastic scattering phenomena near s.p.t.'s.

Universality and scattering phenomena also form two of the themes of the remaining, and major, portion of these lectures (Section 3) which will be devoted to the second of the two areas identified above. I will discuss a recent study establishing that patterns of short range order themselves have universal characteristics, and offering insights into the role of the cluster wall in pattern formation and in the excitation spectrum — not only in one dimension (to which the bulk of earlier work has been restricted) but in higher dimensions also.

1. BACKGROUND

1.1 A Simple Model and the Basic Observables

Much of the essential physics of a s.p.t. is captured by a simple model which has, in fact, formed the basis of the majority of recent analytic and computer-aided studies of the s.p.t. problem. The model is defined by the classical Hamiltonian

$$H = V[\{u(\vec{x}_i)\}] + \sum_i \frac{p^2(\vec{x}_i)}{2M} \qquad (1.1a)$$

The set $\{u(\vec{x}_i)\}$ denote the displacements of a set of N atoms, of mass M, with respect to the sites of a d-dimensional hypercubic lattice, along one specific direction (the coordinates are taken to be scalars) singled out by some strong uniaxial anisotropy; the set $\{p(\vec{x}_i)\}$ are the canonically conjugate momenta. The configurational energy V is defined by

$$V = \sum_i V_s(u(\vec{x}_i)) + \frac{C}{2}\sum_{i,j}^{n.n.} [u(\vec{x}_i) - u(\vec{x}_j)]^2 - E\sum_i u(\vec{x}_i) \qquad (1.1b)$$

The single-site potential V_s embodies the energy of interaction with some underlying sublattice of atoms which do not participate actively in the phase transition; the sublattice is taken to be rigid (the model has no acoustic phonons) and the potential $V_s(u)$ is assumed to have some simple symmetric double-well (and thus anharmonic!) form, with minima at $u = \pm u_0$ and a maximum at the high symmetry site $u = 0$. The second term in eqn. (1.1b) originates in harmonic bonds acting between near-neighbours; the coupling constant C is positive. The remaining term expresses the effect of an ordering field E which will generally be set to zero.

The basic physics of the ordering process that occurs in this model is described by a few key quantities which we now identify, noting in the process the general power law forms of the critical singularities which they exhibit (in dimensionalities satisfying $1 < d < 4$).

We observe firstly that the expectation value $\langle u(\vec{x}_i)\rangle$ is clearly non-vanishing at T = 0 since the favoured configurations are then the two degenerate states in which each atom resides at the bottom of its right (left) well, so that $\langle u(\vec{x}_i)\rangle = +u_0\ (-u_0)$. It is equally clear that, at high temperature where intersite correlation can be ignored, $\langle u(\vec{x}_i)\rangle$ must vanish identically. Accordingly a phase transition is to be expected — its existence can be demonstrated rigorously[7] — at some critical temperature T_c below which there is long range order (ℓ.r.o.) expressed in a non-zero value of the order parameter Q_0 defined by

$$Q_0 = \langle u(\vec{x}_i)\rangle = \frac{1}{N}\langle u(\vec{q}_s = 0)\rangle \tag{1.2}$$

Here $u(\vec{q}_s)$ is one of a set of Fourier coordinates defined by

$$u(\vec{x}_i) = \frac{1}{N}\sum_{\vec{q}} u(\vec{q})e^{-i\vec{q}.\vec{x}_i} \tag{1.3}$$

while $\vec{q}_s$ (zero for the model considered here) is the wavevector characterising the periodicity of the ℓ.r.o. The phase transition is believed to be continuous, the order parameter vanishing smoothly as $T \to T_c^-$ with a power law characterised by an exponent β and an amplitude B,

$$Q_0 \approx B|t|^\beta \tag{1.4}$$

where $t \equiv (T - T_c)/T_c$ is the reduced temperature.

The critically-growing softness of the system against distortions with the profile of the ℓ.r.o is reflected in the behaviour of the isothermal order parameter susceptibility

$$\chi_T = \left(\frac{\partial Q_o}{\partial E}\right)_T = (k_B T)^{-1} \sum_j \langle (u(\vec{x}_i) - Q_o)(u(\vec{x}_j) - Q_o) \rangle$$

$$= (k_B T)^{-1} \{\frac{1}{N} \langle |u(\vec{q}_s = 0)|^2 \rangle - NQ_o^2\} \tag{1.5}$$

which exhibits the critical behaviour

$$\chi_T \approx \begin{cases} C_+^T \, t^{-\gamma} & (T \to T_c^+) \\ C_-^T \, (-t)^{-\gamma} & (T \to T_c^-) \end{cases} \tag{1.6}$$

More generally we may define a wavevector- and frequency-dependent susceptibility $\chi(\vec{q},\omega)$ characterising the response of the ordering coordinates to a space and time dependent field. The isothermal and adiabatic order parameter susceptibilities are recovered as the limits[8]

$$\chi_T = \lim_{q \to o} \lim_{\omega \to o} \chi(\vec{q},\omega)$$

$$\chi_S = \lim_{\omega \to o} \lim_{q \to o} \chi(\vec{q},\omega) \tag{1.7}$$

The spectral function $S(\vec{q},\omega)$ giving the scattering cross section for neutrons or X-rays (in the one phonon approximation) is related to the imaginary part of $\chi(\vec{q},\omega)$ by the fluctuation dissipation result (we denote the time coordinate by τ)

$$S(\vec{q},\omega) = \frac{1}{2\pi N} \int_{-\infty}^{\infty} \langle u(\vec{q},0) u(-\vec{q},\tau) \rangle e^{i\omega\tau} \, d\tau$$

$$= \frac{k_B T}{\pi\omega} \, \mathrm{Im} \, \chi(\vec{q},\omega) + N \, Q_o^2 \delta(\omega) \Delta(\vec{q}) \tag{1.8}$$

so that the net (frequency-integrated) scattering for wavevector transfer $\vec{q}$ is given by

$$I(\vec{q},t) = \int_{-\infty}^{\infty} S(\vec{q},\omega) d\omega = \frac{1}{N} \langle u(\vec{q}) u(-\vec{q}) \rangle$$

$$= k_B T \, \chi(\vec{q},\omega{=}0) + N \, Q_o^2 \Delta(\vec{q}) \tag{1.9}$$

The terms proportional to Q_o^2 in eqns. (1.8) and (1.9) represent the superlattice scattering which signals the existence of ℓ.r.o. The remaining terms represent the fluctuation scattering which is the signature of short range order. The critically-growing spatial and temporal coherence of the s.r.o. (as $T \to T_c^+$) is expressed in the critical narrowing of the fluctuation scattering in q-space and ω-space; the q- and ω-widths are the inverses of the correlation length ξ and correlation time τ which have critical singularities of the form

$$\xi \approx \xi_+ t^{-\nu} \tag{1.10a}$$

$$\tau \approx \tau_+ t^{-z\nu} . \tag{1.10b}$$

The spectrum of basic observables relevant to these lectures is completed by the specific heat (per particle) which exhibits a relatively weak critical singularity; the asymptotic $T \to T_c^+$ behaviour is customarily written in the form

$$\tilde{C}_{E=0} \approx \frac{A_+}{\alpha} t^{-\alpha} + A_o . \tag{1.11}$$

1.2 General Theory of Scaling and Universality

The renormalisation group theory of the critical region provides the foundation for a number of very general statements regarding the behaviour of the model defined above. Two of these — widely believed though by no means rigorously established — are of particular importance. Firstly it is believed[1] that the equilibrium critical properties of any model of the form (1.1) are controlled by a fixed point of the R.G., uniquely prescribed by the system dimensionality and termed the "Ising" fixed point, since the true (fixed-length-spin) Ising model is a special case of (1.1); accordingly all such models are said to belong to the Ising universality class.† Secondly, as regards its time-dependent critical behaviour, the system (1.1) belongs to the universality class of Ising models with conserved energy[11]. Although, in Section 3, we will be deeply concerned with the qualitative character of the excitation spectrum of the model (1.1), our explicit considerations will be restricted to the equilibrium properties. Accordingly we explore here only the

†The evidence suggesting a failure of universality in d = 3, implicit in the results reported by Baker and Kincaid[9], is substantially undermined by the remarkable recent studies of Nickel[10].

first of these two assertions.

Consider the n-point correlation function

$$K^{(n)}(t,E,x) \equiv \langle u(\vec{x}_1) \cdots u(\vec{x}_n)\rangle_c \qquad (1.12)$$

where x is a symbolic representative of the set of position coordinates $\vec{x}_1 \cdots \vec{x}_n$, and c denotes connected part. Given the existence of a unique critical fixed point with two relevant scaling fields proportional respectively to t and E a simple R.G. argument readily establishes that the correlation functions have the homogeneity property[12]

$$K^{(n)}(t,E,x) = (a_0 b^{\lambda_0})^n \tilde{K}^{(n)}(b^{\lambda_1}a_1 t, b^{\lambda_2}a_2 E, b^{-1}a^{-1}x) \qquad (1.13)$$

where b is an arbitrary scale factor. The functions $\tilde{K}^{(n)}$ are universal (i.e. unique to the universality class and thus, in this case, dependent only upon dimensionality). The exponents λ_0, λ_1 and λ_2 also have universal values characteristic of the Hamiltonian flow under the R.G. operation, in the vicinity of the fixed point[5]. The parameters a_0, a_1 and a_2 are non-universal scale factors (i.e. quantities which depend upon the microscopic details of the Hamiltonian (1.1)). Equation (1.13) is predicted to hold asymptotically, in the regime in which t and E are sufficiently small and the inter-site separations (between the positions $\vec{x}_1 \cdots \vec{x}_n$) are sufficiently large. It is an extremely powerful result. In Section 3 we shall see that it guarantees that ordering-coordinate configurations themselves display universal features. In Section 2 we shall explore its consequences for two problems arising in the theory of critical scattering. As a prelude we will first assemble its implications for the critical singularities of the observables identified in the preceding section.

According to equation (1.13) in the case n = 1 the order parameter obeys an equation of state of the form

$$Q_0 = a_0 b^{\lambda_0} \tilde{K}^{(1)}(b^{\lambda_1}a_1 t, b^{\lambda_2}a_2 E) \qquad (1.14)$$

Thus, in zero field, the order parameter has a power law behaviour of the form (1.4) with an exponent and amplitude identifiable as

$$\beta = -\lambda_0/\lambda_1 \qquad\qquad B \doteq a_0 a_1^{-\lambda_0/\lambda_1} \qquad (1.15)$$

The symbol $\doteq$ is introduced to denote equality to within universal multiplicative factors. (i.e. the amplitude B is given by the combination of non-universal scale factors indicated, multiplied by a universal factor determined, in principle, by the function $\tilde{K}^{(1)}$. The order parameter susceptibility follows from (1.14) on differentiating with respect to the field E. In zero field we find a power law singularity of the form (1.6) with

$$\gamma = (\lambda_0 + \lambda_2)/\lambda_1 \qquad C_\pm^T \doteq a_2 a_0 a_1^{-(\lambda_0+\lambda_2)/\lambda_1} \tag{1.16}$$

Next we note that the form of the two-point correlation function

$$\langle u(0)u(\vec{x})\rangle = K^{(2)}(t,x) = (a_0 b^{\lambda_0})^2 \tilde{K}^{(2)}(b^{\lambda_1} a_1 t, b^{-1} a^{-1} x) \tag{1.17}$$

implies a correlation length divergence of the form (1.10a) with

$$\nu = 1/\lambda_1 \qquad \xi_+ \doteq a\, a_1^{-1/\lambda_1} \tag{1.18}$$

Finally (and less trivially) one can determine the character of the specific heat singularity, given the equation of state (1.14)[13]. The result may be written in the form (1.11) with

$$\alpha = 2 - d/\lambda_1 \qquad A_+ \doteq T_c^{-1} a_0^{(\lambda_2-d)/\lambda_0} a_1^{d/\lambda_1} a_2^{-1} \tag{1.19}$$

It is clear, then, that the familiar critical singularities are controlled by the three universal exponents λ_i and the three non-universal scale factors a_i , i = 0,1,2. However neither the three exponents, nor the three scale factors are independent. To establish the relationships which exist between λ_0 and λ_2, and between a_0 and a_2, we utilise the alternative representation of the susceptibility

$$\chi_T = (k_B T)^{-1} \sum_j \{\langle u(\vec{x}_i)u(\vec{x}_j)\rangle - Q_0^2\} \simeq (k_B T_c a^d)^{-1} \int d^d x\, K^{(2)}(t,E,x) \tag{1.20}$$

Making use of eqn. (1.13) with n = 2 one finds a power law singularity in zero field, with

$$\gamma = \frac{2\lambda_0 + d}{\lambda_1} \qquad C_\pm^T \doteq (k_B T_c)^{-1} a_0^2\, a_1^{-(2\lambda_0+d)/\lambda_1} \tag{1.21}$$

Consistency with equations (1.16) then requires

$$\lambda_0 = \lambda_2 - d \qquad a_0 \doteq k_B T_c\, a_2 \tag{1.22}$$

It thus follows, firstly, that all the primary critical exponents may be expressed simply in terms of the two universal flow-exponents λ_1 and λ_2 and, secondly, that all the amplitudes of the primary singularities may be expressed in terms of the two non-universal scale factors a_1 and a_2, together with appropriate factors of the critical temperature T_c and the lattice spacing a. The former result is a familiar corollary of strong scaling theories[14]. The latter result, rather less widely appreciated, is sometimes referred to as two scale factor universality[15].

2. NOVEL UNIVERSAL QUANTITIES

2.1 Adiabatic and Isothermal Susceptibilities

Let us consider the relationship between the adiabatic and isothermal order parameter susceptibilities χ_S and χ_T. According to elementary thermodynamic arguments[14]

$$\frac{\chi_S}{\chi_T} = 1 - \frac{T}{\chi_T \tilde{C}_E} \left(\frac{\partial Q_o}{\partial T}\right)_E^2 \tag{2.1}$$

In the regime $T > T_c$ where $Q_o(T) \equiv 0$ (in zero field) it follows immediately that the two susceptibilities are identical. In the low temperature phase where Q_o is non-vanishing $\chi_S \neq \chi_T$. Making use of equations (1.14), (1.16) and (1.19) we find immediately that in zero field

$$\lim_{t \to 0^-} \frac{T}{\chi_T \tilde{C}_E} \left(\frac{\partial Q_o}{\partial T}\right)_E^2 \equiv r^*_{LP} \tag{2.2}$$

is universal, being independent of all the non-universal factors a_1, a_2, a and T_c. Two important results follow as immediate corollaries.

Firstly, the ratio of the amplitudes characterising the divergence of the adiabatic susceptibility as $T \to T_c^{\pm}$ is itself universal:

$$\frac{C_+^S}{C_-^S} \equiv \lim_{t \to 0^+} \frac{\chi_S(t)}{\chi_S(-t)} = \frac{1}{1 - r^*_{LP}} \lim_{t \to 0^+} \frac{\chi_T(t)}{\chi_T(-t)} = \frac{1}{1 - r^*_{LP}} \frac{C_+^T}{C_-^T} \tag{2.3}$$

We recall, secondly, that a difference between χ_S and χ_T implies (c.f. eqns. (1.7) and (1.8)) the existence of a thermal diffusion central peak, of width $D_T q^2$, in the spectral density $S(q,\omega)$ of the ordering coordinates[8]. In the limit as $T \to T_c^-$ the ratio of the intensity of this central peak to the residual intensity in the spectrum is given by a universal ("Landau-Placzek") constant

$$\lim_{t \to 0^-} \frac{\chi_T(t) - \chi_S(t)}{\chi_S(t)} = \frac{r^*_{LP}}{1 - r^*_{LP}} \equiv R^*_{LP} \tag{2.4}$$

The values of the universal constants identified in equations (2.2), (2.3) and (2.4) are given in Table 1,[16] for various members of the Ising universality class. The d = 2 results are exact. The d = 3 results follow from series expansion studies[17]. The d = 4 results also describe the universality class of d = 3 Ising models with dipolar forces (uniaxial ferroelectrics); they follow from exact R.G. studies[18] capturing the logarithmic corrections to pure power-law singularities, characteristic of borderline dimensionalities. Two points merit specific comment.

Firstly it is clear that the ratio $\chi(t)/\chi(-t)$ strongly depends upon whether the susceptibility is measured under adiabatic or isothermal conditions: in particular, for uniaxial ferroelectrics the familiar "law of 2" obeyed by the isothermal amplitude ratio[19] is replaced by a "law of 8" in the case of the adiabatic amplitudes. This result goes some way towards explaining the deviations from the law of 2 regularly observed in studies of ferroelectrics, where the dielectric susceptibility is almost invariably measured at frequencies which guarantee adiabatic conditions[16].

Table 1. Universal Amplitude Ratios for Ising-like Systems

	r^*_{LP}	C^T_+/C^T_-	C^S_+/C^S_-	R^*_{LP}	R^*_S
d = 2	0	37.69...	37.69...	0	0.0507...
d = 3	0.75	5.07	20.28	3.0	0.014
d = 4†	3/4	2	8	3	$1/(9\pi \ln t)$

†These results are also applicable to uniaxial dipolar systems where ξ (eqn. 2.10) is to be interpreted as the transverse correlation length[18].

Secondly the results establish that, in $d = 3$ Ising systems (with and without dipolar forces), a considerable proportion of the spectral weight of the order parameter fluctuations in the low temperature phase resides (asymptotically as $T \to T_c^-$) in a thermal diffusion central peak. This prediction is likely to prove hard to verify since this scattering will appear merely as a quasi-elastic halo on the dominant superlattice Bragg peak[20] (c.f. eqn. 1.8). The result does, however, make it plain that in these cases the details of the critical fluctuation spectrum immediately above T_c will be correctly captured only by a theory which takes account of energy conservation and the thermal diffusion mode it implies: indeed it is well-established[11] that the critical slowing down in the $d = 3$ Ising system (c.f. eqn. 1.10b) is controlled by the non-linear coupling of the order parameter to the thermal diffusion mode. However, it may actually be of more importance to note that, in the $d = 2$ case, the weight of the spectrum residing in the thermal diffusion central peak is asymptotically negligible as $T \to T_c^-$. Accordingly a non-linear coupling to the thermal diffusion mode, which has frequently[21] been invoked to account for quasi-elastic scattering unexpectedly appearing in the $T \gtrsim T_c$ regime, cannot be the source of the effect of this kind which (c.f. Section 3) does appear to occur in $d = 2$; the results presented in Section 3 suggest that these observations should be understood as a precursor effect of a rather different kind.

2.2 The Onset of Superlattice Scattering

Consider now the temperature-dependence of the cross-section for the scattering of X-rays or neutrons in an experiment probing the behaviour of the ordering coordinates in an Ising-like system. Equation (1.9) gives the net scattering for a specific wavevector transfer $\vec{q}$, in the one phonon approximation. In practice, as a result of the finite instrumental resolution, such an experiment actually measures the scattering associated with a range of momentum transfers in the vicinity of the condensing mode. The observed cross section is thus of the form

$$\mathcal{J}_{\Lambda_R}(t) = \sum_{\vec{q}} I(\vec{q},t)\, F(\vec{q}/\Lambda_R) \tag{2.5}$$

where $F(\vec{q}/\Lambda_R)$ is a resolution function which picks out the contributions of modes lying close to $\vec{q}_s$. Taking $\vec{q}_s$ to be zero†, as it is

† The following arguments are actually more relevant to the case where $\vec{q}_s \neq 0$, when the scattering is not complicated by a sublattice Bragg peak.

for the model (1.1), we may write

$$F(\vec{q}/\Lambda_R) \begin{cases} = 1 & q \ll \Lambda_R \\ = 0 & q \gg \Lambda_R \end{cases} \tag{2.6}$$

so that, recalling eqn. (1.9)

$$\mathcal{G}_{\Lambda_R}(t) = NQ_0^2 + k_B T \sum_{\vec{q}} \chi(\vec{q}, \omega = 0)\, F(\vec{q}/\Lambda_R) \tag{2.7}$$

For an experiment with sufficiently good wavevector resolution there will exist a region of reduced temperatures (below T_c) sufficiently small that the order parameter exhibits the pure power law behaviour (1.4), but still sufficiently large that the superlattice scattering (the first term in 2.7) dominates the fluctuation scattering (the second term). In this regime, then

$$\mathcal{G}_{\Lambda_R}(-t) \simeq N\,B^2(-t)^{2\beta} \tag{2.8a}$$

Above T_c the superlattice scattering vanishes identically; there will exist a comparable region of reduced temperature where one may approximate $\chi(\vec{q},\omega = 0)$ by χ_T (c.f. eqn. 1.7) in the fluctuation-scattering sum, giving

$$\mathcal{G}_{\Lambda_R}(t) \simeq N\,k_B\,T_c\,v_R\,\left(\frac{a}{2\pi}\right)^d C_+^T\,t^{-\gamma} \tag{2.8b}$$

where

$$v_R = \int d^d q\; F(q/\Lambda_R) \tag{2.9}$$

is the resolution volume of the scattering experiment. Combining equations (2.8a) and (2.8b) with equations (1.15), (1.16), (1.18) and (1.22) we find the simple result[22]

$$\frac{\mathcal{G}_{\Lambda_R}(t)}{\mathcal{G}_{\Lambda_R}(-t)} = R_S^*\, v_R\, \xi^d \tag{2.10}$$

where ξ is the correlation length at reduced temperature t, and R_S^* is a universal number whose values for the various Ising universality classes are given in Table 1. Unlike the results established in the preceding section, equation (2.10) rests upon the relationship between a_0 and a_2 prescribed by (1.22): it is thus a direct expression of two scale factor universality, in a form which should be particularly susceptible to experimental test. Physically it asserts that there is a universal link between the amplitudes ξ_+ and C_+^T, which control the growth of short range order above T_c, and the amplitude B, which controls the growth of long range order below the critical point.

This universal relationship is at the heart of a more general result: the evolution in the scattering cross section (2.7) from the pure $(-t)^{2\beta}$ power law form below T_c (when $\Lambda_R\xi \ll 1$) to the pure $t^{-\gamma}$ power law form above T_c (when, once again, $\Lambda_R\xi \ll 1$) is described by the crossover scaling form

$$\mathcal{G}_{\Lambda_R}(t) \simeq B^2|t|^{2\beta}\tilde{\mathcal{G}}(\Lambda_R\xi_+|t|^{-\nu}) \qquad (2.11)$$

where $\tilde{\mathcal{G}}$ is a universal function. The function can be determined[22] with the aid of R.G.-based ε-expansions for the two-point correlation function above and below the critical point[23]. Its form depends weakly on the specific shape of the resolution function $F(\vec{q}/\Lambda_R)$. The results of a calculation assuming a resolution function that prescribes a sharp cut-off on the sphere $|\vec{q}| = \Lambda_R$ are shown in Fig. 1. These calculations allow one to establish the regions within which the scattering cross section may be satisfactorily fitted by pure (2β or $-\gamma$) power laws. Calculations along these lines, but using more realistic resolution functions and improved representations of the correlation function, should ultimately allow one to utilize scattering data throughout the critical region. The temperature-derivative of the scattering intensity is also of interest. It exhibits the singular behaviour of the specific heat — a result which may be established very generally, and which may prove to be of some assistance in the experimental determination of the transition temperature. Although a neutron scattering experiment might seem an excessively elaborate method of investigating the specific heat singularity, it does have one major advantage over calorimetric techniques: it allows one to pick out the contributions to the specific heat originating in the truly critical fluctuations (i.e. those with wavevectors close to $\vec{q}_S$), and thus to discriminate against the contributions from the Brillouin-zone full of non-critical phonon modes which dominate the heat capacity in displacive systems[24]. As discussed in Section 3. an e.p.r. experiment also discriminates against non critical fluctuations; the e.p.r. line width and its temperature-derivative exhibit behaviour very similar to that of the scattering cross section.

3. UNIVERSAL PATTERNS OF SHORT RANGE ORDER

3.1 Introduction

The results presented in the preceding section typify the detail

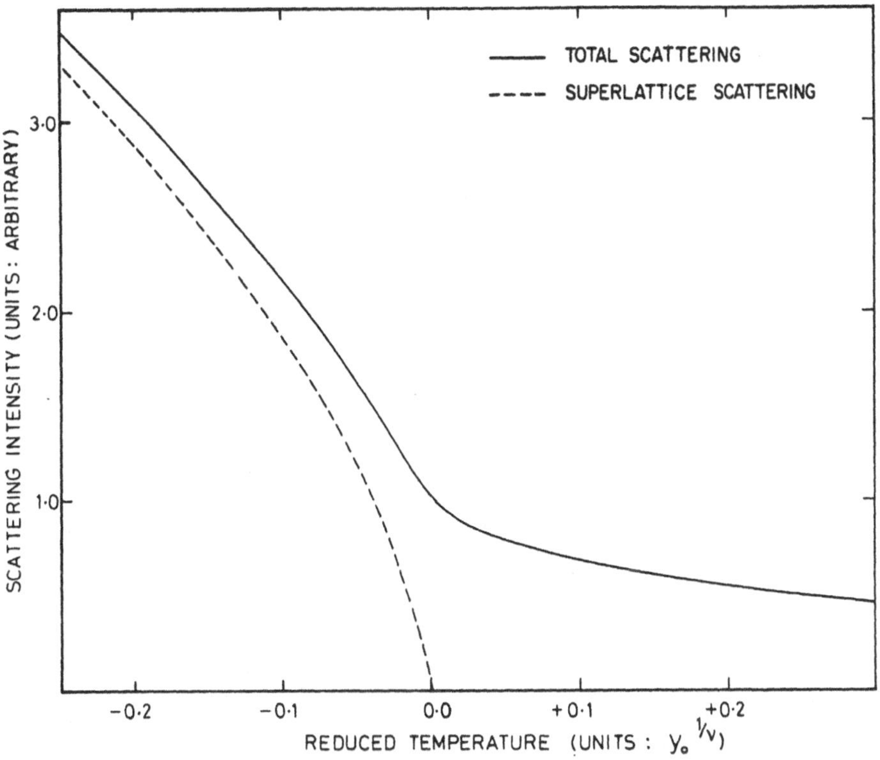

Fig. 1. The temperature-dependence of the one-phonon scattering cross section in an Ising-like system (eqn. 2.11). The cross section is normalised to unity at T_c. The reduced temperature is expressed in units of $(\xi_+ \Lambda_R)^{1/\nu}$ where Λ_R is the radius of the resolution volume.

with which the modern theory of the critical region allows one to characterise critical-point singularities: general R.G. arguments expose a great network of relationships among the basic observables, and explicit realisations of the R.G. (and related techniques) yield actual values for the universal quantities, some of which are now known with great precision[25]. In the course of the remaining portion of these lectures we will see that the contemporary theory may also be used to illuminate the basic physics of the ordering process: we will see that the general R.G. result (1.13) implies that the patterns of short range order themselves have universal features, and we will use an approximate realisation of the R.G. (and other methods) to identify these features explicitly.

To motivate these developments I will first endeavour to identify, more explicitly, the questions about the qualitative character of the ordering process which have begun to emerge in recent years, with the recognition that the picture of this process implicit in classical treatments of the s.p.t. problem is no more trustworthy than the classical predictions for exponent values. In fact the classical framework yields not one picture of the ordering process, but two, corresponding to the fact that the microscopic character of systems (such as 1.1) exhibiting s.p.t.'s has two identifiable extremes, commonly termed "displacive" and "order-disorder"[5].

The microscopic hallmark of a displacive regime is a double-well potential whose depth $E_W \equiv |V_s(u_0)|$ is small on the scale of the thermal energy k_BT

$$g \equiv \frac{E_W}{k_BT} \ll 1 \tag{3.1a}$$

and small on the scale of a typical bond energy $E_B \sim Cu_0^2$:

$$s \equiv \frac{E_W}{E_B} \ll 1 \tag{3.1b}$$

In fact, since $k_BT_c \sim E_B$, at least in dimensionalities $d > 2$, the two conditions are essentially equivalent near the critical point in these cases; in $d = 1$, where the conditions are distinct, it is customary to demand that both should be fulfilled. In the regime thus defined there is, according to eqn. (3.1a) abundant thermal energy for well-to-well motion across the local potential barrier; eqn. (3.1a) implies, moreover, that the overwhelming majority of the Fourier coordinates $u(\vec{q})$ behave quasiharmonically. Accordingly the probability density function (p.d.f.) $P_0(u)$, characterising the distribution of a typical local coordinate u, is essentially Gaussian as indicated in Fig. 2, and as expressed more quantitatively in the value of the parameter

$$G \equiv \frac{3\langle u^2\rangle^2 - \langle u^4\rangle}{2\langle u^2\rangle^2} = -\frac{\Gamma^{(4)}}{2(\Gamma^{(2)})^2} \tag{3.2}$$

where the quantities $\Gamma^{(n)}$ are the cumulants of the p.d.f.; for the displacive p.d.f. $G \simeq 0$.

These microscopic features are usually invoked to justify the

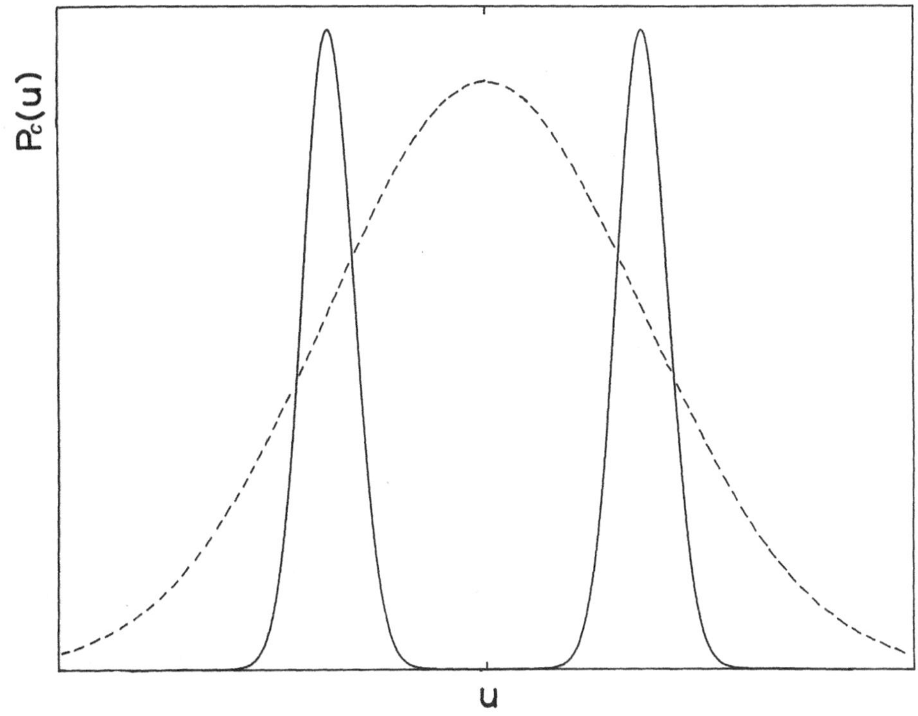

Fig. 2. The p.d.f. of a local ordering coordinate in an Ising-like system in the displacive regime (dashed line) and the order-disorder regime (solid line).

independent mode approximation[26] in which it is assumed that each coordinate $u(\vec{q})$ evolves harmonically in the mean environment created by all the other modes. In this approximation the approach to criticality ($T \to T_c^+$) is signalled by the softening of the modes with wavevectors close to $\vec{q}_s = 0$, which are unstable in the harmonic approximation and are rendered stable by the thermal dressing produced by their interactions with other modes. The growing temporal coherence of the s.r.o. (eqn. 1.10b) is expressed in the vanishing of the soft mode frequency; its growing spatial coherence (eqn. 1.10a) is expressed in the concomitant growth in the population of phonons with wavevectors close to $\vec{q}_s$. The critical scattering spectrum implied by this picture is that of a softening phonon resonance, becoming overdamped and then narrowing critically. The predicted exponent values are those of the spherical model.†

†Strictly, the s.p.t. is first order in this approximation, as discussed in Ref. 5.

The displacive regime thus appears to present a weakly non-linear problem. (It is, in fact, a strongly non-linear problem in disguise.) By contrast the physics of the order-disorder regime is manifestly anharmonic. In this regime the depth of the local potential is large on the scale of the thermal energy, so that g (eqn. 3.1a) is large compared to unity; correspondingly each particle is strongly localised in the vicinity of its well minima $\pm u_0$, and the local coordinate p.d.f. has the strongly double-peaked form suggested schematically in Fig. 2, with a G-value (eqn. 3.2) close to unity. In this case it is clear that the local motion will have two components with quite different time scales: a relatively fast, small-amplitude motion about the well-minima, and a relatively slow, large-amplitude hopping from one well to the other. Both forms of motion have collective counterparts: the counterpart of the first can be envisaged as the collective vibration of the particles about their positions of local equilibrium — essentially the phonons of a quasi-statically disordered system; the collective counterpart of the second mode is to be thought of as the motion of the walls bounding locally ordered clusters — regions within which the particle distortion is predominantly $+u_0$, or predominantly $-u_0$. It is clear that the signatures of both modes should appear in the critical scattering spectrum. However since the existence of the phonon-like mode seems less relevant to the physics of the phase transition the associated in-well degrees of freedom are frequently discarded altogether, while the strongly non-linear inter-well motion is simulated in a semi-phenomenological fashion by the kinetic Ising model[27], (or its quantum analogue, the pseudo-spin model) traditionally solved within an independent site approximation. Within this framework the growing coherence of the s.r.o. (eqns. 1.10a,b) seems to reside in the growth of clusters of local order, and the corresponding growth in the time taken for a cluster wall to diffuse a distance of the order of the maximum cluster size (i.e. the correlation length). The associated scattering takes the form of a central peak, growing in intensity and narrowing critically. The exponent values predicted are those of mean field theory.

In the course of the last ten years the deficiencies of this classical picture have become growingly apparent.

In the first place, of course, the classical predictions (for e.g. exponent values) are now known to be quantitatively incorrect for dimensionalities $d < 4$, the errors increasing the lower the dimensionality, culminating in the well known failure of either (classical) theory to capture the fact that in one dimension there is actually long range order only at zero K.

Secondly it should also be evident that the schizophrenia of the classical view of the ordering process in displacive and order-disorder regimes is hard to reconcile with the spirit of universality.

Thirdly, and for our purposes most significantly, there is now some reason to believe that a system which outside the critical region displays displacive characteristics may, with the approach to criticality, exhibit a crossover to behaviour more reminiscent of the order-disorder regime. The first evidence for such a phenomenon was supplied by neutron-scattering studies[2] of the phase transition in $SrTiO_3$, supposedly the paradigm displacive system: the excitation spectrum of the ordering coordinates, dominated at high temperatures by a softening phonon, was found to display, in addition, a narrow central peak growing critically and ultimately dominant. These observations will be discussed elsewhere[28] and it is not appropriate to dwell on the details here. I will remark only that, while the degree to which these results reflect intrinsic crystal behaviour remains uncertain,[6] they unquestionably supplied the motivation for subsequent theoretical studies which have left little doubt that such a phenomenon does occur intrinsically, at least in some circumstances. Specifically, molecular dynamics studies[3,29,30] of models of the form (1.1) were also found to display a crossover, with the approach to criticality, from a one-feature (soft phonon) excitation spectrum to a two feature (phonon plus central peak) spectrum, in both one and two dimensions. Analytic studies of the $d = 1$ model (1.1) subsequently correlated this effect with a crossover, explicitly identifiable in the equilibrium properties[4,31], from a phonon-dominated regime to a regime in which both phonons and cluster walls have to be taken into account, with cluster wall behaviour controlling the growth in short range order. This programme has more recently been extended to include the excitation spectrum explicitly[32].

These developments have raised many issues deserving of attention: the implications of universality for the order-disorder/displacive classification; the identification of the "basic physics" underlying the progressively more strongly non-classical behaviour the lower the dimensionality; the assessment of the relative importance of the cluster wall and the phonon in the ordering process; and the determination of the extent to which the "central peak" phenomenon is universal. In what follows I shall attempt to show that worthwhile insights, illuminating and interrelating these four issues, are to be found in the patterns of s.r.o. which form as $T \to T_c^+$. We will see that these patterns display universal features provided the system is viewed with a resolution that is sufficiently large on the scale of the lattice spacing. Accordingly the patterns to be studied are those formed not by the local coordinates $u(\vec{x}_i)$ themselves, but by the coarse-grained or block coordinates originally introduced by Kadanoff[33]. The block coordinate

$$u_L(\vec{x}) = \zeta_L \sum_{\{\vec{x}_i\} \varepsilon S(L)} u(\vec{x}_i) \tag{3.3a}$$

describes the instantaneous aggregate of the ordering coordinates, within a region S(L) of linear dimension L centred on $\vec{x}$. (ζ_L is an appropriate scale factor to be defined below). The representation (3.3a) is tractable in "low" dimensions. In "high" dimensions it proves more convenient to define block coordinates by

$$u_\Lambda(\vec{x}) = \frac{\zeta_\Lambda}{N} \sum_{|\vec{q}|<\Lambda} u(\vec{q}) e^{-i\vec{q}.\vec{x}} \tag{3.3b}$$

with ζ_Λ a further appropriate scale factor. With $\Lambda \sim 1/L$ the coordinates (3.3b), like the coordinates (3.3a), form the ingredients of a "picture" of the ordering process whose resolution is of order L.

The link between this coarse-grained "picture" of order-formation and the more practical issue of the nature of the critical scattering spectrum resides in a simple result: the frequency-dependent analogue of eqn. (2.5) describing the cross section for a scattering experiment in which the momentum transfer lies within Λ_R of the condensing mode, may be written in the form

$$S_{\Lambda_R}(\omega) = \sum_{\vec{q}<\Lambda_R} S(\vec{q},\omega) = S_o \int_{-\infty}^{\infty} \langle u_{\Lambda_R}(\vec{x},0) u_{\Lambda_R}(\vec{x},\tau)\rangle e^{i\omega\tau} d\tau \tag{3.4}$$

where S_o is a constant. It follows that such a scattering experiment probes the local correlation function of a block coordinate.

In what follows no attempt is made to calculate the time-dependent block correlation function appearing in eqn. (3.4). Rather, we will study the block coordinate probability density function near criticality. The formal significance of these functions has been appreciated by many authors, notably Patashinskii[34] and Jona-Lasinio[35]. Here we will establish their existence in universal forms whose specific character we will identify for the Ising universality class in dimensions d = 1,2 and 3. The results (presented in more detail elsewhere[36]) will be used, in particular, to illuminate the role of the kink and the phonon in the ordering process, and in the excitation spectrum, whose qualitative character we will infer from the form of the p.d.f., in much the same spirit as we used the double-peaked p.d.f. in Fig. 2 to suggest the character of the local excitation spectrum in an order-disorder system.

3.2 General Theory of the Block Coordinate p.d.f.

Formally the probability density function P_L of a block coordinate[†] u_L is defined by

[†]The following arguments can equally-well be developed in terms of the Fourier representation for the block coordinates, (3.3b).

$$P_L(u_L') = <\delta(u_L' - u_L)> \tag{3.5a}$$

where < > denotes statistical average of the coordinate u_L, and u_L' is a dummy constant. Using the integral representation of the δ function the p.d.f. may be written in the more convenient form

$$P_L(u_L') = \frac{1}{2\pi}\int_{-\infty}^{\infty} dH\, e^{-iHu_L'} <e^{iHu_L}> \equiv \frac{1}{2\pi}\int_{-\infty}^{\infty} dH\, e^{-iHu_L'}\, \hat{P}_L(H) \tag{3.5b}$$

The characteristic function $\hat{P}_L(H) \equiv <e^{iHu_L}>$ has an expansion in the n-point cumulants of the block p.d.f.[37]

$$\hat{P}_L(H) = \exp\{\sum_{n=1}^{\infty} \frac{(iH)^n}{n!} J_L^{(n)}\} \tag{3.6}$$

where, explicitly

$$J_L^{(n)} = \zeta_L^n \sum_{\{\vec{x}_1\}\epsilon S(L)} \cdots \sum_{\{\vec{x}_n\}\epsilon S(L)} <u(\vec{x}_1)\cdots u(\vec{x}_n)>_c \quad . \tag{3.7}$$

When the block size L is sufficiently large the contributions of well-separated sites will dominate the cumulants (3.7) provided t is sufficiently small (ξ sufficiently large); the correlation functions may then be approximated by their scaling forms (1.13) and the sums replaced by integrals giving

$$J_L^{(n)} \approx J^{(n)\infty} = [\zeta_L\, a_0\, (\frac{L}{a})^{(d+2-\eta)/2}]^n\, \tilde{J}^{(n)}(z) \tag{3.8}$$

where $z \equiv L/\xi$ and the functions $\tilde{J}^{(n)}$ are universal. We have made use of equations (1.10a) and (1.18), and have identified the exponent η by

$$\eta = 2 - d - 2\lambda_0 = 2 - \gamma/\nu \tag{3.9}$$

Choosing the scale factor ζ_L such that the p.d.f. has unit variance in the critical $(z \to 0)$ limit,

$$\zeta_L = [a_0\, (\frac{L}{a})^{\frac{d+2-\eta}{2}}]^{-1}\, [\tilde{J}^{(2)}(0)]^{-1/2} \tag{3.10}$$

gives

$$J^{(n)\infty} = \tilde{J}^{(n)}(z)/[\tilde{J}^{(2)}(0)]^{n/2} \tag{3.11}$$

Thus in the specified (large L,ξ) limit the block p.d.f. tends to a universal form[+] P^{∞}, whose cumulants $J^{(n)\infty}$ depend only upon $z \equiv L/\xi$.

3.3 One Dimension: Exact Results

The form of the universal p.d.f. P^{∞} can be determined rigorously and explicitly in d = 1 using transfer operator techniques[36]; here, sacrificing rigour for brevity, I will give only a physically-plausible derivation.

Consider the d = 1 system (1.1). The typical configurations at low temperatures are self-evident: there will be clusters in which all particles reside on a common side of their local potential barrier; their precise locations will deviate from the minima of the locally-favoured wells by virtue of phonon-like fluctuations within the clusters, and by virtue of the walls which delineate the clusters. Accordingly it is natural to decompose the displacements $u(x_i)$ in the form[4]

$$u(x_i) = \sigma(x_i) + y(x_i) \tag{3.12}$$

where $\sigma(x_i) = \pm\sigma_0$ ($\approx \pm u_0$) is an Ising-like variable which identifies the well favoured by the particle at x_i, and $y(x_i)$ is the fluctuation remainder. Now let us identify the various length scales in the problem. There are four — or five if one includes the block size L. First there is the lattice spacing a. Secondly, there is the correlation length ξ, which gives a measure of the typical cluster size and which becomes arbitrarily large at low temperatures. Thirdly there is a further correlation length ξ_{ph} characterising the distance over which the phonon-controlled fluctuations in the y-variables within a cluster are correlated. Fourthly there is the thickness of a cluster wall ξ_c. In practice (for physically reasonable potentials) the latter two lengths are of the same order of magnitude: in the order-disorder limit where s >> 1 (c.f. eqn. 3.1b) both are of the order of the lattice spacing, while in the displacive limit (s << 1) one finds that $\xi_c \sim \xi_{ph}$ with

$$\xi_{ph} = a(C/A)^{1/2} \quad ; \quad A \equiv \frac{1}{2} V_s''(u)\Big|_{u=u_0} \tag{3.13}$$

Thus, for physically reasonable potentials, both lengths are large compared to the lattice spacing in this limit. In fact, for the purposes of this argument it is sufficient to observe that ξ_c and

† The detailed form will depend upon the specific fashion in which the block coordinates are defined (c.f. eqns. 3.3a,b); however the bulk of this dependence will be absorbed by the unit-variance condition leaving forms which, in essential respects, should be definition-independent.

ξ_{ph} are non-critical, so that in the critical ($T \to 0$) limit each becomes small on the scale of ξ. It follows that, in the regime $L, \xi >> a$, the value of a block coordinate in any configuration carrying a non-negligible statistical weight will be entirely determined by the values of the σ-variables within the block; the contributions made by the y-variables will vanish in the spatial average implicit in the concept of a block coordinate. In more physical terms a typical configuration of the $d = 1$ system at low T, when viewed with a resolution large compared to the lattice (more precisely, large compared to ξ_{ph}) will display clusters of complete homogeneous order separated by sharp cluster walls. It follows immediately that the block p.d.f. for any $d = 1$ system will (for large L, ξ) be that of the fixed-length-spin Ising model defined by the partition function

$$Z = \mathop{\mathrm{Tr}}_{\{\sigma(x_i)\}} e^{+\frac{K}{\sigma_o^2} \sum_{i=1}^{N} \sigma(x_i)\sigma(x_{i+1})} \quad ; \quad \sigma(x_i) = \pm\sigma_o \qquad (3.14)$$

where K is a coupling constant which, in the limit as $T \to 0$, gives (one half) the energy required to create a cluster wall (in units of k_BT). Thus, in $d = 1$, the asymptotic form of the block p.d.f. is controlled entirely by the statistical mechanics of cluster walls. The asymptotic form itself is readily obtained from a transfer-matrix calculation of its characteristic function. This calculation yields the universal result

$$P^{\infty}(u_L) = \frac{e^{-\frac{z}{2}}}{2} \Big\{ \delta(1-u_L) + \delta(1+u_L) + \frac{z}{2}\,\Theta(1-u_L^2) \times \Big[I_0\big(\tfrac{z}{2}(1-u_L^2)^{\frac{1}{2}}\big) + \frac{I_1\big(\tfrac{z}{2}(1-u_L^2)^{\frac{1}{2}}\big)}{(1-u_L^2)^{\frac{1}{2}}} \Big] \Big\} \qquad (3.15)$$

where Θ is the step function, I_0 and I_1 are Bessel functions and the choice

$$\zeta_L = a/L\sigma_o \qquad (3.16)$$

has been made, in accordance with the unit variance condition (3.10), given the identification $\eta = 1$. Two limiting cases of (3.15) merit particular attention.

Firstly, in the critical limit, $z \equiv L/\xi \to 0$,

$$P^{\infty}(u_L) \approx P^*(u_L) = \frac{1}{2}\{\delta(1-u_L) + \delta(1+u_L)\} \tag{3.17}$$

so that the fixed point p.d.f. characterising the critical limit has a maximal order-disorder form with a G-value (c.f. eqn. 3.2)

$$G^*(d=1) = 1 \tag{3.18}$$

Secondly, in the limit in which the block size is large compared to the correlation length ($z \to \infty$) one finds

$$P^{\infty}(u_L) \approx \sqrt{\frac{z}{4\pi}}\; e^{-zu_L^2/4} \tag{3.19a}$$

This is a normalised Gaussian with variance

$$\langle u_L^2 \rangle = \frac{2}{z} = \frac{k_B T a \chi_T}{\sigma_o^2 L} \tag{3.19b}$$

in accord with the central limit theorem which is applicable in this regime[38].

Now let us consider the corrections to the universal asymptotic form P^{∞}, eqn. (3.15), arising from the fluctuations in the y-coordinates, eqn. (3.12). The corrections are non-universal. We will examine the form which they assume in systems satisfying the displacive criterion (3.1b). Again we will use a phenomenological argument to establish a result which can be derived rigorously with transfer operator methods. Neglecting the cluster-wall regions, whose weight is small compared to that of the intra-cluster regions, we write the fluctuation energy associated with the y variables in the form

$$\Delta H = \frac{1}{2N} \sum_q \omega^2(q)\, |y(\vec{q})|^2 \tag{3.20}$$

where the y(q) represent ("ordered phase") phonon coordinates. In this approximation the y-variables contribute only to the second cumulant of the block p.d.f. Explicitly this contribution has the form

$$\Delta\Gamma_L^{(2)} = \zeta_L^2 \sum_{\{x_1\}\varepsilon S(L)} \; \sum_{\{x_2\}\varepsilon S(L)} \langle y(x_1) y(x_2) \rangle$$

$$\approx k_B T a \chi_{ph} / \sigma_o^2 L \tag{3.21}$$

where $\chi_{ph} \equiv 1/\omega^2(0) = 1/2A$ is the susceptibility of the phonon coordinates, and we have used eqn. (3.16).

With these phonon-induced corrections taken into account, the characteristic function of the block p.d.f. assumes the form

$$\hat{P}_L(H) \approx \hat{P}^{\infty}(H)\ \hat{P}_{ph}(H) \qquad (3.22)$$

where $\hat{P}^{\infty}(H)$ is the characteristic function in the asymptotic regime (i.e. the Fourier transform of (3.15)) and

$$\hat{P}_{ph}(H) = \exp\{-\frac{H^2 k_B T a \chi_{ph}}{2\sigma_o^2 L}\} \equiv \exp\{-\frac{w^2 H^2}{2}\} \qquad (3.23)$$

Evidently, the effect of the intra-cluster phonons is to smear out the sharp structure inherent in P^{∞}, eqn. (3.15). In particular the δ functions in (3.15) are replaced by Gaussians of variance

$$w^2 = k_B T a \chi_{ph}/\sigma_o^2 L \qquad (3.24)$$

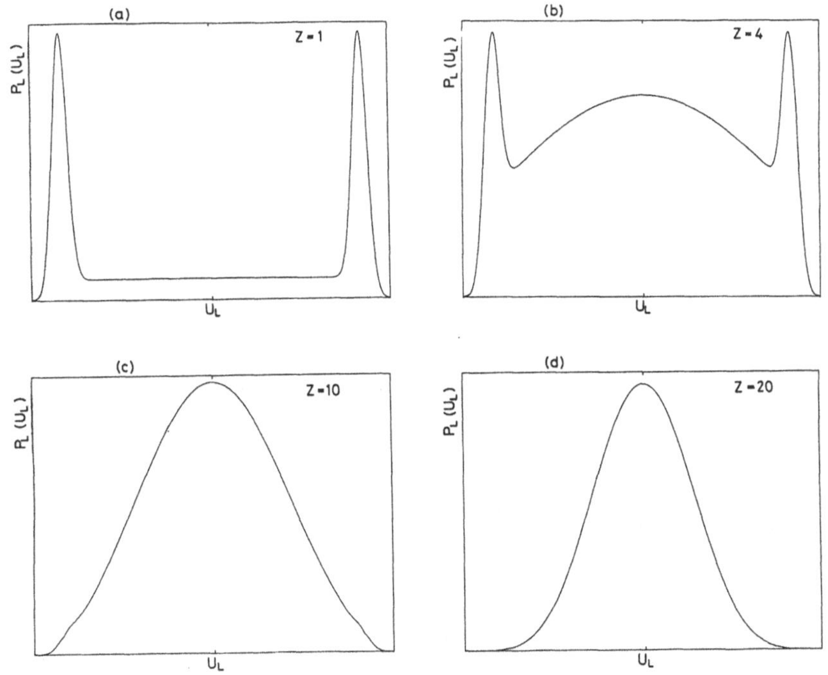

Fig. 3. The p.d.f. of the block coordinate u_L in d = 1 for fixed w = 0.05 (eqn. 3.24) and various values of $z = L/\xi$.

which is clearly an expression of the central limit theorem for the phonon coordinates. The block p.d.f. obtained by Fourier transformation of (3.22) is plotted in Fig. 3 for fixed w = 0.05 and various values of $z = L/\xi$

3.4 Two and Three Dimensions: Approximate Results

In dimensionalities $d > 1$, where ℓ.r.o. sets in at a non-zero T_c, the block coordinate p.d.f. is considerably more interesting and its determination is correspondingly more difficult. An approximate form for the p.d.f. P^*, characterising the critical limit ($a \ll L \ll \xi$), may however be derived with the aid of the realisation of the R.G. known as Wilson's approximate recursion formula[1]. The recursion formula prescribes an explicit non-perturbative relationship between the effective potential V_Λ "seen" by a block coordinate u_Λ, eqn.(3.3b), and the effective potential $V_{\Lambda/2}$ seen by the coordinate of a block with twice the linear dimension. A simple extension of the arguments used to establish this recursion formula (as expounded particularly clearly by Ma[39]) yields a relationship between the p.d.f.'s for the two block sizes. Specifically we write

$$u_\Lambda(\vec{x}) = u^<(\vec{x}) + u^>(\vec{x}) \tag{3.25}$$

where $u^<$ and $u^>$ contain, respectively, the Fourier coordinates with wavevectors less than and greater than $\Lambda/2$. Comparison of equations (3.25) and (3.3b) shows that

$$u^<(\vec{x}) = \frac{\zeta_\Lambda}{\zeta_{\Lambda/2}} u_{\Lambda/2}(\vec{x}) \equiv \alpha_o\, u_{\Lambda/2}(\vec{x}) \tag{3.26a}$$

and

$$\alpha_o = 2^{-\left(\frac{d-2+\eta}{2}\right)} \tag{3.26b}$$

The last step follows from the form of ζ_Λ prescribed by a scaling argument analogous to that used to establish (3.10). It follows that

$$P_\Lambda(u_\Lambda) = \int_{-\infty}^{\infty} du_{\Lambda/2} \int_{-\infty}^{\infty} du^> \, P_{\Lambda/2}(u_{\Lambda/2})\, P_c(\alpha_o u_{\Lambda/2}, u^>)$$

$$\delta[u_\Lambda - (u^> + \alpha_o u_{\Lambda/2})] \tag{3.27}$$

where $P_c(u^<, u^>)du^>$ is the conditional probability that, given a particular $u^<$ as its first argument, the second argument has value between $u^>$ and $u^> + du^>$. Within approximations already inherent in the recursion formula its form is found to be

$$P_c(u^<,u^>) = p_0\exp\{-(u^>)^2 - \frac{1}{2} V_\Lambda(u^< + u^>) - \frac{1}{2} V_\Lambda(u^< - u^>)\} \qquad (3.28a)$$

where p_0 is prescribed by the normalisation condition

$$\int_{-\infty}^{\infty} P_c(u^<,u^>)du^> = 1 \qquad (3.28b)$$

Now from the arguments of Section 3.2 we expect that, in the limit as $T \to T_c$, the block p.d.f. will tend to a fixed point form P^* independent of block size. From equations (3.27) and (3.28) it is clear that this result is fulfilled as a simple corollary of the existence of a fixed point form V^* for the block potential. More explicitly equation (3.27) implies that P^* must satisfy

$$P^*(u_\Lambda) = \int_{-\infty}^{\infty} dy\ P^*(y)\ P_c^*(\alpha_0 y,\ u_\Lambda - \alpha_0 y) \qquad (3.29)$$

where P_c^* is given by (3.28) with the potential assigned its fixed point form. The set of equations (3.28) and (3.29) is closed with the assignment of a value to the exponent η in eqn. (3.26b). For strict internal consistency η should be set to zero. This procedure is satisfactory in d = 3. However, in d = 2, the recursion formula for V_Λ does not then converge to a fixed point[1]. Adopting the expedient[1] of assigning η its known d = 2 Ising value (η = 1/4) convergence to a fixed point is obtained and the results (assessed on the success with which they predict exponent values) are not unreasonable. Implementing these procedures in d = 2 and d = 3, and solving the aforementioned set of equations numerically one obtains for the fixed point p.d.f. the results shown in Figs. 4a,4b. In d = 3 (Fig. 4a) the p.d.f. is singly-peaked, though far from Gaussian, as indicated by its G-value (eqn. 3.2)

$$G^*(d = 3) \approx 0.33 \qquad (3.30)$$

In d = 2, however, the fixed point p.d.f. has a markedly double-peaked character (Fig. 4b) reminiscent of the d = 1 results at low temperatures (Figs. 3a,b); the associated G-value is

$$G^*(d = 2) \approx 0.87 \qquad (3.31)$$

Although the approximate recursion formula is undoubtedly less trustworthy in d = 2 than it is in d = 3 there are two reasons for believing that the structure shown in Fig. 4b is essentially correct.

First of all, molecular dynamics studies by E. Stoll and T. Schneider, presented in a preliminary report of this work[40], reveal a qualitatively similar, though somewhat less sharply peaked, structure. Secondly, utilising exactly established properties of the 2 and 4 spin correlation functions for the planar Ising model[41] it is actually possible to express the parameter G* in a closed form, as an integral which can be evaluated numerically. The result is[36]

$$G^*(d = 2) = 0.90 \pm 0.02 \tag{3.32}$$

in close correspondence with the result obtained from the recursion formula.

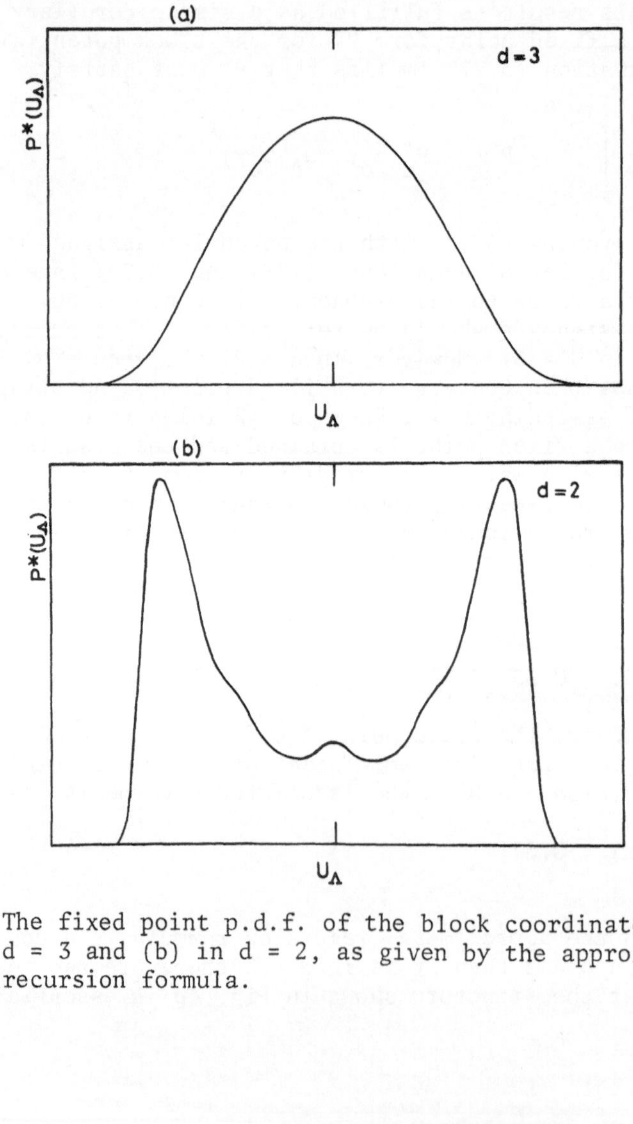

Fig. 4. The fixed point p.d.f. of the block coordinate u_Λ (a) in d = 3 and (b) in d = 2, as given by the approximate recursion formula.

3.5 Discussion

Armed with the results presented in the preceding three sections we now return to the issues identified in the course of Section 3.1, dealing with the ways in which they are illuminated firstly by the existence and secondly by the specific forms of the p.d.f.'s we have studied.

It should be clear that the very existence of universal block-coordinate distribution functions is itself of some significance, suggesting the physically appealing notion that the universality of the critical region effectively resides in the universality of block-coordinate behaviour. This is by no means a new idea: it is, in fact, simply a restatement of the key insight underlying early formulations of scaling theory[33] and the renormalisation group[1]. It is also a perspective illuminatingly emphasized by Jona-Lasinio[35] who discusses the possibility of a link between the stable fixed points of the R.G. and the stable distributions of probability theory. In the structural phase transition context it highlights the inadequacies of the order-disorder/displacive classification scheme inherited from the classical view of s.p.t.'s: any such classification scheme based on the character of the microscopic interactions, is trustworthy in the critical region only as regards the qualitative character of the local behaviour; the qualitatively different forms of collective behaviour suggested by such schemes resolve themselves, in the critical region, into mere differences of length scale.

Turning now to the forms of the fixed-point functions P*, we remark first on the obvious trend displayed qualitatively in Figs. 3 and 4 and quantitatively in the G* values (eqns. 3.30, 3.31, 3.18): it is natural to correlate the progressively more strongly order-disorder character of the p.d.f.'s through the sequence d = 3, 2 and 1 with the corresponding progression, noted earlier, in the "non-classical" character of critical behaviour as attested by exponent values, and by the fluctuation-induced suppression of the critical temperature. Again this is basically a restatement of a widely-appreciated phenomenon — the enhancement of non-linear effects in low dimensions. However the forms of the p.d.f.'s actually allow us to calibrate this trend, as it is displayed in the nature of the short range order and the collective excitation spectrum.

Let us first of all establish that the basic physics of the d = 1 case (clear a priori, given that $T_c = 0$) can be deduced from the form of the corresponding block p.d.f. The universal "picture" of s.r.o. in d = 1 which (as a matter of expediency) we exploited in deriving the form of the p.d.f. is, conversely, implied by that form: the s.r.o. resides in completely and homogeneously ordered clusters separated by sharp domain walls; the length scale associated with the cluster wall thickness and with the coherence of intra-cluster ripples, is non-critical and thus scales out of the asymptotic picture. Presented with the strongly order-disorder character of the block p.d.f. (Fig. 3a) one would naturally expect that the local spectral density of the coordinate in question would reveal traces of two qualitatively different fluctuations: large amplitude fluctuations of the block coordinate between its two most favoured values (asymptotically carrying the entire weight of the spectrum) and small amplitude fluctuations about its most favoured values; the strongly-anharmonic character of the former suggests that it will manifest itself as a heavily-damped central component in the spectrum; in contrast the small amplitude fluctuations should have a resonant character in the low temperature spectrum. Recalling the link (eqn. 3.4) between the block-coordinate spectral density and the excitation spectrum determined in a scattering experiment one thus arrives at the familiar two-frequency-scale scattering spectrum well-established for the d = 1 system at low temperatures[3,4]. The microscopic significance of the two features also follows from the identification of the processes giving rise to the two types of block coordinate motion: the large amplitude motion of a block coordinate (and thus the central peak) originates in the motion of cluster walls across the block length; the small amplitude motion (and thus the resonant side band) resides in intra-cluster phonons.

Consider now the d = 2 problem. Here the critical temperature is finite and the essential character of the s.r.o. and the excitation-spectrum at T_c is not clear a priori; we must rely on the form of the fixed point p.d.f. (Fig. 4b) which, to the author's knowledge, gives the clearest view of the former, and the only firmly-based theoretical clue to the latter. Specifically, comparison of Fig. 4b with the fixed point form in d = 1, eqn. (3.17), shows that, while a qualitatively similar picture of s.r.o. is appropriate in d = 2, some refinements are necessary: certainly Fig. 4b implies the existence of clusters with linear dimensions extending over a range bounded above only by the correlation length, and nested in the fashion proposed by Kadanoff[42] to yield a picture that is scale invariant at T_c. However the smooth form of the p.d.f., to be contrasted with the singular structure in d = 1, is inconsistent with homogeneous intra-cluster order and with sharp cluster walls: a typical configuration must, rather, display fluctuations in the intra-cluster order, and spatially-extended cluster walls. Given the constraint that the picture be scale invariant, it is hard to avoid the conclusion that the phonon-induced ripples in the intra-cluster order, and the cluster walls themselves should extend over distances up to the (order of magnitude of the) correlation length. In other words it appears that the length scales ξ_{ph} and ξ_c, which scale out of the d = 1 problem, are caught up into the two-dimensional problem and are themselves measures of the critical length scale ξ. With certain obvious reservations the physical character of the d = 2 problem at criticality thus appears to be qualitatively similar to that of the d = 1 system at low but finite temperatures.

It seems plausible that this analogy extends to the excitation spectrum: thus Fig. 4b suggests (but does not prove!) that the critical scattering from a two (or quasi-two) dimensional system may universally display traces of two excitations, again associated with large amplitude and small amplitude fluctuations of the coordinates of blocks with linear dimension ~ $1/\Lambda_R$ (eqn. 3.4): the former component must again reside in the motion of the walls of clusters large compared to the block size: the latter component will originate both in the intra cluster phonons and in the motion of the walls of clusters embedded within the block. Since features associated with both excitations should be apparent in the scattering from arbitrarily large blocks (c.f. the critical character of ξ_{ph} and ξ_c alluded to above) it would seem that, in contrast to the d = 1 case, both features should display critical narrowing; however the dominant part of the spectrum seems likely to be associated with cluster wall motion. This picture is in broad accord with the results of molecular dynamics experiments[3,29,30] alluded to in Section 3.1: in these studies the dominant feature in the scattering cross section close to T_c — the central peak — was explicitly traced to the motion of cluster walls, while the additional feature — stemming from the soft phonon, and ultimately manifested as a wing on the central mode — was associated with intra-cluster phonons. Although the results are suggestive of a non-critical soft mode side band they are probably not inconsistent with a critical narrowing of both features.

Turning now to the d = 3 case we find that there is no unambiguous interpretation of the fixed point p.d.f., Fig. 4a. Certainly, its relatively structureless form, taken along with the d = 3 topology which allows interpenetration of clusters, together make the simple cluster picture less compelling in this case. Nevertheless the form of P^* is not actually inconsistent with the interpretation advanced for its d = 2 analogue: one may take the view that the strongly non-Gaussian character of the p.d.f. (c.f. eqn. 3.30) is a manifestation of substructure similar to that shown in Fig. 4b, but unresolved because the intra-cluster fluctuations are now comparable with the mean intra-cluster distortion.

It seems appropriate to digress at this point, to mention a study[43] of the e.p.r. spectrum of $SrTiO_3$ which casts some light on this point and which is cross-linked at several different levels with the phenomena under consideration here. We recall that the line shape of a paramagnetic resonance whose Larmor frequency is linearly-dependent on the value of a local ordering coordinate $u(\vec{x})$ is given by the function[44]

$$P_{\tau_0}(u'_{\tau_0}) = \frac{1}{2\pi}\int_{-\infty}^{\infty} d\tau\, e^{-i\tau u'_{\tau_0}} \langle e^{i\tau u_\tau} \rangle \qquad (3.33)$$

where

$$u_\tau = \frac{\zeta_{\tau_0}}{\tau} \int_0^\tau u(\vec{x},\tau')d\tau' \tag{3.34}$$

with ζ_{τ_0} an appropriate scale factor. These equations have been written in a form which underlines their close (though incomplete) affinity with equations (3.5b) and (3.3a). Equation (3.3a) defines a space-smoothed representation of the displacement field $u(\vec{x},\tau)$: the large wavevector components are erased by the spatial average; accordingly the coordinate u_L is sensitive to the long-wavelength fluctuations. On the other hand eqn. (3.34) defines a time-smoothed representation of $u(\vec{x},\tau)$: high-frequency components are eliminated by the time average; the coordinate u_τ is sensitive to the slow fluctuations. Since the slowest fluctuations are the longest wavelength fluctuations it is reasonable to expect that the two coordinates will display qualitatively similar behaviour, and thus have similar p.d.f.'s† — an expectation borne out by molecular dynamics studies. This brings us to the structural similarity between the defining equation for the e.p.r. spectrum, (3.33), and the characteristic function representation for the block coordinate p.d.f. eqn. (3.5b). In fact this similarity is incomplete since the "space-block" size L is a constant in equation (3.5b), while the "time-block" size τ forms the variable of integration in eqn. (3.33). Nevertheless the e.p.r. spectrum does represent a first approximation to the p.d.f. of the time-smoothed variable u_{τ_0}, where τ_0 is the e.p.r. measuring time.[44]

In the light of these remarks let us consider briefly the measurement of the e.p.r. spectrum of $SrTiO_3$, shown in Fig. 5. The key points to be noted here are: (i) The e.p.r. line shape is strongly non-Gaussian in the few degrees above T_c. (ii) The gross features of the non-linear structure are captured by the double-Gaussian representation shown and (iii) The magnitude of the typical intra-cluster distortion implied by this representation is $\sigma_0 \simeq 0.22°$ at $T = T_c + 2$ K, taking the ordering variable as the angular rotation of oxygen octahedra. By comparison the limiting intensity of the central peak observed in the most recent neutron scattering study[46], when viewed on the scale of the superlattice intensity, implies (through eqn. 3.4) a value $\sigma_0 \approx 0.28°$. Now it must be remarked that there are features

---

†In particular both p.d.f.'s may be strongly anharmonic even when the equilibrium local p.d.f. P_0 is near-Gaussian, so that the behaviour of P_0 — the essential output of a recent study[45] based on a decimation form of the R.G. — is less sensitive to the occurrence of a kink-driven crossover (c.f. footnote 13 in Ref. 40 and note its implication that for a displacive system in d = 2 the local p.d.f. remains singly peaked at T_c).

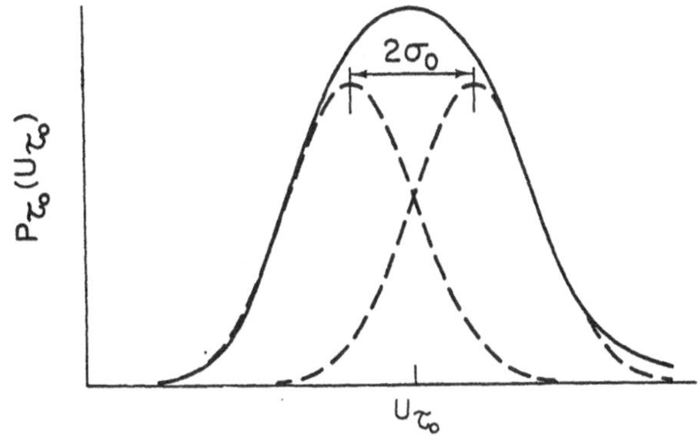

Fig. 5. The line shape function P_{τ_0} (eqn. 3.33) as determined by e.p.r. studies of $SrTiO_3$ at T_c + 2 K, together with its optimum double-Gaussian representation (dashed line).

of both experiments which cannot be accommodated within ideal crystal theory[28]; the apparent saturation of the phonon side band in $SrTiO_3$, it would now appear, may also fall into this category. Nevertheless, returning to our original theme, the correspondence between the two σ_0 values is at the very least an indication that singly-peaked p.d.f.'s may indeed conceal strongly anharmonic substructure. The interpretation of the d = 3 fixed point p.d.f. (Fig. 4a) is thus fraught with ambiguity and the question of the universality of a kink-driven central peak phenomenon in 3 dimensions stands unresolved.

3.6 Summary and Prospects

Finally I will try to summarise the essential insights acquired through the studies described above, and indicate some of the numerous questions which remain to be answered.

Firstly we have seen that patterns of s.r.o. have universal characteristics, independent of the details of the microscopic interactions but dependent upon the system dimensionality. Presumably the character of the s.r.o. should also reflect the other factors (order parameter dimensionality, existence of dipolar forces.....) known to influence the universality class[5] but not yet investigated in this context.

Secondly I have endeavoured to emphasize the link that surely exists between the qualitative character of the s.r.o. and that of the collective excitation spectrum. The full utility and limitations of this link — expressed in eqn. (3.4) — remain to be explored.

We have seen thirdly, and more specifically, that in d = 2, as in d = 1, the patterns of s.r.o. have elements which may be traced sufficiently unambiguously to cluster walls and intra-cluster phonons that both forms of fluctuations may be expected to have signatures in the excitation spectrum; in d = 3, however, neither the s.r.o. nor the excitation spectrum may be resolved in this fashion with comparable conviction. To resolve the three-dimensional problem, and to demonstrate explicitly the universal two-time-scale phenomenon which seems likely to occur in d = 2, an appropriate explicit calculation of the critical scattering spectrum is required.†

The qualifier "appropriate" requires some elaboration, with which I shall conclude. Ideally we require a calculation not only of the limiting form of the critical scattering spectrum, but also of the manner in which it evolves as criticality is approached in a system whose scattering outside the critical region is that of a simple softening underdamped phonon. According to the picture emerging from molecular dynamics studies and substantiated by the calculations presented above, in d = 2 at least, we may expect that the soft phonon evolves into a mere wing on a central component which emerges in the critical region and grows to dominate the spectrum: it seems that both features will remain in the spectrum, even asymptotically. Since the signatures of kink and phonon are certainly more likely to be separately identifiable in d = 2 than in d = 3 it seems more appropriate to utilise an approximation scheme tailored to "low" dimensions (i.e. $d = 1 + \varepsilon$) than high dimensions ($d = 4 - \varepsilon$). Indeed since the non-convergent large order behaviour of the $\varepsilon = 4 - d$ expansion is known to be controlled by kink-like phenomena[47] it is possible that low-order ε expansions may miss significant kink-induced effects altogether. Finally, recalling the concluding insights of Section 2.1 it actually appears less than essential to take account of the influence of energy conservation on the dynamics; it is, it would appear, of greater importance to capture the full extent of the mechanical non-linear effects.

†A number of studies which have already attempted to confront this problem are surveyed in Ref. 6.

References

1. K.G. Wilson, Phys Rev B4, 3174 (1971); K.G. Wilson and J. Kogut, Phys Rep C12, 77 (1974).
2. T. Riste, E.J. Samuelsen, K. Otnes and J. Feder, Solid St. Commun. 9, 1455 (1971); S.M. Shapiro, J.D. Axe, G. Shirane and T. Riste, Phys Rev B6, 4332 (1972).
3. T. Schneider and E. Stoll, Phys Rev Lett 31, 1254 (1973); 35, 296 (1975).
4. J.A. Krumhansl and J.R. Schrieffer, Phys. Rev. B11, 3535 (1975); S. Aubry, J. Chem Phys 62, 3217 (1975); 64, 3392 (1976).
5. A.D. Bruce, Adv. in Physics 29, 111 (1980).
6. A.D. Bruce and R.A. Cowley, Adv in Physics 29, 219 (1980).
7. S. Sarbach, Phys Rev B15, 2694 (1977).
8. H. Thomas, "Structural Phase Transitions and Soft Modes" E.J. Samuelsen, E. Andersen and J. Feder eds., Universitetsforlaget, Oslo, p.15 (1971)
9. G.A. Baker and J.M. Kincaid, Phys Rev Lett 42, 1431 (1979).
10. B.G. Nickel, Proc. Cargese Summer Institute (1980), to be published.
11. R. Bausch and B.I. Halperin, Phys Rev B18, 190 (1978).
12. M.E. Fisher, Proc Nobel Symposium 24, B. Lundquist and S. Lundquist eds. Academic Press, New York, p16 (1974).
13. R.B. Griffiths, Phys Rev 158, 176 (1967).
14. H.E. Stanley, "Introduction to Phase Transitions and Critical Phenomena", Clarendon, Oxford (1971).
15. D. Stauffer, M. Ferer and M. Wortis, Phys Rev Lett 29, 345 (1972); P.C. Hohenberg, A. Aharony, B.I. Halperin and E.D. Siggia, Phys Rev B13, 2986 (1976).
16. A.D. Bruce, Phys Rev Lett 44, 1 (1980).
17. D.D. Betts, J. Phys A12, 2287 (1979).
18. A. Aharony and B.I. Halperin, Phys Rev Lett 35, 1308 (1975).
19. A.I. Larkin and D.E. Khmel'nitskii, Sov Phys JETP 29, 1123 (1969).
20. P. Heller, Int J Magn 1, 53 (1970).
21. J. Feder Solid St Commun 9, 2021 (1974); see also Ref. 6.
22. A.D. Bruce, J Phys C 14, 193 (1981).
23. D.R. Nelson, Phys Rev B14, 1123 (1976); Y. Achiam and J.M. Kosterlitz, J Phys C 10, 4559 (1977).
24. I. Hatta, Y. Shiroishi, K.A. Müller and W. Berlinger, Phys Rev B16, 1138 (1977).
25. J.C. LeGuillou and J. Zinn-Justin, Phys Rev B21, 3976 (1980).
26. E. Eisenriegler, Phys Rev B9, 1029 (1974).
27. R.J. Glauber, J Math Phys 4, 294 (1963); M. Suzuki and R. Kubo, J Phys Soc Japan 24, 51 (1968).
28. K.A. Müller, these proceedings.
29. T. Schneider and E. Stoll, Ferroelectrics 26, 67 (1980).
30. W.C. Kerr, Phys Rev B19, 5773 (1979).
31. J.F. Currie, J.A. Krumhansl, A.R. Bishop and S.E. Trullinger,

Phys Rev B22, 477 (1980); G.F. Mazenko and P.S. Sahni, Phys Rev B18, 6139 (1978).
32. P.S. Sahni and G.F. Mazenko, Phys Rev B20, 4674 (1979).
33. L.P. Kadanoff, Physics 2, 263 (1966).
34. A.Z. Patashinskii, Sov Phys JETP 26, 1126 (1968).
35. G. Jona-Lasinio, Nuovo Cimento 26B, 99 (1975).
36. A.D. Bruce, to be published.
37. H. Cramer, Mathematical Methods of Statistics, Princeton U.P. (1946).
38. G.A. Baker and S. Krinsky, J. Math Phys 18, 590 (1976).
39. S.K. Ma, "Modern Theory of Critical Phenomena", Benjamin, Reading (1976).
40. A.D. Bruce, T. Schneider and E. Stoll, Phys Rev Lett 43, 1284 (1979).
41. A. Luther and I. Peschel, Phys Rev B12, 3908 (1975).
42. L.P. Kadanoff "Phase Transitions and Critical Phenomena" Vol. 5a C. Domb and M.S. Green eds., Academic Press, New York p1 (1976).
43. A.D. Bruce, K.A. Müller and W. Berlinger, Phys Rev Lett 42, 185 (1979).
44. A. Abragham "The Principles of Nuclear Magnetism", Clarendon, Oxford (1961).
45. P.D. Beale, S.K. Sarker and J.A. Krumhansl, Cornell preprint (1981).
46. J. Topler, B. Alefeld and A. Heidemann, J Phys C 10, 635 (1977)
47. D.J. Wallace in "Solitons and Condensed Matter Physics" A.R. Bishop and T. Schneider eds., Springer-Verlag, Berlin p104 (1978).

DYNAMIC CORRELATIONS IN THE ORDERED PHASE OF PEROVSKITES

G. Meissner, N. Menyhárd*, and P. Szépfalusy*,

Theoretische Physik
Universität des Saarlandes
6600 Saarbrücken, Federal Republic of Germany

I. INTRODUCTION

The existence of a non-vanishing average value P of the order parameter below the critical temperature T_c should give rise to qualitatively new features in statics and dynamics of a variety of systems. A crucial role in this context is played by the symmetry of the space spanned by the degrees of freedom of the order parameter of n components in d dimensions.

In systems exhibiting continuous rotational symmetry it has for instance been realized long before the recent interest in critical phenomena that the longitudinal order parameter susceptibility χ_L diverges along the coexistence curve for all temperatures $T < T_c$. This singular behavior of χ_L is induced by the divergence of the transverse susceptibility χ_T due to gapless collective modes (Goldstone excitations) and should exhibit a crossover critical behavior in approaching the critical point T_c (Fig. 1). The crossover behavior near T_c in dynamics has been studied for asymptotically small wave-numbers k and frequencies ω within the relaxational model[1]. There occurs, however, an additional type of 'crossover' in leaving this small ω region and passing over to frequency values typical for the longitudinal excitation branch in and beyond the mean field (MF) region (Fig. 2). While in statics the above mentioned singularities of χ_T and χ_L are caused by large transverse fluctuations dominating at all $T < T_c$, we learned with respect to such high ω-values from investigating isotropic dynamic models in the $n \to \infty$

*permanent address: Central Research Institute for Physics 1525 Budapest, Hungary

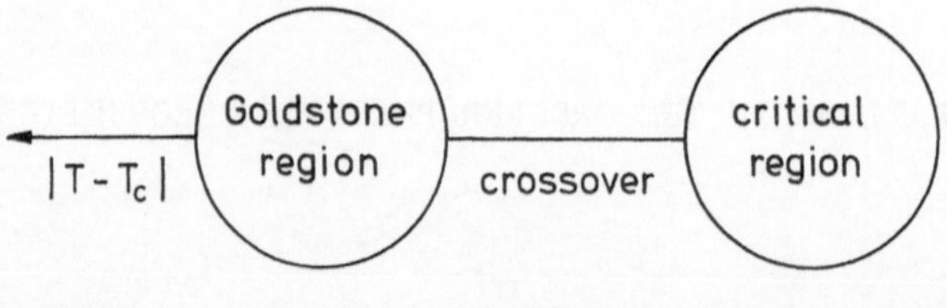

Fig. 1. Crossover in statics of isotropic systems where the momentum field $\pi_\alpha(x_1 \ldots x_d, t)$ conjugate to the order parameter field ϕ_α with $\alpha = 1,\ldots,n$ does not play any role.

limit[2] that a relation $\tilde{\omega}_L \sim P$ between the frequency $\tilde{\omega}_L$ of the longitudinal mode and the order parameter P can be valid in a remarkably large temperature range.

In this seminar we would like to report on inquiring the validity and applicability of similar ideas in systems of finite n including effects of weak anisotropy. In Sec. II a model is introduced appropriate to study dynamic correlations in perovskites below the antiferrodistortive transition. The approximation scheme which leads to an expression for the dynamic longitudinal susceptibility $\chi_L(\underset{\sim}{k} = 0,\omega)$ at temperature $T < T_c$ is indicated in Sec. III. Predictions for the longitudinal frequencies $\tilde{\omega}_L$ and the extent to which they might be detectable experimentally are briefly discussed in Sec. IV.[3]

II. MODEL

First we would like to mention that the role of the momentum field $\pi_\alpha = \partial\phi_\alpha/\partial t$ canonically conjugate to the order-parameter field ϕ_α for dynamic correlations can be accounted for by the generalized Langevin equation

$$\frac{\partial^2}{\partial t^2}\phi_\alpha + \Gamma_o \frac{\partial}{\partial t}\phi_\alpha = -\frac{\delta}{\delta\phi_\alpha} H^\phi + \eta_\alpha \qquad (1)$$

with the damping coefficient Γ_o and the stochastic force η_α.

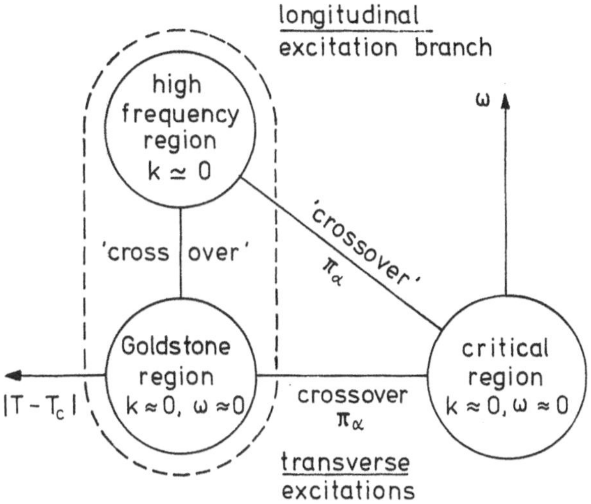

Fig. 2. Crossover in dynamics of isotropic and nearly isotropic systems. Momentum field π_α starts playing a role in the investigated dotted region along transverse and longitudinal branches.

In (1) the effective Hamiltonian $H^\phi = H^\phi/k_BT$ for $\phi_\alpha(\underset{\sim}{x},t)$ of perovskites with n = d = 3 takes the form

$$H^\phi = \int d^3x \; \{\frac{1}{2}r_o \sum_\alpha \phi_\alpha^2 + \frac{1}{2}\sum_\alpha [(\nabla\phi_\alpha)^2 - f_o(\nabla_\alpha\phi_\alpha)^2]$$

$$+ \frac{1}{4!}u_o(\sum_\alpha \phi_\alpha^2)^2 + \frac{1}{4!}v_o\sum_\alpha \phi_\alpha^4 - \sum_\alpha \phi_\alpha E_\alpha\}, \qquad (2)$$

where E_α denotes an external field. In an isotropic solid, only the parameters $r_o = r_o'(T - T_{MF})/T_{MF}$ and u_o would be required where T_{MF} denotes the MF critical temperature. Cubic anisotropy in the lattice structure of perovskites, therefore, is accounted for by the two additional coupling constants f_o and v_o in the bilinear and quartic part of H^ϕ in (2). Because of these symmetry-breaking perturbations in H^ϕ, there are easy axes below T_c for the order parameter $P \equiv \mathrm{Tr}\,\{e^{-H^\phi}\phi_1\}/\mathrm{Tr}\,e^{-H^\phi}$, inspite of its multicomponent nature (n > 1). Therefore, the static transverse susceptibility $\chi_T \equiv r_T^{-1}$ now is no longer divergent for $T < T_c$. In the cubic to tetragonal transition in

$SrTiO_3$, e.g., P then measures static rotations of oxygen octahedra (TiO_6) around cubic axes[4].

III. DYNAMIC CORRELATIONS BELOW T_C

The effect of dynamic correlations on longitudinal dynamic modes $\tilde{\omega}_L$ below T_c in this model of perovskites can be studied, e.g., in the long wavelength ($k \to 0$) limit from zeros of the inverse of the longitudinal susceptibility

$$\chi_L^{-1}(\underset{\sim}{k} = 0, \tilde{\omega}_L) \equiv 0. \tag{3}$$

In order to obtain an approximation scheme for evaluating $\tilde{\omega}_L$ from the l.h.s. of (3) we may start from an usual expression for the inverse of the longitudinal susceptibility

$$\chi_L^{-1}(\underset{\sim}{k},\omega) = -\omega^2 + \omega_s^2 + k^2 - f_o k_L^2 - i\Gamma_o\omega - \Sigma_L(\underset{\sim}{k},\omega) + \Sigma_L(0,0) \tag{4}$$

with the renormalized static excitation frequency[5] $\omega_s^2 = \chi_L^{-1}(\underset{\sim}{k} \to 0,\ \omega = 0) \equiv r_L$ and with the structure

$$\Sigma_L = \tilde{\Sigma}_L - \frac{u_o}{3}\,\frac{\tilde{\Lambda}\,\tilde{\Lambda}'}{1 + \frac{u_o}{6}\,\pi} \tag{5}$$

of the longitudinal self-energy which is formally the same as for isotropic systems[2]. Therefore, we would just like to state that the quantities marked by a tilde are double-irreducible and the appearance of the $\tilde{\Lambda}$, $\tilde{\Lambda}'$- vertices is typical for the ordered phase. The parameter v_o does not appear explicitly in (5), since it has been neglected as compared to u_o. However, v_o does appear through r_T, since the self-energy $\tilde{\Sigma}_L$, the vertices $\tilde{\Lambda}$, $\tilde{\Lambda}'$ and the polarization part π do depend on r_T.

With the small parameters f_o and $|v_o|/u_o$ we have then evaluated Σ_L in tree-approximation and in one-loop approximation, respectively, applying ordinary rules of perturbation theory.

IV. RESULTS AND CONCLUSIONS

The longitudinal mode obtained below T_c from $\chi_L^{-1}(\underset{\sim}{k} = 0, \tilde{\omega}_L) \equiv 0$ is given in one-loop approximation through the relation

$$\tilde{\omega}_L^2 \simeq \frac{1}{3}\, u_o\, P^2 \tag{6}$$

provided the anisotropy is small. This relation which naturally holds in the MF region, i.e., for reduced temperatures $t \equiv |T - T_c|/T_c \gg t_P, t_L$ has been shown to be valid down to t_L and even below. The two reduced Ginzburg temperatures t_P and t_L characterize the onset of the dominance of fluctuations in the equation of state and in r_L, respectively. We have worked out two cases depending on whether (i) $t_L > t_P$ or (ii) $t_L < t_P$. In a crude approximation $t_P/t_L = 36(r_T/r_L)_{MF}$. Whether the exponent β of the order parameter, $P \sim t^\beta$ in Eq. (6), equals its MF value 1/2 or not follows from $t_L \gtrless t_P$, respectively. Besides of the restriction to small anisotropy, relation (6) holding beyond the MF regime is due to the fact that the ω-values of interest, i.e., those which occur along the longitudinal excitation branch in and beyond the MF regime, are high. Hence the fluctuations due to Goldstone singularities which dominate in statics become suppressed.

We have suggested that among perovskites the cubic to trigonal transition in $LaAlO_3$ appears to be suitable for applying the present theory, due to the fact that according to measurements[4] $|v_o|/u_o \ll 1$ and furthermore $f_o \simeq 0$. Unfortunately, in the temperature region near T_c no experimental data of $\tilde{\omega}_L$ exist for this material. $SrTiO_3$ is much wider investigated experimentally, however, it does not fall into the category of small anisotropy, i.e., $|v_o|/u_o \simeq 1/7$ and $f_o \simeq 0.96$ experimentally. Accordingly, it turns out to belong to case (ii), i.e., $t_P > t_L$. Surprisingly, Raman scattering data[6] for the longitudinal mode have given the high-accuracy (3%) evidence for the validity of (6) in a wide range of temperatures, also in the region where $\beta \simeq 1/3$ as being measured by Müller and Berlinger[4]. While reproducing t_P in good agreement with experiments in $SrTiO_3$, we have shown that even if some cancellation of higher order terms takes place on the basis of the present theory, relation (6) can be valid at best with 20-30% accuracy in the temperature range of interest.

ACKNOWLEDGEMENTS

This work was supported in part by SFB 130, Ferroelektrika.

REFERENCES

1. L. Schäfer, Z. Physik B 31, 289 (1978).
2. L. Sasvári and P. Szépfalusy, J. Phys. C 7, 1061 (1974).
3. An account of this work was reported by G. Meissner, N. Menyhárd, and P. Szépfalusy at the 8th MECO-Seminar on Phase Transitions and Critical Phenomena, March 23-25, 1981, Saarbrücken, FRG (unpublished).
4. See, e.g., the lecture of K.A. Müller at this conference.
5. G. Meissner and K. Binder, Phys. Rev. B 12, 3948 (1975).
6. E.F. Steigmeier and H. Auderset, Solid State Commun. 12, 565 (1973).

PHASON LIGHT SCATTERING IN $BaMnF_4$

K. B. Lyons and T. J. Negran

Condensed Matter Physics Department
Bell Laboratories
Murray Hill, N. J. 07974

In an incommensurate phase there exists a lattice distortion with a periodicity incommensurate with that of the ambient lattice. The distortion is the order parameter of the normal—incommensurate phase transition, and may be approximated symbolically as

$$\psi=\psi_o e^{i\vec{k}_o\cdot\vec{r}} e^{i\phi} \quad , \tag{1}$$

where $\vec{k}_o$ is the reciprocal periodicity of the distortion and ϕ is a constant phase. Since $\vec{k}_o$ is incommensurate, a uniform change in ϕ clearly costs no energy. Hence, long wave length fluctuations in ϕ should obey a dispersion law with the property that $\omega \to 0$ as $q \to 0$. Thus, even well below the phase transition, a slow mode may be present which can modify the order parameter dynamics in some ways analagous to the effect of "microdomains"[1] near a commensurate phase transition. That is, the slow mode, the phason, is an intrinsic part of the order-parameter dynamics. The present paper addresses the dynamics of the phasons as observed experimentally by light scattering in $BaMnF_4$.

$BaMnF_4$ undergoes a transition to an incommensurate phase at a temperature T_i=254 K. Recent evidence, including that presented below, suggests the presence of two phase transitions, with the second one occurring at 247 K. The eigenvectors of the soft mode(s) involved remain unknown to this point, but the order parameter wavevector has been shown to be (0.39,0.5,0.5) below 247 K. The phase above T_i is pyroelectric, and hence lacks inversion symmetry.

The measurements reported here employed a single-mode Ar^+ laser operating at 5145 Å, with a power of up to 100 mW incident on the sample. Except as noted, the experiments were carried out in right angle (bb) scattering ($|\vec{q}|=(4\pi n/\lambda)\sin 45^\circ=2.58\cdot 10^4 cm^{-1}$), with $\vec{q}$ in the *ac* plane. Measurements reported below at scattering angles θ=53° and 127° correspond to $|\vec{q}|=1.62\cdot 10^4 cm^{-1}$ and $3.26\cdot 10^4 cm^{-1}$, respectively. We obtained light scattering spectra from two samples, one with faces perpendicular to the crystal axes and one rotated 45° in the *ac* plane. They were cut from neighboring regions of a $BaMnF_4$ boule grown by the Czochralski method. A separate study of the thermal diffusivity employed a third sample cut as a (100) platelet. The samples were mounted in vacuum on a Cu cold finger controlled in temperature to a stability of ±0.005 K. Temperatures were measured by a Pt resistance thermometer inserted in a hole in the Cu block behind the sample. Spectra were obtained simultaneously in Raman and Brillouin scattering by observing light scattered out two different faces of the samples, with both instruments aligned on the same scattering volume. The Raman spectra were analyzed with a Spex 1401 double grating monochromator, while the Brillouin spectra were obtained with the tandem pressure scanned Fabry-Perot (TFP) system previously described.[2] A molecular I_2 reabsorption cell was employed to absorb the elastically scattered component of the light. Computer analysis[2] was used to restore quantitatively the inelastic spectral profile outside the main I_2 absorption which extends out to $\Delta\nu \sim 0.5$ GHz. The instrumental line width of the TFP system was $\Gamma_{inst} \approx 1.8$ GHz (FWHM), while a resolution of $\sim 5 cm^{-1}$ was typically employed in observation of the Raman spectrum.

Initial measurements were carried out with $\vec{q} \parallel \vec{c}$. In this geometry, the main additional spectral feature near T_i is a peak centered at $\omega=0$ with a width which considerably exceeds Γ_{inst}. While its intensity is strongly temperature dependent below T_i, the observed width does not change appreciably over the range (of ~ 10 K) where it is clearly measurable. However, that width is found to depend on $|\vec{q}|$ when the condition $\vec{q} \parallel \vec{c}$ is maintained. The spectra were studied for $\theta=127°$, $90°$, and $53°$, with resulting deconvolved widths of $\Gamma_{CP}=4.4$, 3.2, and ~ 1.6 GHz (FWHM), respectively. (The latter value is less accurate due to the difficulty in extracting such a narrow line width from the spectrum seen through the I_2 cell.) The values obtained are consistent with $\Gamma_{CP}=Dq^2$, where $D=0.14 \pm 0.02 cm^2/sec$.

We shall first consider whether this scattering may be due to entropy fluctuations. Although such a high value for $D_{th}=\Lambda/\rho C_p$ (where Λ is the thermal conductivity, ρ is the density, and C_p is the specific heat) may be unlikely, we cannot exclude it a priori. Therefore, we carried out a direct measurement of the thermal diffusivity on long time scales, employing the pyroelectric response of the sample itself as a probe of small temperature variations. The results[3] showed that the thermal diffusivity (for $\vec{q} \parallel \vec{c}$ in the neighborhood of T_i is $0.005 cm^2/sec$, which is far too small to explain the observed width. We can therefore reject unequivocally the possibility that the scattering is due to the entropy fluctuation peak.

Since $BaMnF_4$ is noncentrosymmetric above T_i, direct phason scattering may be Raman allowed (i.e.: even at $q=0$) below T_i. Clearly, the converse statement is true—that is, in a centrosymmetric system direct phason scattering is not Raman allowed.[3] The latter argument rests on the fact that if $\vec{k}_o$ is truly incommensurate, then it is always possible to pick a lattice point with inversion symmetry to any desired degree of accuracy in the incommensurate phase. Thus, the long wave length properties of the system will reflect centrosymmetry and the amplitude and phase modes are respectively of even and odd parity. Hence the phason will have no Raman activity, although we note it may still be Brillouin allowed for $q \neq 0$.

In order to describe our spectra it is necessary to take phason damping into account. Little theoretical work directly applicable to the present situation has appeared, but Bhatt and McMillan have published a model for damping of incommensurate excitations in the normal phase ($T>T_i$) of a layered dichalcogenide near its CDW phase transition.[4] In order to apply their results, two extensions are necessary. First $\vec{k}_o$ lies at the Brillouin zone boundary for $BaMnF_4$. Second, we must extend the results to the case $T<T_i$. We perform these extensions here by two simple conjectures, with the hope that these conjectures will spark theoretical investigations of their veracity. For the first problem, we simply replace $\vec{k}_o$ in the Bhatt-McMillan theory by the *incommensurate component* $\vec{k}_o^I$ (the a-component) of $\vec{k}_o$ in the $BaMnF_4$ case. For the second, we observe that the phason dynamics may be relatively independent of temperature below T_i. Hence, we use the equations of Bhatt and McMillan evaluated *at* T_i to describe the dynamics of the *phase* mode (not the amplitude mode) *below* T_i. Bhatt and McMillan give results for both the underdamped and overdamped cases—but we consider only the latter here, since the peak we observe is clearly overdamped. For this case, their result for the phason line width Γ_p may be expressed as

$$\Gamma_p = D_{\parallel} q_{\parallel}^2 + D_{\perp} q_{\perp}^2 + A(T-Ti) \quad (2)$$

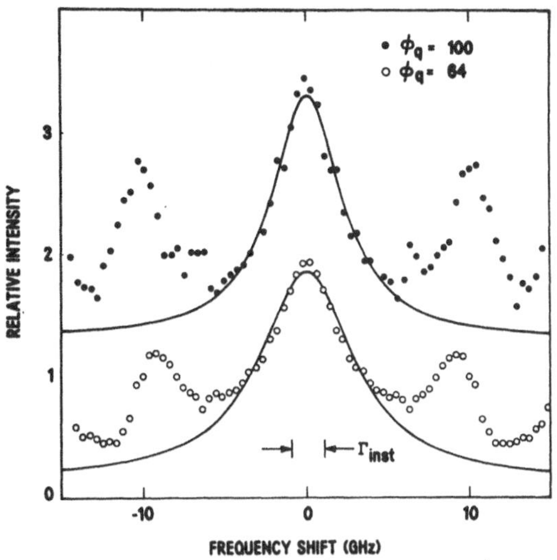

Fig. 1. Spectra observed for two crystal orientations.

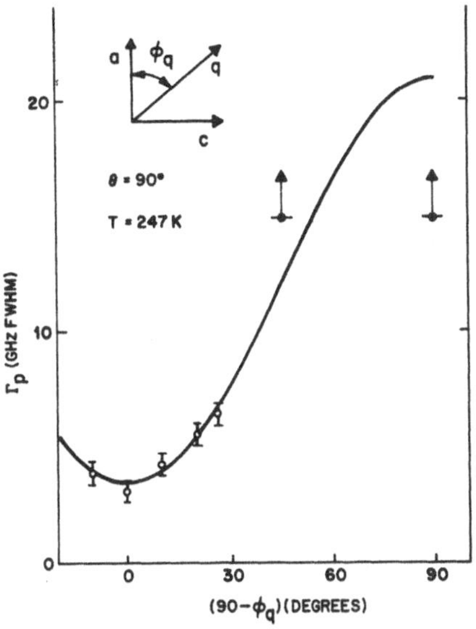

Fig. 2. Phason linewidth as a function of crystal orientation.

where $q_{\parallel,\perp}$ represent the components of $\vec{q}$ parallel and perpendicular to $\vec{k}_o^l$, $D_{\parallel,\perp}$ are independent constants, and we set $T=T_i$. The initial experimental results described above yield $D_\perp=0.14\text{cm}^2/\text{sec}$.

In order to evaluate the possible anisotropy expressed in (2), we measured the central peak line shape as a function of the direction of $\vec{q}$, with $|\vec{q}|$ held constant ($\theta=90°$). For $\vec{q}\parallel\vec{a}$ and for $\vec{q}$ at 45° between $\vec{a}$ and $\vec{c}$, the central component is not clearly distinguishable from the background provided by other scattering processes. If the integrated intensity is independent of $\vec{q}$, we have estimated that a width $\Gamma_p \gtrsim 15$ GHz would have precluded an unambiguous measurement. This implies a minimum value of the ratio $D_\parallel/D_\perp >> 1$. To check this, we measured spectra for smaller deviations of $\vec{q}$ from $\vec{c}$ in the *ac* plane. Two such spectra are shown in Fig. 1, and the deconvolved line widths resulting from the analysis are shown in Fig. 2 where we plot Γ_p against ϕ_q, the latter being the angle formed by $\vec{q}$ and $\vec{a}$. Clearly, the variation observed is qualitatively consistent with (2), and as expected, the ratio $D_\parallel/D_\perp=6.8$ is considerably greater than unity. The fact that the value observed is somewhat smaller than indicated by the result at $\phi_q=45°$ may indicate a slow dependence of the integrated intensity on q.

Furthermore, the model employed relates the anisotropy for the overdamped phason below T_i to that of the underdamped excitation in the normal phase well above T_i, providing that the temperature dependent term may be ignored.[5] The neutron scattering data available[6] well above T_i indicates a ratio $D_\parallel/D_\perp \gtrsim 3$, where the inequality results from the temperature dependent term in (2). Thus, the anisotropy we observe is consistent with the neutron scattering data on the soft mode branch above T_i.

If the scattering results from direct phason scattering, it should vanish at T_i, at least as ψ_o^2. The same is true if it results from coupling of the phason and acoustic modes. This indicates that T_i must be considerably above the temperature T_m where the scattering intensity is maximal. Therefore, it is of interest to study the temperature variation of the intensities of Raman lines which become active below T_i. Such measurements, as shown in Fig. 3, show

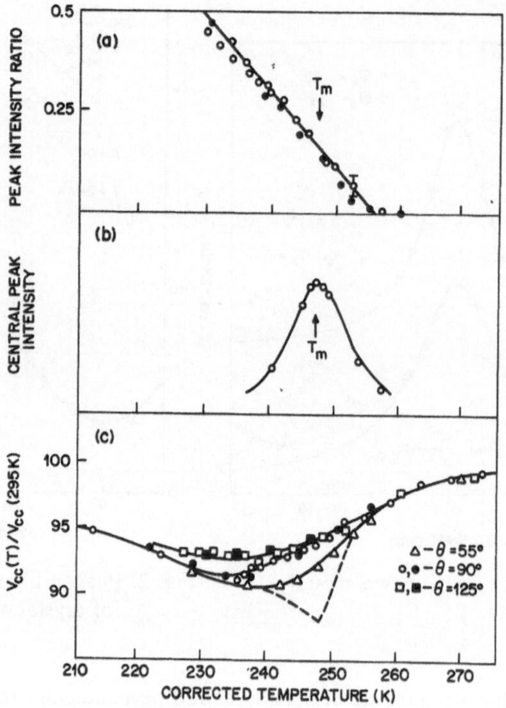

Fig. 3. Three indicators of the phase transition extracted from the light scattering spectra. Part (a) shows the intensity of the 155cm^{-1} peak in the Raman spectrum, (b) shows the intensity of the phason scattering, and (c) shows the relative acoustic velocity V_{cc}, referred to the room temperature value.

clearly that the Raman intensity extrapolates to zero not at T_m but at ~254 K, which is also where the central peak intensity disappears. Thus, we identify T_i~254 K based on this interpretation. It is important to note that the Raman and central peak spectra are obtained simultaneously and all temperatures shown are corrected for laser heating.

We also observed a dispersion of some 4% in the LA mode for $\vec{q} \parallel \vec{c}$, as shown in Fig. 3c, which is evident from a comparison of ultrasonic data[7] with our Brillouin scattering data at frequencies of ~14, ~20, and ~26 GHz. The dispersion is a maximum near T_m, where its relaxation frequency apparently reaches a minimum of some 10 GHz. This lies in contrast to the stronger dispersion in the TA mode observed by Bechtle and Scott, with a relaxation frequency below 1 GHz. It seems, therefore, that the processes responsible for these two relaxation phenomena are quite different.

The sharpness of the ultrasonic anomaly at 247 K (and the associated dispersion) suggests the presence of a second phase transition. This in fact could be related to the decrease in central peak intensity below T_m. Additional evidence for a second phase transition is provided by the recent specific heat measurements of Scott et al[8] who observe a smaller peak in the specific heat at 247 K, subsidiary to a stronger asymmetric one at 254 K. It is therefore of interest to conjecture what the nature of this pair of phase transitions may be.

Based on an x-ray diffraction study which indicates a commensurate phase below 247 K, Scott et al[8] have suggested that the lower phase transition may actually represent lock-in. We note that such an interpretation could explain the disappearance of phason scattering below 247 K. However, the neutron scattering experiments of Cox et al[6], carried out on samples from the same source as ours, clearly show $\vec{k}_o$ to be incommensurate and slightly temperature dependent[9]. Thus, we doubt that the lock-in interpretation is correct. We would suggest, on the contrary, that a model recently proposed by Natterman and Przystawa[9] may be applicable. They find that, within a Landau model explicitly including Umklapp terms in the free energy, there may be a narrow temperature range where $\vec{k}_o$ lies in the *ab* plane. The second phase transition then represents a *partial* lock-in to an incommensurate structure which remains incommensurate along $\vec{a}$ only. The applicability of such a model has been controversial[9] and must remain speculative at this time, but we note the report by Toledano at this meeting that such a structure is allowed by symmetry in the case of $BaMnF_4$.

In conclusion, we have interpreted our light scattering observations in terms of an overdamped phason mode for $T < 254$ K. The scattering may be direct or may be due to coupling of the phason to the acoustic modes. If the latter is true, the resulting asymmetric acoustic mode lineshapes may be observable in appropriate geometries. In either event, the extension of phason damping theories to the case of $BaMnF_4$ is an important problem deserving of theoretical attention. The weight of the evidence now suggests that there are two closely spaced phase transitions in $BaMnF_4$, and we have speculated that the lower one may be of the "partial lock-in" variety. The precise nature of the two phase transitions and the complete spectrum of incommensurate excitations in the various phases are problems which await further experimental investigation.

Acknowledgements

The crystals for this investigation were provided by H. J. Guggenheim. We would like to acknowledge helpful conversations with R. N. Bhatt, P. A. Fleury, D. J. Lockwood, and J. F. Scott. We note in addition useful discussions of the results at the Geilo school itself, notably those with A. D. Bruce, H. Z. Cummins, and R. Pynn. We wish to express our appreciation also to D. J. Lockwood, J. F. Scott, and their coworkers, for sharing their results in preprint form during the course of our respective investigations.

1. A. D. Bruce, see paper elsewhere in this volume.
2. K. B. Lyons and P. A. Fleury, J. Appl. Phys. 47, 4898 (1976).
3. K. B. Lyons, T. J. Negran, and H. J. Guggenheim, Phys. Rev. B, to be published.
4. R. N. Bhatt and W. L. McMillan, Phys. Rev. B 12, 2042 (1975).
5. K. B. Lyons, T. J. Negran, and H. J. Guggenheim, J. Phys. C: Sol. St. Phys. 13, L415 (1980).
6. D. E. Cox, S. M. Shapiro, R. A. Cowley, M. Eibschutz, and H. J. Guggenheim, Phys. Rev. B 19, 5754 (1979).
7. I. J. Fritz, Phys. Lett. 51A, 219 (1975).
8. J. F. Scott, F. Habbal, and M. Hidaka, Phys. Rev. B, to be published.
9. T. Natterman and J. Przystawa, private communication.

note that such an interpretation could explain the temperature-dependent scattering below [illegible] [illegible] [illegible] [illegible] [illegible] [illegible] [illegible] [illegible] [illegible] [illegible] [illegible] [illegible] Fleury [illegible] may be smaller [illegible]. They find that within a Landau model [illegible] [illegible] there may be a narrow temperature range where [illegible] lies in [illegible] [illegible] [illegible] [illegible] [illegible] [illegible] [illegible] [illegible] [illegible] [illegible] [illegible] [illegible] [illegible] such a structure as allowed by symmetry in the case of [illegible].

In conclusion, we have introduced [illegible] [illegible] [illegible] [illegible] [illegible] [illegible] of the phason [illegible] [illegible] [illegible] [illegible] [illegible] [illegible] [illegible] [illegible] [illegible] [illegible] [illegible] [illegible] [illegible] [illegible] [illegible] suggests that there are two [illegible] [illegible] [illegible] [illegible] [illegible] [illegible] [illegible] [illegible] of the two phase transitions [illegible] [illegible] [illegible] [illegible] further experimental investigation.

Acknowledgments

The samples used in this experiment were provided by [illegible]. We wish to acknowledge helpful conversations with R. A. Bruce, P. A. Fleury, [illegible] [illegible] [illegible] [illegible] [illegible] [illegible] [illegible] [illegible] [illegible] [illegible] [illegible] [illegible] [illegible] [illegible] [illegible] [illegible]

1. A. D. Bruce, see paper elsewhere in this volume.
2. K. B. Lyons and P. A. Fleury, J. Appl. Phys. [illegible] (1976).
3. [illegible]
4. [illegible]
5. [illegible]
6. [illegible]
7. [illegible], private communication.

NONLINEAR EXCITATIONS IN SOME ANHARMONIC LATTICE MODELS

Helmut Büttner

Phys. Institut, Universität Bayreuth,
Bayreuth, W.-Germany

1) INTRODUCTION

In recent years there has been a growing interest in anharmonic lattice models, because of their relation to structural phase transitions, and to soliton dynamics and thermodynamics. It is not intended to discuss the many interesting results of the vast literature in these fields, since there exist already very good reviews[1 2] or conference reports[3 4]. But instead I shall study some new aspects of an anharmonic ferroelectric model, the basic properties of which were investigated in two recent papers[5 6]. In addition some results[7] for the Toda-lattice[2] will be discussed and compared with those found by Schneider and Stoll[8]. In these two examples the nature of nonlinear excitations will be clarified and their contribution to various physical quantities will be investigated.

2) NONLINEAR EXCITATIONS IN A MODEL FERROELECTRIC*

The origin of this model is the investigation by Migoni et al.[9], where a nonlinear shell model was used to describe the phonon-spectrum and the giant Raman scattering in some incipient ferroelectrics within the selfconsistent phonon approximation. The physical fact behind this model is the instability of the free doubly negative oxygen-ion and the stabilization in the perovskite structures, which lead to a nonlinear electron-ion-coupling.[13] In the meantime, the selfconsistent phonon description

---

*Most of the results were worked out in collaboration with H. Bilz, Max-Planck-Institut, Stuttgart, W.-Germany.

in a simplified model[10] has been extended to a variety of systems, e.g. SbSI[11] and $KTa_{1-x}Nb_xO_3$[12]. The nonlinear excitations of the model were first studied in Ref. 5. The original lattice was a diatomic chain, with local nonlinear electron-ion interactions. In the following we will study a monatomic chain[6], because most of the relevant excitations can be found in this system and it is much easier to investigate in a stability analysis. A part of the chain is shown in Fig. 1.

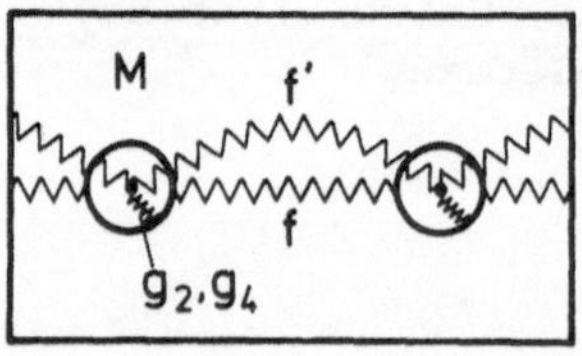

Fig. 1. Part of the one-dimensional lattice-model, with harmonic nearest-neighbour coupling between the electron shells f and the ions f'; the local electron-ion coupling is described by a harmonic coupling g_2 and an anharmonic force-constant g_4.

With M being the mass of the ions, m that of the electrons, g_2, f and f' being harmonic couplings and g_4 a fourth order coupling, the Hamiltonian is written as

$$H = \sum_n \frac{m}{2} \dot{v}_n^2 + \sum_n \frac{M}{2} \dot{u}_n^2 + \frac{g_2}{2} \sum_n (v_n - u_n)^2 + \frac{g_4}{4} \sum_n (v_n - u_n)^4 + $$
$$+ \frac{f}{2} \sum_n (v_n - v_{n-1})^2 + \frac{f'}{2} \sum_n (u_n - u_{n-1})^2 \tag{2.1}$$

As discussed in Ref. 5 to a good approximation one can assume an adiabatic motion of the electrons and the classical equations of motion are:

$$M \ddot{u}_n = g_2 w_n + g_4 w_n^3 + f'(u_{n+1} + u_{n-1} - 2u_n)$$
$$0 = -g_2 w_n - g_4 w_n^3 + f(w_{n+1} + w_{n-1} - 2w_n + u_{n+1} + u_{n-1} - 2u_n) \quad (2.2)$$

where instead of the electron displacement v_n the difference $w_n = v_n - u_n$ is used. This system of coupled difference-differential equations has a large variety of solutions of which certain static and stationary ones will be discussed. Its relation to the well-known ϕ^4-model[1,14,15] is studied in Ref. 5 (see also Ref. 16).

Static solutions to (2.2) can be divided into homogeneous and inhomogeneous states. The simplest one is the paraelectric groundstate with no displacements of the electrons against the ions:

$$u_n = v_n = 0 \quad (2.3)$$

The homogeneous ferroelectric state has a finite difference-displacement

$$w_n^2 = w_o^2 = - g_2/g_4 \quad (2.4)$$

and is only possible for an attractive $g_2 \leqslant 0$ (assuming a positive g_4 for the local double minimum potential). Its energy per particle is found from (2.1):

$$E_o/N = - g_2^2/4g_4 \quad (2.5)$$

The most important inhomogeneous solution is periodic with

$$w_n = A \sin(2\pi n/3)$$
$$u_n = B \sin(2\pi n/3) \quad (2.6)$$

and amplitudes

$$A^2 = - 4g_2(1 + 3ff'/g_2/(f + f'))/3g_4$$
$$B = - A f/(f + f') \quad (2.7)$$

In this state of the chain, every third ion is not displaced against the electrons, while the two in between are displaced in opposite directions. The unit cell of the chain is therefore tribled. Since g_2 is negative it can be seen from (2.7) that for attractive ion-ion interaction ($f' < 0$) the single amplitudes are larger in this state than in the homogeneous case (2.4). This attractive ion-ion coupling can be viewed as coming from the interaction with neighbouring chains[6]. The corresponding energy is

$$E_p/N = - g_2^{\,2}(1 + 3ff'/g_2/(f + f'))^2/6g_4 \tag{2.8}$$

and is lower than E_o/N for negative f', if

$$(1 + 3ff'/g_2/(f + f'))^2 > 3/2 \tag{2.9}$$

The groundstate of the chain is then a state with a periodic variation of the electron-ion displacements and is called a static periodon, which is commensurate to the original lattice with a threefold lattice constant. The stability of these states is considered below. Similar solutions for the original ϕ^4-model have been studied recently by Magyary[16]. In order to find other static solutions it is helpful to write down the equations (2.2) in form of nonlinear recurrence relations[17]. If one eliminates u_n from the first equation these are:

$$\begin{aligned} w_{n+1} &= (1/f + 1/f')(g_2\, w_n + g_4\, w_n^3) + 2w_n - w_{n-1} \\ u_{n+1} &= -(g_2\, w_n + g_4\, w_n^3)/f' + 2u_n - u_{n-1} \end{aligned} \tag{2.10}$$

(Mathematical details for these type of equations can be found in Ref. 17.) The so-called fixed points are the homogeneous solutions. Periodic solutions are called cycles of different length. The inhomogeneous form (2.6) is a cycle of order 3: $w_{n+3} = w_n$. There is also a cycle of order 6. From numerical studies one can find additional solutions not being cycles, but incommensurate with the original lattice. These states have been studied numerically with the following qualitative results (see Fig. 2):

For not too large amplitudes around the paraelectric state (or fixed point (0,0)) there are solutions that are incommensurate on the lattice. In plotting w_{n+1} against $(w_n - w_{n-1})$ the points for increasing n lie on a curve, that becomes more and more 'dense'.

For higher amplitudes a characteristic structure of islands appears, which has an eightfold symmetry indicating a cycle of order 8. It was not possible to find analytic solutions for this period.

At slightly larger amplitudes, before becoming unstable, the static solution appears to be chaotic. The main difference to the incommensurate states is the strong variation of the amplitude. This area of solutions is more pronounced for large negative g_2 (and, or f').

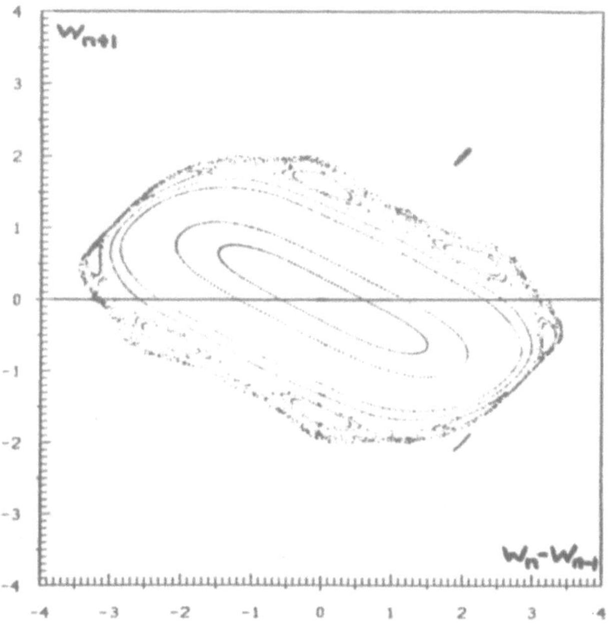

Fig. 2. The electron-ion displacement w_{n+1} as function of $w_n - w_{n-1}$ for 500 lattice points. The various curves belong to different starting-values and parameters $g_2(1/f + 1/f') + 2 = -2$; $g_4(1/f + 1/f') = .25$.

There are similar states in the vicinity of the other fixed points and the cyclic states. The energies of these excitations are not easily accessible, but there seem to be model parameters for which some of them are below the commensurate states[18]. While this manuscript was in preparation a similar discussion of spatial chaotic states on a lattice has been given by Bak[19] in connection with an Ising model.

The energy of all these solutions is proportional to the number of lattice sites, but there are also states with a localized energy, which are most interesting for thermodynamic applications. Besides the well known kink-soliton which describes a transition between the two homogeneous states there is also a kink-like soliton between the two different realizations of the periodons. It is the corresponding excitation above the commensurate state.

Stationary solutions to (2.2) are a much larger manifold and so far only very few are studied in detail[5]. Most of the static periodic states are special limits of more general stationary states. The states in (2.6) can be generalized to

$$w_n = A(q)\sin(\omega t - nqa)$$
$$u_n = B(q)\sin(\omega t - nqa) + C(q)\sin 3(\omega t - nqa) \tag{2.11}$$

where the frequency and the amplitudes are functions of qa:

$$M\omega^2 = \frac{4}{9}(f+f')\sin^2(3qa/2) \tag{2.12}$$

$$A^2 = -\frac{4g_2}{3g_4}\left\{1 + \frac{4f}{g_2}\,\frac{M\omega^2 - 4f'\sin^2(qa/2)}{M\omega^2 - 4(f+f')\sin^2(qa/2)}\,\sin^2(qa/2)\right\}$$
$$= -\frac{4g_2}{3g_4}\left\{1 - \frac{3f}{2g_2}\,\frac{f}{f+f'}\,\frac{(1-\frac{4}{3}\sin^2(qa/2))^2}{1-\frac{2}{3}\sin^2(qa/2)} + \frac{4f'}{g_2}\,\frac{f}{f+f'}\sin^2(qa/2)\right\} \tag{2.13}$$

$$B = -A - (g_2A + 3g_4A^3/4)/4f\sin^2(qa/2) \tag{2.14}$$

$$C = -g_4A^3/16f\sin^2(3qa/2) \tag{2.15}$$

The amplitude of the relative electron-ion displacement A is plotted in Fig. 3 for two different parameter sets.

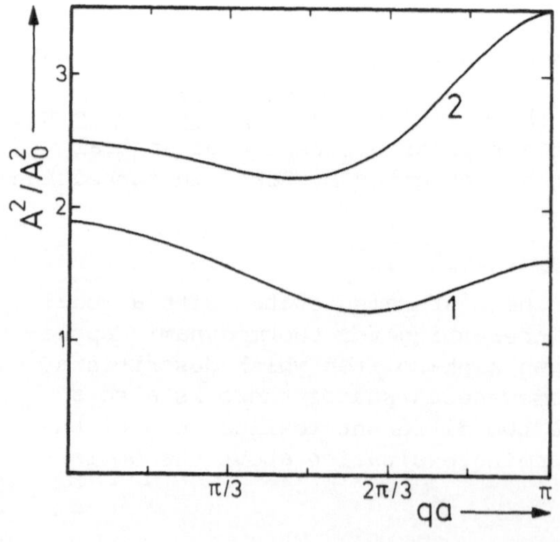

Fig. 3. Periodon amplitude from (2.13) for 1(2): $g_2/f = -2$, (-2) $f'/f = -1/7$, (-1/2)

Notice the minimum at $qa = 2\pi/3$ for $f' = 0$ and how it shifts to smaller q-values for $f' \neq 0$. This may be related to a shift from a commensurate to an incommensurate state[6]. This general stationary periodons will be used later to describe an extension of the selfconsistent phonon approximation. But let us first study the stability of the static periodons.

Stability of static periodons is investigated in the usual way by adding a small perturbation to the original solution.

$$\begin{aligned} w_n &= A \sin(2\pi n/3) + \phi_n(t) \\ u_n &= B \sin(2\pi n/3) + \psi_n(t) \end{aligned} \tag{2.16}$$

and deriving linearized equations of motion for $\phi_n(t)$ and $\psi_n(t)$. Eliminating ψ_n three coupled equations are found (for the three inequivalent ions):

$$M\left\{\frac{g_2}{f}\ddot{\phi}_{3m} - D\ddot{\phi}_{3m}\right\} = D\left\{(1 + f'/f)\, g_2\, \phi_{3m} - f'D\phi_{3m}\right\} \tag{2.17}$$

$$M\left\{\frac{\tilde{g}_2}{f}\ddot{\phi}_{3m\pm1} - D\ddot{\phi}_{3m\pm1}\right\} = D\left\{(1 + f'/f)\, \tilde{g}_2\, \phi_{3m\pm1} - f'D\phi_{3m\pm1}\right\} \tag{2.18}$$

with the difference operator $D\phi_{3m} = \phi_{3m+1} + \phi_{3m-1} - 2\phi_{3m}$ and the effective coupling

$$\tilde{g}_2 = -2g_2 - 9ff'/(f + f') \tag{2.19}$$

The spectrum is found by the Fourieransatz

$$\phi_{3m\pm j} = \phi_j \exp(i\omega t - i3mqa) \qquad j = 0, \pm1 \tag{2.20}$$

from the characteristic equation, which is than solved numerically. In Fig. 4 two qualitative examples of the three phonon branches were plotted. Note the rather large gap for small qa between the upper and lower optical branch. The original periodon is stable if all three frequencies are real. For $qa = 0$ the result can be given analytically

$$\begin{aligned} (M\omega^2)_a &= 0 \\ (M\omega^2)_{ph} &= (2\gamma\tilde{g}_2 + \tilde{\gamma}g_2)/(2\tilde{g}_2 + g_2 + g_2\tilde{g}_2/f) \\ (M\omega^2)_{am} &= \tilde{\gamma}/(1 + \tilde{g}_2/3f) \end{aligned} \tag{2.21}$$

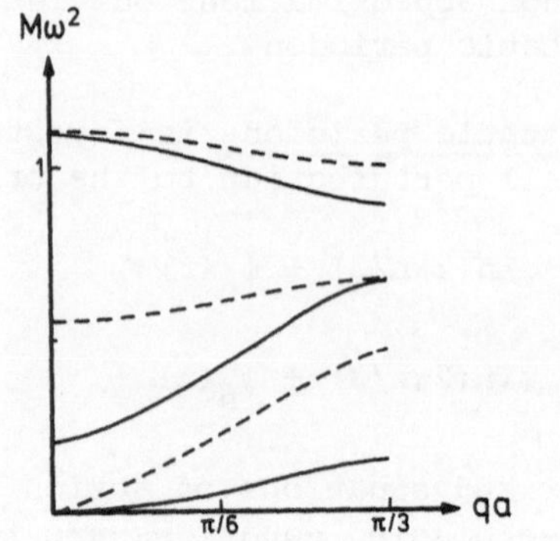

Fig. 4. Qualitative picture of the phonons in the ferroelectri regime for two different parameter sets.

with

$$\gamma = g_2(1 + f'/f) + 3f' \ ; \quad \tilde{\gamma} = \tilde{g}_2(1 + f'/f) + 3f'$$

from which one finds the stability condition (for $|f'| \lesssim f$):

$$2 \lesssim |g_2|/f \lesssim 3 \tag{2.22}$$

The periodon-phonon coupling has as a result not only these linearized phonons, but can also be extended to a selfconsistent phonon approximation. The method is thoroughly discussed in Ref. 6 and consists of the following approximations: There are two additional nonlinear terms neglected in the stability analysis. The first is the contribution of $g_4\,\phi_n{}^3(t)$, which is approximately linearized by its temperature-dependent mean value $<\phi_n{}^2(t)>$:

$$g_4\,\phi_n{}^3(t) \simeq 3g_4\,\phi_n(t)\,<\phi_n{}^2> \tag{2.23}$$

and therefore gives rise to an effective coupling

$$g(T) = g_2 + 3g_4 \langle \phi_n^2 \rangle \tag{2.24}$$

The second nonlinear contribution has the form

$$3g_4 A \sin(2\pi n/3) \phi_n^2(t) \tag{2.25}$$

and is neglected in the following, since the appropriate linearization vanishes.

For this selfconsistent phonon-approximation, where the mean value in (2.24) has to be determined selfconsistently, the spectrum follows from equations (2.17) and (2.18) by replacing g_2 in (2.17) by $g(T)$ and $\tilde{g}_2$ in (2.18) by

$$\tilde{g}_2 \rightarrow \tilde{g}(T) = g(T) - 3g_2(1 + 3ff'/g_2/(f + f')) \tag{2.26}$$

In contrast to the phonons from stability-analysis, the effective coupling $g(T)$ can vary with temperature, but this has only little effect on the spectrum. In the commensurate phase we therefore expect the phonons to show only a weak temperature-dependence (at least near $qa = 0$), which seems to be true in the transition of K_2SeO_4[20]. For further discussion on this and other substances see Ref. 6.

The periodon-phonon coupling can also be present in the paraelectric regime, where the periodon exists as a general time-dependent excitation. In this case the nonlinear term

$$3g_4 w^2_{n,\mathrm{periodon}} \phi_n(t) \tag{2.27}$$

is substituted by an expression with a time averaged solution

$$3g_4 A^2(q) \langle \sin^2(\omega t - nqa) \rangle_{\mathrm{time}} \simeq \frac{3}{2} g_4 A^2(q). \tag{2.28}$$

The physical idea behind this approximation is the fact, that in the paraelectric regime the phonons do not see a static distortion of the lattice, but will couple to the periodons as nonlinear excitations[6]. Therefore in our model there is a single phonon branch with an effective q-dependent coupling

$$g(T,q) = g(T) + \frac{3g_4}{2} A^2(q) \tag{2.29}$$

and the dispersion

$$M\omega^2 = 4f\sin^2 qa/2 \left\{ g(T,q)/(g(T,q) + 4f\sin^2 qa/2) + f'/f \right\} \quad (2.30)$$

For negative g_2 and f' it has a more or less pronounced minimum near $qa = 2\pi/3$. The strength of f' controls the shift away from $qa = 2\pi/3$. A quantitative investigation of this model for various substances can be found in Ref. 6.

The model discussed above is a one-dimensional version of the original three-dimensional lattice[9]. The investigation of two-dimensional systems[21] has given first results and also simple three-dimensional structures are studied for their nonlinear excitations[6].

3) SOME REMARKS ON TODA-LATTICE*

In contrast to the above model the famous Toda-lattice has a nonlinear interaction between neighbouring sites of the chain. It serves as a nonlinear example, where most of the thermodynamic properties can be treated exactly. Since solitary solutions are also known, it is possible to calculate the contributions of these excitations to the thermodynamic properties[7,8,22]. This soliton-gas analogy was also used quite recently by Schneider and Stoll[8] to investigate static and dynamic correlation functions. In the following we would like to comment on some of their results.

We start by reviewing some basic results for the Toda-lattice. The Hamiltonian for this chain can be written as

$$H = \sum_n \left[p_n^2/2M + M\omega^2 \left\{ \exp(-\gamma(q_n - q_{n-1})) + \gamma(q_n - q_{n-1}) - 1 \right\} / \gamma^2 \right] \quad (3.1)$$

with the ion-mass M, the nonlinear coupling γ, and the corresponding harmonic energy $M\omega^2$. The one-soliton solution for the classical equation of motion is usually written for the exponential displacement (with $r_n = q_n - q_{n-1}$)

$$\exp(-\gamma r_n) - 1 = -\frac{\gamma}{M\omega^2} \frac{d}{dr_n} V_{Toda}(r_n) = \sinh^2\alpha \, \mathrm{sech}^2(\alpha n + \omega \sinh\alpha \, t) \quad (3.2)$$

with a parameter α of arbitrary sign. (The first equality is for later use.) This soliton has a certain velocity v_S, momentum P_S and energy E_S:

*Most of the work was done in collaboration with F. G. Mertens, Universtität Bayreuth.

$$v_S = \omega a \sinh(\alpha)/\alpha \tag{3.3}$$

$$P_S = 4M\omega(\alpha \cosh(\alpha) - \sinh(\alpha))/\gamma \tag{3.4}$$

$$E_S = M\omega^2(\sinh(2\alpha) - 2\alpha)/\gamma^2 \tag{3.5}$$

The expression for the momentum P_S should not be confused with the momentum of the soliton-bearing chain.

$$P = \sum_n m\dot{q}_n = \frac{2M}{\gamma} \sinh\alpha \tag{3.6}$$

which is not the canonical variable conjugate to the spatial center of mass of the soliton[7]. The knowledge of canonical variables is most important for the following transformation to action-angle variables, which was found by Flaschka and McLaughlin[23]. Their result can be summarized as follows. There exist a number of discrete variables z_ν (with $|z_\nu| \leqslant 1$) related to the action-variable J_ν and the angle Θ_ν by

$$J_\nu = -(z_\nu + z_\nu^{-1})$$
$$\frac{d\Theta_\nu}{dt} = z_\nu - z_\nu^{-1} \tag{3.7}$$

In addition there are continuous variables $J(\phi)$ and $\Theta(\phi)$ with

$$\frac{d\Theta(\phi)}{dt} = \frac{1}{\pi} \sin\phi \tag{3.8}$$

and the Hamiltonian is transformed (for dimensionless units) into:

$$H = \sum_{\nu=1}^{N_S} \left\{ \frac{1}{2} (z_\nu^{-2} - z_\nu^{2}) + \ln z_\nu^{2} \right\} + \frac{1}{\pi} \int_0^{\pi} d\phi\, J(\phi) \sin\phi \tag{3.9}$$

where the number of solitons N_S is treated as a parameter. From this form of H the soliton contribution to the free energy at low temperatures can be calculated. Taking into account only one-soliton states one finds for the partition function

$$Z_S = \exp\left\{ \int \frac{dJd\Theta}{\Delta\Omega} \ln(1 + \exp(-\beta E_S(J))) \right\} \tag{3.10}$$

where $\Delta\Omega$ is an appropriate phase-space volume[7]. Written with the parameter α from (3.2) one finds for a N-particle chain

$$\frac{1}{N} \ln Z_S = \frac{2}{\Delta\Omega} \int_0^\infty d\alpha \; 4\alpha \sinh(\alpha) \ln\left(1 + \exp(-\beta E_S(\alpha))\right) \tag{3.11}$$

with a free energy contribution at low temperatures:

$$F_S = - \frac{2}{\Delta\Omega} \frac{\pi^2}{12} T^2 + \ldots \tag{3.12}$$

For the phase space volume $\Delta\Omega = 2\pi^2$ this is equal to the first nonlinear contribution in the exact result. This is in contrast to the calculation of Schneider and Stoll[8], who used the chain momentum P for the soliton. The higher order terms in the free energy are strongly influenced by soliton-soliton scattering events, but since we are only interested in the low-temperature regime, this positive result (3.12) encourages to use the soliton-gas analogy in a calculation of static correlations. This was first done in the work by Schneider and Stoll[8], who give some exact results and compare it with calculations using the soliton-gas picture. An interesting function for correlation studies is the exponential displacement (we use the notation of Ref. 8)

$$e_\ell = \exp(q_{\ell-1} - q_\ell) \tag{3.13}$$

with the mean value

$$\langle e_\ell \rangle = 1 \; ; \qquad \langle e_\ell^{\,2} \rangle = 1 + T \tag{3.14}$$

where the mean value is calculated from

$$\langle A \rangle = \int dr \; A(r) \exp\left\{ -\beta \left[\exp(-r) + r - 1 \right] \right\} / \int dr \exp\{ \ldots \tag{3.15}$$

Notice that function e_ℓ is related to the potential derivative as written in (3.2). Therefore the quadratic mean value is related to the virial by

$$\langle (e_\ell - \langle e_\ell \rangle)^2 \rangle = \langle (\frac{dV}{dr})^2 \rangle = \langle (r - V) \frac{dV}{dr} \rangle = \langle r \frac{dV}{dr} \rangle \tag{3.16}$$

where the last equation is valid because $\langle V \; dV/dr \rangle = 0$ for the Toda-potential. Since the virial mean value is T in general we find

$$\langle (e_\ell - \langle e_\ell \rangle)^2 \rangle = T \tag{3.17}$$

which also follows from Schneider and Stoll's direct calculation. The structure factor for the e_ℓ is therefore directly related to the virial and therefore q-independent. This follows from the definition

$$S_{ee}(q) = < \left| \frac{1}{N^{\frac{1}{2}}} \sum_\ell e^{iq\ell} (e_\ell - <e_\ell>) \right|^2 > \tag{3.18}$$

$$= <(e_\ell - <e_\ell>)^2> = T$$

The q-independence of S_{ee} is therefore not a curious result but comes from the fact, that the force-force correlation is related to the virial. Since the same is true for a purely harmonic lattice, namely that $<(dV/dr)^2> = <r\, dV/dr>$, one finds also in this case $S_{ee}(q) = T$. In this respect the Toda-lattice again behaves quite harmonically.

In order to investigate this further, Schneider and Stoll[8] proposed an approximate method to calculate the soliton contribution to $S_{ee}(q)$. They also extended this ansatz to dynamical correlation, but in the following we only discuss the static problem. The product $<e_\ell - <e_\ell>><e_o - <e_o>>$ is written as a sum of soliton contributions

$$\delta e_\ell\, \delta e_o \simeq \sum_\nu \delta e_\ell(Q_\nu)\, \delta e_o(Q_\nu) \tag{3.19}$$

where $\delta e_\ell(Q\ ) = e_\ell(Q_\nu) - 1$ is the soliton (3.2) centered at Q_ν. The mean values are defined as for the free energy calculations by

$$<\delta e_\ell(Q_\nu)> = \frac{1}{Z_S} \sum_{N_S} \frac{1}{N_S!} \sum_\nu^{N_S} \int \prod_\mu dP_\mu\, dQ_\mu\, \sinh^2\alpha_\nu\, \mathrm{sech}^2[\alpha_\nu(\ell-Q_\nu)] \cdot$$

$$\cdot\, \ell n(1+e^{-\beta E_S})$$

$$= \int dPdQ\, \sinh^2\alpha\, \mathrm{sech}^2\, [\alpha(\ell-Q)]\, \ell n\, (1 + \exp(-\beta E_S)) \tag{3.20}$$

In the same way one finds

$$<\delta e_\ell\, \delta e_o> - <\delta e_\ell><\delta e_o> = 32 \int d\alpha\, \sinh^5\alpha\, \ell n(1+e^{-\beta E_S(\alpha)}) \frac{\alpha\ell\, \coth(\alpha\ell)-1}{\sinh^2(\alpha\ell)} \tag{3.21}$$

Our method again differs from that in Ref. 8 in two aspects. The proper soliton-momentum (3.4) is taken into account and the discreteness of the α-values. In doing the Fouriertransformation of (3.21) this results in

$$S_{ee}(q) = 32 \int d\alpha \, e^{-\beta E(\alpha)} \, \frac{\sinh^5\alpha}{\alpha} \, \frac{(\pi qa/2\alpha)^2}{\sinh^2(\pi qa/2\alpha)} \tag{3.22}$$

For very low temperatures (and finite qa) these functions vanish as $\exp(-T^{1/4})$ and for higher temperatures a numerical investigation shows that it is always much smaller than the harmonic term. In contrast to Ref. 8 we therefore conclude, that the solitons do not contribute to the force-force correlation $S_{ee}(q)$.

For a study of dynamical correlation see Ref. 8, 25, 26.

Valuable discussions with H. Bilz and F. G. Mertens are greatly acknowledged. We thank T. Schneider for sending us a preprint of his work prior to publication.

1. R. A. Cowley, Adv. in Physics 29, 1 (1980)
   A. D. Bruce, Adv. in Physics 29, 111 (1980)
2. M. Toda, Suppl. Progr. Theor. Phys. 59, 1 (1976)
3. A. R. Bishop and T. Schneider, ed., Solitons and Condensed Matter Physics, Springer Series in Sol. State Science, Vol. 8 (1978)
4. R. K. Bullough and P. J. Caudrey, ed., Solitons, Topics in Current Physics 17, Springer 1980
5. H. Büttner and H. Bilz, in: Recent Developments in Condensed Matter Physics, Vol. I, ed. Devreese, Plenum, p. 49 (1981)
6. H. Bilz, A. Bussmann-Holder, W. Kress, H. Büttner and U. Schröder, preprint
7. F. G. Mertens and H. Büttner, submitted to Phys. Letters
8. T. Schneider and E. Stoll, Phys. Rev. B, to be published
9. R. Migoni, H. Bilz and D. Bäuerle, Phys. Rev. Lett. 37, 1155 (1976)
10. H. Bilz, A. Bussmann, G. Benedek, H. Büttner and D. Strauch, Ferroelectrics 25, 339 (1980)
11. M. Balkanski, M. K. Teng, M. Massot and H. Bilz, Ferroelectrics 26, 737 (1980)
12. D. Rytz, U. T. Höchli and H. Bilz, Phys. Rev. B22, 359 (1980)
13. A. Bussmann, H. Bilz, R. Roenspiess and K. Schwarz, Ferroelectrics 25, 347 (1980)
14. J. A. Krumhansl and J. R. Schrieffer, Phys. Rev. B11, 3535 (1975)

15. S. Aubry, J. Chem. Phys. 62, 3217 (1975)
16. E. Magyary, preprint
17. I. Gumowski and C. Mira, Recurrences and Discrete Dynamical Systems, Lecture Notes in Mathematics 809, Springer 1980
18. H. Büttner, to be published
19. P. Bak, Phys. Rev. Lett. 46, 791 (1981)
20. J. D. Axe, M. Izumi and G. Shirane, Phys. Rev. B22, 3408 (1980)
21. G. Behnke and H. Büttner, J. Phys. A: Math. Gen. 14, L113 (1981)
22. H. Büttner and F. G. Mertens, Sol. State Comm. 29, 663 (1979)
F. G. Mertens and H. Büttner, in: Recent Developments in Cond. Matter Phys., ed. Devreese, Plenum 1981
23. H. Flaschka, Progr. Theor. Phys. 51, 703 (1974)
D. W. McLaughlin, Journ. Math. Phys. 16, 96 (1975); 16, 1704 (1975)
24. H. Bolterauer and M. Opper, Z. f. Physik, to be published; Phys. Letters 83A, 69 (1981)
25. S. Dietrich, Z. f. Physik, submitted
26. F. G. Mertens and H. Büttner, to be published

SOLITONS IN THE ONE-DIMENSIONAL PLANAR FERROMAGNET $CsNiF_3$

R. Pynn[x], M. Steiner[+], W. Knop[+], K. Kakurai, J.K. Kjems[o]

[x]Institut Laue Langevin, 156X Centre de Tri,
38042 Grenoble Cedex, France
[+]Hahn-Meitner-Institute, Glienicker Str. 100,
D-1000 Berlin 29, West Germany
[o]Risø National Laboratory, DK-4000 Roskilde, Denmark

ABSTRACT

Neutron inelastic scattering has been used to study the spectrum $S_{\perp}(q,\omega)$ of spin fluctuations perpendicular to a magnetic field applied in the easy plane of the one-dimensional ferromagnet $CsNiF_3$. It is shown that the observed central peak can be described in terms of the independent-soliton model proposed by Mikeska. In the geometry used there is no contribution to the observed scattering from the two-magnon processes which have been proposed as an alternative explanation of the central peak found in $S_{\parallel}(q,\omega)$.

INTRODUCTION

The low-temperature magnetic properties of $CsNiF_3$ have been investigated by a number of methods. The consensus [1] is that, at temperatures above ~3 K, the material behaves as a one-dimensional Heisenberg ferromagnet with planar anisotropy. Spins of magnitude S=1 associated with the Ni^{2+} ions are coupled ferromagnetically to form chains along the crystallographic c axis [2]. Below the Néel temperature [3] T_N=2.61 K the small antiferromagnetic interchain coupling results in a three-dimensional magnetic order. However, the interchain coupling is about 10^{-2} of the intrachain interaction and is insignificant at temperatures large with respect to T_N. A crystal-field anisotropy tends to constrain the spins to the plane perpendicular to the chain axis. In view of this picture the Hamiltonian (we add the magnetic field term for later convenience)

$$\mathcal{H} = -2J\sum_{i=1}^{N} \vec{S}_i.\vec{S}_{i+1} + A\sum_{i=1}^{N} (S_i^z)^2 - g\mu_B H\sum_{i=1}^{N} S_i^x \qquad (1)$$

has often been invoked to describe the magnetic properties of $CsNiF_3$. The restriction to nearest neighbour interactions in the first term follows from the observed cosine behaviour of the magnon dispersion relation in the chain direction [2]. This dispersion relation is well described by the linear excitations of the Hamiltonian (1) and, in a classical approximation a fit of the measured magnon frequencies yields [2] $J/k = (11.8 \pm 0.3)$K and $(A/k)_{classical} = (4.5 \pm 2.0)$ K. A quantum mechanical treatment of the harmonic spin waves which result from equation (1) indicates that the value of A should be doubled [4,5]; thus $(A/k)_{q.m.} = (9 \pm 4)$ K. More accurate results for the parameters of the Hamiltonian have been obtained from the measurements of spin waves in a magnetic field. These experiments gave [6] $J/k = (11.5 \pm 0.05)$ K and $A/k = (8.9 \pm 0.2)$ K at a temperature of 4.2 K. Accurate susceptibility measurements [7] have also been interpreted on the basis of the Hamiltonian of eqn (1). In a temperature region between 15 K and 170 K these data are well described by $J/k = (9.9 \pm 0.2)$ K, $A/k = (7.5 \pm 0.2)$ K and $g = 2.26 \pm 0.01$. One concludes that the Hamiltonian of equation (1) is capable of providing a quantitative description of a wide range of properties of $CsNiF_3$ but that the precise values of the parameters used depends somewhat on the property concerned.

It was suggested by Mikeska [8] that part of the dynamic spin response of $CsNiF_3$ in a magnetic field ought to be describable in terms of magnetic solitons. In an appropriate region of temperature and magnetic field, quantum effects can be ignored [9] and the single-ion anisotropy is sufficiently strong to supress spin fluctuations parallel to the chain (z) direction. In a continuum approximation the equation of motion of such a system is the classical Sine-Gordon equation whose solutions comprise free oscillations, solitons and breathers [10]. Solitons, which correspond to a complete 2π rotation of the classical spin direction about the chain axis, contribute to the fluctuation spectrum and hence to the neutron scattering cross section. Provided the solitons do not interact with one another it is possible to show that the contribution to the spin fluctuation spectrum is given by [8,11,12].

$$S^x(q,\omega) = \frac{64}{\pi^2}\frac{\beta}{qc}\exp\{-16mJ\beta\}\exp\left\{-8mJ\beta\,\frac{\hbar^2\omega^2}{c^2\pi^2q^2}\right\}\left\{\frac{\pi^2 q}{2m}\right\}^2 \times\left\{\sinh\left(\frac{\pi^2 q}{2m}\right)\right\}^{-2} \qquad (2)$$

where $\beta = 1/k_BT$; $c = (4JA)^{1/2}$; $m = (g\mu_B H/2J)^{1/2}$

and $$S^\alpha(q,\omega) = \frac{1}{2\pi}\int dz\, e^{i2\pi qz}\int d\omega e^{i\omega t} \langle S^\alpha(z,t)S^\alpha(0,0)\rangle \qquad (3)$$

In the above equations $S^{\alpha}(z,t)$ is the α cartesian spin component at position za (a is the unit cell side in the z direction) along the chain at time t. The wavevector q in the above expressions is parallel to the chain direction and is measured in reciprocal lattice units ($\equiv 2\pi/a$). Equation (2) describes spin fluctuations parallel to the applied magnetic field (x direction). Spin fluctuations in the transverse (y) direction are described by $S^y(q,\omega)$, a quantity which is obtainable from eqn. (2) by replacing the sinh by cosh.

Although the model outlined above is physically appealing it is well to note that its validity depends on the fulfillment of a number of conditions. These are outlined in the following table

Reason for constraint	ref.	constraint	constraint applied to $CsNiF_3$
X-Y behaviour	8	$k_BT \ll (2AJ)^{1/2}$	$T \ll 15$
quantum effects unimportant	8,13	$A/J \ll 8\pi^2 S(S+1)$	$0.21 \ll 79$
quantum effects unimportant	13	$k_BT \gg (2Ag\mu_BH)^{1/2}$	$\sqrt{H}/T \ll 20$
continuum approx. valid	8	$q \ll (A/2\pi^2 J)^{1/2}$	$q \ll 0.15$
continuum approx. valid	8	$g\mu_BH \ll 2k_BA$	$H \ll 5\times10^4$
limit of soliton velocity	12	$h\nu < q\pi S(4JA)^{1/2}$	$\nu < q$
non-interacting solitons	8	$(2g\mu_BHJ)^{1/2}/k_BT \gtrsim 0.5$	$\sqrt{H}/T \gtrsim 10$
one-dimensional behaviour	2	$T \gg T_N$	$T \gtrsim 5$

In this table $\nu(\equiv\omega/2\pi)$ is in THz, q in reciprocal lattice units, T in Kelvin and H in Oersteds. For $CsNiF_3$ the classical value of A (which is half the quantum value [9]) has been used. Some of the conditions are difficult to fulfil simultaneously in an experiment with $CsNiF_3$ and any purported observation of solitons is thus easily contested.

The first experimental evidence for solitons in $CsNiF_3$ was reported by Kjems and Steiner [14]. These authors used neutron scattering techniques to measure the fluctuation spectrum around the (0,0,2) Bragg peak and thus determined $S^x(q,\omega) + S^y(q,\omega)$. The only theory available at the time of the experiment was [8] for $S^x(q,\omega)$ and the results were analysed on this basis. However, for the experimental conditions used there is very little difference between the sinh and cosh in eqn. (2); both are essentially given by half the exponential of the argument. Fairly good agreement between theory and experiment was reported. Unfortunately, as Reiter [15] has pointed out, the calculation of the magnetic field dependence of the integrated intensity of the soliton contribution to

$S(q,\omega)$ is incorrect in ref. [14]. However, the corrected calculation disagrees with experiment only for the lowest field used for which $\sqrt{H}/T \lesssim 8$. Thus one might argue that the discrepancy between experiment and Mikeska's theory represents the effects of soliton interaction (see line 7 of the above table). Apart from this minor disagreement the soliton theory provides a description of the neutron scattering data which is qualitatively good and quantitatively respectable.

Several theoretical attempts have been made to explain the $CsNiF_3$ data [14] by theories which do not involve solitons. Reiter [15] has produced a harmonic spin wave theory which retains the quantum aspects of the Hamiltonian given in equation (1) and which apparently describes the data of Kjems and Steiner [14] at least as well as the soliton theory. However, while Reiter [15] correctly predicts the field dependence of the observed neutron intensity, his lack of temperature dependence for the energy width of the scattering disagrees with experiment. The latter quantity is well reproduced by the soliton theory. It is not clear (at least to us!) that the low-temperature high-field theory of Reiter [15] is applicable to the spin dynamics of $CsNiF_3$ under the experimental conditions used.

In a recent numerical simulation Loveluck et al. [13] have investigated the conditions under which the Hamiltonian given in eqn. (1) can be mapped onto that of a decrete Sine-Gordon model. These authors conclude that the mapping fails because (c.f. first line of table above) out-of-plane spin fluctuations cannot be ignored at the temperatures of the experiments. Loveluck et al. [13] claim that the central peak observed [14] in $CsNiF_3$ is a result of two-magnon (creation plus annihilation) scattering. However, since Loveluck's molecular dynamics calculations based on equation (1) of this paper do not reproduce the observed spectrum it is difficult to know how relevant the comparison of eqn. (1) and the Sine-Gordon model is to experiments with $CsNiF_3$.

Experiment

To add fuel to the controversy described above we have carried out an experiment on $CsNiF_3$ to measure $S^y(q,\omega)+S^z(q,\omega)$ between 1.9 K and 12 K. If the soliton theory is valid $S^y(q,\omega)$ ought to be given by eqn (2) above with sinh replaced by cosh and S^z ought to be small. In addition two magnon difference processes contribute only to $S^x(q,\omega)$ and therefore are not observed in our experiment.

The experiment was performed on the IN12 three-axis spectrometer at the Institut Laue-Langevin. Incident neutrons of wavevector 1.55 $\mathring{A}^{-1}$ were used and a cooled Be filter removed higher order contamination. The monochromator and analyser were PG(002) and horizontal collimations were 40'-30'-30'-60'. Experiments were performed at wavevectors $\vec{Q}=(0.6, 0, -q_c)$ with a magnetic field of 10 kG applied in the x direction. Data were taken at temperatures of 1.9 K, 9 K and 12 K.

Data evaluation

The measured scans (c.f. Figs. 1 and 2) all show a pronounced three-peak structure; two peaks at finite frequencies which correspond to single-magnon creation and annihilation processes and a central peak which is predominantly due to incoherent scattering. The scans at T=1.9 K (c.f. Fig. 1) have been fitted to three Gaussian peaks and a frequency independent background. These fits give several essential pieces of information; the instrumental line-width to be associated with the magnon peaks, the magnitude and width of the incoherent scattering component and the magnitude of the background. At higher temperatures (T=9 K and 12 K) we have attempted to fit the spectra by a model in which the spin-wave peaks are described by Lorentzian functions convolved with the appropriate resolution widths determined at 1.9 K. Both the central peak and the flat background have been constrained to their values at 1.9 K.

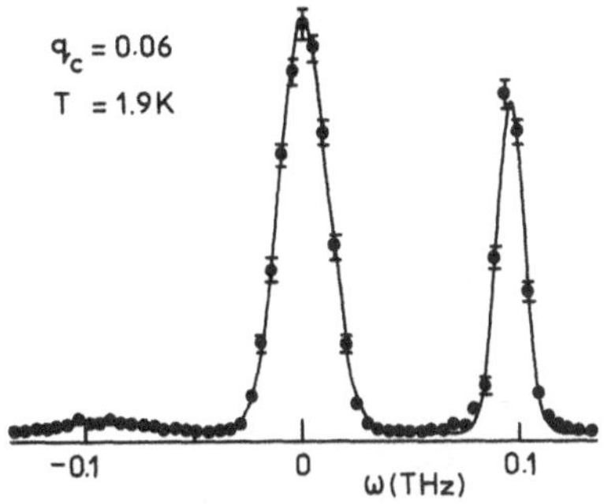

Fig. 1

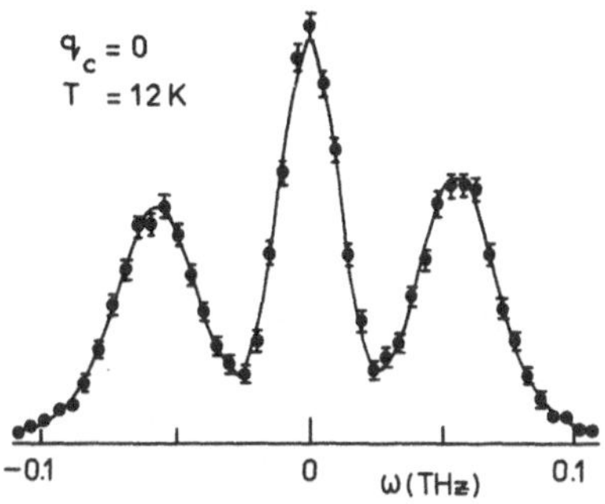

Fig. 2

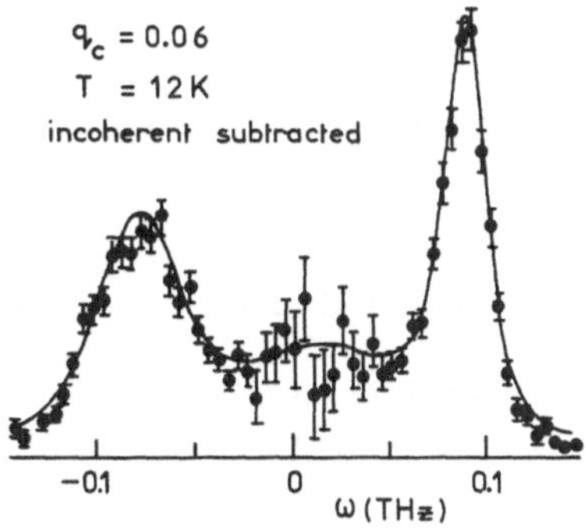

Fig. 3

Figs. 1 and 2: measured constant-Q scans

Fig. 3: constant-Q scan with incoherent background subtracted. The lines through the data points are the result of fits described in the text

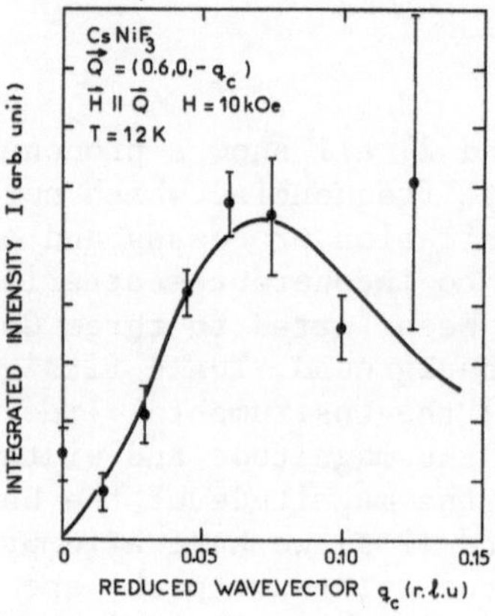

Fig. 4: Integrated intensity of measured 'soliton' scattering. The line is calculated from Mikeska's [8] theory and normalised to the data at q_c=0.045.

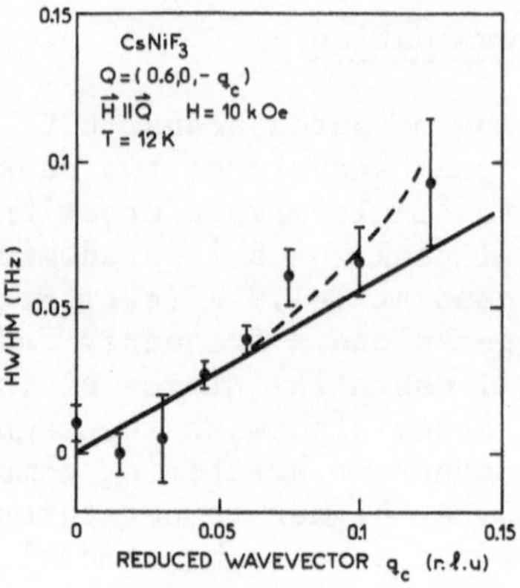

Fig. 5: Half width of measured 'soliton' scattering. The solid line is calculated from Mikeska's [8] theory and the dashed line includes the relativistic corrections of Leung and Huber [12].

As figure 2 shows this fit is very successful at q_c=0; even when the energy gain and loss magnon peaks are not constrained to have the same frequency, this result is produced by the fit. With increasing q_c the quality of the fit deteriorates and at q_c=0.06 and T=12 K for example the best model has a goodness of fit parameter χ^2=5.3. If the frequencies of the magnon creation and annihilation peaks are constrained to be equal the fit becomes even worse ($\chi^2 \sim 7.5$). A close examination of the data indicates that, at finite q_c, there is more intensity around ω=0 than is explained by the incoherent scattering. This is demonstrated in figure 3 which shows data at q_c=0.06 and T=12 K from which the measured incoherent scattering has been subtracted. The additional scattering around ω=0 has been fitted by Gaussian peak in our data analysis.

The integrated intensity of the extra central peak is plotted in figure 4 as a function of q_c. Superimposed on the experimental data is the result obtained from eqn. (2) scaled to fit the data at q_c = 0.045 r.ℓ.u. Evidently the agreement between theory and experiment is good. In the case of the measured energy widths shown in Figure 5 the agreement is equally convincing. In the latter case however the experimental results are very sensitive to the method of background subtraction and the agreement may be fortuitous.

Since two-magnon scattering does not contribute to the measured scattering our results would seem to add experimental support to the contention that solitons do contribute to the spin dynamics of $CsNiF_3$. However, we expect that the publication of our data will stimulate the production of an alternative theory perhaps based on the out-of-plane spin fluctuations. According to Leung and Huber [12] these fluctuations have, in the Sine-Gordon model, a spectrum which is not unlike eqn. (2). However, the predicted dependence of the integrated intensity on q_c is not as observed: in particular the calculated intensity tends to a finite value as q_c tends to zero and decreases with increasing q_c.

REFERENCES

1. M. Steiner, J. Villain and C. Windsor, Adv. Phys. 25, 87 (1976)
2. M. Steiner and B. Dorner, Solid St. Commun. 12, 537 (1973)
3. Lebesque, J. Snel and J.J. Smit, Solid St. Commun. 13, 371 (1973)
4. P.A. Lindgaard and A. Kowalska, J. Phys. C9, 2081 (1976)
5. G. Reiter, PhD Thesis, Stanford University
6. M. Steiner and J. Kjems, J. Phys. C 10, 2665 (1977)
7. J.V. Lebesque and N.F. Huyboom, Comm. Phys. 1, 33 (1976)
8. H.J. Mikeska, J. Phys. C 11, L29, 1978
9. H.J. Mikeska and E. Patzak, Z. Phys. B26, 253 (1977)
10. I. Faddeev and L.A. Takhtadzhyan, Theor. Math. Phys. 21 160 (1974)
11. M. Steiner, J. Appl. Phys. 50, 7395 (1979)
12. K.M. Leung and D.L. Huber, Solid St. Commun. 32, 127 (1979)
13. J.M. Loveluck, T. Schneider, E. Stoll and H.R. Jauslin, Phys. Rev. Lett. 45, 1505 (1980)
14. J.K. Kjems and M. Steiner, Phys. Rev. Lett. 41, 1137 (1978)
15. G. Reiter, Phys. Rev. Lett. 46, 202 (1981)

ON THE POSSIBILITY TO CREATE NONTHERMAL SOLITONS IN A ONE DIMENSIONAL MAGNETIC SINE GORDON SYSTEM

D. Hackenbracht

Institut für Theoretische Physik
Universität Frankfurt, Robert Mayer-Str. 8
6000 Frankfurt/M. (W. Germany)

1. INTRODUCTION

Recently thermally activated sine-Gordon solitons have been observed in the one dimensional magnetic systems TMMC [1,2] and $CsNiF_3$ [3-5] (actually, in $CsNiF_3$ this is a point of discussion[6-9]). In this article we want to discuss the possibility to create nonthermal solitons in a one dimensional sine-Gordon system by a time dependent magnetic field. These nonthermal solitons will appear in addition to the thermally excited ones. This would open e.g. the possibility to study a nonequilibrium system and questions like soliton-soliton interaction (if the density is made high enough) and soliton diffusion (because they are created locally).

The principle idea how magnetic solitons could be created by an external field $\vec{B}_e(z,t)$ is shown in fig.1. If $\vec{B}_e$ is switched on locally, the spins $\vec{S}$ in that region will start to precess around the z-axis, and soliton-antisoliton pairs can emerge under appropriate conditions.

2. THE MODEL

For convenience, we shall only treat the ferromagnetic case and start from the Hamiltonian $H = H_0 + H_1$ with

$$H_0 = \int (dz/a) \ \{\frac{1}{2} Ja^2S^2(\phi_z(z,t))^2 + AS^2\Theta^2(z,t) + \tilde{B}(1-\cos\phi(z,t))\} \qquad (1a)$$

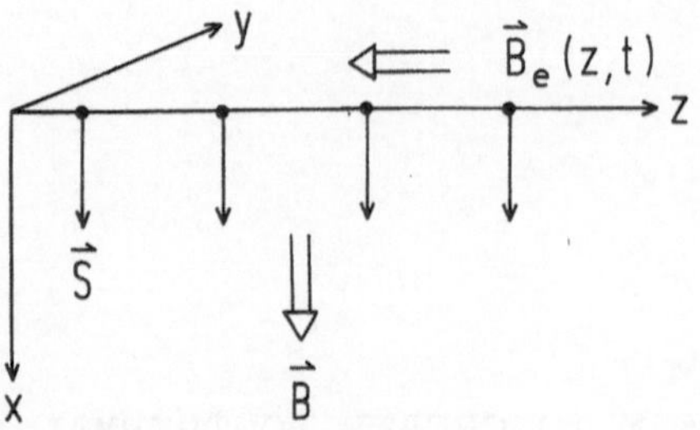

Fig. 1 Geometry of the model

$$H_1 = \int (dz/a)\{\tilde{B}_e(z,t)\Theta(z,t)\} \tag{1b}$$

H_0 describes a one dimensional Heisenberg ferromagnet with coupling J and planar anisotropy A in the symmetry-breaking field $\vec{B}$ parallel to the x-axis (see fig. 1). $\tilde{B} = g\mu_B SB$, a is the lattice constant. The spin $\vec{S}$ has been taken as a classical vector $S(\cos\phi\cos\Theta, \sin\phi\cos\Theta, \sin\Theta)$. Hamilton's equations of motion for H_0 lead to the sine-Gordon equation (SGE) for the angle ϕ, as was shown by Mikeska[6].

H_1 represents the interaction of the additional external field $\vec{B}_e=(0,0,-B_e)$ and the spins (the sign is chosen only for convenience). In H, the Θ-dependent terms have been kept only to lowest order. This limits the strength of B_e ($\tilde{B}_e^2/4A^2S^4 \ll 1$). Instead of directly assuming Θ to be small, we can linearize Θ around $\Theta_e = \tilde{B}_e/2AS^2$ and obtain a SGE with renormalized parameters. For the required moderate fields, the renormalization effects are small. We therefore continue with H as given in (1). Hamilton's equations for H result in a modified SGE

$$\phi_{tt} - \phi_{zz} + \sin\phi = \frac{1}{\hbar\omega_0 S}(\tilde{B}_e(z,t))_t \tag{2}$$

(The variables are dimensionless now, for we scale length and time with $d_0=c_0/\omega_0$ and $1/\omega_0$, respectively, where $c_0{}^2 = 2AJS^2a^2/\hbar^2$ and $\omega_0{}^2 = 2A\tilde{B}/\hbar^2$. If the additional field is switched on at t = 0 by a unit step function, we have $(\tilde{B}_e(z,t))_t = \tilde{B}_e(z)\delta(t)$. The solution of (2) is then reduced to an initial value problem for the SGE with $\phi(z,0) = 0$ and $\phi_t(z,0) = \tilde{B}_e(z)/(\hbar\omega_0 S)$. We have assumed that the system is in the classical ground state ($\phi = 0$, $\Theta = 0$) for t<0, neglecting the thermal fluctuations.

3. RESULTS

We have solved the initial value problem by the inverse scattering method (ISM)[10]. But here we will present only a rather intuitive arguing and refer the reader to a more detailed version of this report[11]. We begin by specifying the spatial shape of $B_e(z)$ (see fig.2). Clearly it has to be inhomogenous, because a localized object like a soliton will not arise from an overall homogenous excitation of the spin system's ground state. Now it can be shown that the field B_e has to exceed a certain value B_c to allow for soliton production. This limiting value already arises from a simple energy consideration for a single spin in the symmetry-breaking field B and the anisotropy field A: When B_e has been switched on, the spin can only arrive at $\phi = \pi$ if $B_e \geq B_c$. Therefore the length ℓ_1 in fig.2 is chosen so that $B_e(z>\ell_1) < B_c$. It is important that only in a part of the chain the conditions necessary for soliton production exist. We would like to mention that the fact that the time dependent field B_e is nonzero all over the probe in a realistic experiment requires some care when using the ISM.

With $\phi_t(z,0) \sim B_e$, we see that at $z = \ell_1$ spins with different initial velocities are coupled: the faster spins

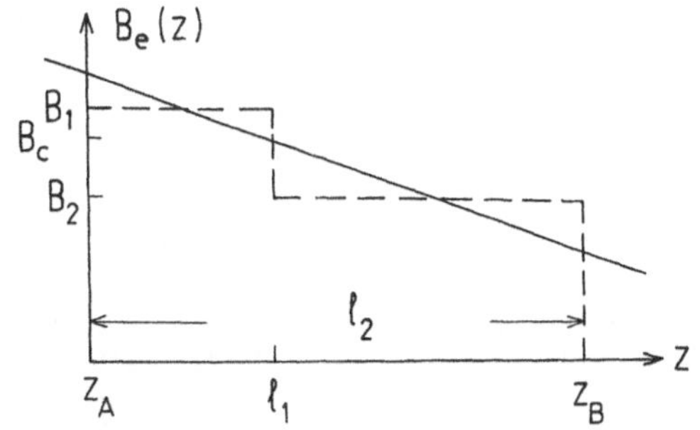

Fig. 2 Spatial variation of the additional field (full line: experimental situation, z_A and z_B mark the chain ends; broken line: approximation used in the calculation).

(left to ℓ_1) have to share their kinetic energy with their slower right neighbours. After some time an antisoliton-like shape will appear, when the spins with $z < \ell_1$, which are not affected by the energy loss, have reached $\phi = 2\pi$. It is important to note that now the region $z < \ell_1$ is no longer homogenous in ϕ and ϕ_t, and this gives rise to the successive creation of antisolitons, all moving to the right. Their total number is proportional to ℓ_1 (when dealing with a chain of finite extension ℓ_2 we have to make sure that $\ell_1 << \ell_2$!). That we only find antisolitons is due to the fact that the additional field B_e introduces a preferred winding sense ($\phi_t > 0$) and direction (due to the gradient) into the system.

4. DISCUSSION

The experimental realization of soliton production by an additional field B_e seems difficult due to the required space and time dependence of B_e. It has to vary along a single chain from a value high enough for soliton production ($B_e > B_c$) to an "undercritical" value. This severely restricts the number of nonthermal solitons, because a single chain is very small compared to the probe. And even for rather long chains the necessary field gradient would be very high. In our caculation, the gradient is approximated by $(B_1 - B_2)/\ell_1$ (see fig.2). As an illustration, we have for $CsNiF_3$ $B_1 \sim 17$KG, $B_2 \sim 14$ KG with B = 1 KG, A = 5K (taken from ref. 6). Taking ℓ_1 = 1mm, the number N of nonthermally excited solitons is $4{,}7 \cdot 10^4$, which is comparable to the number of thermally excited ones ($\sim 4{,}1 \cdot 10^5$ in a probe of 1cm length at 5K). But for realistic lengths of a single chain, we have to take $\ell_1 < 1\,\mu$m! This reduces N by a factor 10^3 and makes gradient unreasonably high. These problems might be overcome when taking into acount the intrinsic inhomogenities of the system. We are now left with the question: What happens if the field is not switched on suddenly? With damping it is possible that the system will only relax to a new equilibrium. When also including the thermal fluctuations (thus leaving the deterministic description) and assuming the time derivative of B_e to be constant, we obtain a driven sine-Gordon system for which Büttiker and Landauer[12] have discussed the nucleation of soliton-antisoliton pairs. But looking again at $CsNiF_3$, we find that the driving force, which is just right-hand side of (2), is very small.

Acknowledgements

This report is the result of the joint work of the author and H.G. Schuster. It is a pleasure to thank Dr. Benner, Dr. Liebmann and Prof. H.J. Mikeska for valuable comments and discussions and Prof. Steiner for drawing our attention to ref.13 after this work has been completed.

REFERENCES

1. J.P. Boucher, L.P. Regnault, J. Rossat-Mignod, J.P. Renard, J. Bouillot, W.G. Stirling Sol.St.Comm. 33,171(1980)
2. J.P. Boucher, J.P. Renard, Phys.Rev.Lett. 45,486 (1980)
3. J.K.Kjems, M. Steiner, Phys.Rev.Lett. 41, 1137 (1978)
4. M. Steiner, J. Appl. Phys. 50,7395 (1979)
5. R. Pynn, M. Steiner, W.Knop, K.Kakuri, J.Kjems, preprint and these Proceedings
6. H.J. Mikeska, J.Phys.C 11, L 29 (1978)
7. E. Allroth, H.J. Mikeska, J.Phys. C 13, L 725(1980)
8. J.M. Loveluck, T. Schneider, E.Stoll, H.R.Jauslin, Phys.Rev.Lett. 45,1505(1980)
9. G. Reiter, Phys.Rev.Lett. 46, 202
10. D.J.Kaup, Stud.Appl.Math. LIV,165(1975)
11. D. Hackenbracht, H.G. Schuster, Z.Phys.B(1981a), to be published
12. M. Büttiker, R. Landauer, Phys.Rev.A 23,1397(1980) and these Proceedings
13. M. Maki,P. Kumar, Phys.Rev.B 14,3920(1976)

TRANSPORT AND FLUCTUATIONS IN LINEAR ARRAYS OF MULTISTABLE SYSTEMS

Markus Büttiker and Rolf Landauer

IBM Research Center
Yorktown Heights, N. Y. 10598, U.S.A.

I. CLASSIFICATION OF MULTISTABLE SYSTEMS

In this paper we study systems which are spread out in one spatial dimension, and where different spatial portions can be in differing states of local stability. Adjacent portions will be separated by transition regions called domain walls, kinks, or solitons. Our considerations are limited to the case in which adjacent members of the one-dimensional array are tightly coupled, and a continuum model applies. We will first attempt to classify a number of cases fitting the preceding description.

The number of competing states of local stability can be variable. There are systems which have two competing states of local stability, with transition regions exemplified by a ferroelectric domain wall. In the absence of an external field, in the simplest case, there will be two equally favored competing states of stability. In the presence of an applied field we will have an unsymmetrical system with a preferred state, toward which the system is driven, resulting in a moving domain wall.

Alternatively, we can consider systems which have an unlimited sequence of states of local stability, typified by the sine-Gordon chain. Such systems were studied in connection with dislocation theory more than fifty years ago. An elegant review of sine-Gordon soliton theory, emphasizing the long history of the subject, including its relationship to nineteenth century differential geometry, has been provided by Seeger (1980). By the mid-sixties the theory of such systems included their statistical mechanics (Seeger and Schiller, 1966).

Our more detailed analytical discussions, following the introductory discussion, will concentrate on the sine-Gordon chain.

Further distinctions can be made according to the detailed dynamics of the array members and their coupling mechanism. The total system can be a conservative Hamiltonian system. Thus if we consider particles in a sinusoidal potential $V = V_0(1 - \cos\theta)$, and coupled to each other by linear mechanisms, causing nearby particles to be near each other, we have the system which gives rise to the mathematician's sine-Gordon soliton, which is a solution of

$$I\partial^2\theta/\partial t^2 = -\partial V/\partial\theta + \kappa\partial^2\theta/\partial x^2. \tag{1.1}$$

Here θ is the displacement of a particle, x the coordinate along the particle chain, I the inertia per unit length, and κ is the coupling constant. This is the system in which kinks or solitons can pass through each other and remain unchanged in the process, and is the subject of Seeger's (1980) review, but will not receive much further attention from us.

The equations of motion for a conservative Hamiltonian can be supplemented by noise sources and by viscosity, leading to systems which are no longer deterministic, but subject to fluctuations. This will be the main area of interest in our later more analytical discussions. In such systems one can study thermal equilibrium, e.g. the fluctuations back and forth for a chain of coupled particles, with each particle in a symmetrical bistable potential well. We shall, later, analyze such fluctuations for the damped and noisy sine-Gordon chain, whose motion is described by

$$I\partial^2\theta/\partial t^2 + \gamma\,\partial\theta/\partial t = -\partial V/\partial\theta + \kappa\,\partial^2\theta/\partial x^2 + \xi. \tag{1.2}$$

V is as defined for Eq. (1.1), γ is a damping constant, and ξ describes the coupling of the chain to a thermal reservoir, with $\langle\xi\rangle = 0$ and

$$\langle\xi(x,t)\xi(x',t')\rangle = 2\gamma kT\delta(x - x')\delta(t - t'). \tag{1.3}$$

The study of fluctuations in one-dimensional models of ferroelectrics (i.e. fluctuations in a bistable symmetrical potential), and the possible relation of these fluctuations to the central peak (Riste et al., 1971; Riste, 1974), were an important reason for the resurgence of interest in the statistical mechanics of solitons (Krumhansl and Schrieffer, 1975; Varma, 1976). We will not, subsequently, discuss the bistable potential case in detail, it is reviewed elsewhere (Bishop et al. 1980; Currie et al. 1980). One could also study relaxation to thermal equilibrium. Consider, for example, the chain of particles in a bistable symmetrical well. Now apply a bias field along the whole chain, favoring one of the two wells. How fast does the relaxation to the new thermal equilibrium, i.e. the switching process, proceed? Our more detailed analytical sections will shed no further light on this problem.

If we apply a bias field F to the sine-Gordon chain, i.e. such that particles move in a potential,

$$V = V_0(1 - \cos\theta) - F\theta, \tag{1.4}$$

(see Fig. 1), then no equilibrium can be reached. The particles will, instead, move with time to lower values of the potential V. For $0 \leq F < V_0$ the potential V has local minima at $\theta_n = \theta_s + 2\pi n$, where $\theta_s = \arcsin F/V_0$, $(0 \leq \theta_s < \pi/2)$. The uniform state at θ_n is stable against small perturbations. Thermal fluctuations are needed to permit motion of the chain. On the other hand, if $F > V_0$ the potential minima are washed out and the chain will simply move along the potential gradient. The interesting regime is, therefore, $0 \leq F < V_0$. A typical chain configuration is shown in Fig. 2. At low temperatures long segments of the chain lie in the potential valleys of V. These segments are connected by kinks and antikinks (transition from one valley of V to an adjacent one with $\partial\theta/\partial x > 0$ are "kinks", whereas "antikinks" have $\partial\theta/\partial x < 0$). We shall describe the transport process in this system, and the fluctuations accompanying it, in detail in Sections III and IV.

First, however, we will continue to classify multistable systems. The coupled parameters in a one-dimensional array need not describe particle positions, but can characterize the state of an open system, i.e. systems which require a continual energy dissipation just to maintain the locally stable state. We shall, in the next Section, describe two detailed examples of this sort.

The modern soliton literature may leave the impression that a moving transition region, between two states of local stability, is necessarily a sine-Gordon soliton, ϕ^4 soliton (i.e. a soliton in a symmetrical bistable fourth order potential), or a close relative of these. In contrast to the soliton work there is a completely independent and separate body of literature, much more experimental in its orientation, and summarized in two recent reviews (Duvall and Graham, 1977; Ahrens, 1980), which, instead, treat transitions induced by shock waves. The sophistication of modern nonlinear dispersive wave propagation

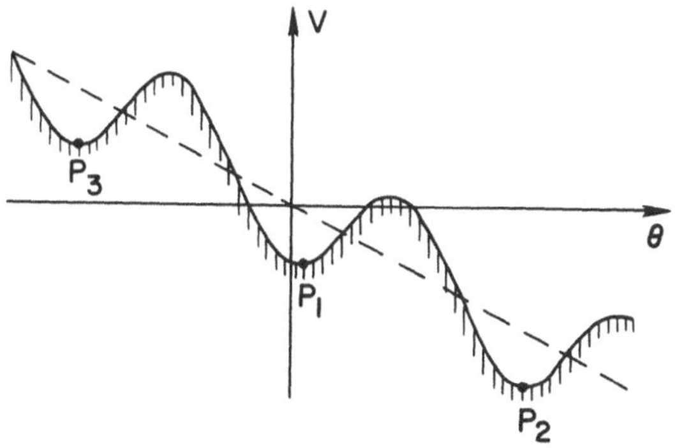

Fig. 1. Single particle potential of Eq. (1.4).

theory seems to have obscured, rather than illuminated, this dichotomy between shock waves and solitons. A detailed comparison between shock waves and solitons has recently been provided by one of the authors (Landauer, 1980) and we shall allude to this subject only with a brief extract from Landauer (1980).

Let us describe a mechanical model of a ϕ^4 chain. The basic unit will be a bistable spring, i.e. one which can be kept at two different positions without force, and has an unstable position in between. One can construct such a device by connecting a particle, through ordinary springs, to the corners of an equilateral triangle. Choose the three springs so that their unstretched lengths are equal and exceed that of the distance between the vertices of the triangle and its center. Furthermore, for conceptual simplicity, think of the particle on a track, confining its motion to a line perpendicular to the triangle, and passing through its center. This system, with motion along the track as the degree of freedom under consideration, is our bistable spring. Alternatively one could invoke a plate or card, buckled under a compressive load, and subject to additional forces in the direction of buckling deflection. The general behavior of a bistable spring is illustrated in Fig. 3.

To construct a system which can exhibit solitons, or domain walls, we must couple a linear string of particles, in multistable potentials, to each other. We can imagine the bistable springs strung out along one direction, and with their particle displacement in a perpendicular direction. Coupling through a torsion element, parallel to the line of bistable elements, but at some distance from it, would be one way of achieving the required coupling.

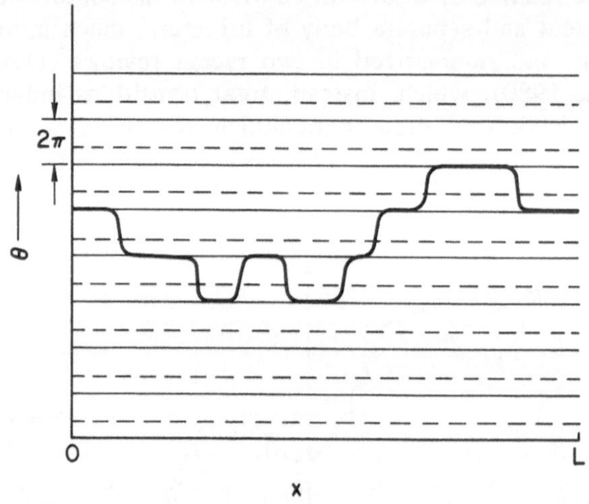

Fig. 2. Typical low temperature configuration of the displacement field $\theta(x,t)$ of the sine-Gordon chain at a given instant of time. Long segments of the chain lying in a valley of the potential shown in Fig. 1 are connected by kinks and antikinks which span the hills of V. Thin horizontal solid lines correspond to minima of V, thin broken lines to maxima of V.

Let us now consider a system which can exhibit shocks. Fig. 4 shows a chain of masses coupled by springs. If the springs stiffen upon compression, then pushing in the plunger with a force that increases with time, launches a shock wave (Al'tshuler, 1965). The later portions of the wave see springs which are already compressed. As a result these later portions see stiffer springs and move more rapidly, causing them to catch up with the earlier portions. This self-steepening produces shocks. Let us now replace the nonlinear springs by the bistable spring shown in Fig. 3. For clarity it is best to think of each bistable spring augmented by a structural member of fixed length, so that both stable positions of the spring give a positive value for the extension of the combined structure.

Consider a semi-infinite chain, with its bistable springs initially at point E in Fig. 3. Then apply a force to the free end increasing with time, past the point D. The initial wave, generated as the force increases up to that at D,

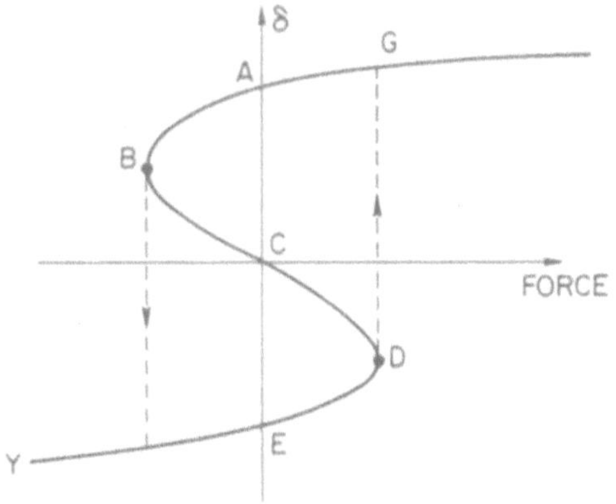

Fig. 3. Extension, δ, of a bistable spring, vs. applied force, F. E and A are points of stability, C of instability. Increasing the force applied to a particle, initially at E, beyond that associated with the point D, will require a transition to the upper branch.

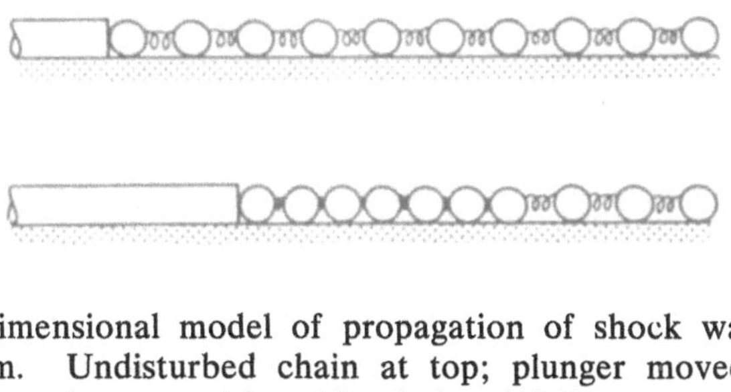

Fig. 4. One dimensional model of propagation of shock wave in an elastic medium. Undisturbed chain at top; plunger moved in at constant velocity at bottom. After Al'tschuler, (1965).

will see increasingly soft springs, and will spread out, rather than form a shock. If the force applied at the end increases monotonically, and the increasing values of force then move into the chain, we cannot follow the characteristic DCBAG, but must make a transition from the lower branch, to the upper one. The motion of this discontinuity can be treated by the usual conservation equations applied to shock propagation. This leads to the conclusion (Al'tshuler, 1965) that a discontinuity moves with the stiffness characterizing the slope of the line $(F_2-F_1)/(\delta_2-\delta_1)$, connecting the endpoints of the shock transition. Thus the transition DG would be stationary. As, however, the force at the input end rises beyond that corresponding to G we obtain a finite slope, and nonvanishing velocity for the transition. Parts of the wave which are generated after the shock has left the input, will move faster than the shock, catch up with it, and coalesce with it. The shock need not be launched on a completely undisturbed chain. As an alternative we can consider a chain of springs which are already compressed and somewhere in the range ED, but with the particles initially at rest.

The reader may have noticed that in the bistable spring model the soliton is a transverse wave and the shock wave is longitudinal. That is a property of this particular example, and is not a characteristic distinction between solitons and shock waves.

II. OPEN SYSTEMS

Thermal Instabilities in Electrical Conductors

As pointed out above, the members of our coupled one-dimensional array can be open systems. The fact that temperature dependent conductors can be multistable has been recognized throughout this century, but that does not prevent a variety of specialized communities from repeated rediscoveries of that fact. Despite the injustice done to history, the later contributors often add new insights. There are even review papers (Dharmadurai, 1980) unaware of the history of their subject.

There are two particularly well known versions of these effects. In the ballast resistor, for which we can provide only a few recent citations (Gray, 1976; Landauer, 1977; Landauer, 1978a; Safranova, 1979; Gurevich and Mints, 1980; Mazur and Bedeaux, 1981), we have an essentially one-dimensional conductor, e.g. a wire, which has a region of temperature in which the resistivity increases relatively rapidly with temperature. The wire is cooled by lateral heat flow to the surrounding ambient, by a mechanism, such as thermal conductivity, which provides heat flow which depends smoothly on the temperature elevation of the wire. If the resistance increase with temperature is sufficiently steep, and a fixed current is fed into the wire, the system is bistable. The wire can be at a temperature only slightly above ambient, leading to a relatively low value of ohmic dissipation, consistent with the slight temperature elevation. Alternatively it can be at a higher temperature, leading to more resistance, and, thus, to the greater heat generation needed for the higher temperature.

The other well known instability involves systems in which the conductance increases rapidly and the electric field, or the voltage, is held fixed. This system, in the 1920's, attracted the attention of some leading scientific intellects, e.g. V. Fock, Th. v. Kárman, and Charles Steinmetz, and by the mid thirties had reached a sophisticated state of understanding (Lüder and Spenke, 1935; Spenke, 1936a; 1936b; Lüder et al., 1936). About a decade ago the people interested in switching in amorphous semiconductors (which is, most likely, a more complex electronic effect) began to rediscover and redevelop these notions. A note (Landauer and Woo, 1972), attempted to explain to this community that Spenke and his coworkers had been there before. The work in this particular area, however, has generally emphasized cylindrical structures. Only recently, to the best of our knowledge, have one-dimensionally extended structures been studied (Kalafati et al., 1979) allowing an opportunity for transition regions of the type emphasized in this paper. We should also note that practical electrical systems are likely to be dominated by end effects, and by inhomogeneities. The theoretician's system with translational invariance, or periodic boundary conditions, doesn't really exist.

Let us briefly discuss the ballast resistor. The basic equation for the time dependence of the temperature, along a wire extended in the x direction, is

$$c_v \partial T/\partial t = -A(T) + i^2 R(T) + \lambda \partial^2 T/\partial x^2. \tag{2.1}$$

c_v is the heat capacity per unit length, A(T) is the cooling rate, λ the heat conductance along the wire, and i is the externally imposed current. A static transition requires $\partial T/\partial t = 0$. If we omit the $\partial T/\partial t$ term from Eq. (2.1), multiply equation (2.1) by $\partial T/\partial x$, integrate along x, and impose the fact that $\partial T/\partial x = 0$ far from the transition region, we find

$$\int_{T_a}^{T_b} (A - i_o^2 R)\,dT = 0. \tag{2.2}$$

T_a and T_b are the temperatures in the wire far away from the transition region. At these limiting temperatures $A = i_o^2 R$. Eq. (2.2) is a condition for i_o, the current which permits a static transition region. As will be shown in more detail, at higher values of current the hot region will be favored and at lower values of current the cold region will be favored, leading in both cases to moving transition regions. Let us assume a temperature profile: $T(x,t) = T(x-ut)$, and substitute this in Eq. (2.1). This yields

$$\lambda d^2T/dx^2 + c_v u\,dT/dx - A(T) + i^2 R(T) = 0. \tag{2.3}$$

This equation is entirely analogous to an equation of motion, in a damped potential

$$m\ddot{q} + \beta\dot{q} + \partial V/\partial q = 0. \tag{2.4}$$

In the analogy T is replaced by q, x is replaced by t, λ by m, $c_v u$ by the damping constant β, and

$$V = \int (- A + i^2 R)dT. \tag{2.5}$$

The force, $-\partial V/\partial q$, vanishes at the initial and final steady states, a and b, where there is a heat balance, without any lateral heat flow. The values of the potential V at a and b will, however, be unequal unless Eq. (2.2) is satisfied. Furthermore at the terminal states, far from a transition region, the velocity $\dot{q} \sim \partial T/\partial x$ vanishes, and there is no kinetic energy, $m\dot{q}^2/2$, near these terminal states. If Eq. (2.2) is satisfied we have "motion" between two states of the same energy, which is possible if and only if the damping β vanishes, i.e. if u = 0.

For definiteness now assume that a is the cold state present at $x = -\infty$ and that b is the hot state present at $x = +\infty$. Let us change i away from i_0, say by increasing it. Consider

$$\frac{\partial}{\partial i} \int_{T_a}^{T_b} (- A + i^2 R)dT. \tag{2.6}$$

This derivative has contributions from the fact that the limits, T_a and T_b, vary with i. These contributions, however, vanish because the integrand vanishes at these limits. Thus the derivative (2.6) arises solely from the heating term, $i^2 R$, and the derivative specified by (2.6) must be positive. Thus, at currents above i_0, we have $V(b) > V(a)$. Motion from a to b, in Eq. (2.4), can therefore only occur if the damping β is negative. This, in turn, corresponds to a negative transition region velocity u, and thus the hot region at $x = +\infty$ is expanding, as we would expect.

Chains of Phase Coupled and Externally Synchronized Oscillators

Consider an electronic oscillator, constructed by using an amplifier with positive feedback through a tuned circuit. The circuit will tend to oscillate at some fixed amplitude, set by the nonlinearities in the amplifier, and at an angular frequency ω_f. We can now attempt to shift this free running frequency by introducing an external signal, i.e. attempting to lock the phase development with time, of the active oscillator, to that of a synchronizing signal. Let ω_s be the angular frequency of the injected signal. The behavior of such an oscillator, under very broad conditions, can be shown to obey an equation given by Adler (1946). Let us denote the instantaneous phase difference between the oscillators' actual phase ϕ and the synchronizing phase ϕ_s by $\theta = \phi - \phi_s$. Denote the difference between the free running unsynchronized oscillator frequency and the synchronizing frequency by Δ. The Adler equation states

$$d\theta/dt = -\Delta_s \sin \theta + \Delta. \tag{2.7}$$

This is obviously equivalent to the equation of motion of a particle in a heavily damped, tilted, sinusoidal potential (see Fig. 1). In the Adler equation the coefficient Δ_s represents the effectiveness of the synchronization signal, and is proportional to it, but also depends on the internal kinetics of the oscillator. If $\Delta_s = 0$ then there is no synchronization, and $d\theta/dt = \Delta$ implies that the oscillator is running at its natural frequency, ω_f. If $|\Delta_s|$ is large enough compared to $|\Delta|$ then Eq. (2.7) will have points of local stability, i.e. the actual oscillator phase is locked to the impressed signal. These points of stability are apart by 2π ; if we let the oscillator phase jump by 2π we end up with a situation indistinguishable from that which existed before the jump.

If we supplement Eq. (2.7) with a noise source whose strength, C, is independent of θ, and whose correlation time is negligible compared to other time constants in the problem, this leads via a Langevin equation to the Smoluchowski equation for Brownian motion in a tilted sinusoidal potential. This equation was put forth and studied in detail in this connection by Stratonovich (1967). The same equation has reappeared elsewhere in similar applications (Lindsey, 1969; Haus and Dyckman, 1978), but has also attracted considerable attention in solid state physics (Bishop and Trullinger, 1978; Schneider et al., 1978), following the original work on Josephson junctions by Ambegaokar and Halperin (1969). As is apparent from Fig. 1, if we are locked into a stable synchronization pattern, say at P_1, and the natural oscillator frequency exceeds that of the injected signal, the oscillator phase in the presence of noise has a

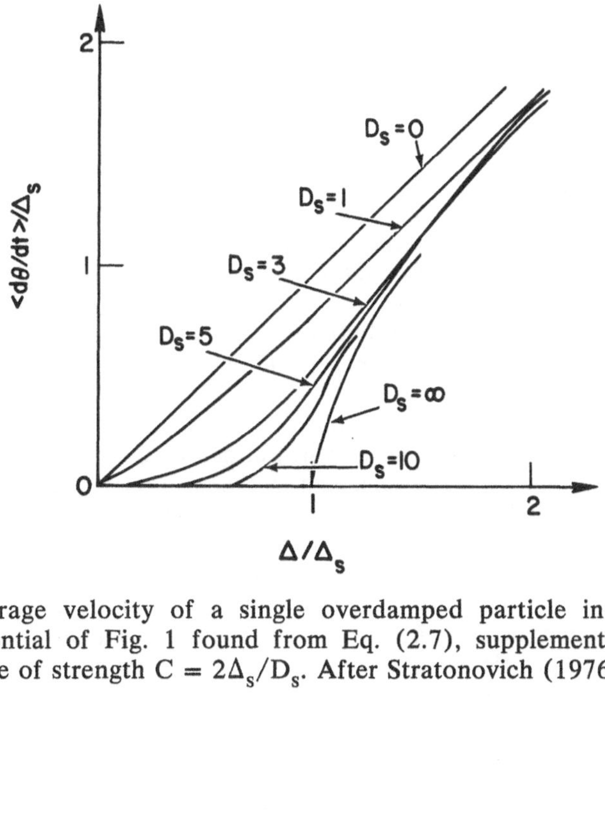

Fig. 5 Average velocity of a single overdamped particle in the effective potential of Fig. 1 found from Eq. (2.7), supplemented with white noise of strength $C = 2\Delta_s/D_s$. After Stratonovich (1976).

nonvanishing probability for jumping ahead by 2π, to the next stable position at P_2. The probability for an uphill jump to P_3 is much smaller. The result for the average motion $\langle d\theta/dt \rangle$ is shown in Fig. 5.

Instead of considering an individual oscillator, we can consider a one-dimensional chain of oscillators, all synchronized by the same external signal, with each oscillator coupled tightly to its neighbors, so that adjacent oscillators cannot be far apart in their phase of oscillation. Each oscillator of the chain is subject to three synchronization signals: the external signal and the signals from its neighbors. The Adler equation for the phase ϕ_n of the n-th oscillator becomes

$$d\phi_n/dt = -\Delta_s \sin(\phi_n - \phi_s) + \omega_f \qquad (2.8)$$
$$-b \sin(\phi_n - \phi_{n+1}) - b \sin(\phi_n - \phi_{n-1})$$

where b describes the effectiveness of the synchronization signals from the adjacent oscillators at n+1 and n-1 on the nth oscillator. (In Eq.(2.8) we assumed that the three signals affecting the nth oscillator are injected in similar ways, and not, for example, at points seeing different phases of the oscillator output.) In the limit of tight coupling we can linearize the corresponding sine functions. The resulting equation of motion for $\theta(x,t)$, of these phase coupled oscillators, is the heavily damped sine-Gordon system (1.2, 1.4) with $I = 0$, $V_0/\gamma = \Delta_s$, $F/\gamma = \Delta$ and $\kappa = a^2 b/\gamma$, where a is the distance between neighboring oscillators. We thus have another example of the heavily damped sine-Gordon system. If the individual oscillators are in the regime shown in Fig. 1, i.e. with states of local stability, then we expect that the chain will have parts along its length lying near such a point of stability, separated by transition regions, leading to an adjacent state of local stability (see Fig. 2).

We can take a finite chain of length L, with one kink in it, and tie the ends together, i.e. impose $\theta(x + L) = 2\pi + \theta(x)$ as boundary condition, and thus lock a kink into the chain. We can tie the ends together without violence because the absolute phase of the oscillators has no significance, the two ends are in undistinguishable physical states. Once we have done this we can generate additional kinks and antikinks, in compensating pairs, preserving the value of their difference. Instead of one initial built in kink we could, of course, have chosen a larger number (see also Seeger and Schiller, 1966, p. 427).

Note, also, that the synchronization signal need not be at the natural frequency of the oscillator, or close to it, but can be near a harmonic or subharmonic of that frequency. The concepts in this section (to the extent that they do not relate to noise) were discussed by Landauer (1955), in connection with degenerate parametrically excited oscillators. These are equivalent, for the purpose of the present discussion, to oscillators synchronized by an injected signal near twice the oscillator's natural free running frequency. In that case a stable synchronized signal shifted by π (at its own frequency) will again bear the same physical relationship to the synchronization signal and will be equally stable. Thus, we have the possibility of a chain, with a single kink, corresponding to a phase shift of π, which can be produced in a real chain by building a 180° twist into the chain, before bringing its ends together.

III. TRANSPORT IN THE DRIVEN SINE-GORDON CHAIN

In this Section we will review the calculation (Büttiker and Landauer, 1979, 1981a) of the average displacement velocity $<\partial\theta/\partial t>$ of particles of the heavily damped and driven sine-Gordon chain (Eqs. 1.3, 1.4 with I = 0). Consider a particle located at x. Each kink moving past this particle, to the right, will decrease the displacement, θ, of the particle by 2π (see Fig. 2). Each antikink moving past a particle, to the right, will increase the displacement by 2π. Hence, if j_K is the kink current density and j_{AK} the antikink current density, we find

$$\partial\theta/\partial t = -2\pi(j_K - j_{AK}). \tag{3.1}$$

In presence of a field, F, kinks propagate with a velocity $-u(F)$ and antikinks with a velocity $u(F)$. The kink currents are given by $j_K = -u(F)m$ and $j_{AK} = u(F)n$, where m and n are the local kink and antikink densities. Thus, we find a velocity (Büttiker and Landauer, 1979)

$$<\partial\theta/\partial t> = 4\pi u n_0 \tag{3.2}$$

where $n_0 = <m> = <n>$ is the average kink density. The two average densities are equal because we have assumed periodic boundary conditions.

The steady state motion of the chain under the action of a field is thus characterized by the propagation velocity u(F) of the kinks and their density n_0. In the remaining part of this Section we will give a short review of the evaluation of these two quantities. In the last Section we will discuss fluctuations of the chain away from the ensemble average motion:

$$<\theta(t)> = \theta_0 + 4\pi u n_0 t. \tag{3.3}$$

A calculation of the quantity $<\partial\theta/\partial t>$ was put forth by Trullinger et al. (1978), Guyer and Miller (1978), and Lee and Trullinger (1980), using a perturbative approach, which does not take essential spatial correlations into account. Schneider and Stoll (1978) have evaluated $<\partial\theta/\partial t>$ by computationally following the time evolution of a chain. For a comparison of the results of these authors with our own approach we refer the reader to Büttiker and Landauer (1981a).

Propagation Velocity of Driven Kinks

To study the transitions in the sine-Gordon chain (Eq. 1.3, 1.4 with I=0) we follow the general approach discussed in connection with the ballast resistor, in Section II, and examine solutions of the form $\theta(x,t) = \theta(z)$, where

$z = x + ut$. Here, u is the propagation velocity of the wave in minus x direction. The waves $\theta(z)$ are found as solutions of

$$\kappa d^2\theta/dz^2 - u\gamma\, d\theta/dz = -dU/d\theta = V_o \sin\theta - F, \tag{3.4}$$

where U is the original potential V of Fig. 1, turned upside down, $U = -V$. This equation describes the motion of a single particle with mass κ in a potential U and subject to damping $\eta = -u\gamma$. We search for solutions which describe a transition from one valley of V (maximum of U) to a neighboring valley of V. That means, we have to find the friction constant, η, such that the dynamical system (Eq. 3.4) exhibits a solution of the following type: A particle started at $z = -\infty$ from a local maximum at U with zero initial velocity has to approach a neighboring maximum of U for $z \to +\infty$ with zero final velocity. We have determined the critical friction η for the occurence of such solutions as a function of the field (Büttiker and Landauer, 1979, 1981a, 1981b). We find that the result $u(F) = -\eta(F)/\gamma$ can be written in the form

$$u(F) = u_o\phi(F/V_o), \tag{3.5}$$

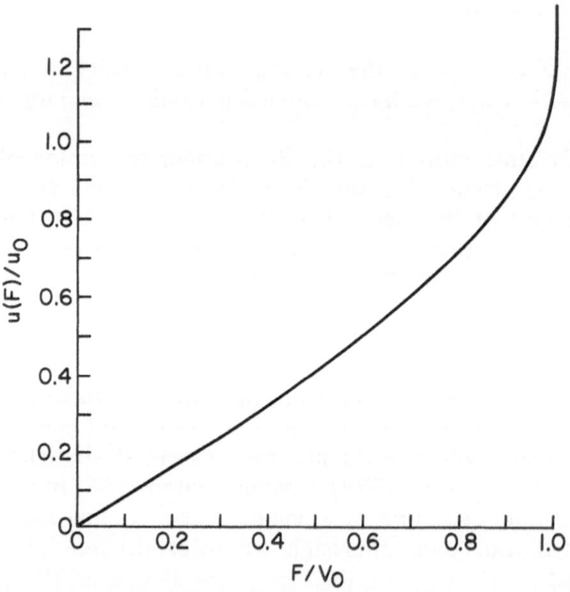

Fig. 6. Propagation velocity u of a kink in the purely viscous sine-Gordon chain. The curve shown defines the parameter independent function $\phi(F/V_o) = u(F)/u_o$. At $F = V_o$, where the minima of the effective potential (Fig. 1) disappear, a driven kink can have any velocity $u \geq 1.19u_o$, where $u_o = (\kappa V_o)^{1/2}/\gamma$.

where $u_o = (\kappa V_o)^{1/2}/\gamma$ is a constant and the function ϕ shown in Fig. 6 is independent of the parameters of Eq. (3.4). For small fields the propagation velocity (3.5) is linear in the field giving rise to a kink mobility

$$\mu = u(F)/F = (\pi/4\gamma)(\kappa/V_o)^{1/2}. \tag{3.6}$$

The propagation velocity increases monotonically with increasing field to a value $u^* = u_o\phi^*$ at $F = V_o$, where $\phi^* \cong 1.19$ in accordance with the result of Urabe (1955), who studied the motion of a damped and driven pendulum. For $F = V_o$ there is a kink structure for every propagation velocity $u > u^*$.

Büttiker and Thomas (1980) have shown that the propagation velocity of kinks in a chain with inertia ($I \neq 0$) can also be expressed with the function ϕ. In that case we find a kink velocity

$$u(F) = u_o\phi(F/V_o)/[1 + \chi^2\phi^2(F/V_o)]^{1/2}, \tag{3.7}$$

where the parameter $\chi = (IV_o)^{1/2}/\gamma$ measures the influence of the inertia. Note, that the small field mobility is unaffected by the inertial term and given by Eq. (3.6). The maximum velocity of a kink in a chain with inertia is given by the velocity of sound of the chain, in the absence of the periodic potential, $u_c = (\kappa/I)^{1/2} = u_o/\chi$. The results found from Eq. (3.7) are in accordance with the data of Marcus and Imry (1980), found by direct computation.

The Steady State Density of Kinks

Consider a sine-Gordon chain in a bias field small enough to allow the particles to have positions of local stability. Assume that the whole chain is at the same position of stability, without any kinks. There will be a lower lying state where each particle is displaced by 2π in the direction of the bias force. In the absence of noise the chain must stay in its original trough. In the presence of fluctuations, however, the chain can make a noise-activated transition, over intervening states of higher energy, to move toward the eventual state of lower energy. Thermal fluctuations can throw a section of the chain into the next lower valley. This transferred section of the chain is terminated by a kink and an antikink. If the transferred region is small, then the attraction between the newly formed kink and antikink outweighs the driving force, and the incipient nucleus will collapse. If the transferred section of chain is long enough, the driving force outweighs the attractive force. This leads to the growth of the transferred region via further separation of the newly formed kink-antikink pair. Thus in the establishment of a region of transferred chain, long enough to grow, we must pass through a saddle point configuration (critical nucleus), i.e. through the minimum energy barrier permitting departure out of the initial valley. Fig. 7 shows a qualitative sketch of the energy of a kink-antikink pair separated by a distance d. Initially the chain lies in one valley (A). Energy must be supplied to pull the kink-antikink pair apart until a critical separation d_{crit} is reached, characterizing the critical nucleus (B). In the range (C) the energy change is dominated by the applied force, rather than by interaction of the original partners, i.e. the transferred region expands with rate 2u(F). Eventually, the kink (antikink) of this pair will recombine with an antikink (kink) created elsewhere. There is an attractive force between kinks and antikinks and since, in the heavily damped case, any kinetic energy is immediately removed, a colliding pair cannot separate again. Thus the location of the recombination events (D) depends not only on the separation d of the original pair but also on the location of the other kinks involved in the recombination.

Thermally activated transitions over a barrier were first treated as a Brownian motion problem, for systems with a one dimensional phase space, by Kramers (1940). This was generalized, for the case of heavily damped motion, to systems with a many dimensional phase space, in which a saddle point has to be crossed, by Brinkman (1956), long before particle physicists discussed such problems. Brinkman's method was rediscovered by Landauer and Swanson (1961) and then again by Langer (1968 and 1969), and will subsequently be called the BLSL method. Since Langer's time, the problem has been rediscovered many times, and tackled with varying success, by people unaware of the earlier efforts. This past year, 1980, has produced some half-dozen papers in this category. Some of these papers do, of course, contribute new insights. For the case of motion in an underdamped potential, not particularly relevant to the later parts of this paper, it should be noted that the discussion by Langer (1969) is not identical to that of Landauer and Swanson (1961).

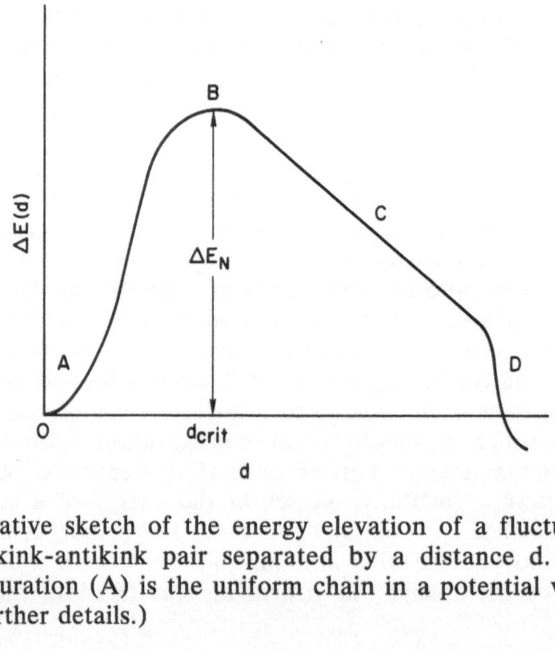

Fig. 7. Qualitative sketch of the energy elevation of a fluctuation consisting of a kink-antikink pair separated by a distance d. The reference configuration (A) is the uniform chain in a potential valley. (See text for further details.)

While it is the many dimensional case that will be needed in this paper, the case treating a single fluctuating degree of freedom continues to be the subject of a great many papers with unnecessary complexity. Alternatively it is rediscovered under a new guise such as *Stochastic Catastrophe Models* (Cobb, 1978). We have, therefore, added an appendix, stressing its simplicity. Somewhat related points were made by Gunther et al. (1979), but without awareness of the fact that the concepts involved have been known for many years.

The BLSL approach has been used to calculate the nucleation rate of kink-antikink pairs (Büttiker and Landauer, 1979, 1981a). This yields a rate per unit time and length of the form

$$j = \Omega\, e^{-\Delta E_N/kT} \tag{3.8}$$

ΔE_N is the energy of the saddle point configuration (nucleus) relative to the initial state (Fig. 8). The evaluation of the pre-exponential factor requires a detailed characterization of all normal modes near the initial state, and also about the saddle point configuration.

Our results for the heavily damped and driven sine-Gordon chain are valid for $\Delta E_N >> kT$ and $F >> n_o kT$, where n_o is the kink density. The former condition is necessary for the applicability of the BLSL approximation. The

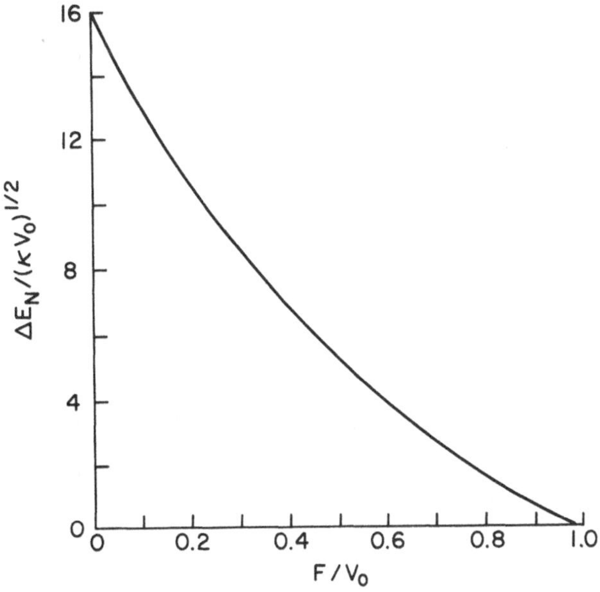

Fig. 8. Energy barrier as function of the applied field. $E_o = 8(\kappa V_o)^{1/2}$ is the equilibrium energy of a kink.

latter condition, however, is specific to our solution of the problem. It ensures that once a configuration has arrived in the region C of Fig. 7 it is driven away from the saddle. This will only be the case if the kink-antikink pair can gain an energy large compared to kT by moving away from each other. The total energy gained by a kink-antikink pair, through their separation, is $2\pi \ell F$, where ℓ is the total distance moved before annihilation. This distance in turn is determined by the mean separation of kinks, i.e. $\ell \sim 1/n_0$.

For fields $F < n_0 kT$ region C of Fig. 7 becomes very flat and even kink-antikink pairs with separation $d >> d_{crit}$ have an appreciable chance of recombining with their original partners. For such small fields, therefore, the successful nucleation of a kink-antikink pair becomes ambiguous and requires carefully consistent definitions of nucleation and recombination. If the driving field becomes very small the shape of the region C is no longer determined by the applied field but by the interaction with other kinks present in the chain. For $F=0$ a critical nucleus in the classical sense does not exist, point B in Fig. 7 goes to infinity. At $F = 0$ the nucleation of kink-antikink pairs can only be discussed sensibly by taking into account the interaction with the additional kinks. These many kink problems have not been addressed by authors (Seeger and Schiller, 1966; Wonneberger, 1980; Skinner and Wolynes, 1980) who have attempted to calculate the nucleation rate of kink-antikink pairs at equilibrium. Skinner and Wolynes (1980) even consider the nucleation of single kinks and single antikinks. Nucleation of transition regions must occur in pairs, except for the introduction of a kink at a free and unconstrained end of a chain.

The complementary process to nucleation is the recombination of kink-antikink pairs. The steady state density of kinks, m_0, and antikinks, n_0, is determined by the balance of these processes. A kink travelling with velocity u sweeps out a range 2udt relative to an approaching antikink, in a time interval dt. Thus the probability that a kink encounters an antikink, in this time, is $2un_0 dt$. Since there are m_0 kinks per unit length the recombination rate j^{rec}, per unit time and unit length, is given by

$$j^{rec} = 2un_0 m_0 = 2un_0^2, \tag{3.9}$$

where we have used periodic boundary conditions $n_0 = m_0$. The steady state density of kinks is, therefore, determined by

$$j^{nuc} - j^{rec} = j - 2un_0^2 = 0, \tag{3.10}$$

where $j^{nuc} = j$ is given by Eq. (3.8). A more detailed discussion by Bennett et al. (1981), subsequently cited as BBLT, has shown that this equation is valid for $F >> n_0 kT$ and for $jL^2/2u >> 1$, i.e. for not too small fields and for large samples (L = length of the chain).

With these results we can now evaluate the average velocity $<\partial\theta/\partial t>$. With Eq. (3.10) the particle velocity of Eq. (3.2) becomes

$$<\partial\theta/\partial t> = 2\pi(2uj)^{1/2}, \tag{3.11}$$

for fields larger than n_0kT. For low fields, where the velocity u is proportional to the field, $u = \mu F$, we can use Eq. (3.2) directly to find (Büttiker and Landauer, 1979)

$$<\partial\theta/\partial t> = 4\pi\, n_{eq}\mu F, \tag{3.12}$$

where

$$n_{eq} = \left(\frac{2}{\pi}\right)^{1/2} \left(\frac{V_o}{\kappa}\right)^{1/2} \left(\frac{E_o}{kT}\right)^{1/2} \exp\left(-\frac{E_o}{kT}\right), \tag{3.13}$$

is the equilibrium density of kinks (Seeger and Schiller, 1966; Currie et al., 1980; Büttiker and Landauer, 1981a). The results of Eqs. (3.11, 3.12) are shown in Fig. 9.

We would like to point out that sine-Gordon models are also used to describe the growth of a crystal-vapor interface. For an extensive list of references we refer to the review article by Weeks (1980). The one dimensional case studied in our paper can serve as a model for situations where an interface as shown in Fig. 2 characterizes the growth. Obviously the strong coupling approximation made in this paper is not very realistic for the crystal growth

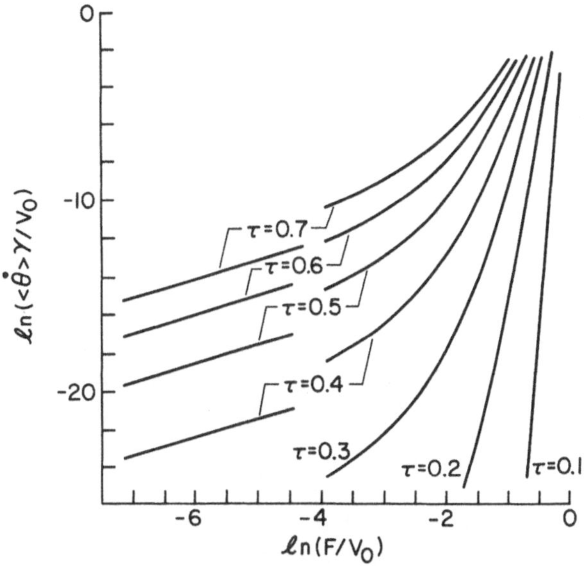

Fig. 9. Velocity of particles in the chain as a function of the field for different temperatures specified by $\tau = kT/(\kappa V_0)^{1/2}$. The straight lines on the left show the extrapolation of Eq. (3.12) to higher fields. The other curves, at larger values of F/V_0, are computational results.

problem. However, the macroscopic features of crystal growth turn out to be largely independent of the details of the microscopic equations. Eq. (3.11) is a well known result in crystal growth theory (Frank, 1974).

IV. FLUCTUATIONS IN THE SINE-GORDON CHAIN

This Section discusses the fluctuations away from the steady state. Our main goal is to understand the long wavelength and long term behavior of the fluctuations. We present two approaches. In the first subsection we discuss *the kink counting approach,* stressing a physical picture. In a second subsection we will present a more formal *hydrodynamic approach* to fluctuations. This will be a simple analytical formulation which yields explicit results and a unified treatment of the fluctuations, both for the equilibrium chain and for the driven chain in the limit of large damping. In a third subsection we will finally compare the long term behavior of the sine-Gordon chain with the behavior of other one-dimensional systems.

The Kink Counting Approach

Let us first investigate the *spatial* fluctuations of a chain at a given instant of time. Consider the instantaneous chain configuration shown in Fig. 2. As we follow along the chain, moving away from a particular position x_0 in the positive x direction, we see that each kink increases the difference in displacement $\theta(x)-\theta(x_0)$ by 2π and each antikink decreases the difference by 2π. BBLT have shown that given N kinks and N antikinks, each geometrical sequence of kinks and antikinks is equally likely. (At equilibrium, $F = 0$, this is a simple consequence of the short range interaction of kinks and antikinks.) Thus as we follow the chain we see an uncorrelated sequence of jumps, each of magnitude 2π. Hence each configuration $\theta(x)$ at a given instant of time is a realization of a diffusion process with x playing the role of time and a diffusion constant (Büttiker and Landauer, 1981b),

$$D_\theta = \frac{1}{2}(2n_0)(2\pi)^2. \tag{4.1}$$

Here, the total kink density, $2n_0$, gives the jump rate per unit length. Thus the mean square difference of two particles a distance x apart grows as

$$<(\theta(x)-\theta(0))^2> = 2D_\theta x. \tag{4.2}$$

Now consider a set of points $x_0 = 0 < .... < x_i < .... x_N = L$ subdividing an interval of length L of the infinitely long chain. The probability that the displacement field increases to $\theta(x_{i+1})$ at x_{i+1}, given a value $\theta(x_i)$ at x_i, for a diffusion process is given by

$$p(\theta(x_{i+1});\theta(x_i)) = \tag{4.3}$$
$$(4\pi D_\theta(x_{i+1}-x_i))^{-1/2} \exp\left[-(\theta(x_{i+1})-\theta(x_i))^2/4D_\theta(x_{i+1}-x_i)\right].$$

The probability of a configuration with values $\theta(x_0),..,\theta(x_i),..,\theta(x_N)$ is obtained by multiplying the probabilities (Eq. (4.3), i = 1, ..., N). Returning to the continuous length scale we find (Büttiker, 1981) the probability density functional for a configuration $\theta(x)$

$$P(\theta(x)) = \exp\left[-\frac{1}{4D_\theta}\int_0^L (\partial\theta/\partial x)^2 dx\right]. \tag{4.4}$$

Eq. (4.4) is also correct for a finite chain of length L subject to periodic boundary conditions, if we restrict the configuration space to functions $\theta(x)$ which are periodic in the length L of the chain. Eq. (4.4) gives rise to a Gaussian distribution for the Fourier amplitudes

$$\theta_{q_n} = L^{-1/2}\int dx \exp(-iq_n x)\theta(x), \tag{4.5}$$

where $q_n = (2\pi/L)n$. The distribution function is

$$P(\theta_{q_n}) = |q_n|(4\pi D_\theta)^{-1/2} \exp[-(q_n^2/4D_\theta)|\theta_{q_n}|^2]. \tag{4.6}$$

The variance of this distribution is the static structure factor of the sine-Gordon chain in the limit of long wave lengths

$$S^{\theta\theta}(q_n) = \langle|\theta_n|^2\rangle = 2D_\theta/q_n^2. \tag{4.7}$$

This result has also been obtained with the transfer integral approach for the equilibrium sine-Gordon chain (Schneider and Stoll, 1980a). However, we wish to stress the much broader range of applicability of the results (4.1 - 4.7). They are independent of the detailed dynamics of the system and apply both to the driven case $0<F<V_0$ as well as to the equilibrium case. The results (4.1 - 4.4, 4.6 - 4.7) depend on the particular function of period 2π specified in the single particle potential V of Eq. (1.4) only via the steady state kink density n_0 (i.e. they are valid for any function of period 2π for which Eq. (1.4) has at most one maximum and minimum in a period).

Let us now turn to the fluctuations of θ, in *time*, for a particle located at x. Our aim is to evaluate the mean square displacement $\langle(\theta(x,t)-\theta(x,0))^2\rangle$ for the chain at *equilibrium*. (Similar quantities for the driven case will be discussed in the next subsection.) As in the driven case the motion of the chain at equilibrium is governed by generation and recombination of kink-antikink pairs. However in the absence of an external force, the motion of the kinks during their lifetime is purely diffusive. The diffusion constant of the kinks is related to the mobility by an Einstein relation (Büttiker and Landauer, 1980) and given by

$$D = \frac{1}{2\pi}\mu kT. \tag{4.8}$$

The factor $(2\pi)^{-1}$ arises because μF is the velocity along the x axis of a kink but the conjugate field to F is θ. A unit displacement of the kink moves a unit length of chain through 2π, under the force F.

First consider a chain of *finite* length L. A given particle is coupled to its neighbors and will, thus, over long times, follow the center of gravity (c.g.) of the chain

$$\theta_{cg}(t) = \frac{1}{L}\int_0^L dx\ \theta(x,t). \tag{4.9}$$

The c.g. of the chain must exhibit a diffusive motion because if we take long enough successive time intervals then the damping forces and the fluctuations assure us that the motion in successive intervals is uncorrelated.

The diffusion constant of the c.g. must be related, through an Einstein relation, to the low field mobility of the c.g., $D_{cg} = \mu_{cg}kT$, as discussed, for example by Guyer (1980). μ_{cg} is the ratio of the chain velocity to the force applied to the whole chain, FL. The low field velocity of the c.g. is given by Eq. (3.12) and thus $\mu_{cg} = 4\pi n_{eq}\mu/L$. Here μ is the kink mobility (Eq. (3.6)) which can be expressed in terms of the kink diffusion constant (Eq. (4.8)), $\mu = 2\pi D/kT$. Thus $\mu_{cg} = 8\pi^2 n_{eq}D/kT\,L$, and using the Einstein relation we find

$$D_{cg} = 8\pi^2 n_{eq}D/L. \tag{4.10}$$

Hence for long times a particle exhibits an average mean square displacement

$$<(\theta(x,t)-\theta(x,o))^2> \cong <(\theta_{cg}(t)-\theta_{cg}(0))^2> \tag{4.11}$$

$$= 2D_{cg}t = 2D_\theta(2Dt/L)$$

with D_θ given by Eq. (4.1), $n_o = n_{eq}$. Apart from numerical factors Eq. (4.11) is identical with the result presented by Gunther and Imry (1980).

Let us now turn to the *infinite* chain at equilibrium, where the c.g. must be stationary, according to Eq. (4.11). We will here simplify an argument presented by Gunther and Imry (1980) for the case of kinks with infinite lifetime, and then extend it to the case of kinks with finite lifetime (Büttiker and Landauer, 1981c). Consider N kinks in a chain of length L, with periodic boundary conditions, and $L^2 >> 2Dt$. Passage of a soliton past a point x will cause θ to change by $\pm 2\pi$, depending on the sign of the soliton (kink vs. antikink) and on its direction of motion. Let there be (1/a) particles per unit distance along the chain. Assume that the soliton diffuses through a distance Δx, in the time t under consideration. The kink will then have caused $|\Delta x|/a$ particles to jump by $\pm 2\pi$. Note that all particles within the range Δx will have been passed an odd number of times during the diffusion, resulting in a net displacement in θ of magnitude 2π. All particles outside of the diffusion distance Δx will have been passed an even number of times, or not at all, causing no net particle displacement.

Separate kinks will diffuse independently, and the two signs for jumps in $\Delta\theta = \theta(x,t) - \theta(x,0)$ will be equally likely. Thus $\Delta\theta$ can be considered to undergo a diffusion, with one jump for each soliton which in the time t has included the position in its net range Δx. The 2N independent kinks and antikinks will provide $2N<|\Delta x|>/a$ jumps of random sign, spread over L/a particles, or $2N<|\Delta x|>/L = 2n_{eq}<|\Delta x|>$ jumps per particle, with $n_{eq} = N/L$. The diffusion in $\Delta\theta$ is determined by the number of random jumps, and their size, 2π. Thus

$$<\Delta\theta^2> = 2n_{eq}<|\Delta x|>\cdot(2\pi)^2 = 2D_\theta<|\Delta x|>, \tag{4.12}$$

with D_θ given by Eq. (4.1). Δx is, itself, determined by diffusive motion. Therefore $<\Delta x^2> = 2Dt$, where D is the kink diffusion constant. Since Δx has a Gaussian distribution, $<|\Delta x|> = (4Dt/\pi)^{1/2}$. Therefore Eq. (4.12), in agreement with Gunther and Imry (1980), becomes

$$<\Delta\theta^2> = 2D_\theta(4Dt/\pi)^{1/2} = 16\pi^{3/2}n_{eq}(Dt)^{1/2}. \tag{4.13}$$

The above results, for infinite soliton lifetime, will turn out to be valid for all soliton lifetimes. We show this by an argument, adapted from BBLT. We follow various ensemble members, classifying them according to the total number of kinks, M. When an ensemble member is subject to a recombination, $M \rightarrow (M - 1)$, there is another ensemble member which has a generation event, $(M - 1) \rightarrow M$, and where the two ensemble members otherwise have the same configuration $\theta(x,t)$, when these events occur. Furthermore, the two compensating events occur with equal probability, at the same spatial position x. (We are in thermal equilibrium, subject to detailed balancing). Thus in the evolution of $\theta(x,t)$, instead of following a given physical ensemble member, we can equally well stay within the class where M is fixed, jumping from one ensemble member, to another, at each annihilation and generation event. After ensemble averaging, the jumping process becomes irrelevant. Thus Eq. (4.13) becomes valid for kink lifetimes short compared to t. This reasoning can also be adapted to rederive the finite chain result of Eq. (4.11).

In the next subsection we will show that the hydrodynamic approach yields exactly the same result. The $t^{1/2}$ dependence was also found by Schneider and Stoll (1980b), however, without specifying the coefficient for $t^{1/2}$.

The Hydrodynamic Approach

With the kink counting approach we had to treat fluctuations in space and time independently. The hydrodynamic approach put forth by BBLT for the driven case, and by Büttiker and Landauer (1980) for the equilibrium case, determines the dynamic structure factor $S^{\theta\theta}(q,\omega)$ defined by

$$<\delta\theta_{q\omega}\ \delta\theta^*_{q'\omega'}> = S^{\theta\theta}(q,\omega)\delta(q - q')\delta(\omega - \omega'). \tag{4.14}$$

Here $\delta\theta_{q\omega}$ is the Fourier transform of the fluctuation

$$\delta\theta(x,t) = \theta(x,t) - <\theta(x,t)> \tag{4.15}$$
$$= (2\pi L)^{-1/2} \sum_n \int d\omega\, \delta\theta_{q_n\omega}\, e^{i(q_n x - \omega t)}.$$

away from the ensemble average. From $S^{\theta\theta}(q,\omega)$ we can, in turn, reestablish the results of the preceeding Section. Our approach is limited to the over-damped case, but will cover both the equilibrium and the driven chain.

The local kink density m(x,t) changes as a consequence of the local kink current j_K and also of the generation and recombination of kinks with rates j^{nuc} and j^{rec}, respectively. The kink balance equation describing these three processes is thus given by

$$\partial m/\partial t + \mathrm{div} j_K = j^{nuc} - j^{rec} + \phi. \tag{4.16}$$

Here, ϕ is the noise associated with the discrete generation and recombination events. Similarly the antikink density is determined by the balance equation

$$\partial n/\partial t + \mathrm{div} j_{AK} = j^{nuc} - j^{rec} + \phi, \tag{4.17}$$

where j_{AK} is the local antikink current. Since kinks and antikinks are generated in pairs, and annihilate in pairs, the r.h.s. of Eq. (4.16) is identical to that of (4.17). Therefore the difference $\rho = m - n$ obeys a continuity equation

$$\partial\rho/\partial t + \mathrm{div} j_\rho = 0, \tag{4.18}$$

where $j_\rho = j_K - j_{AK}$. The excess number of kinks M, over antikinks N, i.e. $N - M = \int \rho dx$, is a constant of the motion. In connection with the sine-Gordon chain, equations of the form (4.16) and (4.17) were studied some time ago (Brailsford, 1961). Frank (1974) used these equations to study crystal growth rates.

Let us first consdier the *driven* case for fields $F > n_o kT$, such that the diffusive motion of the kinks, within their lifetime, can be neglected. BBLT have studied this case in great detail and derived Eqs. (4.17) and (4.18) from a microscopic model. The kink currents and antikink currents describe the drift of the solitons due to the external field, $j_K = -um$ and $j_{AK} = un$, respectively. The recombination rate is given by Eq. (3.9), $j^{rec} = 2umn$. The noise ϕ, which is shot noise, consists of two contributions. Nucleation events occur randomly in space and time with rate j^{nuc}. BBLT showed that in the steady state the recombination events also occur randomly in space and time. Since in the steady state $j^{nuc} = j^{rec}$, the generation events contribute equally and independently to the noise. On the time and length scale of interest, here, the stochastic force has a very short correlation length and time; $<\phi(x,t)\phi(x',t')> = 2j^{nuc}\delta(x - x')\delta(t - t')$.

The hydrodynamic equations (4.16, 4.17) determine the time evolution of the coarse grained displacement field θ, as follows. The spatial variation of θ is related to the imbalance of the kink and antikink densities

$$\partial\theta/\partial x = 2\pi\rho. \tag{4.19}$$

The temporal variation of the field, at a given point, is determined by the difference of the kink and antikink currents at that point

$$\partial\theta/\partial t = -2\pi j_\rho = -2\pi(j_K - j_{AK}), \tag{4.20}$$

as argued in Section III. Using the equations following and including (4.16) and (4.17), we find for the coarse grained displacement field the time evolution equation (BBLT)

$$(\partial^2\theta/\partial t^2 - u^2\partial^2\theta/\partial x^2) + \frac{1}{2\pi}((\partial\theta/\partial t)^2 - u^2(\partial\theta/\partial x)^2) \tag{4.21}$$
$$= 4\pi u(j^{nuc} + \phi).$$

After linearization about the ensemble average motion $<\theta> = \theta_0 + 4\pi u n_0 t$, Eq. (4.21) becomes

$$\partial^2\delta\theta/\partial t^2 + 2u\eta_0\partial\delta\theta/\partial t - u^2\partial^2\delta\theta/\partial x^2 = 4\pi u\phi \tag{4.22}$$

where $\eta_0 = n_0 + m_0$ is the total steady state kink density. Eq. (4.22) has two types of long wavelength modes. One branch with frequency $\omega_1 = -4iun_0$, determined by the lifetime of the kinks, describes the relaxation of fluctuations in the total density η. Additonally we find a diffusive branch of modes with frequency $\omega_2 = -iD_\rho q^2$, where $D_\rho = u/4n_0$, describing the slow decay of fluctuations in the kink-antikink imbalance density ρ via spatial motion of solitons. The long term behavior of the displacement field is determined by the slow modes $\omega_2(q)$. Taking the Fourier transform of the displacement field in Eq. (4.22) and using the correlation function for the noise ϕ, we find for small ω and q

$$S^{\theta\theta}(\omega,q) = \frac{4D_\theta D_\rho}{\omega^2 + D_\rho^2 q^4}. \tag{4.23}$$

At *equilibrium* the motion of the kinks is purely diffusive. In this case the kink current and antikink currents are, $j_K = -D\partial m/\partial x + \xi_K$ and $j_{AK} = -D\partial n/\partial x + \xi_{AK}$ respectively. D is the kink diffusion constant of Eq. (4.8) and ξ_K, ξ_{AK} are fluctuations in the x directed motion of kinks. ξ_K and ξ_{AK} are uncorrelated and from fluctuation dissipation theory we find $<\xi_K(x,t)\xi_K(x',t')> = 2n_{eq}D\ \delta(x-x')\delta(t-t')$, where n_{eq} is given by (3.13). Using the conservation law (4.18), we find with (4.19) and (4.20)

$$\partial\delta\theta/\partial t - D\partial^2\delta\theta/\partial x^2 = 2\pi\xi, \tag{4.24}$$

where $\xi = \xi_{AK} - \xi_K$. Taking the Fourier transform and using the noise strength just cited, we find at equilibrium

$$S^{\theta\theta}(\omega,q) = \frac{4D_\theta D}{\omega^2 + D^2 q^4}, \tag{4.25}$$

where $D_\theta = 4\pi^2 n_{eq}$ is given by Eq. (4.1). Note that for the derivation of Eq. (4.25) we did not have to specify the nucleation and recombination rates. Thus (4.25) is independent of the lifetime of the kinks. Eq. (4.25) has the same form as Eq. (4.23), only the coefficients are different. The total area of the peak, for both Eq. (4.23) and Eq. (4.25), is given by $\int d\omega\, S^{\theta\theta}(\omega,q) = 2\pi S^{\theta\theta}(q) = 4\pi D_\theta/q^2$. Here $S^{\theta\theta}(q)$ is the static structure factor given by Eq. (4.7). Since D_θ is proportional to the steady state density of kinks [see Eq. (4.1)] the total area, in the driven case of Eq. (4.23), increases rapidly with increasing field F, $n_o >> n_{eq}$. All the correlation functions discussed in the previous section can be obtained from Eq. (4.25) in the equilibrium case, and from Eq. (4.23) in the driven case. Taking the Fourier transform of the fluctuations, $\delta\theta(x,t) = \theta(x,t) - <\theta(x,t)>$, and invoking Eq. (4.14) together with Eq. (4.23) or Eq. (4.25) respectively, yields

$$<[\delta\theta(x,t) - \delta\theta(0,0)]^2> =$$

$$\frac{2}{\pi L} \sum_n \int d\omega\, S^{\theta\theta}(q_n,\omega) \sin^2 \frac{1}{2}[q_n x - \omega t], \tag{4.26}$$

where $q_n = (2\pi/L)n$. In particular, by integrating over ω and summing over n we obtain for the difference in displacement between two particles in a chain of finite length L

$$<[\theta(x,0) - \theta(0,0)]^2> = 2D_\theta x(1-(x/L)). \tag{4.27}$$

In Eq. (4.27) we have replaced the deviations, defined in Eq. (4.15), by the total displacement. Eq. (4.27) reduces to Eq. (4.2) for an infinitely long chain. For the mean square displacement we find

$$<[\delta\theta(0,t) - \delta\theta(0,0)]^2> = <[\Delta h_{cg}(t)]^2> + <[\Delta h_R(0,t)]^2>. \tag{4.28}$$

Here, $h_{cg}(t) = \theta_{cg}(t) - <\theta(t)>$ measures the deviation of the center of gravity of the chain away from the ensemble average and $h_R(0,t) = \theta(0,t) - \theta_{cg}(t)$ measures the deviation of a particle at x = 0 from the center of gravity of its chain (see Fig. 10). The differences $\Delta h_{cg}(t) = h_{cg}(t) - h_{cg}(0)$ and $\Delta h_R(0,t) = h_R(0,t) - h_R(0,0)$ are obviously uncorrelated. We find

$$<[\Delta h_{cg}(t)]^2> = 2D_{cg} t \tag{4.29}$$

where the diffusion constant D_{cg} of the center of gravity, for the equilibrium chain, is given by Eq. (4.11), and by $D_{cg} = 16\pi^2 n_o D_\rho/L$, in the driven case. The deviations from the center of gravity have the following poperties. For a finite chain of length L

$$\lim_{t \to \infty} <[\Delta h_R(0,t)]^2> = \frac{2}{3} D_\theta L, \tag{4.30}$$

which is equal to the space average of Eq. (4.27). Thus over long times the particle follows the center of gravity. In the infinite chain the center of gravity follows the ensemble average, $h_{cg} \equiv 0$, and we find

$$<[\delta\theta(0,t) - \delta\theta(0,0)]^2> = <[\Delta h_R(0,t)]^2> = 16\pi^{3/2} n_o (D_\rho t)^{1/2} \tag{4.31a}$$

in the driven case and

$$<[\theta(0,t) - \theta(0,0)]^2> = 16\pi^{3/2} n_{eq} (Dt)^{1/2} \tag{4.31b}$$

at equilibrium. In Eq. (4.31b) we have replaced the deviations, defined in Eq. (4.15), by the total displacements. Note that Eq. (4.31b) agrees exactly with our kink counting derivation, Eq. (4.13). Finally, with both x and t nonvanishing, we find from Eq. (4.26) for the infinitely long chain at equilibrium

$$<[\theta(x,t) - \theta(0,0)]^2> =$$

$$2D_\theta\{(4Dt/\pi)^{1/2} \exp(-x^2/4Dt) + x \,\mathrm{erf}[x/(4Dt)^{1/2}]\}, \tag{4.31c}$$

where erf denotes the error function.

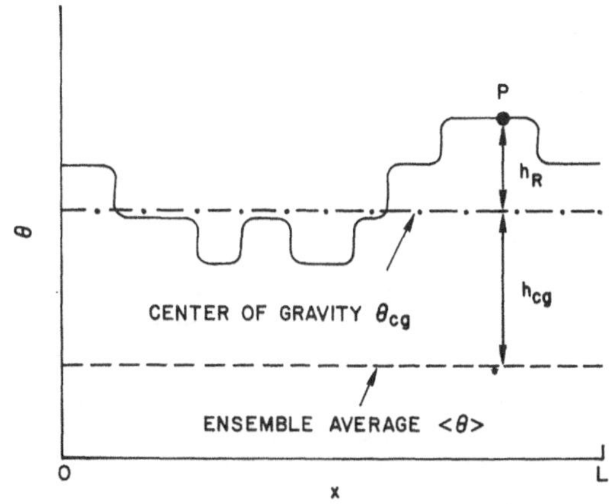

Fig. 10. Fluctuations of a particle away from the ensemble average can be separated into two components; the excursion of the center of gravity away from the ensemble average $h_{cg}(t) = \theta_{cg}(t) - <\theta>$ and the excursion of the particle away from the center of gravity $h_R(x,t) = \theta(x,t) - \theta_{cg}(t)$.

Universality

There is a whole class of one dimensional systems which exhibit the same long term behavior as the sine-Gordon chain, i.e. a group of systems all having a dynamical structure factor of the form Eq. (4.23) or Eq. (4.25), respectively (Büttiker and Landauer, 1981b). All these systems exhibit a constant of the motion $P = \int \rho dx$. Thus the density ρ must obey a continuity equation

$$\partial\rho/\partial t + \mathrm{div} j_\rho = 0. \tag{4.32}$$

In all these systems the current, for slowly varying densities ρ, is given by a diffusion current

$$j_\rho = -D\partial\rho/\partial x + \xi \tag{4.33}$$

where the noise current ξ has the usual properties $<\xi> = 0$ and $<\xi(x,t)\xi(x',t')> = C\delta(x-x)\delta(t-t')$. The field ψ, whose long term behavior is of interest, is related to j_ρ and ρ by

$$\partial\psi/\partial t = -\Lambda j_\rho, \quad \partial\psi/\partial x = \Lambda\rho, \tag{4.34}$$

where Λ is a constant. Eqs. (4.34) are consistent with Eq. (4.32). From Eqs. (4.32 - 4.34) we obtain

$$\partial\psi/\partial t - D\partial^2\psi/\partial x^2 = \Lambda\xi. \tag{4.35}$$

Note that this is not a diffusion equation, which would have the *spatial derivative* of a noise term on the r.h.s. For periodic boundary conditions $\psi(x) = \psi(x+L)$, or $<P> = 0$, we obtain for the structure factor

$$S^{\psi\psi}(q,\omega) = \frac{4D_\psi D}{\omega^2 + (Dq^2)^2} \tag{4.36}$$

where $D_\psi = \Lambda^2(C/4D)$. Below we list a number of systems belonging to this dynamic universality class.

Linear Chain of Harmonically Coupled Particles. This system is described by Eq. (1.2-1.4) for $V_0 = 0$, and has no sinusoidal potential. The equilibrium version, $F = 0$, was described by Imry and Gavish (1974). The fluctuations $\psi(x,t) = \theta(x,t) - <\theta(x,t)>$, with $<\theta(x,t)> = \theta_0 + (F/\gamma)t$, obey an equation of motion of the form (4.35) supplemented by a term $(I/\gamma)\partial^2\psi/\partial t^2$. The diffusion constant is given by $D = \kappa/\gamma$. For small ω and q this yields a structure factor of the form Eq. (4.36), independent of I/γ, with $D_\psi = kT/2\kappa$.

Thus the sine-Gordon chain and the linear damped chain of harmonically coupled particles can be mapped onto one another. Eqs. (4.23, 4.25) demonstrate that the heavily damped sine-Gordon chain, at low temperatures, belongs to the universality class, characterized by a structure factor of the form (4.36). At very high temperatures the periodic potential $V_0(1 - \cos\theta)$ becomes unimportant, taking us to the linear chain of harmonically coupled particles, and the fluctuations of the sine-Gordon chain will also be determined by a structure factor of the form (4.36). Thus, we would expect for the infinitely long chain $<(\psi(x,t)-\psi(x,0))^2> \propto t^{1/2}$, over the whole temperature range. Indeed this is the result of a scaling approach by Schneider and Stoll (1980b) for the heavy damping limit of the equilibrium sine-Gordon chain. But the result (4.36) for the linear chain is quite independent of the value of damping. We also suggest that the long term behavior of the heavily damped sine-Gordon chain is the same as for the lightly damped case, and is always determined by a structure factor of the form (4.36). This conjecture is in contrast with results obtained by Schneider and Stoll (1980b) for a lightly damped sine-Gordon chain. They found from molecular dynamics studies that $<(\psi(x,t)-\psi(x,0))^2> \sim t^{2/3}$. It is, however, questionable whether these simulations were carried out over times long enough to bring the system into the asymptotic region, where (4.36) is valid.

Hopping on a One Dimensional Lattice. Consider N particles on a one dimensional lattice with spacing a. A lattice site can be occupied by only one particle. The particles have a hopping rate W to adjacent empty sites. This system was computationally simulated by Richards (1977), and a hydrodynamic approach of the form Eq. (4.32-4.34) was subsequently given by Alexander and Pincus (1978) and justified in detail by Alexander and Holstein (1978). ψ is the deviation of a particle from a uniform reference configuration, in which the particles are a distance $1/n_0$ apart. n_0 is the equilibrium concentration of particles. Then Eqs. (4.32-4.34) hold with $\Lambda = -1/n_0$, $\rho = n - n_0$ and a diffusion constant $D = a^2W$. Therefore spatially nonuniform particle density distributions, ρ, relax with an effective diffusion constant D equal to that of a single particle on an otherwise unoccupied lattice. We find that $D_\psi = (1 - an_0)/2n_0$. It is interesting that Eqs. (4.32-4.34) hold for all densities $0<n_0<1/a$ of particles and one finds that the mean square particle displacement $<(\psi(x,t)-\psi(x,0))^2>$ grows as $t^{1/2}$ independent of n_0. That the low density and high density limit behave in the same way is simply a consequence of the fact that these cases can be mapped onto one another by exchanging particles in the low density limit with holes in the high density limit. This system can be regarded as a simple model of a kink gas (no antikinks); one identifies kinks with particles in the low density limit and with holes in the high density limit. That kinks repel each other is taken into account in this model by permitting a lattice site to be occupied only by one particle.

Reptation of Polymers. deGennes (1971) has studied the reptation of polymers with a set of equations of the form (4.32-4.34). Motion of monomers on a polymer is caused by N diffusively moving defects, each storing a length Λ. The number of defects is conserved. In the absence of defects the monomers are located at $x = ma$, m integral, along a curvilinear path. $\psi(x,t)$ denotes the deviation, from ma, of the mth monomer in presence of defects and Eq. (4.32-4.34) hold with $\rho = n$, where n is the local defect concentration. The diffusion constant of the defects is given by $D = \mu kT$, where μ is the defect mobility. We find that $D_\psi = n_o \Lambda^2/2$.

In summary: we have shown that there is a class of one dimensional systems whose dynamical behavior is governed by diffusing entities. These entities exhibit some form of interaction which inhibits unhindered diffusion. These systems exhibit a structure factor of universal functional form, for small q and ω.

APPENDIX: THE SIMPLICITY OF TRANSPORT IN THE SMOLUCHOWSKI EQUATION

Consider first a classical gas of noninteracting particles, overdamped in their motion., and spatially nonuniform in one direction. There are many papers on this subject, treating particles crossing single potential barriers, and also particles in sinusoidal potentials. Our key point: We invoke electrochemical potentials as the driving force for particle motion. This is hardly a new point, and is implicit, but not explicit, in Kramers' original paper (1940) on the thermally activated barrier crossing process. It is *the* prevalent method in the semiconductor device literature (Shockley, 1950). After all, an electron moving through a p region, in an npn structure, is crossing a potential barrier. Electrochemical potentials occur in many other places, e.g. in the literature dealing with the effect of activity coefficients on diffusion (Le Claire, 1949). The concept was particularly emphasized by Swanson (1957).

Consider particles with mobility μ and producing a flux

$$j = -\mu kT \frac{dn}{dx} - \mu n \frac{dU}{dx}, \tag{A.1}$$

as a result of a concentration gradient and also as a result of a field induced drift term. In Eq. (A.1) we have implicitly used the Einstein relation, $D = \mu kT$. We can rewrite (A.1)

$$j = -\mu n\, kT \frac{d \log n}{dx} - \mu n \frac{dU}{dx} = -\mu n \frac{d}{dx}(U + kT \log n) = -\mu n \frac{d}{dx}\psi \tag{A.2}$$

where ψ is the chemical potential of the particles. In equilibrium ψ is constant and its derivative vanishes. For small deviations from equilibrium we can, to first order in the deviations, take the n above, which multiplies $d\psi/dx$, as the equilibrium particle density $n_o(x)$. In the one-dimensional case, and in the steady state, j must be independent of x, and

$$\frac{d\psi}{dx} = -j/\mu n_o(x). \tag{A.3}$$

The total drop in ψ, therefore, can be calculated by simply integrating this equation.

$$\psi(B) - \psi(A) = -j\int_A^B dx/\mu\, n_o(x). \tag{A.4}$$

The extent to which the drop in ψ is composed of concentration gradients vs. applied fields is immaterial. We are essentially dealing with resistances in series and the driving force must be largest where the "conductivity" $\mu n_0(x)$ is smallest. Note that μ can be a function of x, it need not be constant. Furthermore, if j is kept inside the integral sign, it can also be a function of x, allowing for an arbitrary distribution of sources and sinks. Thus we can inject particles in one well, remove them from an adjacent well, and ask how much deviation from the equilibrium population distribution between the two wells is needed, to sustain the assumed flux over the barrier. We shall have occasion to refer again, later, to the possibility of simultaneously injecting and removing particles.

So far, in the utilization of Eq. (A.4), we have assumed small deviations from equilibrium. If, however, we are dealing with a linear diffusion problem, we can simply scale up the currents and concentration gradients, and thus describe large departures from equilibrium *exactly*. Note, however, that it is the concentration gradients, and concentration differences, that scale linearly, not the deviations in ψ. Thus, utilizing $\delta\psi = kT\delta n/n$, Eq. (A.4) becomes

$$kT\left(\frac{\delta n(A)}{n_0(A)} - \frac{\delta n(B)}{n_0(B)}\right) = -j\int_A^B dx/\mu\, n_0(x) \tag{A.5}$$

where $\delta n(A)$ and $\delta n(B)$ are the deviations from the equilibrium concentrations $n_0(A)$ and $n_0(B)$ maintained at A and B, respectively. Eq. (A.5) is now valid for large values of $\delta n(A)$, $\delta n(B)$, and j, as well as for small values. Note that $n_0(x)$, in Eq. (A.5), is independent of j, *and does not reflect the perturbed concentration profile and its shifted minimum.* If we consider the application of a large force, with or without simultaneous application of concentration changes, we can again return to Eq. (A.5) by choosing n_0 to be the equilibrium distribution characteristic of the potential *in the presence of the applied force.* Note that we obtained Eq. (A.5) indirectly, via Eq. (A.4), in order to make contact with concepts used in other fields. It is, of course, not necessary to do this.

For a high potential barrier, near which n_0 is very small compared to its value elsewhere, it has become common to invoke the approximation used by Kramers, in which the spatial variation of n_0 is represented only through the second derivative of the potential at the maximum. We can see, however, from the above result, that in the steady state case it is trivial to go beyond that.

If the particles are charged and interacting, the above equations are still valid. Now, however, the density perturbations and the field variation are no longer independent, but are related through Poisson's equation. It can then be shown (Landauer, 1978b) that over large distances the drop in the applied voltage, multiplied by the particle's charge, must equal the drop in ψ.

If, instead of a one-dimensional continuum problem, we have a ladder of points, and the particles make transitions only between adjacent points, we can still write

$$j = g_{n,n+1}(\delta\psi_n - \delta\psi_{n+1}), \tag{A.6}$$

where $\delta\psi_n$ is a *small* deviation, in the electrochemical potential from its equilibrium value, at the nth point, and the "conductance" $g_{n,n+1}$ depends upon the jump rates. Then the total change in ψ, across a ladder of points, is just the sum of terms $j/g_{n,n+1}$, one per step.

Let us now go on to point out that the same general point applies to more general dynamic open systems, and to small departures from their steady state. Once again we can discuss a discrete ladder or, alternatively, a continuum system. To emphasize the analogy with the equations already written down we will discuss the continuum case. We assume that there is a probability distribution $\rho(q)$, whose flux is j, with q denoting the stochastic variable of interest. We assume

$$j = \rho v(q) - D\partial\rho/\partial q \tag{A.7}$$

where the first r.h.s. term represents the deterministic laws of motion, and the diffusive term represents the fact that ensemble members are not all compelled to follow that law. We will not pause to ask when the above equation will be a good approximation; that is a more subtle question than the things discussed in this Appendix. We refer the reader to another paper (Landauer and Woo, 1973) for a discussion of $D\partial\rho/\partial q$ vs. $\partial(D\rho)/\partial q$.

From (A.7) we find that the steady state distribution function, at $j = 0$, is

$$\rho_{ss} = c \exp\left(\int (v/D)dq\right). \tag{A.8}$$

If we now assume $\rho = \beta(q)\rho_{ss}$, then

$$j = -D(\partial\beta/\partial q)\rho_{ss} = -D\beta^{-1}(\partial\beta/\partial q)\rho, \tag{A.9}$$

or $j = -D(\partial \log \beta/\partial q)\rho$. Thus if we let $\psi = \log \beta$ then $j \doteq -D(\partial\psi/\partial q)\rho$ or

$$\psi(q_2) - \psi(q_1) = -\int_{q_1}^{q_2} (j/D\rho)dq. \tag{A.10}$$

If we are close to the steady state, then, to first order in j, we can replace ρ by ρ_{ss} in the above equation. Just as in the transition from (A.4) to (A.5) we can eliminate ψ, and thus find

$$\frac{\delta\rho(q_2)}{\rho_{ss}(q_2)} - \frac{\delta\rho(q_1)}{\rho_{ss}(q_1)} = -\int_{q_2}^{q_1} (j/D\rho_{ss})dq \tag{A.11}$$

without restrictions to small departures from ρ_{ss}.

Let us now break away from the steady state case for some more general remarks. Many authors discuss the "first passage time" question: How long does a particle now at x_0 take to get to x_1, for the first time? While this question exists for all the variants we have discussed above, let us use the specific language appropriate to a continuous coordinate x, and to motion in an overdamped potential, U. Let us insert a current j at point x_0, and remove the particles the moment they reach the location x_1. Thus if $x_1 > x_0$ we will have a solution of the equations, given at the beginning of this Appendix, in which ψ is constant, for $x < x_0$. At x_0 we must match on to a solution which corresponds to the injected current, and which vanishes at x_1. This matching requires continuity in ψ, or n, at x_0. The particles injected at x_0 will build up a total integrated density proportional to the time they spend in the space under consideration. Thus

$$j\tau = \int n dx, \tag{A.12}$$

where the integral is evaluated from the solution we have described, and gives us the desired averaged first passage time τ. If x_0 is near the bottom of a bistable potential, and x_1 is at the peak of the barrier, then $1/\tau$ is *not* the escape rate. A particle reaching the peak of the barrier has an equal probability of falling back into its original well and continuing on into the new well. This makes the escape rate $1/2\tau$. (This particularly simple explanation for the factor 1/2 was pointed out to us by B. J. Matkowsky and Z. Schuss.)

There are many other variations on the basic problem discussed above. The case where the noise intensity depends upon the state of the system is covered by the preceding discussion of more general dynamic systems, and was originally treated by Landauer (1962). The role of such nonuniform noise, in the determination of relative stability, has become widely appreciated in recent years, and is now frequently discussed with the aid of new labels: *Multiplicative noise* vs. *additive noise, external noise* vs. *internal noise.*

The genuinely time-dependent problem, which departs from the steady state that we have discussed, is certainly more complex, and not without interest or importance. It is, however, the steady state result which is often the one that is really needed. Furthermore in the case of a long lived metastable state, the longest of the relaxation times found in the time-dependent problem, can easily be found from the solution of the steady state problem described above, where we inject particles into one well and remove them from the adjacent well (Landauer and Swanson, 1961).

REFERENCES

Adler, R., 1946, Proc. I.R.E. 34, 351.
Ahrens, T. J., 1980, Science 207, 1035.
Alexander, S., and Holstein, T., 1978, Phys. Rev. B 18, 301.
Alexander, S., and Pincus, P., 1978, Phys. Rev. B 18, 2011.

Al'tshuler, L. V., 1965, Sov. Phys. Usp. 8, 52.
Ambegaokar, V., and Halperin, B. I., 1969, Phys. Rev. Lett. 22, 1364.
Bennett, C. H., Büttiker, M., Landauer, R. and Thomas, H., 1981, J. Stat. Phys. 24, 421.
Bishop, A. R., and Trullinger, S. E., 1978, Phys. Rev. B 17, 2175.
Bishop, A. R., Krumhansl, J. A. and Trullinger, S. E., 1980, Physica D, 1, 1.
Brailsford, A. D., 1961, Phys. Rev. 122, 778.
Brinkman, H. C., 1956, Physica 22, 149.
Büttiker, M., 1981, Phys. Lett. 81A, 391.
Büttiker, M., and Landauer, R., 1979, Phys. Rev. Lett. 43, 1453.
Büttiker, M., and Landauer, R., 1980, J. Phys. C. 13, L325.
Büttiker, M., and Landauer, R., 1981a, Phys. Rev. A 23, 1397.
Büttiker, M., and Landauer, R., 1981b, in *Physics in One Dimension,* J. Bernasconi and T. Schneider, eds., Springer, Heidelberg, p. 87.
Büttiker, M., and Landauer, R., 1981c, Phys. Rev. Lett. 46, 75.
Büttiker, M., and Thomas, H., 1980, Phys. Lett. 77A, 372.
Cobb, L., 1978, Behavioural Science 23, 360.
Currie, J. F., Krumhansl, J. A., Bishop, A. R. and Trullinger, S. E. 1980, Phys. Rev. B, 22, 477.
deGennes, P. G., 1971, J. Chem. Phys. 55, 572.
Dharmadurai, G., 1980, Phys. Stat. Sol. A 62, 11.
Duvall, G. E., and Graham, R. A., 1977, Rev. Mod. Phys. 49, 523.
Frank, F. C., 1974, J. Cryst. Growth, 22, 233.
Gray, K. E., 1976, J. Low Temp. Phys. 23, 679.
Gunther, L., and Imry, Y., 1980, Phys. Rev. Lett. 44, 1225.
Gunther, L., Revzen, M. and Ron, A., 1979, Physica 95A, 367.
Gurevich, A. V., and Mints, R. G., 1980, JETP Lett. 31, 49.
Guyer, R. A., 1980, Phys. Rev. B 21, 4484.
Guyer, R. A., and Miller, M. D., 1978, Phys. Rev. A 17, 1774.
Haus, H. A., and Dyckman, H. L., 1978, Int. J. Electron. 44, 225.
Imry, Y., and Gavish, B., 1974, J. Chem. Phys. 61, 1554.
Kalafati, Yu. D., Serbinov, I. A., and Ryabova, L. A., 1979, JETP Lett. 29, 583.
Kramers, H. A., 1940, Physica 7, 284.
Krumhansl, J. A., and Schrieffer, J. R., 1975, Phys. Rev. B 11, 3535.
Landauer, R., 1955, *Moebius Strip Coupling of Bistable Elements.* Reprinted in 1976 as IBM Research Report RC 6093, available from the author or his laboratory.
Landauer, R., 1962, J. Appl. Phys. 33, 2209.
Landauer, R., 1977, Phys. Rev. A 15, 2117.
Landauer, R., 1978a, Physics Today 31, 23.
Landauer, R., 1978b, in *Electrical Transport and Optical Properties of Inhomogeneous Media,* J. C. Garland and D. B. Tanner, eds., American Institute of Physics, New York, Sec. 9.
Landauer, R., 1980, J. Appl. Phys. 51, 5594.
Landauer, R., and Swanson, J. A., 1961, Phys. Rev. 121, 1668.
Landauer, R., and Woo, J. W. F., 1972, Comments on Solid State Physics 4, 139.

Landauer, R., and Woo, J. W. F. 1973, in *Synergetics,* H. Haken, ed., B. G. Teubner, Stuttgart.
Langer, J. S., 1968, Phys. Rev. Lett. 21, 973.
Langer, J. S., 1969, Ann. Phys. (N.Y.) 54, 258.
LeClaire, A. D., 1949, in *Progress in Metal Physics,* Vol. I, B. Chalmers, ed., Interscience, New York, p. 306.
Lee, K. C., and Trullinger, S. E., 1980, Phys. Rev. B 21, 589.
Lindsey, W. C., 1969, Proc. IEEE 57, 1705.
Lüder, H., Schottky, W., and Spenke, E., 1936, Naturwiss. 24, 61.
Lüder, H., and Spenke, E., 1935, Physik Z. 36, 767, also appeared in Z. Techn. Physik 11, 373 (1935).
Marcus, P., and Imry, Y., 1980, Solid State Commun. 33, 345.
Mazur, P., and Bedeaux, D., 1981, J. Stat. Phys. 24, 215.
Richards, P. M., 1977, Phys. Rev. B 16, 1393.
Riste, T., 1974, *Anharmonic Lattices, Structural Transitions and Melting,* Nordhoff, Leiden.
Riste, T., Samuelsen, E. J., Otnes, U., and Feder, J., 1971, Solid State Commun. 9, 1455.
Safronova, O. I., 1979, Sov. Phys. Semicond. 13, 1217.
Seeger, A., 1980, in *Continuum Models of Discrete Systems,* Ê. Kröner and K.-H. Anthony, eds., University of Waterloo Press, Waterloo, p. 253.
Seeger, A., and Schiller, P., 1966, in *Physical Acoustics* Vol. III, W. P. Mason, ed., Academic Press, New York, p. 361.
Schneider, T., and Stoll, E., 1978, in *Solitons and Condensed Matter Physics,* A. R. Bishop and T. Schneider, eds., Springer Verlag, New York, p. 326.
Schneider, T., and Stoll, E., 1980a, Phys. Rev. B 22, 5317.
Schneider, T., and Stoll, E., 1980b, Phys. Rev. B 22, 395.
Schneider, T., Stoll, E. and Schlup, W. A., 1978, Helv. Phys. Acta, 51, 103.
Shockley, W., *Electrons and Holes in Semiconductors,* D. Van Nostrand, New York, 1950, Sec. 12.4, p. 302.
Skinner, J. L., and Wolynes, P. G., 1980, J. Chem. Phys., 73, 4015.
Spenke, E., 1936a, Arch. Elektrotech. 30, 15.
Spenke, E., 1936b, Wiss. Veröff, Siemens-Werk., 15, 92.
Stratonovich, R. L., 1967, *Topics in the Theory of Random Noise,* Vol. II, Gordon and Breach, New York, Chapt. 9, p. 222.
Swanson, J. A., 1957, IBM J. Res. Develop. 1, 39.
Trullinger, S. E., Miller, M. D., Guyer, R. A., Bishop, A. R., Palmer, F., and Krumhansl, J. A., 1978, Phys. Rev. Lett. 40, 206; 1603.
Urabe, M., 1955, J. Sci. Hiroshima Univ., Ser. A 18, 379.
Varma, C. M., 1976, Phys. Rev. B 14, 244.
Weeks, J. D., 1980, in *Ordering in Strongly Fluctuating Condensed Matter Systems,* T. Riste, ed., Plenum, New York, p. 293.
Wonneberger, W., 1980, Physica 103A, (Utrecht), 543.

NON-LINEAR THERMAL CONVECTION

Enok Palm

Department of Mathematics
University of Oslo
Blindern
Norway

1. INTRODUCTION

If a horizontal fluid layer is heated from below or cooled from above, the heat will be transported through the fluid by conduction alone, if the heating is very weak. If, however, the amount of heating is increased, the conduction state becomes unstable and a convective motion is set up.

The first experimental investigations on thermal convection dates back to Thomson (1881) and Bénard (1900). The experiments by Bénard in particular have attracted great attention and are today considered classical in fluid mechanics. The theory of thermal convection was initiated by Rayleigh (1916). He assumed that the amplitude of the motion was infinitesimal such that the equations could be linearized. He thus derived the critical temperature gradient (or in modern language, the critical Rayleigh number) for the onset of convection together with the wavenumber for the marginal stable mode.

Thermal convection occurs in various forms in nature. The experiments by Bénard, which showed a pattern of regular hexagons, was early noticed by meteorologists who poined out the resemblance between certain cloud forms and the cells formed in the experiments. The motion in clouds is, however, very complicated due to the release of latent heat. The concept of thermal convection has also become very important in the study of the earth's interior (see Turcotte and Oxburgh, 1972) and also in branches of astrophysics (Spiegel 1971, 1972).

We shall here consider a fluid layer, bounded by two horizontal boundaries. Usually we assume that there are no restricting lateral walls, but when comparing the theory with experiments, we have to comment on the effect of lateral walls. When all fluid properties are constant, the theory gives that the steady, stable solution consists of a regular pattern of two-dimesional rolls. To prove this, it is necessary to apply the non-linear equations. We shall also see that if the fluid properties are not constant, we may end up with a pattern consisting of hexagons.

Closely connected with thermal convection in a fluid is the thermally induced motion in a porous media. In this case the friction force $\nu \nabla^2\vec{v}$ is replaced by Darcy's law. A porous media is, from a mathematical viewpoint, somewhat simpler than an ordinary fluid. However, in porous media the hydrodynamic dispersion effect may be important and this effect complicates the mathematics considerably.

2. THE BASIC EQUATIONS AND BOUNDARY CONDITIONS

The mathematical analysis is considerably simplified by introducing the Oberbeck-Boussinesq approximation. In the equations of motion the density is then considered to be constant except as far as it modifies the gravity term. Furthermore, in the equation of state the density is assumed independent of the pressure (the anelastic approximation) and a linear function of the temperature whereas the velocity is supposed to be non-divergent. For a critical discussion, se Spiegel and Veronis (1960), Mihaljan (1962) and Velarde and Perez-Gordon (1976).

The governing equations then become

$$\rho_0(\frac{\partial\vec{v}}{\partial t} + \vec{v}\cdot\nabla\vec{v}) = -\nabla p + \rho\vec{g} + \nu\nabla^2\vec{v} \tag{2.1}$$

$$\rho = \rho_0[1 - \alpha(T - T_0)] \tag{2.2}$$

$$\nabla\cdot\vec{v} = 0 \tag{2.3}$$

In addition we need the equation of heat. Neglecting the viscous dissipation, which usually is very small, we have

$$\frac{\partial T}{\partial t} + \vec{v}\cdot\nabla T = \kappa\nabla^2 T \tag{2.4}$$

Here ρ is density, ρ_0 standard density, $\vec{v}$ velocity, t time, p pressure, g acceleration of gravity, ν kinematic viscosity, α coefficient of expansion, T temperature, T_0 standard temperature, κ thermal diffusivity, and ∇^2 the Laplacian.

The Oberbeck-Boussinesq approximation applied on the equation of motion (2.1) physically means that the gravity force is acting on a fluid particle of constant density. With other words, the dynamic effect of the density variation is retained whereas the kinematic effect is neglected. This kinematic effect is usually very small and, furthermore, will only lead to quantitative errors. It should be mentioned that in geophysical and astrophysical applications the approximations involved in (2.2) and (2.3) may not be fulfilled. The results obtained by (2.1)-(2.4) are, however, still valid if the density and temperature in these equations are interpreted as potential density and potential temperature, respectively. The equations cease to be quantitative correct when the vertical length scale of the motion is larger or of the same order as the length scale for the vertical variation of density without motion.

The temperature T may appropriately be written

$$T = T_0 - \frac{\Delta T}{h} z + \theta \tag{2.5}$$

where ΔT is the temperature difference between the lower and upper horizontal boundaries, h is the depth of the fluid layer and z denotes the vertical coordinate measured positive upwards with $z=0$ corresponding to the lower boundary. The horizontal coordinates will be denoted by x,y. $T_0 - \Delta T z/h$ is the temperature distribution by pure conduction and θ is the deviation from this linear distribution due to the motion. Correspondingly the pressure is divided into two parts

$$p = p_s + \hat{p} \tag{2.6}$$

where p_s is defined by

$$\frac{\partial p_s}{\partial z} = -g\rho_0 (1 + \frac{\alpha \Delta T}{h} z) \tag{2.7}$$

Thus p_s is the static pressure due to the two first terms in (2.5).

To get the equations in a dimensionless form we set

$$\begin{aligned} (x,y,z) &= h(x',y',z'), \quad t = h^2 t'/\kappa \\ \theta &= \kappa\nu\theta'/\alpha g h^3 \quad \hat{p} = \kappa^2 \rho_0 p'/h^2 \end{aligned} \tag{2.8}$$

Eliminating ρ by (2.2) and dropping the primes, we then obtain

$$\frac{\partial \vec{v}}{\partial t} + \vec{v}\cdot\nabla\vec{v} = -\nabla p + P\theta\vec{k} + P\nabla^2\vec{v} \tag{2.9}$$

$$\frac{\partial \theta}{\partial t} + \vec{v}\cdot\nabla\theta = \nabla^2\theta + Rw \tag{2.10}$$

$$\nabla \cdot \vec{v} = 0 \tag{2.11}$$

Here $\vec{k}$ is the vertical unit vector and w the vertical velocity. P and R denote the Prandtl number and Rayleigh number, respectively, defined by

$$P = \frac{\nu}{\kappa}, \quad R = \frac{\alpha g \Delta T h^3}{\kappa \nu} \tag{2.12}$$

In most theoretical analysis the fluid layer is assumed to be of infinite horizontal extent. We shall, however, shortly discuss the effect of vertical boundaries later. The horizontal boundaries are either rigid walls or free surfaces. In the first case the no-slip condition yields a velocity of zero at the respective boundaries. In the latter case, with the effect of capillarity neglected, both the normal stress and shearing stress are zero at the free surface. It is customary to simplify this boundary condition by requiring that the shearing stress be zero and the free surface horizontal. This last condition is identical to putting the vertical velocity zero at the boundary. From the equation of continuity (2.11), vanishing shearing stress leads to $\partial^2 w/\partial z^2 = 0$. The boundary conditions for the momentum then becomes

$$w = w_z = 0 \quad \text{at rigid walls} \tag{2.13}$$

$$w = w_{zz} = 0 \quad \text{at free surfaces} \tag{2.14}$$

where in (2.13) we have applied the equation of continuity (2.11). The thermal boundary condition is usually well described by a relation of the form

$$\theta + \lambda \theta_z = 0 \tag{2.15}$$

where λ is a constant which must be known from the experimental set up. In theoretical studies λ is usually assumed to be zero which corresponds to boundaries being perfect heat conductors. On the other side, $\lambda \to \infty$ corresponds to perfect insulating boundaries.

3. THE LINEAR THEORY

a. The horizontally unbounded layer

To study the onset of the convection, we assume that the motion is of small amplitude. In (2.9) and (2.10) we may then neglect the non-linear terms $\vec{v} \cdot \nabla \vec{v}$ and $\vec{v} \cdot \nabla \theta$. Elimination of the horizontal velocity and the pressure in (2.9) and (2.11) gives

$$P^{-1}\nabla^2\dot{w} = \nabla_1^2\theta + \nabla^4 w \tag{3.1}$$

where the dot denotes time differentiation and ∇_1^2 is the horizontal two-dimensional Laplacian. Equation (3.1) together with the linearized equation

$$\dot{\theta} = \nabla^2\theta + Rw \tag{3.2}$$

determines w and θ.

For the purpose of illustration we assume that the horizontal boundaries are both free surfaces. This assumption simplifies the mathematics without essential physical effects being lost. The boundary conditions are then satisfied for solutions of the form

$$w = A(t)e^{i\vec{k}\cdot\vec{r}}\sin\pi z \tag{3.3}$$

$$\theta = B(t)e^{i\vec{k}\cdot\vec{r}}\sin\pi z \tag{3.4}$$

where

$$\vec{r} = (x,y),\quad |\vec{k}| = a \tag{3.5}$$

Introducing (3.3)-(3.5) into (3.1) and (3.2) and eliminating $B(t)$, we obtain for $A(t)$

$$P^{-1}(a^2+\pi^2)\ddot{A} + (1+P^{-1})(a^2+\pi^2)^2\dot{A} + [(a^2+\pi^2)^3 - Ra^2]A = 0 \tag{3.6}$$

which formally is the equation for damped oscillations. The "viscous force" always acts as a stabilizing force whereas the "elastic force" acts as a stabilizing or destabilizing force according to whether

$$(a^2 + \pi^2)^3 - Ra^2 \gtrless 0 \tag{3.7}$$

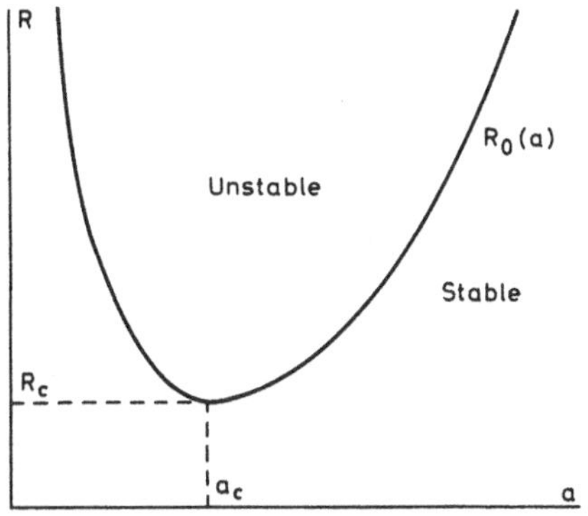

Figure 1. Stability diagram for the onset of convection.

Equation (3.7) determines the critical Rayleigh number R_0 for the onset of convection, which is a function of the wavenumber a, but independent of the Prandtl number. The variation of R_0 with a is displayed in Fig. 1, which shows that for values of R less then R_c, A(t) decays and no convection takes place. For $R = R_c$ the state of conduction is marginally unstable, and a convective motion with wavenumber a_c is set up. For $R > R_c$ a continuous spectrum of modes becomes unstable. It is easily deduced that

$$R_c = \frac{27\pi^4}{4}, \quad a_c = \frac{\pi}{\sqrt{2}} \tag{3.8}$$

a result first obtained by Rayleigh (1916).

For other boundary conditions the calculations become somewhat more complicated, although the main result is the same (Reid and Harris, 1958). The diagram corresponding to Fig. 1 has approximately the same form, and the critical Rayleigh number and wavenumber are given by

rigid-rigid boundaries: $R_c = 1707.7$, $a_c = 3.117$

rigid-free boundaries: $R_c = 1100.6$, $a_c = 2.68$.

The solution of (3.6) is of the form $\exp(\sigma t)$ where σ is real. Also in the case of rigid-rigid boundaries and rigid-free boundaries will σ be real. This is due to the fact that the problems are self-adjoint. Owing to the real value of σ, the motion will not be oscillatory.

Returning to (3.6) we note that for small supercritical Rayleigh numbers, the first term in the equation may be cancelled. Introducing the analytical expression for R_0, we may then write (3.6) as

$$\dot{A} - \sigma A = 0 \tag{3.9}$$

where

$$\sigma = \frac{(a^2+\pi^2)\Delta R}{(1+P^{-1})R_0}, \quad \Delta R = R - R_0 . \tag{3.10}$$

Hence A(t) increases exponentially with the time untill the non-linear terms become important and (3.9) ceases to be valid. These non-linear terms lead to that A(t) may approach a steady finite value. There is, however, another reason for taking into account non-linear effects. By linear theory a continuous spectrum of waves becomes unstable for supercritical Rayleigh numbers. Therefore, as long as the linear equation governs the motion, the observed pattern is very complicated. Observations reveal, however, that the motion has a marked tendency toward a simple cellular form.

This ordering tendency obviously must be attributed to the non-linear terms.

b. The effect of lateral walls

In all experiments in thermal convection there will be lateral boundaries confining the fluid layer horizontally. One effect of these boundaries is that the horizontal wavenumber spectrum cannot be continuous, as was the case for the horizontally unbounded layer. Experimentally, the effect of these boundaries is noticed on the form of the observed planform. Thus Stork and Müller (1972) observe that the cells in containers of rectangular parallelepiped form always are straight rolls parallel with the shortest side. In the case ov convection in a right circular cylinder, the experimental results are more complicated to interprete. Koschmieder (1966, 1974) reports a general strong tendency for formation of cells which are concentric circles. Stork and Müller (1975) find, however, that concentric circles are only a stable pattern for a small range of aspect ratios (defined as a characteristic lateral length of the container divided by the depth of the fluid layer) and small super-critical Rayleigh numbers. For other aspect ratios they observe cells which have a tendency for being parallel with a diameter

The influence of the lateral confinement of the fluid layer in the linear theory has been discussed in several papers. We mention here the papers by Davis (1967), Segel (1969) and Charlson and Sani (1970, 1971). The theoretical findings of Davis is in very good agreement with the experiments by Stork and Müller (1972). Also many of the calculations by Charlson and Sani are well confirmed by experiments (Koschmieder 1966, Stork and Müller 1975). For completeness we mention here that Charlson and Sani (1974) also have examined the non-linear stability.

Davis considers convection in a rectangular box with moderate aspect ratios. His main results are summarized as follows

(a) The preferred mode is always some number of finite rolls (two non-zero velocity components dependent on three spatial variables) with axes parallel to the short side (square boxes excepted).
(b) When the depth is the smallest dimension, finite rolls of near-square cross-section are predicted. Otherwise narrower finite rolls appear.
(c) The critical Rayleigh number decreases rapidly to the value of 1708 as the horizontal dimensions increase so that most experiments, which use thin layers, would appear to have onset occur at about $R_c = 1708$.

It may be worth mentioning that Charlson and Sani (and in fact also Davis) base their calculations on an extremal principle

$$R_c^{-\frac{1}{2}} = \text{Max} \frac{2 \langle \theta w \rangle}{\langle \nabla \vec{v} : \nabla \vec{v} + \nabla \theta \cdot \nabla \theta \rangle} \tag{3.11}$$

Here $\langle\ \rangle$ denotes integration over the entire fluid layers and the admissible functions v and θ must fulfill the boundary conditions. In the present context the principle was first deduced by Sani (1963). As pointed out by Normand, Pomeau and Velarde (1977), (3.11) is a special case of a variational principle for the lowest eigenvalue valid for problems governed by an equation of the form

$$LV + \lambda MV = 0 \tag{3.12}$$

in which V is some vector function. L and M are two real, self-adjoint positive definite linear operators where M is of lower order than L.

We shall here take the effort to show that (3.11) may be derived from an extremal principle which takes into account the time-development of the motion. The same maximum principle, in a non-linear form, will be applied in the next section. Let us go back to equations (2.9-2.11). We introduce in these equations

$$\theta = R^{\frac{1}{2}} \theta' \tag{3.13}$$

and linearize them. The equations then take the form

$$P^{-1} \frac{\partial \vec{v}}{\partial t} = -P^{-1} \nabla p + R^{\frac{1}{2}} \theta \vec{k} + \nabla^2 \vec{v} \tag{3.14}$$

$$\frac{\partial \theta}{\partial t} = \nabla^2 \theta + R^{\frac{1}{2}} w \tag{3.15}$$

$$\nabla \cdot \vec{v} = 0 \tag{3.16}$$

where we for simplicity have cancelled the prime in θ. Multiplying (3.14) with $\partial \vec{v}/\partial t$ and (3.15) with $\partial \theta/\partial t$, adding the equations and integrating over the fluid layer, applying (3.16) leads to

$$\frac{\partial}{\partial t} R^{\frac{1}{2}} \langle \theta w \rangle + \langle \frac{\partial \vec{v}}{\partial t} \cdot \nabla^2 \vec{v} \rangle + \langle \frac{\partial \theta}{\partial t} \nabla^2 \theta \rangle = P^{-1} \langle (\frac{\partial \vec{v}}{\partial t})^2 + (\frac{\partial \theta}{\partial t})^2 \rangle \tag{3.17}$$

By applying the boundary conditions for $\vec{v}$ and θ (either $\theta = 0$ or $\partial \theta/\partial n = 0$ where $\partial/\partial n$ denotes the derivative normal to the boundary), (3.17) may be written

$$\frac{\partial}{\partial t} V = P^{-1} \langle (\frac{\partial v}{\partial t})^2 + (\frac{\partial \theta}{\partial t})^2 \rangle \tag{3.18}$$

where

$$V = R_c^{\frac{1}{2}}\langle\theta w\rangle - \tfrac{1}{2}\langle\nabla\vec{v}:\nabla\vec{v}\rangle - \tfrac{1}{2}\langle\nabla\theta\cdot\nabla\theta\rangle \tag{3.19}$$

and R is chosen as the critical Rayleigh number R_c. It follows from (3.18) that V is ever increasing with time and reaches its maximum value when the motion becomes steady. It is easily seen from (3.14)-(3.16) that for the steady motion $V=0$. From (3.19)

$$\frac{2\langle\theta w\rangle}{\langle\nabla\vec{v}:\nabla\vec{v}\rangle+\langle\nabla\theta\cdot\nabla\theta\rangle} = \left[\frac{2V}{\langle\nabla\vec{v}:\nabla\vec{v}\rangle+\langle\nabla\theta\cdot\nabla\theta\rangle} + 1\right]R_c^{-\frac{1}{2}} \tag{3.20}$$

Since $\langle\nabla\vec{v}:\nabla\vec{v}\rangle+\langle\nabla\theta\cdot\nabla\theta\rangle$ are always positive, it is concluded that the term on the left hand side has its maximum value when the motion is steady, i.e. when $\vec{v}$ and θ has attained the time-independent values corresponding to R_c. For this latter motion the left hand side of (3.20) is equal to $R_c^{\frac{1}{2}}$, since $V=0$. Hereby follows the formula (3.11).

4. NON-LINEAR CONVECTION

4.1. Perturbation approach

a. The amplitude equations. When for supercritical Rayleigh numbers the amplitude A increases in accordance with (3.9), non-linear self-interaction of the first-order mode becomes important, and second-order modes, proportional to A^2, are excited. Then second-order modes and the first-order mode interact, resulting in third-order terms proportional to A^3, and so on. We assume for the moment that in the physical system a kind of symmetry exists so that by reversing all the fluid velocities, an equivalent pattern is obtained. The amplitude equation will then only contain terms of odd power of A and takes the form

$$\dot{A} = \sigma A - bA^3 + \ldots \tag{4.1}$$

which is the Landau-Hopf equation. Here σ is the linear amplification factor (3.10) and b is a constant. In the most problems in thermal convection b is positive, which will be assumed here. Neglecting terms of higher order than third order, we notice that, in contrast to (3.9), the solution of (4.1) tends to an equilibrium state for large values of time. When several first-order modes are taken into account, the problem becomes considerably more complicated. The various modes will interact and instead of (4.1) we get a coupled system of non-linear equations.

Formally the amplitude equation (4.1) may be derived by using a perturbation expansion. We write

$$\vec{v} = \sum_n \varepsilon^n \vec{v}_n, \quad \theta = \sum_n \varepsilon^n \theta_n, \quad R = R_0 + \sum_n \varepsilon^n R_n \tag{4.2}$$

Furthermore, it may be shown that

$$\frac{\partial}{\partial t} = O(\varepsilon^2) \tag{4.3}$$

Introducing this in (2.9-2.11) and utilizing that the equations must be fulfilled for all powers of ε independently, we obtain an infinite set of inhomogeneous linear differential equations. The unknown R_n are successively determined by the conditions of solvability of the equations.

A procedure of this form, assuming the motion to be steady, was first applied in the theory of thermal convection by Gorkov (1957) and, independently and more completely, by Malkus and Veronis (1958). Important contributions were made in the papers by Busse (1962) and Schlüter, Lortz and Busse (1965). These papers are restricted to small supercritical Rayleigh numbers, that is, only terms up to the third order in the amplitude are taken into account. On the other hand, within this frame they were able to determine a whole manifold of steady solutions at once. The first order vertical velocity, for example, was written in the rather general form

$$w = \sum_{k=-N}^{N} C_k e^{i(m_k x + n_k y)} f(z) \tag{4.4}$$

where to assure real solutions $(m_{-k}, n_{-k}) = -(m_k, n_k)$, $C_k = C^*_{-k}$ and for normalization $\Sigma |C_k|^2 = 1$. Furthermore, it is assumed that $m_k^2 + m_k^2 = a^2$, that is, the analysis of the steady motion is restricted to a single horizontal wavenumber a. It turnes out that the conditions of solvability restrict the C_k in (4.4), but there is still an infinite number of possibilities. To find which of these solutions will be realized in an experiment, it is necessary to introduce stability considerations. Assuming that the infinitesimal disturbances have a time dependence of the form $\exp(\tilde{\sigma} t)$, the stability problem leads to a set of linear homogeneous differential equations with homogeneous boundary conditions where $\tilde{\sigma}$ is the eigenvalue. Since the coefficients in these equations are power series in ε, it is reasonable to develop also $\tilde{\sigma}$ and the disturbance in such series. Therefore we have

$$\tilde{\sigma} = \sum_{n=1}^{\infty} \tilde{\sigma}_n \varepsilon^n \tag{4.5}$$

and a similar power series for the disturbance.

The results obtained by Schlüter et al. by this procedure are discussed in the next section in connection with application of an extremal principle. We shall then, in contrast to the papers mentioned above, consider the non-steady version of the amplitude equation (4.1).

<u>b. The preference for two-dimensional rolls</u>. It is of interest to note that for sufficiently small supercritical Rayleigh numbers the motion is governed by an extremal principle. This was first shown by Busse (1967). The principle is only true when modes of the same wave number are considered and has only been proved to hold to the third order in the amplitude. The same extremal principle was derived by Palm (1972) by applying the time-dependent amplitude equations. He also found the physical quantity which attained its maximum value.

The first order approximation (4.4) of the vertical velocity w is then written

$$w = \sum_n A_n(t) e^{i(\vec{k}_n \cdot \vec{r} + \psi_n)} f(z) + \text{complex conjugate} \tag{4.6}$$

where $A_n(t)$ is a real amplitude and ψ_n an arbitrary phase. Since we do not have lateral walls, ψ_n is assumed independent of time. Furthermore,

$$|\vec{k}_n| = a \tag{4.7}$$

The amplitude equation up to the third power of A_n will take the form

$$\frac{dA_n}{dt} = \sigma A_n - A_n \sum_{m \neq n} b_{mn} A_m^2 - b A_n^3 \tag{4.8}$$

Due to the lateral homogeneity of the problem $b_{mn} = b_{nm}$ and b is independent of n. The equation (4.8) may therefore be written

$$\frac{dA_n}{dt} = \frac{\partial V}{\partial A_n} \tag{4.9}$$

where

$$V = +\tfrac{1}{2}\sigma \sum_n A_n^2 \div \tfrac{1}{4} \sum_{m \neq n} b_{mn} A_m^2 A_n^2 \div \tfrac{1}{4} b \sum A_n^4 \tag{4.10}$$

Hence in the n-dimensional amplitude space dA_n will always be directed towards increasing values of V. The potential V will therefore continually increase until it (eventually) attains a maximal value corresponding to the motion becoming steady.

We shall now apply the maximum principle to prove the important result obtained by Schlüter et al. (1965) that among the infinite number of possible steady state solutions only a solution composed of one single mode in (4.6) may be stable. Since $\vec{k}_n$ in (4.6) has an arbitrary direction, this solution corresponds to a periodic set of two-dimensional rolls with arbitrary orientation. To prove the statement we assume for the moment that a steady, stable solution really exists that consists of more than one mode. This solution must then be composed of at least two modes with non-zero amplitudes, for example A_m and A_n. Since V has an extremal value for this solution, we have

$$\frac{\partial V}{\partial A_n} = 2\frac{\partial V}{\partial A_n^2} A_n = 0, \quad \frac{\partial V}{\partial A_m} = 2\frac{\partial V}{\partial A_m^2} A_m = 0 \tag{4.11}$$

Furthermore, applying (4.11) we obtain

$$\frac{\partial^2 V}{\partial A_n^2} = 4\frac{\partial^2 V}{\partial (A_n^2)^2} A_n^2, \quad \frac{\partial^2 V}{\partial A_m^2} = 4\frac{\partial^2 V}{\partial (A_m^2)^2} A_m^2$$

$$\frac{\partial^2 V}{\partial A_m \partial A_n} = 4\frac{\partial^2 V}{\partial A_m^2 \partial A_n^2} A_m A_n \tag{4.12}$$

and thereby we have

$$\frac{\partial^2 V}{\partial A_n^2} \cdot \frac{\partial^2 V}{\partial A_m^2} - \left(\frac{\partial^2 V}{\partial A_m \partial A_n}\right)^2 = 4A_m^2 A_n^2 (b^2 - b_{mn}^2) \tag{4.13}$$

To insure that V has a maximum value, the considered point must be of elliptic type, that is, the right side of (4.13) must be positive. A numerical calculation (Schlüter et al., 1965) shows, however, that for all values of m and n

$$b_{mn} > b \tag{4.14}$$

Therefore, the supposed stable solution does not exist, and we conclude that the only possible steady, stable solution is the two-dimensional roll solution.

It is readily shown that V really has a maximum value for the two-dimensional roll solution. This is most easily seen by examining the stability of the steady solution

$$bA_n^2 = \sigma \tag{4.15}$$

for disturbance satisfying (4.8). We end up with

$$\delta\dot{A}_m = (1 - \frac{b_{mn}}{b})\sigma\delta A_m \tag{4.16}$$

which shows that, due to (4.14), the roll solution is stable for this kind of disturbances.

It follows that the roll-solution will, as far as (4.9) govern the motion, be established for <u>all</u> initial values. This solution therefore posesses global stability. It is noticed that V is a Liapunov function since, according to (4.9), $dV/dt < 0$.

The roll-solution may, however, be unstable for other kind of disturbances, i.e. disturbances with wave numbers different from that of the steady solution.

We abstain from going into details here and refer to Schlüter et al. (1965) and to the article by Busse (1971). The main results of these investigations are shown schematically in Fig. 2, where the shaded region corresponds to stable non-linear solutions (in the form of two-dimensional rolls). Comparing with Fig. 1, we note that only a fraction of the unstable area in the linear case gives rise to a stable non-linear motion. On the other hand, for given supercritical Rayleigh numbers the realized steady motion is not unique. It is obvious that the orientation of the two-dimensional rolls is arbitrary. The figure shows, however, that also the wave number may vary within certain limits.

The boundary curve c in Fig. 2 is somewhat dependent on the Prandtl number. For large Prandtl numbers the curve is determined by disturbances in the form of two-dimensional rolls, oriented normal to the steady two-dimensional roll ("cross-roll instability").

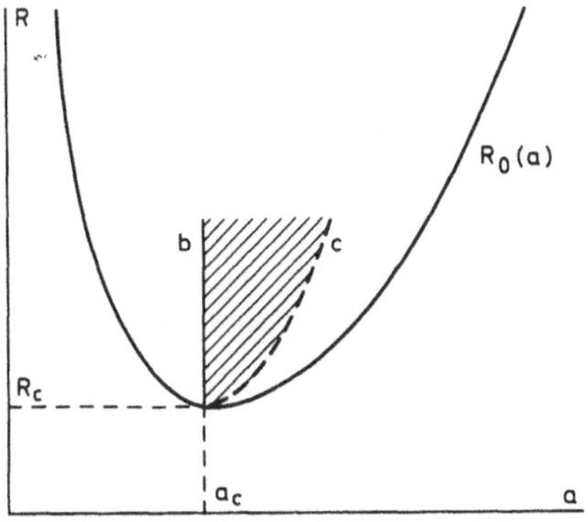

Figure 2. Stability diagram for two-dimensional rolls, small supercritical Rayleigh numbers.

For smaller Prandtl numbers (less than about 1.1 in the case of rigid boundaries and 3.5 in the case of free boundaries) the marginal disturbances are disturbances with wave numbers close to and parallel with the wave number of the steady solution ("Eckhaus instability", see Eckhaus, 1965). The boundary curve denoted by b is determined by disturbances in the form of two-dimensional rolls oblique to the steady solution ("Zigzag instability").

Returning back to the maximum principle (4.9)-(4.10), it may be of interest to note that this is a non-linear extention of the maximum principle (3.18)-(3.19), leading to (3.20).

Applying the non-linear version of (3.14)-(3.16) it may after some manipulation be shown that the functional V which is maximum when the motion is steady is given by

$$V = \left(R^{\frac{1}{2}} + \frac{\Delta R}{R^{\frac{1}{2}}}\right)\langle\theta w\rangle - \tfrac{1}{2}\langle(\nabla\theta)\rangle^2 - \tfrac{1}{2}\langle\nabla\vec{v} : \nabla\vec{v}\rangle \qquad (4.17)$$

valid to the forth order in the amplitude. Comparing with (3.19) we see that (4.17) contains a new term $(\Delta R/R^{\frac{1}{2}})\langle\theta w\rangle$. The two last terms in (4.17) are the generalized dissipation whereas the first term expresses the conversion of potential energi.

c. The tendency toward hexagonal cells. As mentioned in the introduction, a hexagonal pattern is often observed in experiments. From a mathematical point of view, this tendency must be linked with the occurence of a second order term in the amplitude equation (4.1). To discuss the formation of this second order term, let us in (4.6) pick out 3 modes $\vec{k}_1$, $\vec{k}_2$, $\vec{k}_3$ rotated 120^0 to each other, together with their complex conjugates. Since $\vec{k}_1 + \vec{k}_2 + \vec{k}_3 = 0$, we notice that the product of two modes, generates the third mode. Therefore, as far as the horizontal variation is concerned, we end always up with a second order term. This is, however, not true when we consider the z-dependence. To obtain that the conditions for a second order term is fulfilled also in the vertical direction, a kind of asymmetry must be present. This can be obtained, for example, by the viscosity being a function of temperature (Palm, 1960). To see this, let us consider the case of free-free boundaries. The variation of w with z in (4.4), is then given by $f(z) = \sin \pi z$. For simplicity we assume that the kinematic viscosity ν as a function of temperature may be written as

$$\nu = \nu_0 + \Delta\nu \cos \mu(T - T_0) \qquad (4.18)$$

Here μ is constant and, for the moment, we are working with dimensional quantities. The temperature T and T_0 are defined by (2.5). Applying (2.5), assuming that $\Delta\nu/\nu_0 \ll 1$ and that μ approximately satisfies $\mu\Delta T = \pi$ we may write (4.18) as

$$\nu = \nu_0 + \Delta\nu \cos \pi z + \mu\Delta\nu\theta \sin \pi z \tag{4.19}$$

where the various quantities have been now put into dimensionless form again. When (4.19) is introduced in the equation of motion, the last term in (4.19) will give rise to a term which fulfills all the requirements for a second order term to occur.

It is easily seen from the new amplitude equation (for simplicity we have put the phases $\psi_n = 0$) that a possible steady solution corresponds to the amplitudes A_1, A_2, A_3 all are equal. But this is exactly the hexagon solution.

The theory of the preferred mode has been advanced further in works by Segel and Stuart (1962), Segel (1965), Stuart (1964), Palm and Øiann (1964), Palm, Ellingsen and Gjevik (1967), and, independently, in works by Busse (1962, 1967a). In the two last-mentioned papers other fluid properties such as the coefficient of expansion and the thermal diffusivity are also allowed to vary with the temperature, and in addition, a very broad class of disturbances is considered. Restricting attention to the case when only ν is non-constant, it is found that the motion is upward in the center of the polygons when ν decreases with temperature (as for water) and downward in the center of the polygons when ν increases with temperature (as for gases). Furthermore, it is also obtained that hexagons are the only steady, stable planform for small supercritical Rayleigh numbers, less than R_2, say. For R larger than R_2, but less than R_3, both hexagons and two-dimensional rolls are stable, steady solutions. For R larger than R_3 only the roll solution is stable. The values of R_1, R_2 and R_3 is computed by Palm, Ellingsen and Gjevik (1967). For ordinary fluids the Rayleigh number region where hexagons are found to be steady and stable is very small.

The theoretical results above have some qualitative confirmation in the experiments by Hoard, Robertson and Acrivos (1970). There is, however, a quantitative disagreement between the experimental and theoretical findings, which may be due partly to the influence of the lateral cylindrical walls in the experiments, and partly to the fact that the Rayleigh numbers in the actual experiments are outside the region where the theory is strictly valid.

It has been assumed above that the kinematic viscosity is a function of the temperature. As seen from equation (4.19) the kinematic viscosity is then a function of the vertical coordinate z and of θ defined by (2.5). It turnes out that it is only the θ dependence that causes the tendency towards hexagonal cells. Hence if the kinematic viscosity is a function of the vertical coordinate only, there is, to the order considered in the theory, no tendency towards formation of a hexagonal pattern.

The cellular motion observed in the experiments by Thomson (1881) and Bénard (1900) was not caused by an effect as described above. Indeed the gravity was of no importance in their experiments, the surface tension (being a function of temperature) was the driving mechanism. This was first suggested by Block (1956). The first theoretical study of convection driven by surface tension was given by Pearson (1958).

A hexagonal pattern may also occur when the fluid properties are constant whereas the boundary conditions are asymmetric. We consider only amplitude equations up to the third order. Introducing the ε-expansion (4.2) in the equations (2.9)-(2.11), we obtain for the first and second order equations, respectively

$$-\nabla p_1 + P\theta_1\vec{k} + P\nabla^2 v_1 = 0$$
$$\nabla^2\theta_1 + R_0 w_1 = 0 \qquad (4.20)$$
$$\nabla\cdot\vec{v}_1 = 0$$

$$-\nabla p_2 + P\theta_2\vec{k} + P\nabla^2\vec{v}_2 = \vec{v}_1\cdot\nabla\vec{v}_1$$
$$\nabla^2\theta_2 + R_0 w_2 = \vec{v}_1\cdot\nabla\theta_1 - R_1 w_1 \qquad (4.21)$$
$$\nabla\cdot\vec{v}_2 = 0$$

We have here applied that $\partial/\partial t = O(\varepsilon^2)$. Defining the scalar product by

$$R_0\langle\vec{v}'\cdot\vec{v}''\rangle + P\langle\theta'\theta''\rangle \qquad (4.22)$$

we note that the operator on the left hand side in (4.20) and (4.21) is self-adjoint. The condition for solvability of (4.21) then becomes

$$R_0\langle\vec{v}'(\vec{v}_1\cdot\nabla\vec{v}_1)\rangle + P\langle\theta'\vec{v}_1\cdot\nabla\theta_1\rangle = PR_1\langle w_1\theta_1\rangle \qquad (4.23)$$

where $\vec{v}'$, θ' denote an arbitrary solution of (4.20), fulfilling the boundary conditions. It has been shown by Schlüter et al. (1965) that the left hand side in (4.23) is zero, provided the velocity normal the boundaries is zero (impenetrable walls). Therefore R_1 is zero and no second order term exist in the amplitude equation. Hence, to the order of magnitude considered, there is no tendency for formation of hexagons on a fluid with constant materiel properties when $w = 0$ at the two horizontal boundaries.

We saw above that if ν is a function of temperature, a pattern of hexagons would be formed. If, however, ν is a function of z, there is no tendency for formation of hexagons. Let us now assume that some material properties are a function of z. The left hand side in (4.20) and (4.21) will then be changed. If the operator on the left side is self-adjoint, the left hand side of (4.23) is, however, still zero. Therefore, if the considered operator is self-adjoint, material properties being a function of z will not lead to hexagons. It may be shown that for $\nu = \nu(z)$, the operator is self-adjoint, and no hexagons will be formed.

The case $\nu = \nu(z)$ does not occur in an ordinary fluid. However, it may be of interest for convection in a porous medium or in a fluid with small scale turbulence. If, on the other hand, the thermal diffusivity is a function of z, the operator is not self-adjoint and hexagons will occur. This is due to the fact that also the temperature gradient for pure conduction now becomes a function of z. This will also be true if this temperature gradient is non-steady. Hence, there is always a tendency for hexagons to form before the motion has become steady.

4.2. Numerical approaches

a. Stability analysis by numerical methods. A thorough study of the convective motion and its stability behaviour for small and supercritical Rayleigh numbers have been performed by Busse and his collaborators, using the Galerkin method. It turns out that the stability behaviour of the roll motion is very dependent on the Prandtl number. It should be pointed out that large Prandtl number means that the convection of heat $(\vec{v} \cdot \nabla \theta)$ is much more important than the convection of momentum $(\vec{v} \cdot \nabla \vec{v})$. For small Prandtl numbers it is opposite. The stability diagram for very large Prandtl numbers is shown in Fig. 3. As in Fig. 2 the attached letters b and c denote zig-zag instability and cross roll

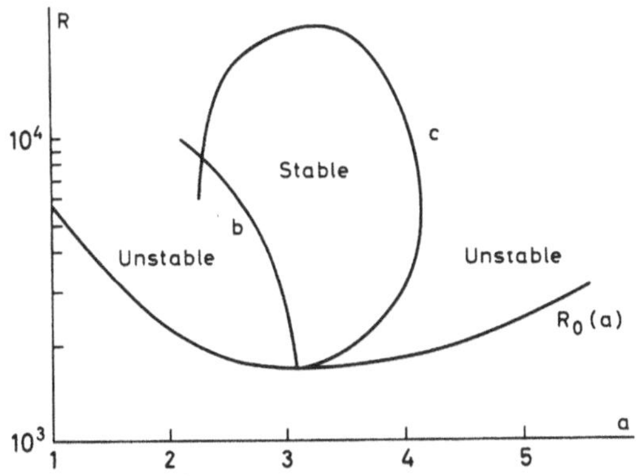

Figure 3. Stability diagram for two-dimensional rolls, moderate supercritical Rayleigh numbers, and rigid boundaries (based on Busse & Whitehead 1971).

instability, respectively. It is noted from the figure that rolls may be stable for Rayleigh numbers up to about 23.000. It does not, however, follow from the analysis that rolls are the only stable solution.

The stability diagram has changed considerably for $P = 7$ (water). It is seen from Fig. 4 that the stable region for two-dimensional rolls is bounded by curves denoted b,c,d and e. The letters d and e denote two new kind of instabilities, by Busse denoted as skewed varicose instability and knot instability, respectively.

In Fig. 5 is shown the stability diagram for $P = 0.71$ (air). It is seen that the region for stable rolls are now much smaller than for higher Prandtl numbers. We also note that one of the bounding curves corresponds to oscillatory instability. Hence we expect that the two-dimensional roll pattern will for increasing values of the Rayleigh number be replaced by an oscillatory motion. We also see that one of the bounding curves corresponds to Eckhaus instability.

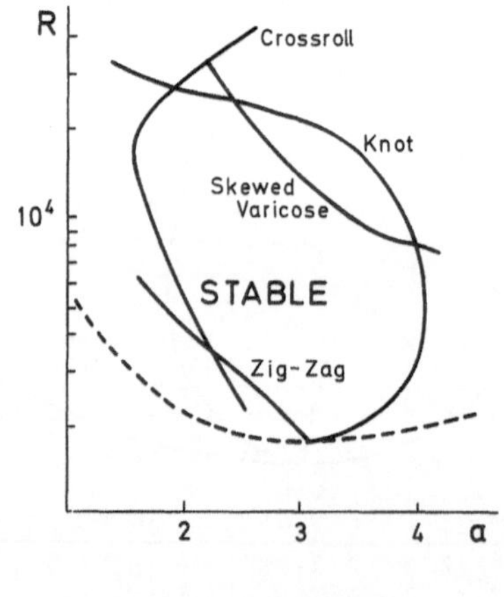

Figure 4. Stability diagram for convection rolls for Prandtl number $P = 7$ (after Busse and Clever 1978). The dashed curve indicates the critical Rayleigh number for the onset of convection.

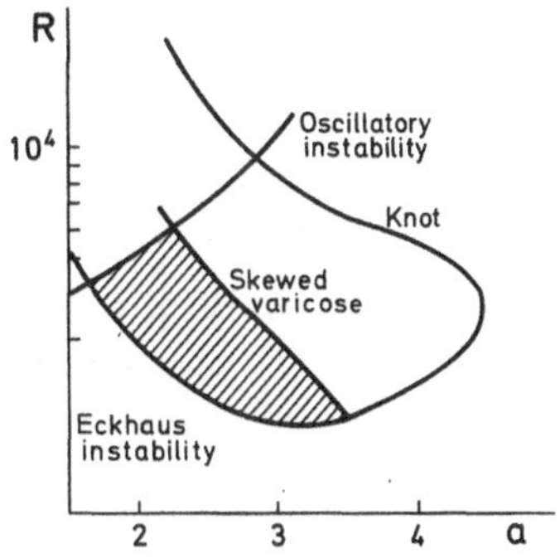

Figure 5. Stability diagram for convection rolls in air, P = 0.71 (after Busse and Clever 1978). Shaded area is stable.

4.3. Comparison with experiments

The theoretical predictions for the critical Rayleigh number have been confirmed in many experiments. The convective motion obtained in the theory for small supercritical Rayleigh numbers has been confirmed in the experiments by Bergé and Dubois [Bergé (1976), Dubois (1976) and Dubois and Bergé (1977)]. They have performed a set of carefully arranged experiments for a large Prandtl number fluid in a rectangular container with the Rayleigh number varying from R_c to about $10R_c$. The amplitudes of the first, second and third harmonics were determined for several Rayleigh numbers. The results have been compared with the theoretical results by Normand, Pomeau and Velarde (1977) obtained by an amplitude expansion and by Busse (1965) by a Galerkin procedure. Generally, a very good agreement between the experiments and the theory is found.

For higher Rayleigh numbers, the relations between the theoretical predictions and the experimental results is more complicated. As we have seen, a regime consisting of rolls is, according to theory, stable for Rayleigh numbers up to about 23.000. This result has been confirmed experimentally by Chen and Whitehead (1968) and Busse and Whitehead (1971), also for large aspect ratios. It is important to point out that in their experiments they use controlled initial conditions. By generating small two-dimensional temperature perturbation before the Rayleigh number of the convection layer is raised above the critical value, rolls with a prescribed wave length are induced.

If, on the other side, the motion is allowed to develop from arbitrary initial conditions, it seems impossible to obtain in experiments a regular roll pattern for large aspect ratios, even if the experiments are run for a very long time. We refer in this connection to the experiments by Ahlers and Behringer (1978). For large aspect ratios and small supercritical Rayleigh numbers they find in a high Prandtl number fluid no signs for the fluid to line up in a regular, two-dimensional pattern. The motion turns out to be very disorderly and reminds more of turbulence than of a regular pattern.

The theory and the experiments by Chen and Whitehead (1968) and Busse and Whitehead (1972) show that a two-dimensional roll motion is a possible, steady and stable motion. On the other hand, experiments also suggest that the steady state of a two-dimensional roll pattern cannot be obtained from arbitrary initial conditions for large aspect ratios. It seems reasonable that in this case it must be difficult for the fluid to find its way to a steady state. There is a continuous infinity of possible equilibrium states, corresponding to rolls orientated in every possible horizontal direction and with the same ab initio probability for being formed.

No theory exists for convection of still higher Rayleigh numbers. Experiments show that for large Prandtl number fluids, the motion becomes 3-dimensional, but steady. For increasing Rayleigh numbers the motion becomes oscillatory and finally turbulent. For a small Prandtl number fluid the motion may go directly to an oscillatory state when the Rayleigh number is increased, without passing any steady 3-dimensional motion.

4.4. The mean field theory

In applications, especially in geophysics and astrophysics, the Rayleigh number is usually so large that it is impossible to obtain any detailed informations of the motion. In many problems, however, also informations of the order of magnitude of some mean quantities may be of great interest. Of special importance are informations of the (horizontally) mean heat transport, expressed by the Nusselt number Nu. One interesting approach to this problem is the optimum theory. We shall abstain from discussing this theory here and refer to the review article by Busse (1978).

An alternative approach is to use some simplified version of the governing equations, and hoping that the revised and simpler equations are a fair approximation to the correct ones. To obtain equations which are essentially simpler than the original ones, it is necessary to neglect some of non-linear effects. The problem is that for high Rayleigh numbers, the non-linear terms are usually very important. A set of simplified equations which have been applied with some success are the equations based on the mean field approximation. The equations are

$$-\nabla p + \theta'\vec{k} + \nabla^2 v = 0 \tag{4.24}$$

$$-\frac{\partial\bar{\theta}}{\partial z} w + \nabla^2\theta' = 0 \tag{4.25}$$

$$\frac{\partial\bar{\theta}}{\partial z} = \overline{w\theta'} - \langle w\theta' \rangle - R \tag{4.26}$$

where the bar indicates the horizontal average and the brackets $\langle\ \rangle$ denote average over the entire fluid layer. In these equations we have introduced

$$\theta = \bar{\theta} + \theta' \tag{4.27}$$

Equation (4.26) is exact. In (4.24) the acceleration term is neglected. This may be a good approximation for high Prandtl number fluids, but does not work for small Prandtl numbers. The most serious approximation is done in the heat equation where the non-linear convection term is neglected.

These equations have been solved numerically by Herring (1963, 1964) with the horizontal dependence described by one horizontal wave length. The Nusselt number dependence on the Rayleigh number found from these equations agrees fairly well with the results obtained by other methods. In Herring's work the horizontal wave number is chosen such that the heat transport is maximised. For stress-free boundaries he obtains

$$Nu = 0{,}325 R^{1/3} \tag{4.28}$$

5. CONVECTION IN A SMALL PRANDTL NUMBER FLUID

Since the transition to turbulence in this case takes place for relatively small Rayleigh numbers, it is of interest to study this flow in detail. In this section which is based on a work by Palm and Tveitereid (1981), it is assumed that the fluid layer is of infinite horizontal extent and that the horizontal boundaries are stress-free and perfect conductors. The equations are now scaled with $h, h^2/\nu$ and $\Delta T/RP^{-1}$ as units for length, time and temperature, respectively. The governing equations then become

$$\frac{\partial \vec{v}}{\partial t} + \vec{v}\cdot\nabla\vec{v} = -\nabla p + \vec{k}\theta + \nabla^2\vec{v} \tag{5.1}$$

$$P\left(\frac{\partial \theta}{\partial t} + \vec{v}\cdot\nabla\theta\right) = Rw + \nabla^2\theta \tag{5.2}$$

$$\nabla\cdot\vec{v} = 0 \tag{5.3}$$

The velocity $\vec{v}$ is, according to (5.3), dependent on two scalar quantities. As such we choose V and φ which are closely connected with the vertical velocity w and the vertical vorticity ζ and defined by

$$w = -\nabla_1^2 V \tag{5.4}$$

$$\zeta = -\nabla_1^2 \varphi \tag{5.5}$$

The velocity $\vec{v}$ is then given by

$$\vec{v} = \vec{\delta}_1\varphi + \vec{\delta}_2 V \tag{5.6}$$

where

$$\vec{\delta}_1 = \left(\frac{\partial}{\partial y}, -\frac{\partial}{\partial x}, 0\right) \quad \vec{\delta}_2 = \left(\frac{\partial^2}{\partial x\partial z}, \frac{\partial^2}{\partial y\partial z}, -\nabla_1^2\right)$$

The boundary conditions are

$$V = \frac{\partial^2 V}{\partial z^2} = \frac{\partial \varphi}{\partial z} = \theta = 0 \quad \text{at} \quad z = 0,1 \tag{5.7}$$

The steady two-dimensional roll solution is for $P \ll 1$ given by (Schlüter et al., 1965)

$$V = 2A\cos\alpha x \sin\pi z \tag{5.8}$$

$$\varphi = 0$$

where

$$(\alpha A)^2 = \frac{6\Delta R}{P^2 R}, \quad \Delta R = R - R_c \tag{5.10}$$

It was shown by Busse (1972) that a disturbance of the form

$$B(t)\sin\alpha x\, e^{i\beta y}\sin\pi z \tag{5.11}$$

where β is small, is the mode which first becomes unstable. We therefore write the disturbance as a sum of terms of the form

$$e^{im\alpha x}e^{in\beta y}\begin{matrix}\sin k\pi z\\ \cos k\pi z\end{matrix} \tag{5.12}$$

where m,n and k are positive integers or zero. We only retain terms for which

$$m \leqq 1,\ n = 1,\ k \leqq 2 \tag{5.13}$$

After some straightforward calculations we obtain an equation for B

$$\ddot{B} + 2\sigma\dot{B} + \omega^2 B = 0 \tag{5.14}$$

where

$$2\sigma = \beta^2(1 - \tfrac{7}{8}(\alpha A)^2) \tag{5.15}$$

and

$$\omega^2 = (\beta\alpha A\pi)^2 \tag{5.16}$$

It is found that the mode proportional to $\cos\alpha x$ always decays. It is of interest to note that the form (5.14) is valid independent on how many modes (5.12) is taken into account. In fact, the existence of a second order equation for $B(x)$ is only dependent on the assumption that β is small. From (5.15) we note that the roll solution becomes unstable when

$$(\alpha A)^2 > \tfrac{8}{7} \tag{5.17}$$

This result may be compared with the numerical result by Busse (1972). The agreement is fairly good, the difference being about 20%. We have also computed 2σ by taking into account the next approximation, in which case we also retain the mode $m = 2$ and $k = 0$. We then obtain exactly the numerical result by Busse.

Let us now extend the analysis to the non-linear regime. To get a consistent approximation we must extend the number of modes such that

$$m \leqq 1,\ n \leqq 2,\ k \leqq 2 \tag{5.18}$$

We want to examine the behaviour of $B(t)$ for large values of t. It is obvious that $B(t)$ will be oscillatory, i.e. the motion will approach a limit cycle. But what kind of oscillatory motion, travelling waves or standing oscillations or a combination of these two types? We consider only the case when the non-linearity is weak, i.e. σ, and thereby B, are small. $B(t)$ may then be written as

$$B(t) = a(t_2)e^{i\omega t_0} + b(t_2)e^{-i\omega t_0} \tag{5.19}$$

where we intend to apply a two-time scale analysis. Here t_0 is the rapid varying time-coordinate whereas t_2 denotes the slowly varying time. It is straightforward to show that $a(t_2)$ and $b(t_2)$ fulfills the following equations

$$\frac{\partial a}{\partial t_2} + i\omega(C_1 a\bar{a} + C_3 b\bar{b})a = -\tfrac{1}{2}\beta^2\{1 - \tfrac{7}{8}(\alpha A)^2\}a - \beta^2\{C_2 a\bar{a} + C_4 b\bar{b}\}a \quad (5.20)$$

$$\frac{\partial b}{\partial t_2} - i\omega(C_1 b\bar{b} + C_3 a\bar{a})b = -\tfrac{1}{2}\beta^2\{1 - \tfrac{7}{8}(\alpha A)^2\}a - \beta^2\{C_2 b\bar{b} + C_4 a\bar{a}\}b \quad (5.21)$$

where C_1, C_2, C_3 and C_4 are known constants, and a bar denotes complex conjugates. It is of interest to note that (5.20), (5.21) are the equations determining $a(t_2)$ and $b(t_2)$ also when more modes than given by (5.18) are taken into account, except for a trivial change in the first term on the right hand side. (5.20) and (5.21) are readily solved for large values of t by introducing

$$a = \rho e^{i\varphi} \qquad b = re^{i\psi} \quad (5.22)$$

We obtain

$$\frac{\partial \rho}{\partial t_2} = -\tfrac{1}{2}\beta^2\{1 - \tfrac{7}{8}(\alpha A)^2\}\rho - \beta^2(C_2\rho^2 + C_4 r^2)\rho \quad (5.23)$$

$$\frac{\partial r}{\partial t_2} = -\tfrac{1}{2}\beta^2\{1 - \tfrac{7}{8}(\alpha A)^2\}r - \beta^2(C_2 r^2 + C_4\rho^2)r \quad (5.24)$$

Hence,

$$\frac{\partial \rho}{\partial t_2} = \frac{\partial V}{\partial \rho} \quad , \quad \frac{\partial r}{\partial t_2} = \frac{\partial V}{\partial r} \quad (5.25)$$

where

$$V = -\tfrac{1}{4}\beta^2\{1 - \tfrac{7}{8}(\alpha A)^2\}(\rho^2 + r^2) - \beta^2\{\tfrac{1}{4}C_2(\rho^4 + r^4) + \tfrac{1}{2}C_4 r^2\rho^2\} \quad (5.26)$$

The solution of (5.20) and (5.21) for $t \to \infty$ is therefore obtained by finding the maximum values of V. It is noted that these maximum values are obtained for all initial states of a and b. It is easily shown that if

$$C_2 < C_4 : \quad \begin{array}{l} b(\text{or } a) \to 0 \\ a(\text{or } b) \to a_0 e^{-i\omega(C_1 a_0\bar{a}_0)t_2} \end{array} \quad (5.27)$$

where

$$a_0\bar{a}_0 = -\tfrac{1}{2}\{1 - \tfrac{7}{8}(\alpha A)^2\}/C_2$$

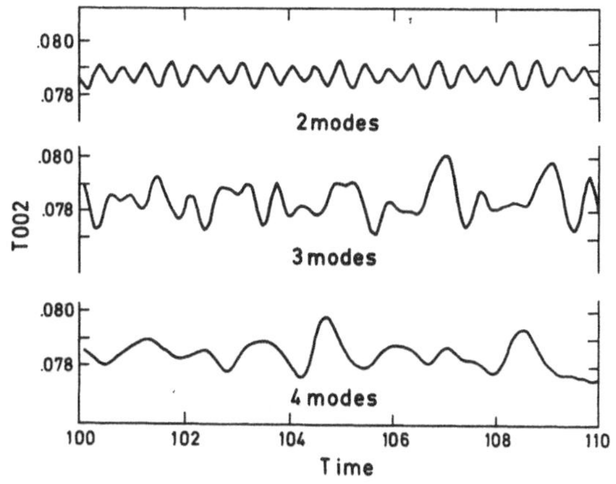

Figure 6. T_{002} denotes the temperature mode $m=n=0$ and $k=2$, and expresses the horizontally mean heat flow. 2,3,4 modes correspond to $n=2,3,4$. $P=0.7$, $R=800$, $\beta=0.07\pi=0.1\alpha$.

and if

$$C_2 > C_4 : \quad \begin{array}{l} {a \atop b} \rightarrow a_0 e^{\mp i\omega(C_1+C_3)a_0\bar{a}_0 t_2} \\ a_0\bar{a}_0 = -\tfrac{1}{2}\{1-\tfrac{7}{8}(\alpha A)^2\}/(C_2+C_4) \end{array} \tag{5.28}$$

In the first case the solution corresponds to travelling waves whereas in the last case the solution corresponds to standing oscillations. With the approximations involved in (5.18), the calculations give that $C_2 < C_4$, i.e. the motion is standing oscillatious.

To examine this result more closely, we have applied numerical methods. We then consider modes for which

$$m \leqq 1, \quad n \leqq 4, \quad k \leqq 2 \tag{5.29}$$

and β is not supposed small. The same assumption (5.29) has been applied by Lauglin and Martin (1975) in their discussion of convection in a small Prandtl number fluid. Some of the results are shown in Figs. 6, 7 and 8. It is noted that $P=0.7$ and $R=800$

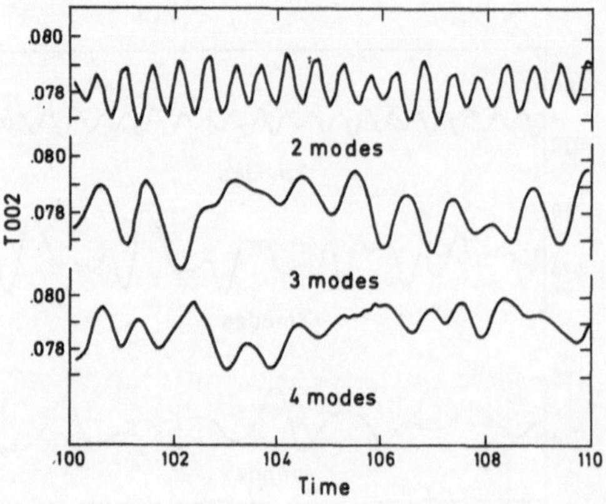

Figure 7. The only difference between Fig. 6 is that now $\beta = 0.094\pi = 0.13\alpha$.

in all three cases. In Figs. 6 and 7 β is relatively small and we obtain a motion which for two modes ($n = 2$) is close to regular standing waves, as expected. For 3 and 4 modes ($n = 3$ and $n = 4$) the motion is much more chaotic. The main impression is that in these cases we have something like a strange attractor. In Fig. 8, however, the situation is totally changed. We now have travelling waves and the solution is very stable. In this case the roll motion and the heat transport are completely steady. The explanation for the differen behaviour is partly that in the last case the higher modes are linearly stable and they therefore travel with the same velocity as the first mode. This is, however, not true for small β where several modes may be linearly unstable.

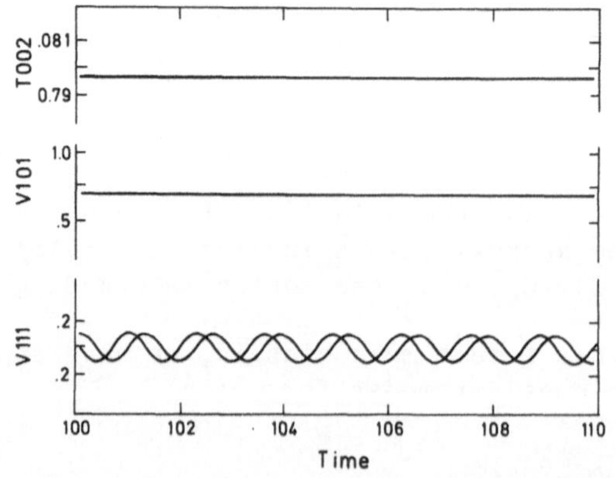

Figure 8. V_{101} and V_{111} denote the amplitudes for the roll motions and the disturbance $m = n = k = 1$, respectively. The two curves for V_{111} describe the real and imaginary parts. $P = 0.7$, $R = 800$, $\beta = 0.375\pi = 0.53\alpha$

REFERENCES

Ahlers, G. and Behringer, R.P., 1978, Evolution of turbulence from the Rayleigh-Bénard instability, Phys. Rev. Lett. 40:712
Bénard, H., 1900, Les tourbillons cellulaires dans une nappe liquide, Rev. Gén. Sci. Pures Appl. 11:1261-71, 1309-28
Bergé, P., 1976, J. Phys. (Paris) 37, Colloq. C1, 23
Block, M.J., 1956, Nature 178:650
Busse, F., 1962, Ph.D. thesis (Munich) unpublished
Busse, F., 1967, On the stability of two-dimensional convection in a layer heated from below, J. Math. and Phys. 46:140
Busse, F., 1967, The stability of finite amplitude cellular convection and its relation to an extremum principle, J. Fl. Mech. 30:625
Busse, F., 1971, Instability of continuous systems, ed. H. Leipholz, 41-47, Berlin, Heidelberg, New York:Springer, 422 pp.
Busse, F. and Whitehead, J.A., 1971, Instabilities of convection rolls in a high Prandtl number fluid, J. Fl. Mech. 47:305
Busse, F., 1972, The oscillatory instability of convection rolls in a low Prandtl number fluid, J. Fl. Mech. 52:97
Busse, F., 1978, Instabilities of convection rolls in a fluid of moderate Prandtl number, J. Fl. Mech. 91:319
Charlson, G.S. and Sani, R.L., 1970, Thermoconvective instability in a bounded cylindrical fluid layer, Int. J. Heat Mass Transfer 13:1479
Charlson, G.S. and Sani, R.L., 1971, On thermoconvective instability in a bounded cylindrical fluid layer, Int. J. Heat Mass Transfer 14:2157
Charlson, G.S. and Sani, R.L., 1975, Thermoconvective flows in a cylindrical fluid layer, J. Fl. Mech. 71, part 1:1
Chen, M.M. and Whitehead, J.A., 1968, Evolution of two-dimensional periodic Rayleigh convection cells of arbitrary wave-number, J. Fl. Mech. 31:1
Davies, S.H., 1968, Convection in a box: on the dependence of preferred wave number upon the Rayleigh number at finite amplitude, J. Fl. Mech. 32:619
Dubois, M., 1976, J. Phys. (Paris) 37, Colloq. C1, 137
Dubois, M. and Bergé, P., 1978, Experimental study of the velocity field in a Rayleigh-Bénard convection, J. Fl. Mech. 85:
Eckhaus, W., 1965,"Studies in non-linear stability theory", Berlin, Heidelberg, New York, Springer, 177 pp
Gorkov, L.P., 1957, Sov. Phys. JETP 6:311
Herring, J.R., 1963, J. Atmos. Sci. 20:325
Herring, J.R., 1964, J. Atmos. Sci. 21:277
Hoard, C.O., Robertson, C.R., Acrivos, A., 1970, Experiments on the cellular structure in Bénard convection, Int. J. Heat Mass Transfer 13:849
Malkus, W.V.R. and Veronis, G., 1958, Finite amplitude cellular convection, J. Fl. Mech. 4:225

Mihaljan, J.M., 1962, Astrophys. J. 136:1126
Normand, C., Pomeau, J. and Velarde, M.G., 1977, Convective instability; A physicist's approach, Rev. Mod. Phys. 49:581
Palm, E., 1960, On the tendency towards hexagonal cells in steady convection, J. Fl. Mech. 8:183
Palm, E. and Øiann, H., 1964, Contribution to the theory of cellular thermal convection, J. Fl. Mech. 19:353
Palm, E., Ellingsen, T. and Gjevik, B., 1967, On the occurence of cellular motion in Bénard convection, J. Fl. Mech. 30:651
Palm, E., 1972, A note on a minimum principle in Bénard convection, Int. J. Heat Mass Transfer 15:2409
Palm, E. and Tveitereid, M., 1981, To be published
Pearson, J.R.A., 1958, On convection cells induced by surface tension, J. Fl. Mech. 4:489
Rayleigh, Lord, 1916, Phil. Mag. 32:529
Schlüter, A,. Lortz, D. and Busse, F., 1965, On the stability of steady finite amplitude convection, J. Fl. Mech. 23:129
Segel, L.A., 1965, The non-linear interaction of a finity number of disturbances to a layer of fluid heated from below, 21:359
Segel, L.A., 1969, Distant sidewalls cause slow amplitude modulation of cellular convection, J. Fl. Mech. 38:203
Segel, L.A. and Stuart, J.T., 1962, On the question of the preferred mode in cellular thermal convection, J. Fl. Mech. 13:289
Spiegel, E.A., 1971, Ann. Rev. Astron. Astrophys. 9:323
Spiegel, E.A., 1972, Ann. Rev. Astron. Astrophys. 10:260
Spiegel, E.A. and Veronis, G., 1960, Astrophys. J. 131:442
Stork, K. and Müller, U., 1972, Convection in boxes: experiments, J.Fl. Mech., 54:599
Stork, K. and Müller, U., 1975, Convection in boxes: an experimental investigation in vertical cylinders and annuli, J. Fl. Mech. 71:231
Thomson, J., 1881, Proc. Glasgow Phil. Soc. 13:464
Turcotte, D.L. and Oxburgh, E.R., 1972, Ann. Rev. Fl. Mech. 4:33
Velarde, M.G. and Perez Gordon, R., 1976, J. Phys. (Paris) 37:177

NEMATIC INSTABILITY INDUCED BY AN ELLIPTICAL SHEAR

Elisabeth Guazzelli

E.S.P.C.I. Lab. H.M.P.
10, rue Vauquelin
75231 Paris Cedex 05

This paper corresponds to a 16 mm movie of 15 mn.[1] Nematic liquid crystals offer a large variety of convective instabilities [2]. Very little is known about the non linear behavior of these instabilities but there are some indications that a sequence of bifurcations similar to isotropic materials should arise [3].

We consider here the evolution of spatial structures in a nematic instability induced by an elliptical shear.[4] The advantage of such a system is manifold : small thicknesses imply the possibility of observing a large number of structures (large aspect ratio - 200 rolls) ; the strong birefringence permits a direct study of the structure by the diffraction of a laser beam which has been extended to sample the entire surface of the cell ($\sim 1\ cm^2$).

I - Experimental apparatus (Figure 1)

The nematic is sandwiched between two rectangular plates with the director $\vec{n}$ (parallel to the optical axis) perpendicular to the parallel boundary plates. The thickness of the cell is d (= 30 to 150 µm). The two plates move perpendicularly at frequency $\omega/2\pi$ ($\sim 10^2$ HZ), by imposing displacements of the form $X(t) = X_0 \cos \omega t$, $Y(t) = Y_0 \sin \omega t$ respectively to the lower and upper plate. This induces an elliptical shear in the bulk of the cell.

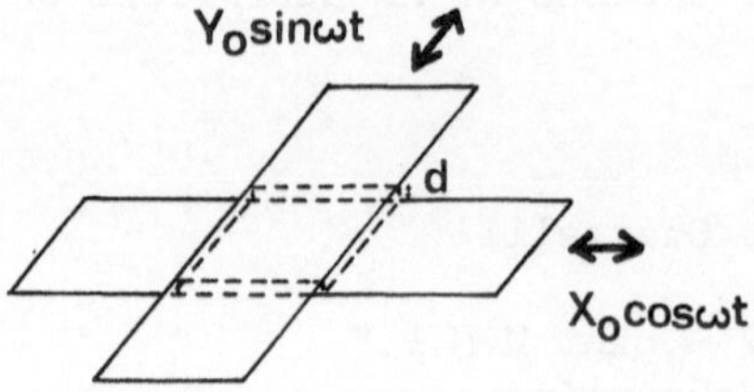

Figure 1

II - Instability mechanism (Figure 2)

Figure 2.1 : without shear :

The director $\vec{n}$ is perpendicular to the boundary plates.

Figure 2.2 : for low shear rates :

The applied velocity field causes the director $\vec{n}$ to move at a frequency ω on an elliptical cone which has its axis perpendicular to the glass plates. The distortion of the director is uniform in a plane parallel to the plates.

Figure 2.3 : As the shear exceeds a certain linear threshold :

The director $\vec{n}$ rotates around a tilted axis. The spatial period of this tilting, Λ, is ~ 2 d. This mean permanent distortion in each point of the sample is caused by the coupling of the director at each instant with a secondary hydrodynamic flow induced by the distortion of the director.

Fig. 2.1

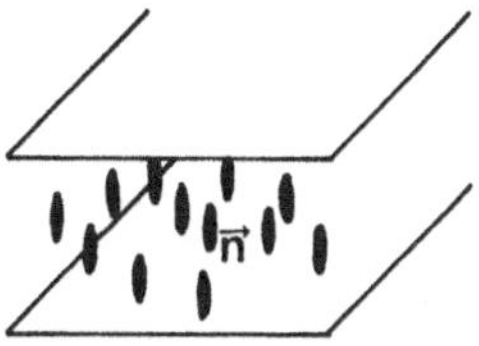

Fig. 2.2

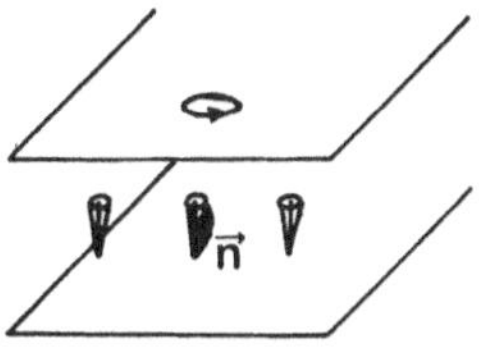

Fig. 2.3

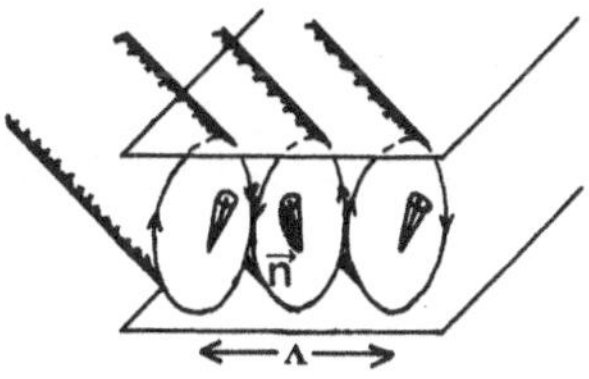

III - Modification of the convective structure on increasing the control parameter N

- Control parameter : A dimensionless number $N = X_o Y_o \omega/D$ can be constructed to characterize the instability threshold (D is the diffusivity of the nematic orientation which damps out the distortion).

- Instability diagram in the X_o-Y_o axis (Figure 3)

The experiments are performed for increasing values of N keeping a constant ellipticity (constant $\frac{X_o}{Y_o}$)

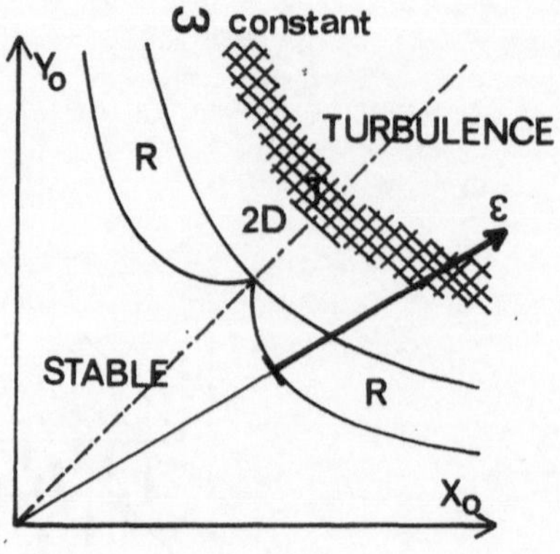

Fig. 3

- Figure 4.1 : Roll instability

Above a first linear threshold N_1, a very regular pattern of Rayleigh Bénard like rolls (Fig. 4.1.a) develops ($\sim$200 in the whole sample). The diffracted pattern is characteristic of a thick index grating (Fig. 4.1.b).

- Figure 4.2 : Square instability

Above a larger threshold N_2, a square pattern develops (Fig. 4.2.a) ($\sim 4.10^4$ cells in the sample). The diffracted image made up of spots of limited extention suggests the possibility of a quasi long range order for this 2D periodic structure (Fig. 4.2.b).

- Figure 4.3 : Partial melting

For larger values of N, arcs of diffraction develop around the original points (Fig. 4.3.b). This feature corresponds, in real space, to clusters of crystallized square pattern slightly disoriented from one another and separated from each other by grain boundaries (Fig. 4.3.a)

The occurrence of a finite density of grain boundaries induces the melting of the positional order but not of the orientational order [5].
Such a description suggests a possible analogy with the loss of ordering of 2D thermodynamic systems. [6]

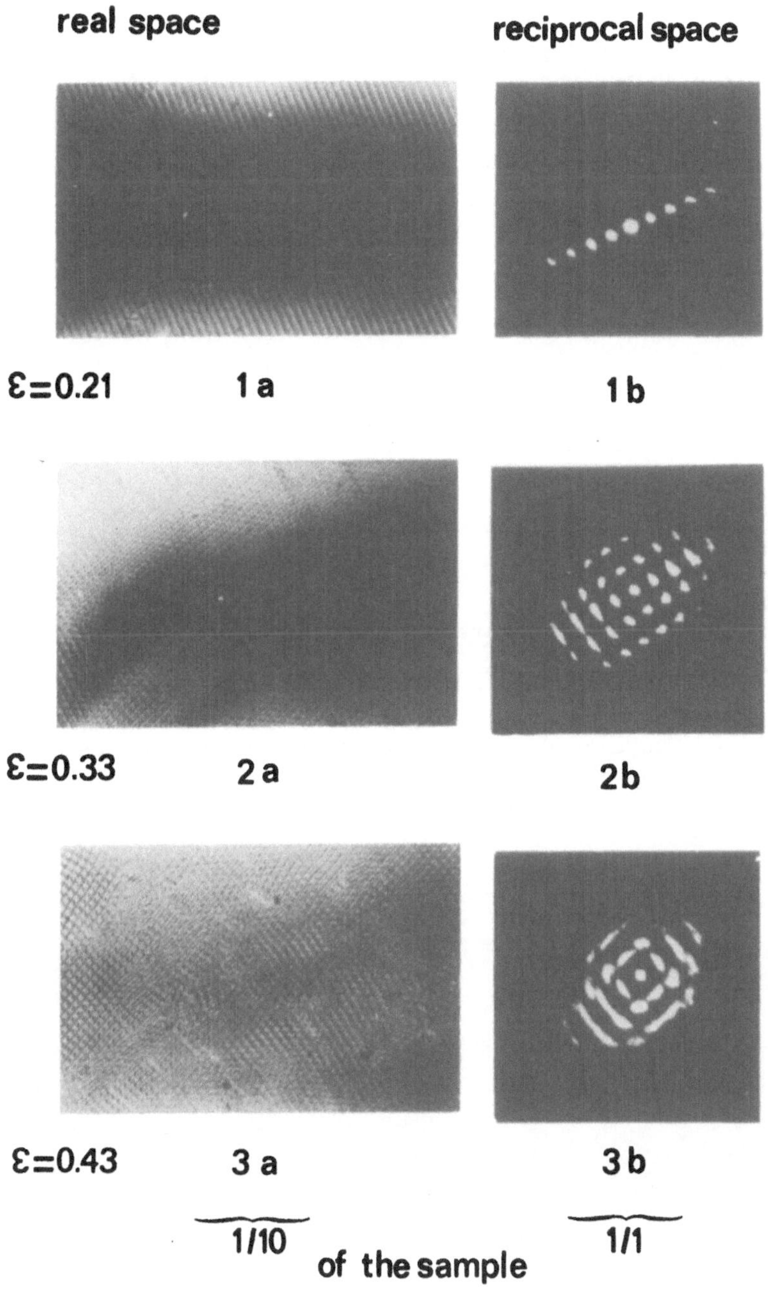
real space
reciprocal space
ε=0.21
1 a
1 b
ε=0.33
2 a
2b
ε=0.43
3 a
3 b
1/10
of the sample
1/1

Fig. 4

IV - The time evolution of the convective structure at fixed N

Such topological objects, as dislocations and grain boundaries play an important role in the transition to turbulence and their dynamics is a manifestation of spatial chaos.

Figure 5 : The deformation field of an edge dislocation in the roll structure :

The Burgers vector of the dislocation corresponds to a pair of rolls (because of the continuity of the velocity field)[7]

Fig. 5

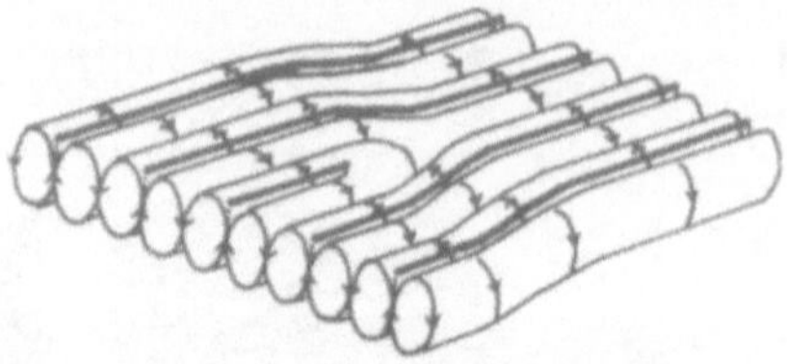

- Roll structures at fixed N

Figure 6 : Dynamics of the defects :

The motion of an edge dislocation is both gliding and climbing. Two edge dislocations of opposite signs can interact either by annihilating (Figure 6) or by passing each other and accelerating.

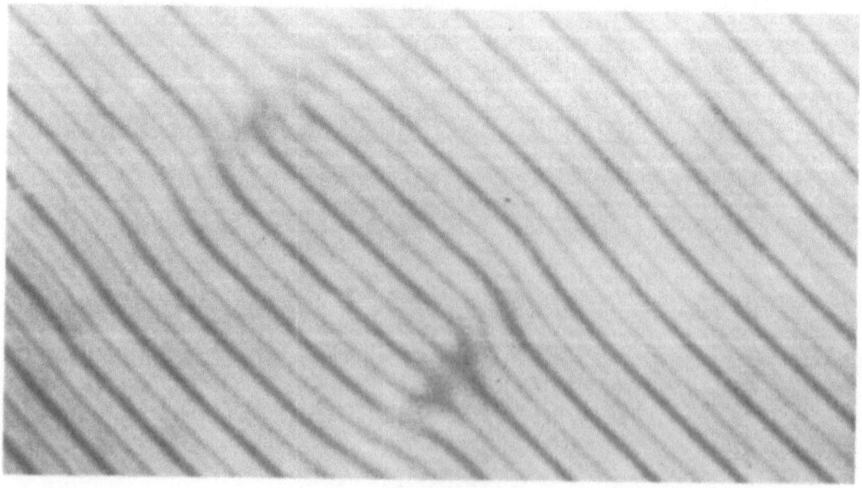

Fig. 6

Dynamics of a dust particle :

The deformation field induced by a dust particle is analogous to that induced by a dislocation loop. A dust particle can eject a dislocation by an heterogeneous mechanism of nucleation and can modify the trajectory of a dislocation.

Figure 7 : Edge of 0,1 rad across the cell :

A variation of the thickness of the cell induces a variation of the wavelength of the rolls ($\Lambda \sim d$). Under these conditions an array of dislocations of the same sign moves across the cell at the same velocity from the small to the large end.

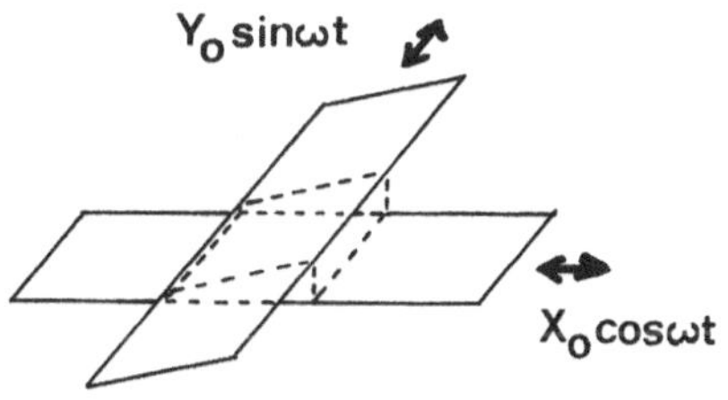

Fig. 7

- Square structures at fixed N

The transition between the roll and the square structure

The dislocations which exist between this square and roll structure allow the wavelength to ajust from the square to the roll structure.

Dynamics of defects :

Motions appear faster than in the roll case.

- Partial melting at fixed N

Clusters of crystallized square patterns are separated from each other by grain boundaries of locally sixfold symmetry which move quickly.

In the sequence showing the instability structure the time scale is contracted 50 times.

References

(1) A copy of the film can be obtained from us at ESPCI

(2) E. Dubois Violette, G. Durand, E. Guyon, P. Manneville, P. Pieranski, in Liquid Crystals, Ed L. Liebert, Academic Press (1978)

(3) H. B. Moller, T. Riste, Phys. Rev. Lett. 34, 996, (1975)

(4) P. Pieranski, E. Guyon, Phys. Rev. Lett. 39, 1281, (1977)

E. Dubois Violette, F. Rothen, J. de Phys. 10, 1039, (1978)

(5) E. Guazzelli, E. Guyon, C.R.A.S. II 292, 141, (1981)

(6) J. M. Dreyfus, E. Guyon, J. de Phys. 42, 283, (1981)

D. R. Nelson, B. I. Halperin, Phys. Rev. B 19, 2157 (1979)

(7) E. Guazzelli, E. Guyon, J. E. Wesfreid, Symmetry and broken symmetry, Ed. N. Boccara, IDSET Paris (1981)

Acknowledgments

The film has been produced with D. Leonard during his stage at E.S.P.C.I. The physics presented in the film is the result of a collaboration with E. Dubois Violette, E. Guyon, P. Manneville, Y. Pomeau, J. Prost, J.E. Wesfreid. Its technical realisation owes much to the skill of M. Clement.

INSTABILITIES AND FLUCTUATIONS

Vittorio Degiorgio

Istituto di Fisica Applicata
Università di Pavia
27100 Pavia, Italy

ROLE OF FLUCTUATIONS[1,2]

Self-organizing systems present generally a critical value of the control parameter λ_c below which the system is in the disordered phase (below threshold) and above which the system goes into the ordered phase. The buildup of the ordered phase is triggered by a spontaneous fluctuation. From a mathematical point of view, one notes that the macroscopic dynamic equations for self-organizing systems are intrinsically homogeneous, i.e. $\rho = 0$ must be a solution (ρ is the order parameter). Therefore if the system is initially at $\rho = 0$, it stays there forever and no self-organization takes place unless a spontaneous fluctuation is occurring. The fluctuation may be originated by thermal noise, by quantum noise (e.g., from the spontaneous emission of light), or by external disturbances.

The role of fluctuations is to probe the stability of the system, and to drive the system into new states which are more stable than the initial one. Of course, not any fluctuation is useful to perform this task. Normally, fluctuations have a large spectrum in frequency ω and wave vector k. Only the fluctuation components having the appropriate frequency and wave vector (ω_c and k_c characteristic of the mode which goes unstable) are capable to drive the system into the new stable state. As far as the effect of the fluctuation amplitude is concerned one should distinguish between continous transitions (second order) and discontinous ones (first order, bistable systems, metastable states). In the latter case, there is usually a potential barrier (activation energy) and the fluctuation must be large enough to overcome the barrier, whereas in the former case there is no threshold for the fluctuation amplitude which triggers the transition.

At steady state, fluctuations act as a disturbance which limits the degree of order of the system. They usually become less and less important as the system is driven away from the threshold point. However, fluctuations in the transient behavior may be large even if the system is for above threshold.

The aim of these lectures is to present an intuitive description of some aspects of fluctuations near critical points and near instability points in driven systems. The description is introduced by a general elementary discussion of random processes, and is illustrated by several experimental examples.

DESCRIPTION OF RANDOM PROCESSES[3,4]

A random process $x(t)$ is a process in which the variable x does not depend in a completely definite way on the independent variable t, as in a causal process; instead one gets in different observations different functions $x(t)$. The random process $x(t)$ is completely described by the following set of probability distributions:

$W_1(xt)dx$ = probability of finding x in the range $(x, x+dx)$ at time t.

(1)

$W_2(x_1t_1; x_2t_2)dx_1dx_2$ = joint probability of finding x in the range (x_1, x_1+dx) at time t_1 and in the range (x_2, x_2+dx_2) at time t_2.

And so on.
The set of functions (1) form a kind of hierarchy since each function W_n must imply all the previous W_k with $k < n$. They describe successively the random process in more detail.

To determine the function W_k experimentally, one needs a great number of records $x(t)$ obtained from a great number of experiments on "similarly prepared" systems. In most applications (and especially for the Brownian motion problems), however, it is possible to make a simplification because the processes are stationary in time. This means that the underlying "mechanism" which causes the fluctuations does not change in course of time. The functions W_k can now be experimentally determined from one record $x(t)$ taken over a sufficiently long time. One can then cut the record in pieces of length T (where T is long compared to all "periods" occurring in the process), and one may consider the different pieces as the different records of an ensemble of observations. In computing average values one has in general to distinguish between an ensemble average and a time average. However, for a stationary process these two ways of averaging will always give the same result, and one can, therefore, use either of them. From now on I will only consider stationary processes.

The set of probability distributions (1) leads immediately to a method of classifying the random processes.

A random process is called a purely random process when the successive values of x are not correlated at all. This means that:

$$W_2(x_1t_1;\ x_2t_2) = W_1(x_1t_1)\ W_1(x_2t_2)$$

and analogously for the higher W_n. All the information about the process is then completely contained in the first distribution function W_1. When t is discrete, it is easy to give examples, but for continuous t, the purely random process can only be considered as a kind of limiting case; in any actual example, the x_1 and x_2 will surely be correlated when the time interval t_2-t_1 is small enough.

In the next more complicated case, all the information about the process will be contained in W_2. Such processes are called Markov processes. For a more precise definition it is useful first to introduce the notion of conditional probabilities. I will write $P_2(x_1/x_2,t)dx_2$ for the probability that given x_1 one finds x in the range (x_2,x_2+dx_2) a time t later. Of course, one finds P_2 from W_2 according to

$$W_2(x_1;x_2t)= W_1(x_1)P_2(x_1/x_2,t) \tag{2}$$

A Markov process can now be defined more precisely by stating that for such a process the conditional probability that x lies in the interval (x_n,x_n+dx_n) at time t_n, given that x is equal to $x_1,x_2\ldots x_{n-1}$ at the times t, $t_2\ldots t_{n-1}$ (where $t_n>t_{n-1}\ldots\ t_2\ t_1$) depends besides on x_nt_n only on the value of x at the previous time t_{n-1}, that is,

$$P_n(x_1t_1,x_2t_2\ldots y_{n-1}t_{n-1}/x_nt_n) = P_2(x_{n-1}t_{n-1}/x_nt_n) \tag{3}$$

Clearly, all the W_n for $n > 2$ can be found, when only W_2 is known. One derives for istance easily from (3) that:

$$W_3(x_1t_1,x_2t_2,x_3t_3) = \frac{W_2(x_1t_1,x_2t_2)W_2(x_2t_2,x_3t_3)}{W_1(x_2t_2)} \tag{4}$$

and so on. Therefore, W_2 or P_2 completely describes the process.

The next class of processes will be completely described by giving W_3. Very often, when a process is not a Markow process one can still consider it as a kind of "projection" of a more complicated Markov process. Besides x, one then considers another dependent variable z (which may be, for instance, dx/dt or it may be a coordinate of another system), and it may be that for the two variables x,z combined, the process is then a Markov process. The $W_2(x_1x_2t)$ which one obtains by integrating $W_2(x_1z_1,x_2z_2t)$ over z_1 and z_2 will then in general not be a Markov process, and one can say that this is due to the fact one has not given a complete enough description of the process.
Often, the measurement does not give the full probability distribution, but only some appropriate average, like the moments

$$M_k = \langle x^k \rangle = \int x^k W_1 dx \tag{5}$$

and the correlation functions

$$G(t_1,\ldots,t_k) = \langle x(t_1)\ldots x(t_k)\rangle = \int x_1\ldots x_k W_k dx_1,\ldots dx_k \tag{6}$$

where $x_1 = x(t_1)$, and so on.

As a particular example, we shortly discuss the Brownian motion of a free particle. The standard mathematical model for analyzing the process is the Langevin equation

$$m\dot{v} = -fv + F(t) \tag{7}$$

where m is the mass of the particle and v its velocity. The second term on the left-hand side denotes the viscous drag, f being the friction coefficient. F(t) is a rapidly varying Gaussian force having the properties

$$\begin{aligned} &\langle F(t)\rangle = 0 \\ &\langle F(t)F(t')\rangle = \Gamma\,\delta\,(t-t') \end{aligned} \tag{8}$$

The second property is strictly related to the assumption that f is time-independent. Since the corresponding power spectrum has a constant value for all values of ω, one says that F(t) has a white-noise spectrum. From further analysis it is found that

$$\langle v(0)v(t)\rangle_{eq} = \frac{\Gamma}{2mf}\exp(-\gamma t) \tag{9}$$

where $\gamma = f/m$. We observe that in contrast to $F(t)$, the velocity fluctuations persist for some time, more precisely they have a correlation time γ^{-1}. Eq. (9) evaluated for $t = 0$ gives $<v^2> = \Gamma/2mf$. Since $1/2\, m<v^2> = 1/2\, k_BT$ in our one-dimensional problem, we find

$$\Gamma = 2\, f\, k_BT. \tag{10}$$

The power spectrum of the velocity fluctuations has a Lorentzian shape, and is given by

$$S(\omega) = \frac{2\gamma k_BT}{m} \frac{1}{\gamma^2+\omega^2} \tag{11}$$

Since the velocity fluctuates both in magnitude and direction, the position of the Brownian particle also fluctuates following a random walk pattern. The nature of the random walk was first investigated by Einstein who showed that the mean square displacement of the particle position $x(t)$ is, for long times,

$$<\Delta\, x^2(t)>_{eq} = <(x(t)-x(o))^2>_{eq} = 2Dt \tag{12}$$

where D is the diffusion coefficient, given by

$$D = \int_0^\infty dt<v(0)v(t)>_{eq} = \frac{k_BT}{f} \tag{13}$$

This result is important because it links a transport coefficient describing dissipation with a quantity connected with fluctuations.

In order to fully characterize the stochastic process defined by the Langevin equation, one should evaluate the conditional probability $P(x_o,0/x,T)$. This can be accomplished by solving the Fokker-Planck equation associated with Eq. 7. An example of this procedure will be shown later on in connection with an elementary description of the laser statistics.

PHASE TRANSITIONS[5]

Phase transitions in thermodynamic systems are well known to most of the participants to this school, and are also discussed by other lecturers. I want here to recall some aspects connected with the role of critical fluctuations.

A simple description of the critical region is provided by Landau's theory which assumes that free energy G can be expanded in a power series of the order parameter ρ. I consider, for simplicity, a real scalar order parameter. The expansion takes the following form

$$G = G_o - H\rho + a(T-T_c)\rho^2 + b\rho^4 + c(\nabla\rho)^2, \tag{14}$$

where H is the field conjugate to the order parameter, a,b and c are constants.
The most probable value of ρ is determined by minimizing G. The result is

$$\rho = 0 \qquad \text{for} \qquad T>T_c \tag{15}$$

$$\rho = \pm \left(\frac{aT_c}{2b}\right)^{\frac{1}{2}} \varepsilon^{\frac{1}{2}} \qquad \text{for} \qquad T<T_c$$

with $\varepsilon = |T-T_c| / T_c$.
More generally, one can derive the full probability density $W_1(\rho)$ as

$$W_1(\rho) \sim e^{-G/k_B T} \tag{16}$$

Landau makes the assumption that the most probable value of ρ coincides with the mean value. This is only true if $W_1(\rho)$ is symmetric around the maximum, as it happens, for instance, with a Gaussian probability distribution.
An important quantity is the spatial correlation function

$$g(\vec{r},\vec{r}\,') = \left\langle \left[\rho(\vec{r}) - \langle\rho(\vec{r})\rangle\right] \left[\rho(\vec{r}\,') - \langle\rho(\vec{r}\,')\rangle\right] \right\rangle \tag{17}$$

Within the frame of the mean-field theory, it can be shown that, for H = 0,

$$g(\vec{r},\vec{r}\,') = \frac{\exp(-|\vec{r}-\vec{r}\,'|/\xi)}{|\vec{r}-\vec{r}\,'|} \, \frac{k_B T}{8\pi c} \tag{18}$$

with

$$\xi = (c/aT_c)^{\frac{1}{2}} \varepsilon^{-\frac{1}{2}} \qquad \text{for} \qquad T>T_c$$

$$\xi = (c/2aT_c)^{\frac{1}{2}} \varepsilon^{-\frac{1}{2}} \qquad \text{for} \qquad T<T_c \tag{19}$$

The function $g(\vec{r},\vec{r}')$ indicates how the deviation of ρ from its average at one point in the material is correlated to the similar fluctuations in neighboring regions. The characteristic range ξ of the correlation function measures the size of a region in which a coherent fluctuation occurs. Eqs. (17 - 19), show that, as we get closer to the critical point, these fluctuations cover more space.

It is also useful to recall that the susceptibility $\chi = (\partial\rho/\partial H)_T$ is given by

$$\chi = (k_B T)^{-1} \int ds\, g(\vec{s}) \tag{20}$$

where $\vec{s} = \vec{r}-\vec{r}'$.
By using Eq.(20), $\chi_{H=0} = (2c)^{-1}\xi^2$.

It is known that in many cases of physical interest the Landau theory does not work. Ginzburg has suggested the following criterion for the validity of Landau theory: fluctuations in the order parameter over distances comparable with ξ must be small in comparison with the order parameter itself[6]. That is,

$$\langle\delta\rho^2\rangle_\xi = g(\vec{r},\vec{r}')\,\Big|_{|\vec{r}-\vec{r}'| = \xi} \ll \langle\rho\rangle^2 \tag{21}$$

This inequality can be written as

$$\frac{k_B T_c \chi_o}{4\pi\xi_o^3}\,\varepsilon^{\frac{1}{2}} \ll \frac{aT_c}{2b}\,\varepsilon \tag{22}$$

where $\xi_o = (c/2aT_c)^{\frac{1}{2}}$ and $\chi_o = (2aT_c)^{-1}$. Since the left-hand side goes to zero as $\varepsilon \to 0$ more slowly than the right-hand side (see Fig. 1), there will be a cross-over at ε_c. When $\varepsilon < \varepsilon_c$, the Landau theory breaks down. The quantity ε_c turns out to be quite large ($\varepsilon_c \simeq 10^{-2}$) for the ferromagnetic transition, for the classical gas-liquid transition and for the consolution point of binary mixtures. In connection with the latter system, it is interesting to note that recent light-scattering measurements on macromolecular[7] and micellar[8] solutions are in agreement with the Landau theory. Ishimoto and Tanaka[7] have measured χ, ξ and the coexistence curve near the consolution point of a water-protein binary mixture. I have computed ε_c by using their data: the result is $\varepsilon_c \simeq 3\times10^{-4}$. This could explain why their measurements, performed in the range $\varepsilon \gtrsim 10^{-3}$, find a mean-field behavior of the critical mixture. The micellar solution studied in Ref. 8 is made of water plus a nonionic surfactant. The surfactant aggregates spontaneously in aqueous solution to form micelles having an aggregation number of the order of 100.

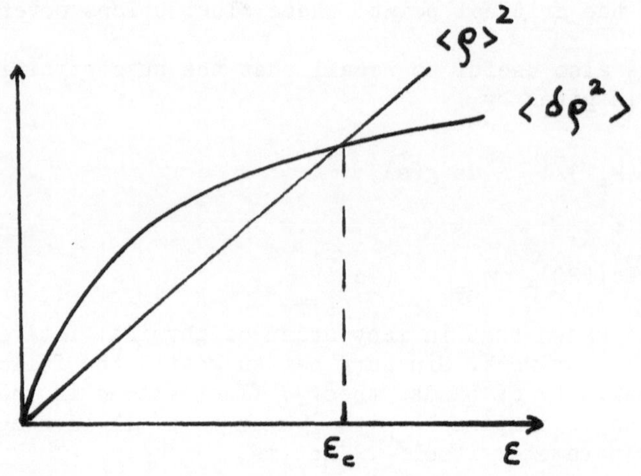

Fig. 1 Application of the Ginzburg criterion to a mean-field second-order phase transition: the order parameter square and the mean square order-parameter fluctuation as function of $\varepsilon = (T_c - T)/T_c$.

The experimental results are shown in Figs. 2, 3. The calculation of ε_c is more difficult in the case of the micellar solution because the coexistence curve is not symmetric and therefore an approach similar to that of the Flory-Huggins theory of polymer solutions must be followed to describe this system. Furthermore accurate measurements of the coexistence curve are lacking.

According to the scaling laws formulation, $<\delta\rho^2> \sim \varepsilon^{-\gamma' + d\nu'}$ where d is the dimensionality of the system, and γ', ν' are the critical exponents for χ and ξ, respectively, in the ordered phase. Furthermore $<\rho> \sim \varepsilon^{\beta}$. Therefore the Ginzburg criterion is always satisfied provided that $d\nu' - \gamma' \geqslant 2\beta$. Scaling predicts (or assumes?) indeed that $d\nu' - \gamma' = 2\beta$, as obtained by combining the two equalities $d\nu' = 2 - \alpha = \gamma' + 2\beta$. The physical meaning of the equality is that the mean square fluctuation averaged over a correlation volume goes to zero as $T \to T_c$ remaining always proportional to the square of the mean order parameter. This self-consistency argument about fluctuations can be used, of course, to derive scaling laws[9].

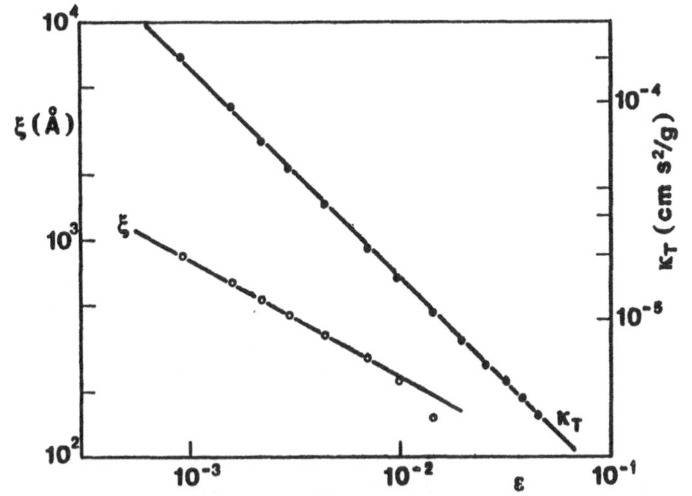

Fig. 2 The osmotic isothermal compressibility κ_T and the correlation length as function of ε along the critical isochore of the H_2O-$C_{12}E_6$ micellar solution (see Ref. 8).

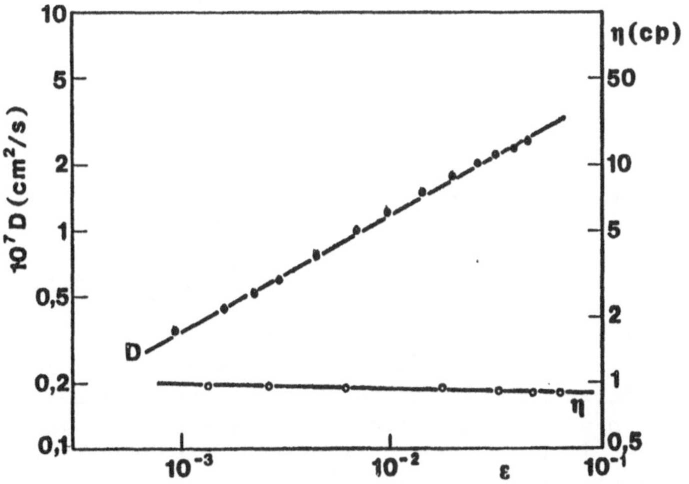

Fig. 3 The mass diffusion coefficient D and the macroscopic shear viscosity η as a function of ε along the critical isochore of the H_2O-$C_{12}E_6$ micellar solution. The solid line relative to D represents the Einstein-Stokes relation (see Ref.8).

THE LASER INSTABILITY[10,11]

A laser consists typically of a dilute collection of active atoms in an optical cavity which is excited by an external source of energy. If the pump power fed into the cavity is weak, the laser acts as an ordinary lamp. However, if both the active atoms and the external energy source are chosen properly, there is a critical pump power above which the device will emit a highly directional optical beam with a very long coherence time. The threshold value of the pump power is determined by the characteristics of the active atoms and by the energy losses of the optical cavity.

This behavior of the laser can be seen as a transition from a disordered state below threshold, where the atoms emit waves with random phases independently of each other, to an ordered state above threshold, where stimulated emission is predominant.

The laser is not a system in thermal equilibrium but rather one in which a stationary ordered state is created and maintained by an energy flow through the device. There is therefore no a priori reason why the concepts and techniques developed in equilibrium thermodynamics to study phase transitions should be useful in the laser case. Particular efforts have been devoted to the study of the region around threshold, because it is the behavior in this region that provides us with a better understanding of the laser transition. From the experimental point of view, it was a very fortunate coincidence to have available a specific system, the helium-neon laser, in which the transition region is rather easily accessible for accurate measurements. The investigation of the threshold region revealed many features qualitatively similar to those encountered in second-order phase transitions near their critical points, such as a large increase both in the magnitude and in the decay time of amplitude fluctuations of the laser field.

The analogy was put on a quantitative basis by comparing the laser theory with the so-called "mean-field" (or Landau) treatment of second-order phase transitions[12]. The origin of this similarity becomes evident when we recall that the usual treatments of laser behavior are self-consistent-field theories. In the laser analysis each atom is a radiating dipole in the electromagnetic field emitted by all the other atoms. The radiation field produced by this set of radiating atoms is then calculated in a self-consistent fashion. The physics of the laser is similar in this way to that of a ferromagnet in which each spin sees a mean magnetic field due to all the other spins and aligns itself accordingly, thus adding its contribution to the average magnetic field.

The laser can be schematized, as any oscillator can, as a positive-feedback amplifier. The gain at optical frequencies is provided by the pumped active medium and the positive feedback is obtained by the two parallel mirrors that form the optical cavity.

This cavity can sustain many modes. However, only the mode with the largest net gain will oscillate in the region near threshold. Because we are mainly interested in this region, our discussion will be confined to a single-mode laser.

The laser electric field can be written as $E(t)e^{i\omega t}$, where ω is the angular frequency of the laser mode and $E(t)$ is a complex quantity with temporal variations much slower than those of the oscillating exponential factor. The dynamical response of the system near threshold is well described by the following equation of motion which expresses the competition between the gain g and the losses,

$$\frac{dE}{dt} = (g_o-\gamma)E - \beta|E|^2E \tag{23}$$

The laser gain is written in Eq.(23) as $g(E)=g_o-\beta|E|^2$, where both g_o and β are proportional to the unsaturared population inversion σ, the difference between the number of atoms in the upper laser level and that in the lower level. The threshold condition, $g_o=\gamma$, defines the threshold population inversion σ_t.

The steady-state behavior of the laser field is easily derived from Eq.(23). Below threshold ($\sigma<\sigma_t$) the only solution is E=0. Above threshold ($\sigma>\sigma_t$) the only stable solution is

$$E = E_o = A\left(\frac{\sigma-\sigma_t}{\sigma}\right)^{\frac{1}{2}} \tag{24}$$

where A is a proportionality constant. Note that only the modulus of the field is assigned; the phase ρ is arbitrary.

So far we have considered the laser field as a deterministic variable; however, fluctuations can not be neglected in the threshold region. The most important source of noise in the laser is spontaneous emission from the upper to the lower laser level. A quantum theory that intrinsically accounts for fluctuations has been developed[10]. A simplified approach can be introduced as follows. By adding a random noise source $\Gamma(t)$ to Eq. (23), a non linear Langevin equation is obtained. Let $\Gamma(t)$ be a complex stationary random process with the properties

$$\langle\Gamma(t)\rangle = 0, \quad \langle\Gamma(0)\Gamma(t)\rangle = Q\delta(t) \tag{25}$$

The associated Fokker-Planck equation is

$$\frac{\partial P}{\partial t} + \mathrm{div}_E \{ \left[g(E) - \gamma \right] EP \} = \frac{Q}{4} \nabla_E^2 P \qquad (26)$$

A first important result of the theory is the expression for the steady-state probability distribution P(E) of the laser field near threshold. The function P(E) can be put into the very suggestive form (compare with Eq. (16))[12]

$$P(E) \sim \exp \left[- \frac{G(E)}{k\sigma} \right] \qquad (27)$$

where k is proportional to the spontaneous emission rate per atom and G(E) is a laser "potential" defined by the relation dE/dt = - dG/dE. From Eq.(23) I find

$$G(E) = - \frac{g_o - \gamma}{2} E^2 + \frac{\beta}{4} E^4 + \mathrm{const} \qquad (28)$$

The behavior of G(E) is shown in Fig.4. Note that G(E) does not depend on the phase ϕ of the laser field.

Below threshold P(E) is a bell-shaped function peaked at E=0. Well below threshold the quartic term in G(E) can be neglected, so that P(E) reduces to a gaussian function of E. Above threshold P(E) is maximum for the value $E=E_o$, given by Eq. (24). It is then clear that the steady-state solution found when fluctuations are neglected gives the most probable value (not the mean value) of the laser elctric field. The lowest-order moment of P(E) having an operational meaning is the average intensity $I=\langle |E|^2 \rangle$. Recent experimental results giving I as a function of the reduced normalized population inversion, $\varepsilon = \sigma/\sigma_t - 1$, are shown in Fig.5.

We can infer the magnitude of field-amplitude fluctuations qualitatively by considering the potential curves in Fig.4 and using the fact that energy fluctuations due to spontaneous emission depend very weakly on $\sigma - \sigma_t$ in the threshold region, $\varepsilon \ll 1$. For instance, it is immediately seen that the same change in the laser potential at the steady state produces a change in the field amplitude that is very large at threshold but relatively small above threshold, where the presence of a strong restoring force has a stabilizing effect on the field amplitude. More quantitatively, it can be shown that the mean-square fluctuation is proportional to $|\sigma - \sigma_t|^{-1}$.

Up to now our attention has been focussed on P(E), in view of the analogy with the second-order phase transitions. It should be recalled, however, that measurements of the statistical properties of laser beams give the photon count distribution p(n), and not

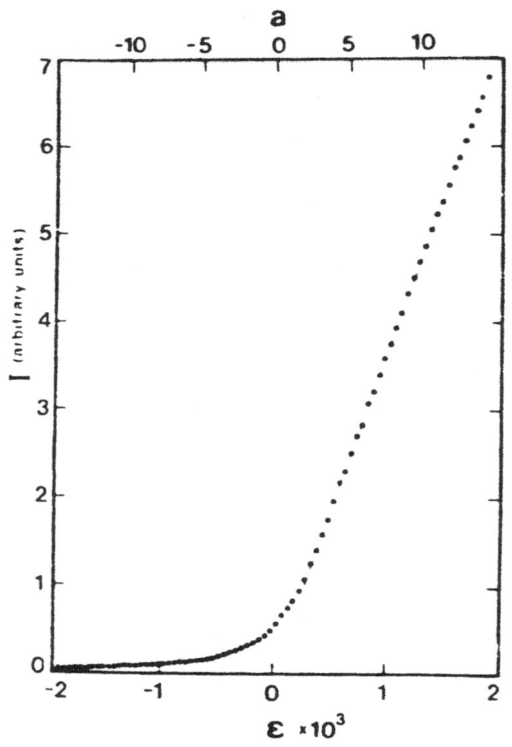

Fig. 4. The laser potential G(E), the minima of which give the steady states of the laser. The magnitudes of the fluctuations of E are found by considering energy fluctuations $k\sigma_t$ (see Eq. 27) above the minima of G(E) and deriving the corresponding changes in the field amplitude.

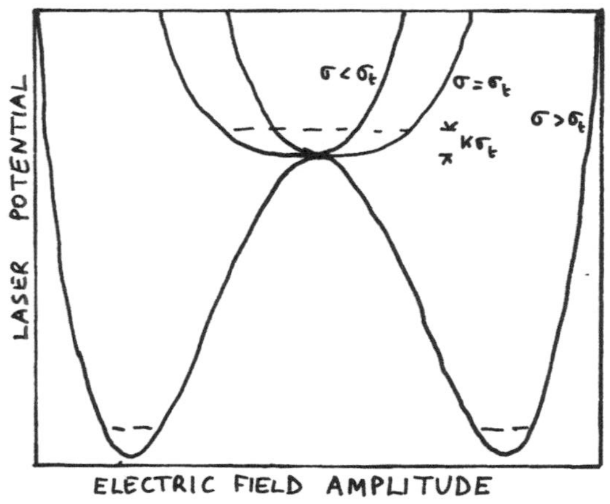

Fig. 5. The average intensity I of the laser field as a function of the reduced population inversion $\varepsilon = \sigma/\sigma_t - 1$. Above threshold, apart from a narrow region close to $\varepsilon = 0$, I is proportional to ε (see Ref. 11).

P(E). The distribution p(n) is directly related to P(E) by the equation[11]

$$p(n) = \int \frac{(\eta|E|^2)^n}{n!} e^{-\eta|E|^2} P(E) d^2E \tag{29}$$

where η is a conversion efficiency. A gaussian probability density for the field (for a laser well below threshold) gives a Bose-Einstein photon distribution; a delta-function-like probability density (for a laser well above threshold) produces a Poisson distribution. The gradual change in the photon probability distribution as the laser field goes from below to above threshold through a succession of stationary states has been measured and compared successfully with theory[11].

Turning now to the dynamics of laser-field fluctuations, we recall that the relaxation time of the amplitude fluctuations τ'_c comes out to be inversely proportional to the curvature of the laser potential at the minimum, therefore varying as $|\sigma - \sigma_t|^{-1}$. The actual behavior of τ'_c shows that very close to threshold the simple power-law dependence on $\sigma - \sigma_t$ breaks down, and a rounding-off effect appears. A similar fact can be observed in the plot of the average intensity I (Fig.5). Both of these effects are due to the finiteness of the volume occupied by the laser mode; they are fully described by the theory.

Phase fluctuations show slowing down below threshold. Their correlation time would diverge in a linear theory, but the effects of the finiteness of the laser volume is to give a finite correlation time at threshold. Above threshold the correlation time for phase fluctuations increases asymptotically as the output power of the laser. This can be intuitively understood by considering Fig.6.

Few other aspects of the laser problem are worth mentioning:

a) Third-order intensity correlations $G_3(\tau_1, \tau_2)$ have been measured and successfully compared with those calculated by using the Markovian laser model[13]. As we said at the beginning of these lectures, G_3 is calculated as an average over the joint probability W_3. For a Markovian system, W_3 can be derived from W_2 (W_2 is obtained from the Fokker-Planck equation), as shown by Eq.(4).

b) The theory was extended to the case of the laser with injected signal[12]. This signal generates an additional polarization S in the active medium. The quantity S is the variable conjugate to the order parameter E and therefore plays the same role as the external magnetic field in the ferromagnet. With $S \neq 0$, the laser transition is first-order. Note, however, that there are no metastable states because the laser field is a two-dimensional variable (amplitude and phase), whereas the feedback mechanism stabilizes only the amplitude[14].

$\delta\varphi$: phase shift due to single spontaneous emission event

$$\delta\varphi \div \frac{1}{E_o} \div \frac{1}{P_{out}^{1/2}}$$

Random walk on the circumference:

$$\langle\Delta\varphi^2\rangle = N \langle\delta\varphi^2\rangle \div \sigma t \frac{1}{P_{out}}$$

Coherence time $\tau_c \longrightarrow \langle\Delta\varphi^2\rangle^{1/2} \sim \pi \longrightarrow \tau_c \div \frac{P_{out}}{\sigma}$

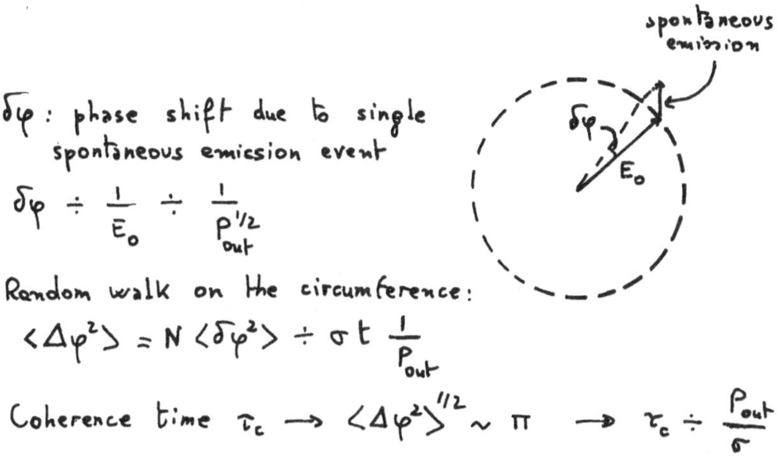

Fig. 6   Phase diffusion in the laser above threshold.

c) Very interesting experiments have recently been performed by Mandel and coworkers[15] on a two-mode laser. The fields E_1 and E_2 of the modes behave as mutually compled random processes governed by nonlinear Langevin equations. The joint probability density $W(E_1,E_2,t)$ obeys a four-dimensional Fokker-Planck equation. I cannot discuss here all the experimental and theoretical results. I will only mention that the potential relative to one mode exhibits two dips corresponding to two metastable states, so that this system allows both the calculation and the measurement of the probability distribution of the first-passage times between bistable states under the influence of quantum fluctuations.

THE TRANSITION TO CHAOS IN OPTICS

I will shortly discuss in this chapter two recent experiments which show limit cycles and chaotic behavior in optical systems driven by a stationary pump.

It is known[16] that the single-mode laser equations for a homogeneously broadened transition are the same as the equations of the Lorenz model for the Rayleigh-Bénard instability (and also for a similar model concerning the Soret-driven instability in a two-component liquid layer[17]). Recently, Casperson[18] considered a laser with a inhomogeneously broadened transition. He was able to show by a numerical study that also in this case there is a critical value of the pump parameter above which the time-independent steady-state solution is unstable and the laser output shows undamped intensity pulsations. The experimental requirements to be met in order to see the instability are high gain and long atomic relaxation times (longer than the photon lifetime in the laser cavity). Casperson[18] performed the experiment with a single-mode Xenon laser, operating at a wavelength $\lambda = 3.51\,\mu m$, which exhibits high gain ($g_o \sim 20\ m^{-1}$) and atomic relaxation times ($T_1 \simeq T_2 \simeq 10^{-6}$ s) much longer than the photon decay time ($\gamma^{-1} \simeq 10^{-9}$ s). Let me recall, incidentally, that γT_2 plays here the role of the Prandtl number. The results, shown in Fig.7, present also some evidence of a period doubling bifurcation.

The second experiment was performed by Gibbs et al.[19] with an optically bistable device, i.e. a device which exhibits two distinct states of optical trasmission[20]. Recently Ikeda et al.[21] have pointed out that ring-cavity bistable-devices are described by difference equations and may show instabilities with periodic and chaotic dynamics. A simplified scheme of the apparatus is shown in Fig.8. A He-Ne laser beam passes through an electrooptic modulator and a polarizer. The transmitted light is detected by a photodiode whose output signal is delayed and then fed back to the electrodes of the modulator crystal. When the composite response time of modular and detector is much shorter than the delay time t_R in the feedback line, the phase γ of the transmitted optical field satisfies approximately the difference equation

$$\phi(t) = \alpha\, P_{in} \left\{ \frac{1+\beta\cos\left[\phi(t-t_R)\right]}{2} \right\}$$

where α and β are constants. In agreement with the predictions of Ref. 21, Gibbs et al.[19] have found, by increasing P_{in}, periodic and chaotic outputs.

INSTABILITY TRANSIENTS

Thermodynamic systems and stationary open systems display fluctuations of macroscopic size only near critical points. The situation may be drastically different if one considers, instead of a stationary system, a system which is suddenly brought to an unstable state and is let to evolve toward an equilibrium state or a steady-state. Particularly interesting is the case of an unstable state characterized by the absence of systematic forces, in which

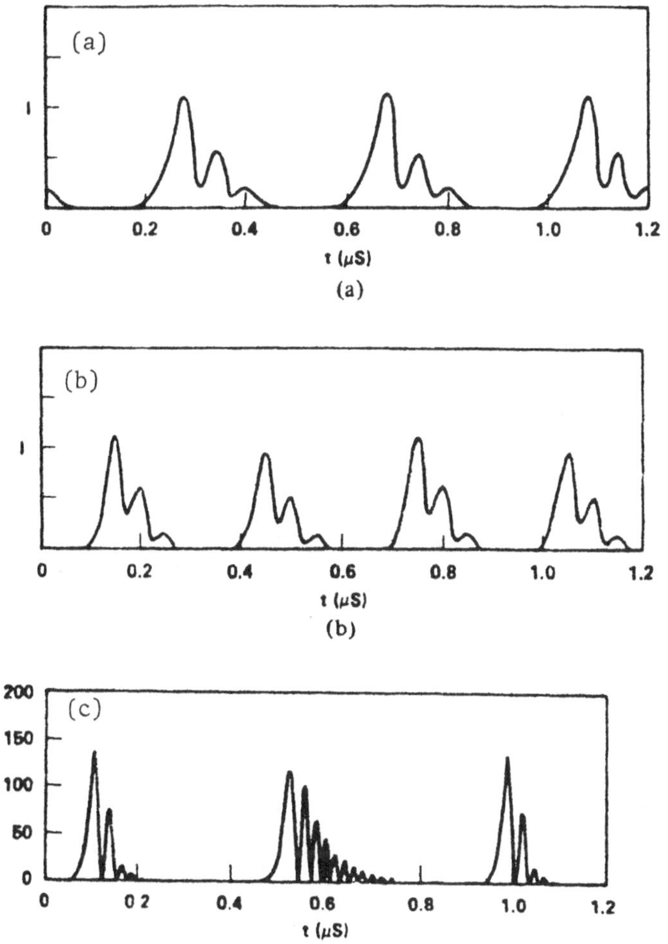

Fig. 7 Plots of the pulsation instability for a single-mode laser. Curves a) and b) are experimental data with discharge currents 40 and 50 mA respectively. Curve c) is a theoretical result. The plots are taken from Ref. 18.

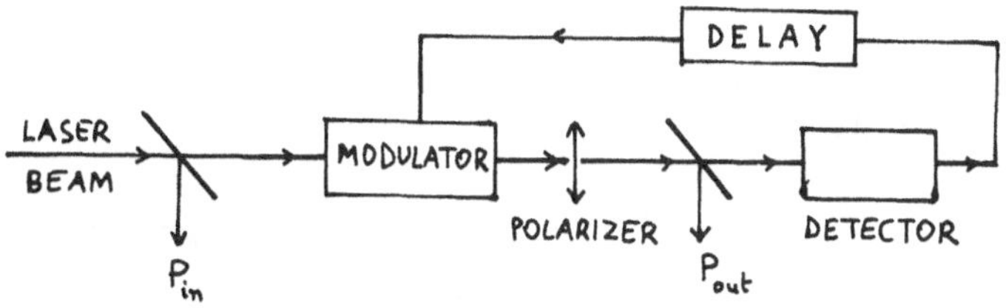

Fig. 8 Scheme of the experiment on chaotic behavior of an optically bistable system (see Ref. 19).

case the decay of the system must be initiated by a fluctuation. Once the observable deviates from its unstable value an amplification process sets in which drives the system toward the ordered stable state. Since fluctuations in the initial state are large (typically, they follow a Gaussian statistics) and the amplification process can be initially described as a linear process, the fluctuations during the transient evolution grow considerably (amplified Gaussian noise) before saturation effects enter into the process and stabilize the system at the final low-noise ordered state. Physical examples of such a process are the switch-on of a laer, the cooperative decay of excited atoms (superfluorescence), spinodal decomposition, the build-up of a steady convection pattern in a Rayleigh-Bénard cell above threshold, and so on.

Detailed experiments were performed on the transient statistics of a single-mode laser (a Q-switched He-Ne laser)[22]. Fig.9 shows an example of the evolution of the photon count probability distribution from the initial Bose-Einstein to the final Poisson distribution. By calculating the evolution of the first moment $<n>$ and of the variance $<\Delta n^2>$, one can see that the variance presents a huge peak approximately in correspondence with the inflexion point of the $<n>$ curve.

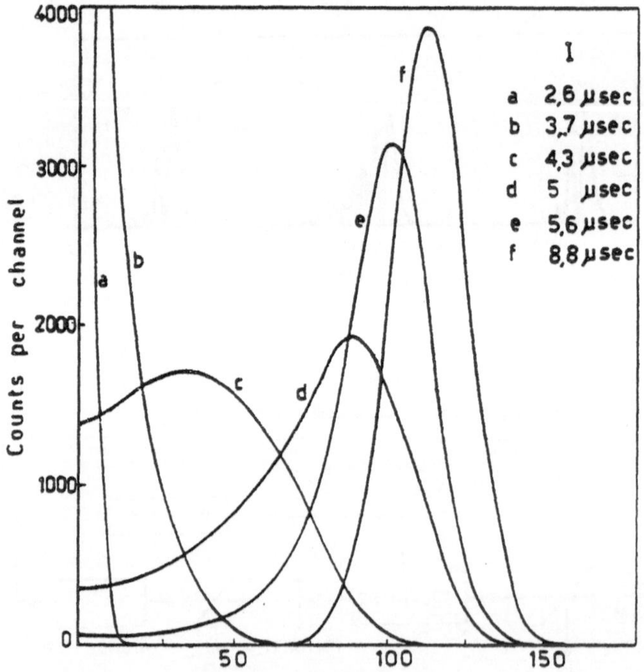

Fig. 9. Evolution of the photocount distribution during the transient of a Q-switched He-Ne laser (see Ref. 22).

The experimental results are well described by the quantum statistical laser theory[10] which, however, does not give analytical results for the transient statistics and, furthermore, does not allow an intuitive picture of the process. An important clue toward a simpler theoretical approach was given by the experimental observation that, far from threshold, all single transients have approximately the same shape and the main statistical effect is a random time-jitter of the leading edge of the transient with respect to the switching time. This observation suggested to treat the problem with the following assumption: a single transient can be represented as a deterministic evolution from a statistically defined initial condition[22].

In the case of the laser transient the quantity which is experimentally determined is the intensity $I=|E|^2$. The dynamic equation for I is easily derived from Eq.(23):

$$I = \alpha I - 2\beta I^2 \tag{30}$$

where $\alpha=2(g_o-\gamma)$. By assuming an initial intensity I_o at t=0, the integration of Eq.(30) gives

$$I(t,I_o) = \frac{I_o I_S}{I_S e^{-\alpha t} + I_o \; 1-e^{-\alpha t}} \tag{31}$$

where $I_S=\alpha/2\beta$ is the asymptotic value of I for long times t. If the initial macroscopic state is well below threshold, the field E_0 is a Gaussian variable, and therefore I_o is a statistical variable following an exponential distribution function

$$p(I_o) = (1/\bar{I}_o)\, e^{-I_o/\bar{I}_o} \tag{32}$$

Note that this distribution represents the classical limit, for high average photon numbers, of the Bose-Einstein distribution.

The time-dependent k-th moment of the intensity distribution is therefore given by

$$\langle I^k(t)\rangle = \int_0^\infty I^k(t,I_o)p(I_o)dI_o \tag{33}$$

It is straightforward to derive analytical expressions for $\langle I\rangle$ and $\langle I^2\rangle$, for instance. I only recall here that the relative variance $v=\langle\Delta I^2\rangle/\langle I\rangle^2$ can be expressed as an universal function of the scaled variable z, defined as

$$z = \frac{I_S\, e^{-\alpha t}}{I_o\left[1-e^{-\alpha t}\right]} \tag{34}$$

The expression for v is

$$v = \frac{z(1-H(z)) - H^2(z)}{(1-H(z))^2}$$

where

$$H(z) = z\, e^z \int_z^{+\infty} \frac{e^{-y}}{y}\, dy$$

Several authors have subsequently produced more sophisticated treatments of the transient statistics[23]. When the final state is well above threshold, all the approaches converge to the simple result given above. Some of the proposed approaches are however able to describe as well the case of a final state close to threshold. There latter approaches are very interesting, but too complex to be summarized here in few words.

It is worth mentioning that the same assumption used for the laser transient can be used to describe the statistics of superfluorescence pulses[24]. Here again one finds that the time-dependent normalized moments of the pulse intensity can be expressed as universal functions of a scaled variable which represents a nonlinear transformation of the time variable.

It is a very general property of instabilities that the single-mode dynamics is described in the threshold region by an equation, like Eq. (23), containing a cubic nonlinearity. The treatment developed for the laser should therefore be applicable to the transient behavior of hydrodynamic instabilities. I will come back to this point in the next chapter.

The transient behavior of systems in thermal equilibrium (e.g., the so-called spinodal decomposition) is intrinsically multimode, which would require a generalization of the treatment outlined above. Some attempts in this direction have recently appeared[25].

FLUCTUATIONS IN HYDRODYNAMIC INSTABILITIES

Several theoretical discussions of intrinsic fluctuation effects near the threshold of the Rayleigh-Bénard instability (RBI) have appeared in the literature in the last few years.

In principle, the calculation procedure is simple. First of all, the set of nonlinear dynamic equations for the appropriate variables has to be written. Secondly, random forces are added to the deterministic equations. The random forces are Gaussian and are δ-correlated in space and time. Their mean square amplitude is determined from the fluctuation-dissipation theorem by assuming local thermodynamic equilibrium, as it was done in the simple example concerning Brownian motion. Finally, the conditional probability density is found by solving the multidimensional Fokker-Planck associated to the set of stochastic dynamic equations.

The first explicit results have been obtained by linearizing the hydrodynamic equations in the pretransitional region[26]. The mean square velocity fluctuation in the critical mode can be expressed as

$$<|\delta v|^2> \sim Q/\lambda$$

where Q is the mean square amplitude of the random force and λ is the relaxation rate of the linear fluctuations. In general Q contains the contribution of the random forces associated with both the momentum conservation and the energy conservation equations. For typical liquids the second contribution is smaller by six order of magnitudes with respect to the first one. When the random force is thermal, Q is given by[27]

$$Q = \frac{k_B T}{\nu^2 \rho d} \frac{P^2}{1+P} \left(\frac{3}{R_c}\right)^{\frac{1}{2}}$$

where $P=\nu/\chi$ is the Prandtl number, ν is the kinematic viscosity, χ the thermal diffusivity, R_c the critical Rayleigh number, ρ the density and d the gap between the plates. A general expression of λ which considers also off-critical modes is

$$\lambda \sim \varepsilon \left[1+ \xi^2 (k-k_c)^2\right]$$

where $\varepsilon=|R-R_c|/R_c$ represents the distance from threshold and $\xi \sim d\varepsilon^{-\frac{1}{2}}$ is the correlation length. The fact that λ is proportional to ε is referred to as critical slowing down.

A nonlinear statistical theory of the RBI was developed by Graham[28] by using a generalized thermodynamic potential similar to that used in the laser case. This theory is too complex to be synthetized here in few words. I only recall that rounding off effects are found very close to threshold, that is the transition is not sharp for a finite system. However, unlike the single-mode laser case, this region seems too narrow to be experimentally accessible.

No experimental observation of intrinsic fluctuations near the RBI threshold has been so far reported. Light-scattering experiments like those performed on critical fluids are not feasible in this case because only the fluctuations of wave vector k_c are here enhanced, and k_c is of the order of 1 cm^{-1}. Consequently the required scattering angle is exceedingly small. Lekkerkerker and Boon[29] have suggested that critical fluctuations may appreciably influence the Brownian motion of a particle suspended in the fluid, so that a light scattering experiment could a modification of the particle diffusion coefficient as R approaches R_c.

I have made recently a proposal based on transient rather than steady-state experiments[18]. Indeed, if the system is prepared in a state very close to threshold (R very close to R_c) and the heat current is changed at t = 0 in a step to a value above threshold, the convective flux will build up in the liquid layer following (not too far above threshold) the same dynamic equation written above for the laser case. The transient evolution contains a magnified information about the statistics of the initial state, as shown above, and can therefore be used, in principle, to study the critical fluctuations. This approach was recently tried experimentally by Ahlers et al.[27] who have measured the transient behavior of the convective heat flow following a jump in the heat current across a liquid helium layer heated from below. These authors have inserted a stochastic forcing field in the amplitude equation with the cubic nonlinearity and have attempted to fit the experimental curves. They have found, however, that a deterministic model with a constant forcing field yields a better fit than the stochastic model. Their conclusion is that some unavoidable imperfection of the system prevents the observation of intrinsic fluctuations.

REFERENCES

1. H. Haken, "Synergetics", Springer Verlag, Berlin (1980)
2. G. Nicolis and I. Prigogine,"Self Organization in Non Equilibrium Systems", Wiley-Interscience, New York (1977)
3. R.L. Stratonovich, "Topics in the Theory of Random Noise", Gordon and Breach, New York (1963) vol.1.
4. "Noise and Stochastic Processes", edited by N. Wax, Dover, New York (1964)
5. L.P. Kadanoff et al., Rev. Mod. Phys. 39: 395 (1967)
6. Besides the general reference 5, see: J. Als-Nielsen and R.J. Birgeneau, Am.J.Phys. 45: 554 (1977)
7. C. Ishimoto and T. Tanaka, Phys.Rev.Lett. 39: 474 (1977)
8. M. Corti and V. Degiorgio, Phys.Rev.Lett. 45: 1045 (1980)

9. V. Degiorgio and M. Giglio, unpublished.
10. M. Sargent, M.O. Scully and W. Lamb, "Laser Physics", Addison-Wesley, Reading (1974); see also Ref. 1 and the papers mentioned therein.
11. F.T. Arecchi and V. Degiorgio in: "Laser Handbook", edited by F.T. Arecchi and E.O. Schulz-Dubois, North-Holland, Amsterdam, vol. 1: 191 (1972). This review paper contains references about the experimental works on laser statistics by Armstrong and Smith, Freed and Haus, Mandel and coworkers, Pike and coworkers, and others.
12. V. Degiorgio and M.O. Scully, Phys.Rev. A 2: 1170 (1970); R. Graham and H. Haken, Z. Physik 237: 31 (1970)
13. M. Corti and V. Degiorgio, Phys.Rev. A 14: 1475 (1976)
14. L.A. Lugiato, Lett.Nuovo Cimento 23: 609 (1978)
15. R. Roy, R. Short, J. Durnin and L. Mandel, Phys.Rev.Lett. 45: 1486 (1980)
16. H. Haken, Phys.Lett. 53A: 77 (1977). The existence of an instability in the laser equations was discussed by H. Risken and K. Nummedal (J.Appl.Phys. 39: 4662 (1968)) independently from Lorenz work.
17. V. Degiorgio, Phys.Rev. A 20: 2193 (1979)
18. L.W. Casperson, IEEE J.Quantum Electron. 14: 756 (1978); Phys. Rev. A 23: 248 (1981)
19. H.M. Gibbs, F.A. Hopf, D.L. Kaplan and R.L. Shoemaker, Phys. Rev.Lett. 46: 474 (1981)
20. A complete set of references about optical bistability, including the works by Gibbs, McCall, Bonifacio, Lugiato and others, can be found in the March 1981 issue of the IEEE J. of Quantum Electron.
21. K. Ikeda, H. Daido and O. Akimoto, Phys.Rev.Lett. 45: 709 (1980)
22. F.T. Arecchi and V. Degiorgio, Phys. Rev. A 3: 1108 (1971)
23. F. Haake, Phys. Rev. Lett. 41: 1685 (1978); M. Suzuki, Phys. Lett. 67A: 339 (1978); F. De Pasquale and P. Tombesi, Phys. Lett. 72A: 7 (1979); F.T. Arecchi and A.Politi, Phys. Rev. Lett. 45: 1219 (1980); L.A.Pokrovsky, Physica 105A: 105 (1981)
24. V. Degiorgio, Opt.Commun. 2: 362 (1971). For the theory of superfluorescence, see R. Bonifacio, P. Schwendimann and F. Haake, Phys. Rev. A 4: 302 (1971)
25. K. Kawasaki, M.C. Yalabik and J.D. Gunton, Phys.Rev. A 17: 455 (1978)
26. V.M. Zaitsev and M.I. Shliomis, Sov.Phys. JETP 32: 866 (1971)
27. G. Ahlers, M.C. Cross, P.C. Hohenberg and S. Safran, preprint
28. R. Graham, Phys. Rev. A 10: 1762 (1974)
29. H.N. Lekkerkerker and J.P. Boon, in: "Fluctuations, Instabilities, and Phase Transitions" edited by T. Riste, Plenum, New York, p.205 (1975).

STEADY STATES, LIMIT CYCLES AND THE ONSET OF TURBULENCE.
A FEW MODEL CALCULATIONS AND EXERCISES.

Manuel G. Velarde

Departamento de Física Fundamental-UNED
Apartado Correos 50 487
Madrid -3 (Spain)

1. INTRODUCTION: FROM STEADY STATES TO CHAOS IN THE TIME EVOLUTION OF NON LINEAR SYSTEMS

1.1. BACKGROUND

In these *Notes* I restrict consideration to the study of some *outstanding aspects* of the *time evolution* of non-linear systems. The fascinating problems of onset of "coherent structures" and other aspects of real space turbulence and fully developped turbulence are completely left out here. My choice has been dictated by the seemingly fertile ideas put forward in the past few years, and the major importance that I think such ideas have,at present, for the understanding of key mechanisms in the onset of turbulence and chaos in nonlinear sciences/17,18,32,37,41,43,47,48/.

To provide (almost) selfcontained and constructive *Notes* I start discussing features concerning steady states,limit cycles and their stability (chapter 2) before embarking in the study of model examples (chapters 3 and 4) for the onset of turbulence,here restricted to "weak" turbulence and the appearance of (strange) non periodic attractors. Chapter 1 gives a reasonable background and a decent summary of Landau-Hopf ideas. Turbulence in experiments as in theory is *what it is or what we should dig out of model calculations and well controlled experiments*. However, a feature that has reached a fairly universal consensus is that all time correlations decay at infinity (t large). Another important property of systems capable of exhibiting chaotic or turbulent behavior is their sensitivity to changes in initial data, a feature related to instability in phase space and the appearance of "strange" attractors in flows with negative divergence (with,however, expanding directions). Moreover, according to experiments turbulent or chaotic states clearly show sensitive dependence on the preparation of the system. Selection of a state of chaos seems to be externally controllable. The symmetry of the ensemble of available possibilities can be broken by the preparation of the system. Thus asking questions about routes to chaos although might ultimately not be the major problem in the study of turbulence appears, however, as an interesting line of thought. Whether a system evolves or is driven smoothly or suddenly to chaos seems to be a fascinating problem.

The field is so vast that I must cut a few trees and forget about the forest, with the hope of cutting relevant trees. Moreover, I concentrate in model problems and sketchy caricatures of the original Navier-Stokes equations. Whether or not the results found

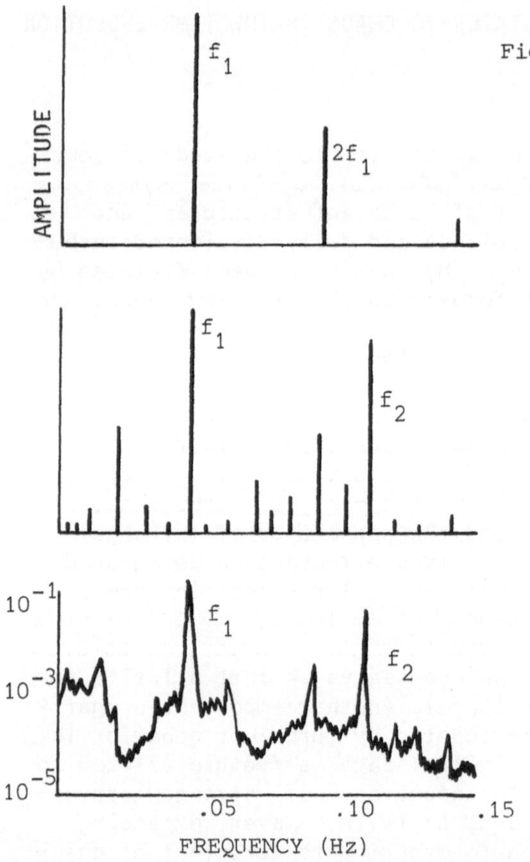

Fig. 1. Power spectra in Rayleigh-Bénard convection(sketch) (a)time-periodic(limit cycle) state. (b)quasiperiodic flow(f_1, f_2,...are not rationally related) (c)'chaotic'state.There is broadband continuous spectrum,with f=0 included. Taken from /16/.

Amplitude(db,log scale) vs frequency(Hz)

with such models or caricatures are relevant to experiment is at present an irrelevant point. If it turns out that more realistic models share features with the models I deal here with, then my *Notes* would have achieved the least modest aim of providing through a *PEDESTRIAN APPROACH* an understanding of complicated dynamics in nonlinear systems.

1.2. LANDAU PICTURE OF THE ONSET OF (QUASIPERIODIC GAUSSIAN) TURBULENCE

Consider the Rayleigh-Bénard problem or the Taylor-Couette flow,... in the case of *time-independent* boundary conditions.For low enough values of the external constraint(thermal gradient,differential rotation,etc.) the system is driven to a steady state. If V denotes the amplitude of the flow velocity,say,we have

$$V = V_0 \ (X,R) \tag{1}$$

where X denotes some space variable and R is a dimensionless measure of the constraint (the Rayleigh number, etc).

Upon increasing the value of the constraint the steady state V_0 might eventually become unstable. Let us denote by $\delta\ V_0(X,t,R)$ some time-dependent, albeit small disturbances upon the steady value V_0 . If V_0 is unstable at $R \geqslant R_1$ we have

$$V = V_0 \ (X,R) + \delta V_0(X,R,t) \tag{2.a}$$

where to a first approximation

$$\delta V_0 \sim V_\sigma \ (X,R) \ \exp\ (\sigma,t) \tag{2.b}$$

Here σ is a complex quantity whose real part determines the instability of V_0 . If $\mathrm{Re}\ \sigma = \gamma > 0$ then the steady state will be *unstable* (a small disturbance analysis,*i.e.*, a linear or local stability analysis merely gives a sufficient condition for instability).

Assume that $\mathrm{Im}\ \sigma \equiv \omega_1 \neq 0$ at $R \geqslant R_1$. Then according to the Hopf bifurcation theorem we expect a $2\pi/\omega_1$ -periodic(limit cycle) at the onset of instability with corrections to ω_1 at $R > R_1$.

Further increase of R past R_1 would eventually destabilize the sustained (nonlinear) limit cycle. Let R_2 be the point of instability of this first limit cycle, and let ω_2 be the frequency of disturbances on the cycle that will grow at $R \geqslant R_2$. Note,however, that a small disturbance analysis like the one sketched before does not permit the stability analysis of ω_1. It is intuitively appealing to proceed with small disturbances upon the ω_1-limit cycle but there are counterexamples to such intuitive idea. This is a point where Landau was unfortunalety unaware of the relevant mathematical theory(see/2,3,21/and sections 2.3,4.2.1) .Eigenfrecuencies of the (time-dependent) linearized operator in real space for disturbances upon time-dependent solutions simply do not define stability of the limit cycle in contradistinction with the case of disturbances upon a steady state!

Then if a new time-periodic solution appears at $R \geqslant R_2$ we have

$$V(X,R,t) \sim \phi(\omega_1,\omega_2,R), \quad R \gtrsim R_2 \qquad (3)$$

(ω_2 is not rationally related to ω_1)

Thus upon further increase of R, Landau conjectured that asymptotically at large values of the constraint ($R \to \infty$) the resulting flow would depend on a collection of frequencies

$$V(X,R,t) \underset{R \to \infty}{\sim} \phi(\omega_1,\omega_2,\ldots,\omega_n) \underset{n \to \infty}{} \qquad (4)$$

where the ω_i are *not* rationally related to each other ! Then ϕ defines a quasiperiodic function.

Note that at R_2 we have ω_1 and ω_2, two frequencies with relative phases *not* determined. These phases depend-in principle-on conditions at the time that the system reaches $R = R_2$ but precise control of these conditions seems beyond control in an experiment !

Note also that the Landau (turbulent) flow is *not* sensitive to change in initial data. Two trajectories which start close remain close in the course of time. A change in initial data brings the system from, say $(\omega_1 t, \omega_2 t, .., \omega_n t)$ to $(\omega_1 t + \alpha_1, \omega_2 t + \alpha_2, .., \omega_n t + \alpha_n)$ for α_n small. The Landau (quasiperiodic) flow yields sharp peaks at ω_1, ω_2,... and their harmonics, while the spectrum of a system which is sensitive to initial conditions yields a broadband 'flat' (noiselike) spectrum. Note, however, that if the frequencies are sufficiently close it may be possible to interpret the spectrum as that of a broadband chaotic flow. But this is a problem of instrumental resolution. A quasiperiodic flow has quasiperiodic (time) correlations and quasiperiodic motions do not phase mix ! All frequencies in the Landau picture *are in different directions* of the phase space of the system and do not *therefore* superpose incoherently. Presently available data show (almost) unambiguously the decay of correlations and broadband flat spectrum in flows accepted as turbulent. Figure 1 illustrates this point. It has been taken from reference /16/. Here lays an upsetting inadequay of Landau's conjecture.

Landau's turbulence leads to gaussian statistics and people say is inadequate to account for experimental data. Thus the usual dogma in the physicists' community that chaos or turbulence arises either from interaction of an infinite number of degrees of freedom or from an external noise, chosen gaussian as the randomness could be related to the infinite number of degrees of freedom via the central limit theorem, is just over !

N.B.: DEFINITION

$f(X)$, $-\infty < X < \infty$, is a quasiperiodic (almost periodic) function if for each $\varepsilon > 0$ there exists $\delta(\varepsilon)$ such that each interval of length $\delta(t)$ contains at least *one* number T for which

$$|f(X + T) - f(X)| < \varepsilon$$

If $\varepsilon = 0$, $f(X)$ is T-periodic. Note that for small values of ε it may be necessary to make $\delta(t)$ very large !

Although Landau's theory is mathematically incorrect for the stability analysis of time-dependent (ω_i) solutions we have seen that its major failure-at present-comes from the inadequacy to account for the available experimental properties of turbulent flows. Better suited alternative theories have been put forward in the past few years. A sketchy account of some of them follows.

i. Landau-Hopf:

S P QP

R --> infinity.

Following steady states (S), periodic (limit cycle)states (P) and quasiperiodic states (QP), "chaos" or "turbulence" is expected at $R \to \infty$ with quasiperiodic time-correlation functions.

ii. Lorenz:

S "chaos"

"Chaos" may arise through inverted bifurcation from steady states at $R < R_\infty$ in the form of "strange"attractors.Time-correlation functions decay.

iii. Ruelle-Takens-Newhouse-Arnold:

S P1 P2 P3+"chaos"

Following steady motion and three successive (limit cycle) Hopf bifurcations, there is a strange attractor. If after two frequencies there appears a third one, then the latter is necessarily accompanied of broadband noise.

iv. May-Feigenbaum-Collet-Eckmann-Koch:

S P 2P 4P ... "chaos"

A period-doubling cascade leads to chaos at $R < \infty$. Curiously enough the state of chaos has a high degree of "disorganization".

v. Pomeau:

S ... P "intermittency + "chaos"

Destabilization of a limit cycle may end in "intermittency" and chaos at $R < \infty$

A description, with exercises, of items ii, iv and v form the major part of these *Notes*.

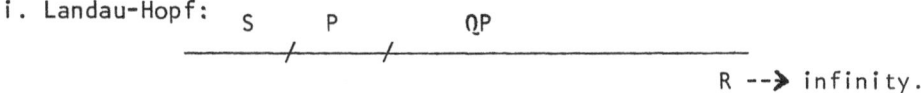

1.3. LORENZ MODEL, DISCRETE MAPS AND RELATED MATTERS

1.3.1. A point of philosophy

The Navier-Stokes equations are rather intractable. Moreover, without arguing about their relevance to physics and more specifically to the study of turbulence I ought to confess we can forget about them here. One of the reasons to support this áttitude is that drastic truncated modal approximations to the Navier-Stokes equations are providing quite a fertile field of research in fluid physics and mathematical physics even if correlation of theory with experimental data is far from acceptable. On the other hand mathematical life and phenomenological arguments in phase space of dimension three seem to have the essential features of higher dimensional problems provided the variables are nonlinearly coupled.This feature became clear when the inadequacy of Landau's theory was accepted as a fact and some rather abstract mathematical ideas reached popularity in the physics community thanks to the work of Lorenz, Arnold, Ruelle and others.

Yet, for specific results, truncated modal approximations to the Navier-Stokes equations demand the use of large computers and a lot of c.p.u. time. Discrete maps, like the 2d Poincaré map for a 3d Lorenz model have the nice feature of permitting a long way in the study of the dynamics of nonlinear systems with the use of commonly available progammable pocket calculators. Discrete maps share common features with truncated modal approximations to the Navier-Stokes equations themselves. An optimistic view of our disregard of the latter is that discrete maps plus Lorenz-like models plus computer analysis plus an updated knowledge of experimental results put all together add up to the Navier-Stokes equations Hopefully all information brought by this approach could guide our fellow mathematicians in their rigorous study of these or any other related equations.

Note also that a discrete map is what naturally comes from data output if one follows the data in the sequential way as it comes from the experiment. Take the Nusselt number (convected heat flux) or the measured values of external constraints (thermal gradient in Rayleigh-Bénard convection) as time proceeds in steps. A discrete map is trivially generated: $N(t_1),N(t_2),\ldots$ Whether or not a stroboscopic approach to the sequence brings useful information is a matter of fortune. It just happens that some experiments about onset of turbulence have indeed shown striking properties, like period-doubling cascades as predicted for simple one-or two-dimensional maps. Until very recently all data on Nusselt numbers and other quantities have traditionally been described by continuous curves which interpolate and extrapolate data from experiment. I shall not, however, give any more argument in favor of discrete maps. Their interest will be apparent in the course of my lectures.

1.3.2. Lorenz Model /26/

Consider the Rayleigh-Bénard problem .For the readers of these Proceedings Prof.Palm has given all the necessary background (see also/38, 49/). At a given value of the Rayleigh number, R = Rc, steady convection develops in the form of two-dimensional (X,Z) rolls with wave-number k = kc. As the flow is 2d we can introduce a velocity potential (stream function) $U = \partial\Psi / \partial Z$, $W = -\partial\Psi/\partial X$ where (U,O,W) are the three components of the flow velocity. The convective states can be analyzed in the form of a series expansion

$$\Psi(X,Z,t) = \sum_{m=0}^{\infty} \sum_{n=1}^{\infty} \Psi_{mn} \sin(\pi k m X) \sin(\pi n Z) \quad (1)$$

$$\Theta(X,Z,t) = \sum_{m=0}^{\infty} \sum_{n=1}^{\infty} \Theta_{mn} \cos(\pi k m X) \sin(\pi n Z) \quad (2)$$

where Θ accounts for the temperature field across the layer. We disregard here the boundary conditions. (1) and (2) correspond to stress-free conducting boundaries, the simplest mathematical case.

To a first approximation we have

$$\Psi(X,Z,t) = A_1 \sin \pi k X \sin \pi Z \quad (3)$$

$$\Theta(X,Z,t) = A_3 \cos \pi k X \sin \pi Z + A_2 \sin 2 \pi Z \quad (4)$$

The second term in the r.h.s. of (4) corresponds to the distortion of the linear temperature profile in the motionless state prior to Rc. It comes from the $\underset{\sim}{v}.\mathrm{grad}\,\Theta$ term in the Navier-Stokes equations.

Thus the original Navier-Stokes equations, etc. which are partial differential equations bring a set of ordinary differential equations for the flow amplitudes A_i (i = 1,2,3). We have

$$dA_1/dt = - P[\, kRA_3 + \pi^3 B^2 A_1 \,] / \pi B \quad (5a)$$

$$dA_2/dt = \frac{1}{2} A_1 A_3 - 4\pi^2 A_2 \quad (5b)$$

$$dA_3/dt = -A_1 (A_2 + \pi k) - \pi^2 B A_3 \quad (5c)$$

where $B = 1 + k^2$, k = kc. $P = \nu/\kappa$ (Prandtl number) with ν and κ the kinematic viscosity and thermometric conductivity, respectively. $R = \alpha g\, \Delta\Theta\, d^3/\nu\kappa$ is the Rayleigh number with α, g and d the thermal expansion coefficient, the acceleration of the gravitational field and the vertical gap across the layer, respectively. The set of equations (5) define the Lorenz model for the time evolution of a horizontal single-component liquid layer heated from below just *in the neigborhood of Rc* ($R \gtrsim Rc$).

1.3.3. Two-component Lorenz model/51/

When the fluid layer contains a binary mixture where Fourier (heat), Fick(mass) and Soret (thermal diffusion) transport phenomena take place the natural extension is the two-component Lorenz model:

$$dA_1/dt = - P [kR(A_3 + SA_5) + \pi^3 B^2 A_1]/\pi B \tag{6.a}$$

$$dA_2/dt = \frac{1}{2} A_1 A_3 - 4\pi^2 A_2 \tag{6.b}$$

$$dA_3/dt = - A_1 (A_3 + \pi k) - \pi^2 B A_3 \tag{6.c}$$

$$dA_4/dt = \frac{1}{2} A_1 A_5 - 4\pi^2 r (A_4 - A_2) \tag{6.d}$$

$$dA_5/dt = - A_1 (A_4 + \pi k) - \pi^2 r B (A_5 - A_3) \tag{6.e}$$

where $r = D/\kappa$ is the (inverse) Lewis number, i.e., the ratio of mass to thermal diffusivity. $S = \alpha \Delta\Theta/\gamma \Delta N_1$ accounts for the Soret cross transport effect (mass transport induced by the imposed thermal gradient). γ is the expansion coefficient related to changes in the mass fraction of either of the two-components. A_4 and A_5 are the amplitudes of the impurity field N_1 in the binary mixture. By analogy with equation (4) we have.

$$N_1(X,Z,t) = A_5 \cos \pi k X \sin \pi Z + A_4 \sin 2\pi Z \tag{7}$$

As a matter of fact the set of equations (6), i.e., the two-component Lorenz model provides also the mathematical description of a laser with saturable absorber /1/

2. STEADY STATES, LIMIT CYCLES AND PHASE TRANSITION PICTURE

2.1. STEADY STATES AND THEIR STABILITY

2.1.1. Exercise: Rayleigh problem at large Prandtl number($P \gg 1$)/7/

Consider eqs (5) of preceeding Section. If $P \gg 1$ then A_1 is the quantity with fastest variation. Thus in the time scale over which $dA_1/dt = 0$, the other quantities A_2 and A_3 have appreciable time variation. Defining $\Delta_T = A_2 + \pi k$, $y_3{}^2 = A_1 A_3/2$ the problem reduces to

$$A_1 = - A_3 Rk/\pi^3 B^2 \tag{1}$$

$$\dot{\Delta}_T = y_3 - 4\pi^2 (\Delta_T - \pi k) \tag{2}$$

$$\dot{y}_3 = y_3 \{R \Delta_T - \pi^5 B^3/k \} k/\pi^3 B^2 \tag{3}$$

where a dot over a quantity denotes time-derivation. Loosely speaking, $A_1 A_3$, defines the convective heat flow (Nusselt number). We take $\varepsilon = (R-Rc)/Rc$ with $Rc = \pi^4 B^3/k^2$. Then eq. (3) becomes

$$\dot{y}_3 = y_3 \{ \Delta_T (\varepsilon + 1) - \pi k \} \pi B/k \tag{4}$$

The system (2) and (4), together with (1) constitutes an approximate description of the Lorenz model. Steady solutions are either the motionless state

$$y_3 = 0 \quad ; \Delta_T = 0 \tag{5}$$

or else a steady convective regime with wave number k,

$$y_3^* = -4\pi^3 k\ \varepsilon/(\varepsilon+1) \quad ; \Delta_T^* = \pi k/(\varepsilon+1) \tag{6}$$

To study the stability of these solutions and more specifically of the steady convective regime(6) we proceed to a disturbance analysis in the linearized approximation.We take

$$y_3 = y_3^* + Y_3 \quad \text{and} \quad \Delta_T = \Delta_T^* + \delta_T$$

Thus the linear evolution problem to study is

$$\dot{\delta}_T = y_3 - 4\pi^2\ \delta_T \tag{7}$$

$$\dot{Y}_3 = -\ 4\pi^4 B\varepsilon\delta_T \tag{8}$$

Solutions are sought as exponentially growing in time with time constant σ .We write

$$Y_3 = Y_{3_0} \exp \sigma t \quad , \quad \delta_T = \delta_{T_0} \exp \sigma t \tag{9}$$

The determinantal (characteristic)equation is

$$\sigma^2 + 4\pi^2\sigma + 4\pi^4 B\varepsilon = 0 \tag{10}$$

which leads to

$$\sigma = 2\pi^2(-1 \pm \sqrt{1-B\varepsilon}\ ) \tag{11}$$

It appears that for $\varepsilon > 1/B(B=3/2, k_c=\sqrt{2}/2)$ σ is complex with

negative real part whereas if $\varepsilon < 1/B$,σ is real and negative.In either case we have stability of the steady state in at least a small enough supercritical region where the three-mode scheme (1) is assumed to be a useful approximate description of Rayleigh convection.Thus we can safely say that the steady convective regime bifurcates from the motionless state with the right sign, i.e., bifurcates continuously as a stable solution.

2.1.2. Exercise: Two-component Rayleigh problem at vanishingly small Lewis number ($r << 1$) /7,50/

If we now consider the two-component Rayleigh problem at finite Prandtl number but vanishing small Lewis number ($r << 1$) the problem is then reduced to the following set of equations

$$dA_1/dt = - \{ RSkA_5 + \pi^3 B^2 A_1 \} \; P/ \; B\pi \qquad (1.a)$$

$$dA_4/dt = A_1 A_5/2 - 4\pi^2 r A_4 \qquad (1.b)$$

$$dA_5/dt = - A_1 (A_4 + \pi k) - \pi^2 B r A_5 \qquad (1.c)$$

If we restrict to liquid mixtures of large enough Prandtl number ($P >> 1$) we have

$$\dot{y} = y\{\Delta_s (\varepsilon+1) - \pi k\} \; r\pi B/k \qquad (2)$$

$$\dot{\Delta}_s = y - 4\pi^2 r(\Delta_s - \pi k) \qquad (3)$$

where we have introduced the following quantities

$$y = A_1 A_5/2, \; \Delta_s = A_4 + \pi \; k, \quad \tilde{R} = RS/r, \quad \tilde{R}_c = \pi^4 B^3/k^2$$

Loosely speaking $A_1 A_5$ denotes the convective mass flux.

Clearly, the system (2), (3) is exactly the same as in the preceeding section, up to a factor r. Thus we have the steady solution

$$y^* = -4\pi^3 k \; r/(\varepsilon+1); \quad \Delta_s^* = \pi \; k/(\varepsilon+1) \qquad (4)$$

The stability of this steady convective regime is determined by the solutions of the characteristic equation

$$\sigma^2 + 4\pi^2 r\sigma + 4\pi^4 B\varepsilon r = 0 \qquad (5)$$

We have

$$\sigma = \{ 2\pi^2(-1 \pm\sqrt{1-B\varepsilon/r}) \} r \tag{6}$$

to be compared with (11) in the preceeding section.

We see that for $\varepsilon > r/B$ ($B = 3/2$ for $k_c = \sqrt{2}/2$) the decay of disturbances upon the steady state (4) is oscillatory, whereas for $\varepsilon < r/B$ the decay is monotonic. In either case, there is at least a small neighbourhood of (4), around the transition point, where (4) is stable, *i.e.*, the convective state in the two-component problem (1) bifurcates with the right sign and should be experimentally observable. This is the case, for instance, when the Rayleigh number is positive (respectively negative) and the Soret separation is also positive (negative). See also /7,15,44,46/.

2.2. PHASE TRANSITION PICTURE: AN ILLUSTRATION OF PHASE COEXISTENCE TRICRITICAL POINTS, TRIPLE POINTS, ETC. IN CONVECTIVE INSTABILITY

In this section we present a straightforward extension of the phenomenological Landau picture of phase transitions to the description of convective states and their stability. We consider the two-component Lorez model when all time derivatives are set to zero. The probelm is reduced to a pictorial representation of a sixth-order polynomial in the order parameter, A_1, say. For convenience in the notation we write $V \equiv A_1$. Then we have the "potential" /4,23,52,54/

$$\Phi(V) = \Phi_0 + \frac{A}{2} V^2 + \frac{B}{4} V^4 + \frac{C}{6} V^6 \tag{1}$$

where Φ_0 is some constant that can be disregarded and A, B and C all depend on R, S, r, ... They do not depend, however, on the Prandtl number. Specifically

$$A = 64\pi^4 r^2 \, [27\pi^4/4 - R(1+S+S/r)]/3 \tag{2.a}$$

$$B = 16[27\pi^4(1+r^2)/4 - R]/9 \tag{2.b}$$

$$C = 1 \tag{2.c}$$

The extrema of the potential (1) are

$$V = 0 \text{ (always)} \tag{3.a}$$

$V = 0$ corresponds to the state of rest, i.e., the motionless (or disordered) state. We also have the following four solutions

$$V^2 = 8[\, R - Q(1+r^2)\,]\,/9 \pm 8\,[\,\{R-Q(1+r^2)\,]\,/9\}^2 - $$
$$-\pi^4 r^2 [\, Q - R(1+S+S/r)\,]\,/3]^{1/2} \qquad (3.b)$$

with $Q = 27\pi^4/4$.

We shall sequentially describe our findings according to the values of R and S at given values of r (see Figure 2). Later on we consider the three-dimensional plot {R,S,r} (see also /54/).

Region I of Fig.2 corresponds to the phase V = 0, i.e., to the rest. A typical plot of the potential (1) is illustrated in Figure 3. Thus the conductive phase is the only available state and it is stable. Conduction is the only heat transport at rest.

The phase V = 0 looses stability in Region II where a steady convective state, i.e., an ordered phase appears and it is stable.

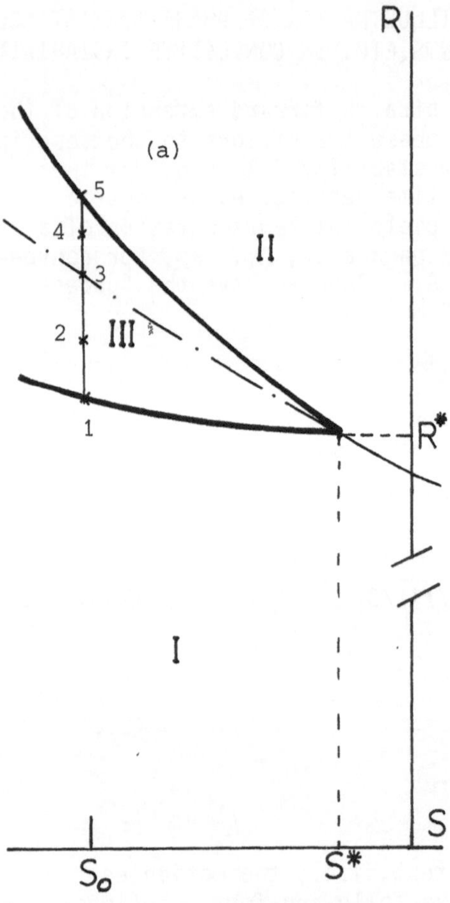

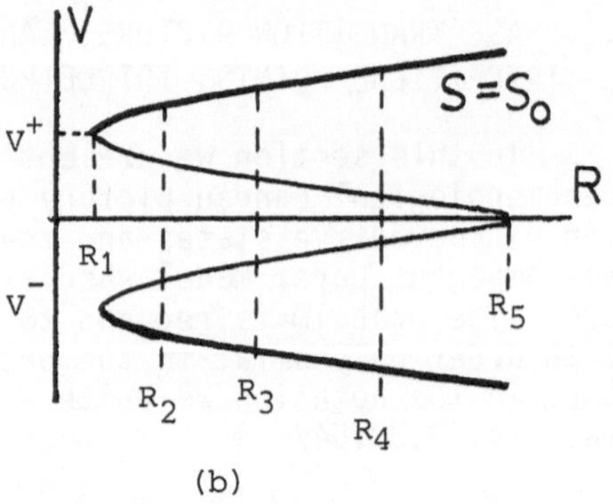

Fig. 2 . (a) R,S(S,negative) phase diagram displaying the region of nonlinear stability of the motionless state(I), of the steady convection(II) and of coexistence of the two (III). The dotted line is a line of 'triple' points.

(b) An inverted bifurcation showing the values of the order parameter as function of Rayleigh number in the coexistence region(III).

(R^*, S^*) is a 'tricritical' point at $r = 10^{-2}$.

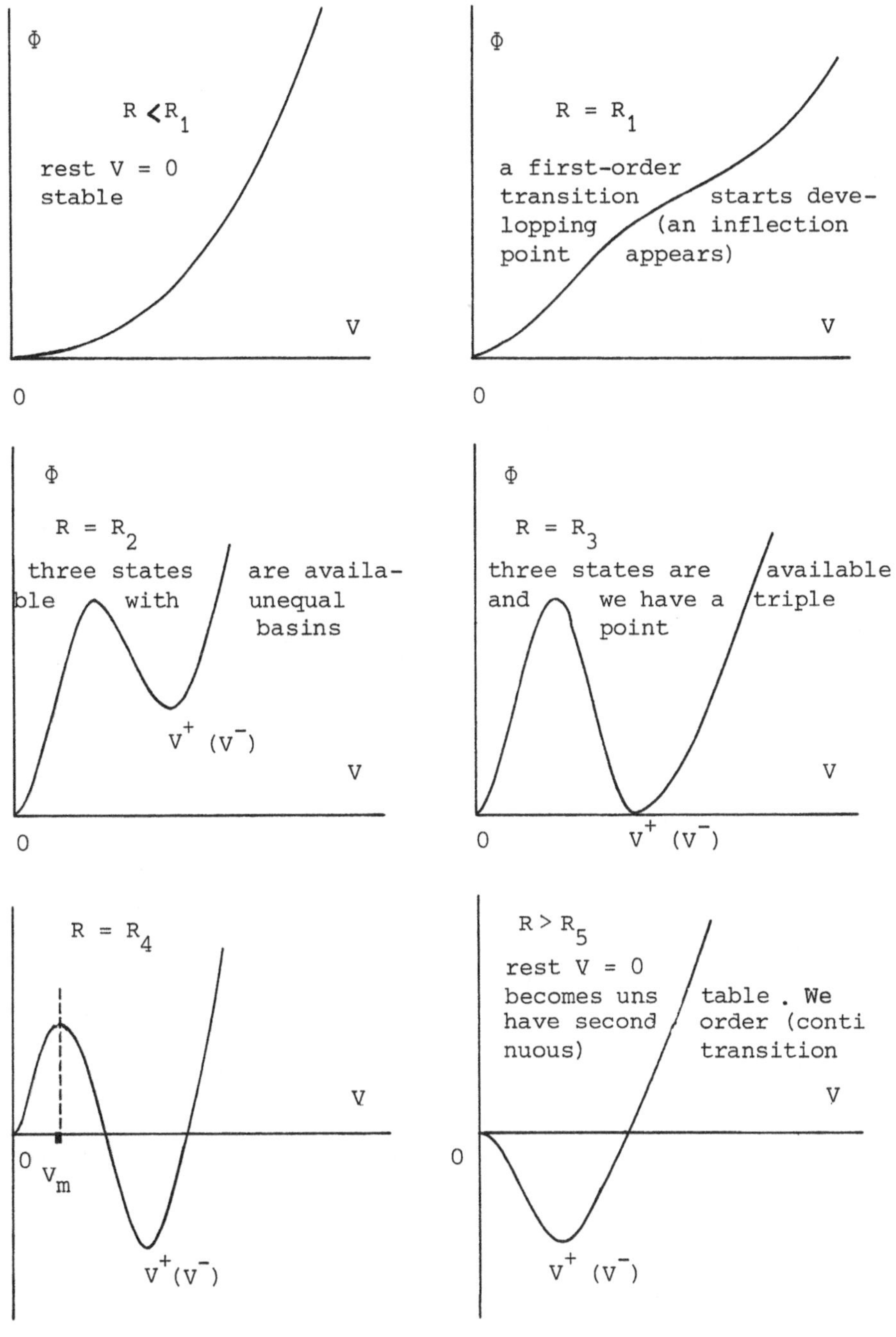

Fig. 3 . The available possibilities('*phases*') and the shape of the Landau "potential" for the two-component Lorenz model(sixth-order polynomial): $R_1 < R_2 < R_3 < R_4 < R_5$.

In region III five steady solutions are available. Their stability depends on the values of the parameters R and S (and eventually r). For illustrative purposes we discuss the expected phases at given S and varying R (see Figure 2). We take five different values of the Rayleigh number (R_i, i=1,...,5).

1. Along the vertical line $S = S_0$, for $R < R_1$ the square root in (3.b) has complex solutions only. At $R = R_1$ the square root vanishes and we have two nonvanishing, V^+ and V^-, solutions of $V^2 = 8[R - Q(1+r^2)]/9$. $V^{\pm}$ are on inflexion points of the potential (see Fig. 3). According to the notation introduced by Ginzburg *et al.* /14/ this point is to be considered as a point of a spinodal line. The location of R_1 is

$$R_1 = \frac{27\pi^4}{4}\{(1-r^2)-2r(1+r)S+2[r(1+r)S[r(1+r)S+r^2-1]]^{1/2}\} \quad (4)$$

which corresponds to the onset of finite amplitude instability as depicted in Fig. 2.b.

2. Slightly above R_1 we have five solutions (V = 0 and four real nonvanishing values of V). The corresponding potential possesses three minima and two maxima as illustrated in Fig. 3. In Figure 2.b the convective solutions are located along the vertical line $R = R_2$. Two convective steady states are stable (V^+) and the other two are unstable (V^-). The (+) and (-) correspond to the ± sign in equation (3.b) respectively.

3. At $R = R_3$ we have a *triple* point. Illustration of a triple point is given in Figure 3. (see also Fig. 2.b.). We have

$$A_{triple} = 3B^2/16C \text{ and}$$

$$R_3 = \frac{27\pi^4}{4}\{1 - \frac{5}{3}r^2 - \frac{8}{3}(1+r)rS +$$

$$+[[\frac{5}{3}r^2-1+\frac{8}{3}(1+r)rS]^2 - [1+r^4-\frac{10}{3}r^2]]^{1/2}\} \quad (5)$$

It should be noted that the line of triple points originates at the *tricritical* point (A = B = 0, C > 0)

$$R^* = 27\pi^4(1+r^2)/4 > 0 \quad (6.a)$$

$$S^* = -r^3/(1+r)(1+r^2) < 0 \quad (6.b)$$

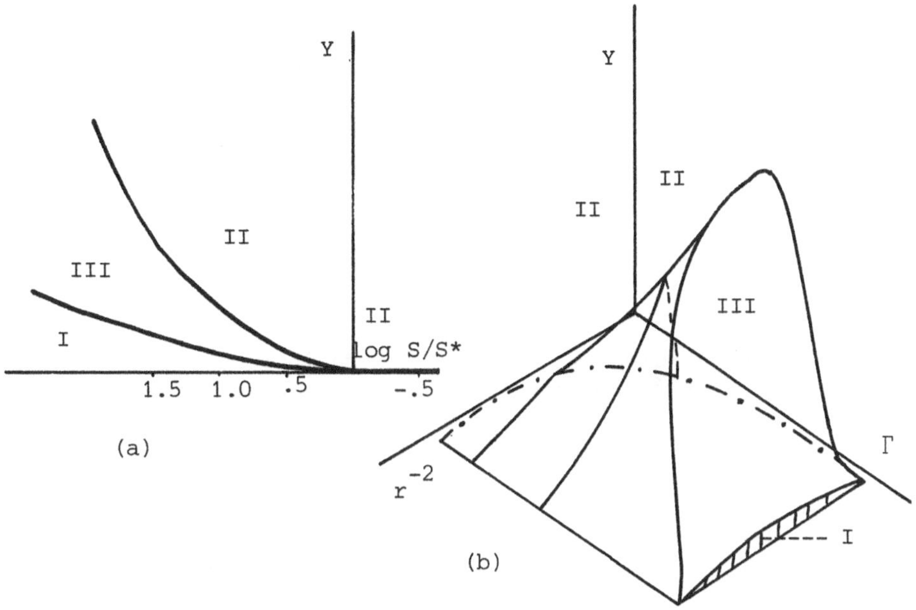

Fig. 4. 'Phase diagram' of two-component Rayleigh-Bénard convection and laser with absorber(units are arbitrary).
(a) Y-S plane(Bénard problem)according to definitions in the main text. Regions I and II contain all states of rest and steady convection,respectively. Region III is the region of 'coexistence' of 'phases', where limit cycle behavior appears.The origin is the 'tricritical' point.
(b) Three-dimensional 'thermodynamic' diagram,(Y,r,Γ).
The broken-dotted line is a line of 'tricritical'points.
Regions I,II and III contain the states decribed in (a).

Fig. 5. Another 'phase diagram' for the same problems described in Fig.4. Here we use the Prandtl(P) number and the Soret(S) separation factor in the two-component Lorenz model.Regions 1 and 3 contain steady states. Region 2 is a region of limit cycle behavior only.

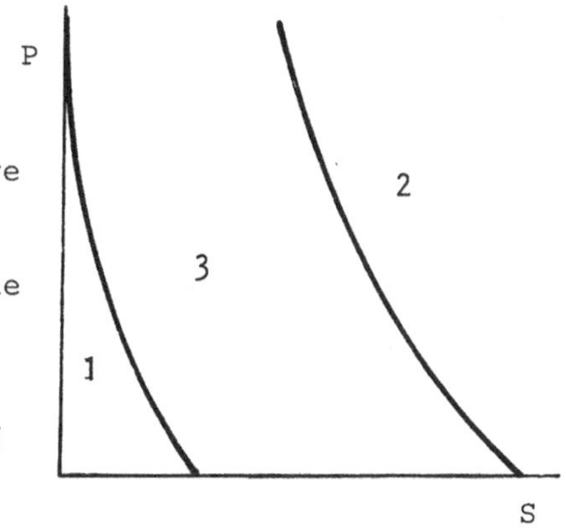

The form of the potential at the tricritical point is concave above with the flattenning corresponding to $A = B = 0$, and $C = 1 > 0$.
4. At $R=R_4$ five solutions are still available. Their relative stability is illustrated in Fig.3.. (see also fig.2 .b).
5. At $R=R_5$ we pass from a minimum at $V=0$ ($R \lesssim R_5$) to a maximum at $V=0$ ($R \gtrsim R_5$). The crossover appears as a consequence of the motion of V_m towards $V=0$ when R is varied. At $R=R_5$, V_m collapses with $V=0$. Although the motionless state is an available solution it loses stability at $R = R_5$. The only expected state is steady convection as illustrated in Fig.2b. In the notation of Ginzburg /14/ R_5 belongs to a spinodal line. We have $A=0$ and $\partial^2\Phi/\partial V^2=0$. Its ordinate is

$$R_5 = 27\pi^4/4(1+S+S/r) \tag{7}$$

It should be noted that if the line defined by equation (7) is not crossed [$S < -r/(1+r)$] the state of rest is always a minimum of the potential. Its stability depends on the Rayleigh number and it can be very weak (too shallow the basin of attraction).

If now we use $\Gamma=1-RSk^2/\pi^4 r(1-r)(1+k^2)^3$ and its conjugate $Y=(\partial\Phi/\partial C)_{R,r}$ a three-dimensional plot (Fig.4) provides a compact account of all above described "phases". Moreover, limit cycle behavior is predicted in the coexistence region as described below.

2.3. LIMIT CYCLES AND THEIR STABILITY

Consider again the two-component Lorenz model (see Introduction Section 1.3.3, Eqs (6)). Linearization of the equations is easily obtained. We just drop the A_iA_j terms. Assume now solutions of the form $A_i = A_{io}\exp(\sigma t)$. The linearized problem leads to the following algebraic system.

$$\begin{bmatrix} -\pi^2 BP-\sigma & 0 & -PRk/\pi B & 0 & -PRkS/\pi B \\ 0 & -4\pi^2-\sigma & 0 & 0 & 0 \\ -\pi k & 0 & -\pi^2 B-\sigma & 0 & 0 \\ 0 & 4\pi^2 r & 0 & -4\pi^2 r-\sigma & 0 \\ -\pi k & 0 & \pi^2 rB & 0 & -\pi^2 rB-\sigma \end{bmatrix} \begin{bmatrix} A_{1_0} \\ A_{2_0} \\ A_{3_0} \\ A_{4_0} \\ A_{5_0} \end{bmatrix} = 0 \tag{1}$$

Overstability (leading to Hopf bifucation) corresponds to finding solutions with $\mathrm{Re}\,\sigma = 0$, $\mathrm{Im}\,\sigma = \omega \neq 0$. From Eq.(1) we obtain

$$\omega^2 = \pi^4 B^2(rP + P + r) - PRk\ (S+1)/B \tag{2}$$

with

$$R = R_{0V} = (27\pi^4/4)\,[\ (P+Sc)(Sc+1)(P+1)/Sc^2\ (PS+P+1)] \tag{3}$$

where $S_C = P/r = \nu/D$ is the Schmidt number.

The (nonlinear)bifurcated time-periodic(Hopf) solution can be obtained by means of a double time scale.We pose $t = t(\tilde{t},\tau)$

$$\frac{d}{dt} = \frac{\partial}{\partial \tilde{t}} + \left(\sum_{i=1}^{\infty} \varepsilon^i R_i\right)\frac{\partial}{\partial \tau} \tag{4}$$

Where $t = t$, $\tau = (R - R_{oy})t$. We also pose

$$R = R_{oy} + \sum_{i=1}^{\infty} \varepsilon^i R_i \tag{5.a}$$

$$A(\tilde{t},\tau,\varepsilon) = \sum_{i=1}^{\infty} \varepsilon^{i+1} A^{(i)} \tag{5.b}$$

where A denotes a five-component vector

$$A^{(i)} = (x_i, y_i, z_i, u_i, v_i) \tag{5.c}$$

$A^{(o)}$ formally coincides with the vector $A_{i\,o}$ $(i = 1,..,5)$ introduced earlier .In the following we shall disregard the tilde on $\tilde{t}$.

Using (5) in the nonlinear problem this is reduced to a hierarchy of linear equations ordered according to powers in ε,

order ε

$$\frac{\partial}{\partial t}\begin{pmatrix} x_0 \\ y_o \\ z_o \\ u_0 \\ v_o \end{pmatrix} = \frac{\partial}{\partial t} A^{(o)} = L \begin{pmatrix} x_o \\ y_0 \\ z_o \\ u_o \\ v_0 \end{pmatrix} \tag{6}$$

where L coincides with the matrix operator in Eq.(1) with the only restriction that $R = R_{ov}$ and that σ does not appear.

order ε^2

$$\frac{\partial}{\partial t}\begin{pmatrix} x_1 \\ y_1 \\ z_1 \\ u_1 \\ v_1 \end{pmatrix} = \frac{\partial}{\partial t} A^{(1)} = LA^{(1)} + N_1[A^{(0)}] - R_1 \frac{\partial}{\partial \tau} A^{(0)} \quad (7)$$

with

$$N_1[A^{(0)}] = \begin{bmatrix} -P[R_1 k(z_0 + v_0 S)]/\pi B \\ x_0 z_0/2 \\ -x_0 y_0 \\ x_0 v_0/2 \\ -u_0 x_0 \end{bmatrix} \quad (8)$$

and so on.

The solution $A^{(0)}(t,\tau)$ can be written in the following form

$$A^{(0)}(t,\tau) = a(\tau)\,\phi(t) + \text{c.c.} = a(\tau)\, e^{i\omega t}\Phi + \text{c.c.} \quad (9)$$

with $\phi(t)$ is the eigenvector of $(i\omega I - L)$ where I is the identity. Φ is the corresponding eigenvector of L .We have

$$\Phi = \begin{bmatrix} -1/\pi k \\ 0 \\ (\pi^2 B + i\omega)^{-1} \\ 0 \\ [1 + \pi^2 rB/(\pi^2 B + i\omega)]/(\pi^2 rB + i\omega) \end{bmatrix} \quad (10)$$

Notice that all t (fast time scale) dependence is carried by the factor $\exp(i\omega t)$ while the amplitude $a(\tau)$ evolves on the slow time scale.

To proceed further we introduce a suitable inner product in the space of solutions.Let $f(t,\tau)$ and $g(t,\tau)$ be two such solutions.We define

$$< f(t,\tau) \mid g(t,\tau) > = \lim_{T \to \infty} T^{-1} \int_0^T [f^*(t,\tau)]^t g(t,\tau)\, dt \qquad (11)$$

where the supercript denotes transpose.
The adjoint problem to Eq.(7) has the following solutions

$$\psi(t) = \Psi e^{i\omega t} \qquad (12)$$

with

$$\Psi = \begin{pmatrix} 1 \\ 0 \\ -PR_{0v}k[\,1/\pi B + \pi rS/(-\pi^2 rB + i\omega)]/(-\pi^2 B + i\omega) \\ 0 \\ R_{\rho v}Ks/\pi B(-\pi^2 rB + i\omega) \end{pmatrix} \qquad (13)$$

The existence of a nontrivial solution of Eq.(8) demands the following condition (Fredholm's alternative)

$$\left< \psi \middle| N_1[A^{(0)}] - R_1 \frac{\partial}{\partial \tau} A^{(0)} \right> = 0 \qquad (14)$$

It follows that $R_1 = 0$. This result could have been guessed from the very beginning. The original nonlinear problem does not distinguish between the alternative ways of heating, from below or from above. Thus at a given location in the fluid layer, positive and negative values of A_1 (the convective velocity, say) are equally probable.

The solution of the Eq.(7) can be expressed as

$$A^{(1)} = a^2(\tau)\, A \exp(2i\omega t) + \text{c.c.} + B\, |a(\tau)|^2 \qquad (15)$$

with

$$A = \frac{1}{2}(2i\omega I - L)^{-1} \begin{pmatrix} 0 \\ \phi_1\phi_2 \\ 0 \\ \phi_1\phi_5 \\ 0 \end{pmatrix} \equiv A_i \qquad (16)$$

and

$$B = -L^{-1}\begin{pmatrix} 0 \\ \Phi_1 \mathrm{Re}\Phi_3 \\ 0 \\ \Phi_1 \mathrm{Re}\Phi_5 \\ 0 \end{pmatrix} \equiv B_i \tag{17}$$

To proceed further we ought to use the equation to order ε^3, and once more Fredholm's alternative. I sketch the procedure.

<u>order ε^3</u>

$$\frac{\partial}{\partial t}\begin{pmatrix} x_2 \\ y_2 \\ z_2 \\ u_2 \\ v_2 \end{pmatrix} = \frac{\partial}{\partial t} A^{(2)} = LA^{(2)} + N_2[A^{(0)}, A^{(1)}] - R_2 \frac{\partial}{\partial \tau} A^{(0)} \tag{18}$$

with

$$N_2[A^{(0)}, A^{(0)}] = \begin{pmatrix} -P\, R_2 k(z_0 + v_0 S) \\ (x_0 z_1 + x_1 z_0)/2 \\ -(x_0 y_1 + x_1 y_0) \\ (x_0 v_1 + x_1 v_0)/2 \\ -(u_0 x_1 + u_1 x_0) \end{pmatrix} \tag{19}$$

We now insert $A^{(0)}$ and $A^{(1)}$ in Eq (18). The terms in the r.h.s. are functions of $\exp(i\omega t)$. The expression for $N_2[A^{(0)}, A^{(1)}]$ is rather involved and I skip it. The other term is

$$-R_2 \frac{\partial}{\partial\tau} A^{(0)} = -R_2 \frac{\partial}{\partial\tau} a(\tau) \exp(i\omega t) \begin{bmatrix} \Phi_1 \\ \Phi_2 \\ \Phi_3 \\ \Phi_4 \\ \Phi_5 \end{bmatrix} + \text{c.c.} \tag{20}$$

Fredholm's alternative yields an evolution equation for the slow varying amplitude.We have

$$R_2 \left[\frac{d}{d\tau} < \psi|\Phi > + < \Psi |M| \Phi > \right] a(\tau) - \lambda a(\tau) |a(\tau)|^2 = 0 \tag{21}$$

with

$$M = \begin{bmatrix} 0 & 0 & kP/\pi B & 0 & kPS/\pi B \\ 0 & 0 & 0 & 0 & 0 \\ 0 & 0 & 0 & 0 & 0 \\ 0 & 0 & 0 & 0 & 0 \\ 0 & 0 & 0 & 0 & 0 \end{bmatrix} \tag{22}$$

and

$$\lambda a(\tau) | a(\tau)|^2 = < \Psi|N_2'[A^{(0)}, A^{(1)}] > \tag{23}$$

where

$$N'[A^{(0)}, A^{(1)}] = \begin{bmatrix} 0 \\ \frac{1}{2} (x_0 z_1 + x_1 z_0) \\ -(x_0 y_1 + x_1 y_0) \\ \frac{1}{2}(x_0 v_1 + x_1 v_0) \\ -(x_0 u_1 + x_1 u_0) \end{bmatrix} \tag{24}$$

Thus, we have

$$\frac{d}{d\tau} a(\tau) = \nu\, a(\tau) + \Lambda\, a(\tau)\, |a(\tau)|^2/R_2 \tag{25}$$

where

$$\Lambda = \lambda / < \Psi|\Phi > \tag{26}$$

and

$$\nu = -< \Psi|\, M\, |\, \Phi > / < \Psi|\Phi > \tag{27}$$

We can now write the amplitude as

$$a(\tau) \equiv \rho\,(\tau)\, \exp\, i\omega(\tau) \tag{28}$$

Then Eq.(25) yields

$$\frac{d}{d\tau}\rho(\tau) = \mathrm{Re}\,\nu\,\rho(\tau) + \mathrm{Re}\Lambda\,\rho^3(\tau)/R_2 \tag{29.a}$$

$$\rho\,\frac{d\,\omega}{d\,\tau} = \mathrm{Im}\,\nu\,\rho + \mathrm{Im}\,\Lambda\,\rho^3/R_2 \tag{29.b}$$

It follows

$$\rho(\tau) = \rho(o)\ \rho(\infty)\ e^{\tau \mathrm{Re}\nu}\,[\rho^2(\infty)+(e^{2\tau \mathrm{Re}\nu} - 1)\rho^2(o)]^{-1/2} \tag{30.a}$$

$$\omega(\tau) \underset{\tau\to\infty}{\sim} (\ \mathrm{Im}\,\nu - \mathrm{Im}\,\Lambda\,\mathrm{Re}\,\nu\ /\ \mathrm{Re}\,\Lambda\ )\ \tau$$

where as introduced earlier $\tau = (R - R_{0V})t$.

We know from linear stability analysis, Eq.(1), that for $R < R_{0V}$ the steady solution ($A_1 = 0$) is stable, i.e., $\rho(\tau)$ decays to zero as t goes to infinity. Thus stability of the motionless steady state demands

$$\mathrm{Re}\ \nu > 0 \tag{31}$$

which merely states the transversality condition in Hopf's bifurcation theory. The limit cycle and the state of rest are available but due to (31) the latter is the only stable phase. Note that

$$(R - R_{0V})\ \mathrm{Re}\ \nu \tag{32}$$

corresponds to the time constant for a disturbance to decay on the limit cycle.

Another way of writing Eq(29.a) is

$$\frac{d}{d\tau}\rho(\tau) = \mathrm{Re}\,\nu\,\rho(\tau)\,[\,1-(-\mathrm{Re}\,\Lambda/R_2\mathrm{Re}\nu)\rho^2] \tag{33}$$

and then

$$\rho(\infty) = (-\mathrm{Re}\,\Lambda/\,R_2\mathrm{Re}\,\nu)^{-1/2} \tag{34}$$

When $\mathrm{Re}\,\nu > 0$ and $\mathrm{Re}\,\Lambda < 0$ then $R_2 > 0$ and $R > R_{0V}$. Thus if $\mathrm{Re}\,\Lambda$ is negative the limit cycle $A(t)$ bifurcates supercritically, i.e., it is stable, whereas if $\mathrm{Re}\,\Lambda$ is positive the limit cycle bifurcates to the wrong side (subcritical) and it branches as an unstable solution. The amplitude and (angular) frequency of the limit cycle are respectively

$$A(t) = 2\,[\,-\mathrm{Re}\,\Lambda\,(R-R_{0V})/\mathrm{Re}\nu]^{1/2}\;\mathrm{Re}[\,\Phi\exp i\tilde{\omega}t\,] \;+$$

$$+\;[|R - R_{0V}|\mathrm{Re}\,\Lambda/\mathrm{Re}\,\nu\,]\{\,-L^{-1}\begin{pmatrix}0\\ \Phi_1\mathrm{Re}\,\Phi_3\\ 0\\ \Phi_0\mathrm{Re}\Phi_5\\ 0\end{pmatrix} \;+$$

$$+\;\mathrm{Re}[\,(2i\,\,I-L)^{-1}\begin{pmatrix}0\\ \Phi_1\Phi_3\\ 0\\ \Phi_1\Phi_5\\ 0\end{pmatrix}\exp i\tilde{\omega}t\,]\} \;+$$

$$\mathcal{O}(\,|R - R_{0V}|^{3/2}) \tag{35.a}$$

and

$$\tilde{\omega} = \omega + (R - R_{0V})(\mathrm{Im}\,\nu - \mathrm{Im}\,\Lambda\,\mathrm{Re}\,\nu\,/\,\mathrm{Re}\Lambda) \tag{35.b}$$

where we have for convenience of normalization set $R_2 = 1$. The quantities R_{0V} and ω belong to the linear stability analysis and they have been computed by several people(see,for instance, the review paper by Schechter *et al.*/44/). The only quantities that need to be evaluated are ν and Λ .This must be done with the computer and the actual numerical results are of no interest for my purposes here.Note that with this method we construct the limit cycle and delineate its stability. See also /1,42,44,52/.

3. STRANGE ATTRACTORS

3.1. FRACTAL DIMENSTION : AN ILLUSTRATION#

Consider the contour line of the square box in Fig.6 .Replace every side by a new segment following the construction sketched in Fig.7.After a large number of iterations we have Mandelbrot's snow flake.

Each iteration (figure 7) replaces a set, *i.e.*, the side or, more generally, a curd or a meridian intercept of the domain by N subsets (the offspring number, here N = 8) that are similar to the original in a known contraction ratio r < 1 (here r = 1/4).Then the Mandelbrot's fractal dimension of the transformed object is

$$D = \lg N / \lg (1/r) \qquad (1.a)$$

or for iterations with varying N_i and r_i

$$D = < \lg N_i > / < \lg(1/r_i) > \qquad (1.b)$$

Mandelbrot's snow flake has fractal dimension D = lg 8/lg 4 = 3/4 = = 1.5.This is Hausdorff's dimension which is, generally, larger than the topological dimension of the object /27/.

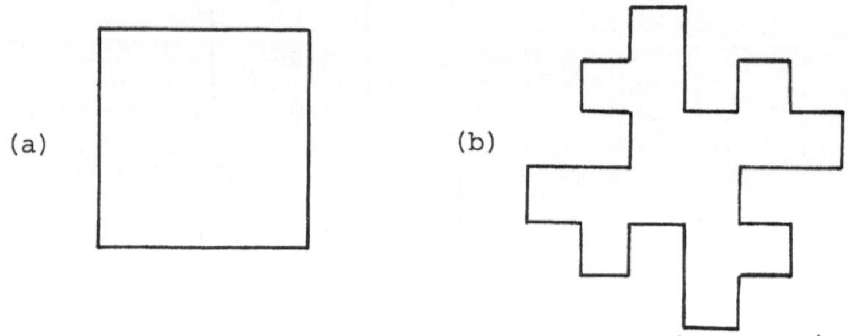

Fig.6. Mandelbrot's snowflake:(a)original set (the line),(b)transformed object (the tortuous line) after one step in the process.

Fig.7 Offspring number (N=8) and contraction ratio(inverse number of parts,4) in Mandelbrot's snow flake.(a)original set, (b) transformed object.

3.2. LYAPUNOV CHARACTERISTIC EXPONENTS / 34-36,39,45,55/

Consider (Figure 8) at $t = 0$ a basic(unit) cell(volume Ω_0) that after the nonlinear transformation F becomes at $t = \tau$, $F(\Omega_0)$. Let d be the phase space dimension (Lorenz model, d = 3; two-component Lorenz model, d = 5). Let m_- be the number of contracting sides (here $m_- = 2$). At time $t = \tau$, define a new *unit cell* Ω_0' of side

$$l_0'(\tau) = [\prod_{\alpha=1}^{m_-} F(l_0)]^{1/m_-} = [\frac{l_0}{2} \times \frac{l_0}{3}]^{1/2} = l_0/\sqrt{6} \quad (1)$$

After n iterations of the transformation we have at $t = n\tau$ a unit cell $\Omega_0^{(m)}$, of side

$$l_0^{(n)} = [\prod_{\alpha=1}^{m_-} F(l_0^{(n-1)})]^{1/m_-} \quad (2)$$

Note that $\Omega_0 \geqslant F(\Omega_0) \geqslant \Omega_0' \geqslant F(\Omega_0') \ldots$

The new *units* following the action of F are constructed using Schmidt orthogonalization procedure. There is no need of normalization to unit length although it is advisable in view of the computational scheme to be described further below. There is also the following sequence of volume ratios

$$\frac{F(\Omega_0)}{\Omega_0}, \frac{F(\Omega_0')}{\Omega_0'}, \ldots\ldots, \frac{F(\Omega_0^{(n)})}{\Omega_0^{(n)}}, \ldots$$

We now define a growth rate λ, $\frac{F(l_0)}{l_0} = e^{\lambda\tau}$, or more properly $F_\alpha(l_0)/(l_0) = e^{\lambda_\alpha \tau}$. The latter notation indicates that the defined growth rate is (e_α) direction-dependent. Thus

$$\lambda_\alpha(\tau) = \frac{1}{\tau} \lg [F_\alpha (l_0)/l_0] \quad (3)$$

where $\alpha = 1,2,3$ for the Lorenz model, and $\alpha = 1,2,\ldots,5$, for the two-component Lorenz model. As a matter of principle we are interested in the asymptotic values at $t \to \infty$ (or $n \to \infty$, for fixed τ). We define

$$\tilde{\lambda}_\alpha = \lim_{n \to \infty} \frac{1}{n} \sum_{i=1}^{n} \lambda_\alpha(n\tau) \quad (4)$$

which incorporates the successive $\tau, 2\tau, \ldots, n\tau, \ldots$ iterations.

In our case (figure 8).

$$\frac{F(\Omega_0)}{\Omega_0} = \frac{(l_0/2)(l_0/3)(2l_0)}{l_0^3} = 1/3 \quad (5.a)$$

or more generally

$$\frac{F(\Omega_0)}{\Omega_0} = \frac{\prod_{\alpha=1}^{d} \{F_\alpha(1_0)\}}{1_0^d} \tag{5.b}$$

$e_\alpha \equiv (e^1, e^2, e^3)$ in the Lorenz model. Thus

$$\frac{F(\Omega_0)}{\Omega_0} = \exp\left[\sum_{\alpha=1}^{d} \lambda_\alpha \tau\right] \tag{6}$$

is the contracting ratio of a d-volume at time $t = \tau$. We write $\sum_{\alpha=1}^{d} \lambda_\alpha = \lambda^{(d)}(\tau)$. Let m_+ and m_0 denote the number of expanding and invariant directions, respectively.Using equation (3) we now define

$$K^\pm(\tau) = \frac{1}{m_\pm} \sum \lambda_\alpha^\pm(\tau) \tag{7}$$

where the + and - superscripts in (7) have their obvious meaning. We have

$$m_+ K^+(\tau) + m_- K^-(\tau) = \sum_{\alpha=1}^{d} \lambda_\alpha(\tau) = \lambda^{(d)}(\tau) \tag{8}$$

where no superscripts + or - are on λ. Note that if a direction is invariant,as $F_\alpha(1_0)/1_0=1$,we have $\lambda_\alpha(\tau)= 0$.

We can also define the following ratios

$$r'(\tau) = \frac{1_0'(\tau)}{1_0}, \quad r''(2\tau) = \frac{1_0''}{1_0'}, \; \ldots\ldots, r^{(n)} = \frac{1_0^{(n)}}{1_0^{(n-1)}} \tag{9}$$

where $r^{(n)}$ is at time instant $t = n\tau$. We have

$$r^{(n)}(\tau) = \left[\prod_{\alpha=1}^{m_-} F(1_0^{(n-1)})\right]^{1/m_-} / \, 1_0^{(n-1)} \tag{10}$$

Note that $m_- \leqslant 3$, or $m_- \leqslant 5$ in the Lorenz models here considered.

Thus

$$r^{(n)}(\tau) = e^{\tau K_-(\tau)} \tag{11}$$

On the other hand, after the <u>n</u> iterations

$$\frac{1_0^{(n)}}{1_0}(t = n\tau) = \frac{1_0^{(n)}}{1_0^{(n-1)}} \cdots\cdots \frac{1_0^{(1)}}{1_0} = r^{(n)} \cdots\cdots r^{(1)} \tag{12}$$

or else

$$\prod_{i=1}^{n} r^{(i)} = e^{n\tau K_-(\tau)} = r(T) \tag{13}$$

where now $T = n\tau$.

The ratio $F(\Omega_0)/\Omega_0' = (l_0^3/3)/(l_0/\sqrt{6})^3$ is the offspring or subsets number obtained from Ω_0 after the transformation. We call

$$N_i = F(\Omega_0)/\Omega_0' = \frac{F(\Omega_0)}{\Pi\, l_0^{(i)}} = \frac{e^{\tau\sum_{\alpha=1}^{d}\lambda_\alpha(\tau)}\,\Omega_0}{r^{(i)d}\,\Omega_0} = \frac{e^{\lambda^{(d)}(\tau)}}{r^{(i)d}} \qquad (14)$$

where the superscript i has the obvious interpretation.

Then, Mandelbrot's fractal dimension of the transformed object is (see Section 3.1)

$$D = \langle \lg N_i \rangle / \langle \lg 1/r_i \rangle \qquad (15.a)$$

Here

$$\mathrm{Lg}\, N_i = \tau\,[\lambda^{(d)}_{(\tau)} - d\, K_-(\tau)] \qquad (16.a)$$

$$\langle \lg N_i \rangle = \lim_{n\to\infty} \frac{1}{n}\, \tau\,[\lambda^{(d)}_{(\tau)} - dK_-(\tau)] =$$

$$= \tau[\langle\lambda^{(d)}\rangle - d\langle K_-\rangle] \qquad (16.b)$$

and

$$\langle \lambda^{(d)} \rangle = m_- \langle K_+ \rangle + m_- \langle K_- \rangle \qquad (16.c)$$

On the other hand

$$\langle \lg 1/r_i \rangle = -\tau \langle K_- \rangle \qquad (17)$$

Thus the fractal dimension (15) is

$$D = m_0 + m_+ \{1 + |\frac{\langle K_+\rangle}{\langle K_-\rangle}|\} \qquad (15.b)$$

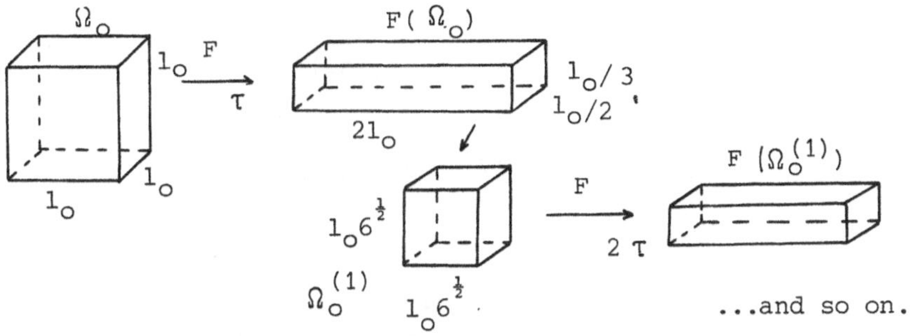

Fig. 8 . A unit cell and its "evolution" with the "flow" F.

3.3. EXERCISE: LORENZ MODEL/36,45/

We consider the system introduced in section 1.3.2., Eqs (5)

$$\frac{d}{dt}\begin{bmatrix} A_1 \\ A_2 \\ A_3 \end{bmatrix} = \begin{bmatrix} -\pi^2 PBA_1 & -PRkA_3/\pi B \\ \frac{A_1A_3}{2} & -4\pi^2 A_2 \\ -A_1(A_2+\pi k) & -\pi^2 BA_3 \end{bmatrix} = F(A) \qquad (1)$$

where $B = 1 + k^2$. Note that with $\tau = \pi^2 Bt$, $A_1 = \pi^2 B\sqrt{2}\, X$, $A_2 = -\pi kZ/\gamma$, $A_3 = -\pi k\sqrt{2}\, Y/\gamma$, $b = 4/B$ and $\sigma = P$, we reproduce the usual Lorenz equations in X,Y,Z, variables. We consider the values used by Shimada and Nagashima /45/, $\sigma = 16$, $b = 4$ and $\gamma = 40$. The parameter b has a rather unusual value as it corresponds to a vanishing wavenumber (k=0) but this has no relevance for my purpose here.

It should be noted that (1) defines a contracting flow in phase space. We have

$$\text{div } F = \sum_{i=1}^{3} \frac{\partial dA_i/dt}{\partial A_i} = -(\sigma + b + 1) = -21 < 0 \qquad (2)$$

Let $X_0 = (A_{1_0}, A_{2_0}, A_{3_0})$ and $e^1 \equiv \{e_1, e_2, e_3\}$ denote an initial point, a basic length, surface or d-volume in the neighborhood of X_0 respectively.

We apply F to the initial phase point X_0 and either F or $L = \partial F/\partial A$ to the extreme points of the unit length, etc. The latter gives to a first approximation the same transformed value that F does. We have $X_1 = F(X_0)$, $\hat{X}_1 = L(X_0 + e_1)$, and so on. The unit length e_1, is transformed into l_1; the unit surface $||e_1 \wedge e_2||$ is transformed into $||a \wedge b||$ whereas the unit volume $||e_1 \wedge e_2 \wedge e_3||$ goes into a new box (a,b,c) of volume $||a \wedge b \wedge c||$ (Forget about the actual notation which refers to exterior product).

We now introduce the 3-dimensional Lyapunov exponents

$$\frac{l_1}{e_1} = e^{\lambda(e^1, X_0)\tau}, \quad e^1 \equiv e_1 \qquad (3)$$

$$\frac{||a \wedge b||}{||e_1 \wedge e_2||} = e^{\lambda(e^2, X_0)}, \quad e^2 \equiv (e_1, e_2) \qquad (4)$$

$$\frac{||a \wedge b \wedge c||}{||e_1 \wedge e_2 \wedge e_3||} = e^{\lambda(e^3, X_0)}, \quad e^3 \equiv (e_1, e_2, e_3) \qquad (5)$$

Strictly speaking these d-dimensional exponents are defined for $t \to \infty$ ($t = n\tau$, $n \to \infty$) as indicated in Section 3.1. According to the definitions there are $\binom{3}{1} = 3$ exponents in (3), $\binom{3}{2} = 3$ exponents in (4), and $\binom{3}{3} = 1$ in (5). Thus we have seven exponents.

After one iteration we proceed to a second one, following Schmidt's orthonormalization procedure, as described in Section 3.2. In practice, each iteration generates seven characteristic exponents.

According to Benettin *et al.*/5/ if at each iteration we choose at random the orientation of the basic unit, the above introduced d-dimensional Lyapunov exponents converge to their maximum values and do not depend on the initial point X_0. Let λ_1^m be the maximum of $\lambda(e_1, X)$. Then, λ_1^m is one of the three values $\{\lambda(e_1,X_0) \equiv \lambda_1, \lambda(e_2,X_0) \equiv \lambda_2, \lambda(e_3,X_0) \equiv \lambda_3\}$. These values $(\lambda_1,\lambda_2,\lambda_3)$ are the one-dimensional exponents. Thus for the unit surface the Lyapunov exponents are $\{\lambda_1+\lambda_2, \lambda_1+\lambda_3, \lambda_2+\lambda_3\}$. The three-dimensional exponent is $(\lambda_1+\lambda_2+\lambda_3)$ and its value should coincide with the divergence of the flow, div $F = \lambda_1+\lambda_2+\lambda_3$. Here we know that it has the value -21. If we order the one-dimensional exponents $\lambda_1 \geqslant \lambda_2 \geqslant \lambda_3$, then after a suitably large number of iterations we have

$\lambda(e^1,X_0) \to \lambda_1$, $\lambda(e^2,X_0) \to \lambda_1+\lambda_2$, $\lambda(e^3,X_0) \to \lambda_1+\lambda_2+\lambda_3$, Shimada and Nagashima have conducted 2^{12} changes of basis and 100 iterations at each change. They quote the following values

$$r = R/Rc = 40, \quad R = 26300.45, \quad Rc = 27\pi^4/4 = 657.5$$

$$\lambda_1 = 1.37, \quad \lambda_2 = 0, \quad \lambda_3 = -22.37 \tag{6}$$

Thus we have one expanding direction, one invariant direction and one contracting direction. The three-dimensional Lyapunov exponent is

$$\lambda(e^3,X_0) = \lambda_1+\lambda_2+\lambda_3 \simeq -21$$

which provides a check on the numerical procedure.

As the flow is contracting after iterations any initial object goes into a zero-volume attractor. Following the result obtained in Section 3.2. the fractal dimension of this attractor is

$$D = 1 + 1 \left\{ 1 + \frac{1.37}{22.37} \right\} \simeq 2.06 \tag{7}$$

N.B: Note that the largest Lyapunov exponent for steady states is negative, whereas for limit cycles or quasiperiodic orbits it vanishes.

3.4. EXERCISE: TWO-COMPONENT LORENZ MODEL /55/

We again consider Eqs(6), Section 1.3.3. We have

$$\text{div } F = -\pi^2 (BP + 4 + B + 4r + rB) < 0 \tag{1}$$

Consideration is restricted to the following numerical values

$$P = 10, \; k = \sqrt{2}/2, \; B = 3/2, \; r = 10^{-2} \text{ and div } F = -202.82$$

As here the phase space has dimension d = 5 the cell *has five directions*. There are the following Lyapunov characterisctic exponents

one-dimensional: $\lambda(e^1, X_0)$; $\binom{5}{1} = 5$

two-dimensional: $\lambda(e^2, X_0)$; $\binom{5}{2} = 10$

three-dimensional: $\lambda(e^3, X_0)$; $\binom{5}{3} = 10$

four-dimensional: $\lambda(e^4, X_0)$; $\binom{5}{4} = 5$

five-dimensional: $\lambda(e^5, X_0)$; $\binom{5}{5} = 1$

Now we have the volume contracting ratio

$$||\lambda_1 e_1 \wedge \lambda_2 e_2 \wedge \lambda_3 e_3 \wedge \lambda_4 e_4 \wedge \lambda_5 e_5|| / ||e_1 \wedge e_2 \wedge e_3 \wedge e_4 \wedge e_5|| = e^{\left(\sum_{i=1}^{5} \lambda_i\right)\tau} \quad (3)$$

where the λ_i are the one-dimensional Lyapunov exponents. We can order them: $\lambda_1 \geqslant \lambda_2 \geqslant \lambda_3 \geqslant \lambda_4 \geqslant \lambda_5$.

After a suitably large number of changes of basis and iterations (in our case 2,000 and 20,000, respectively) we can write the d-dimensional exponents as

$$\lambda(e^1, X_0) = \lambda_1,\ \lambda(e^2, X_0) = \lambda_1 + \lambda_2, \ldots$$

$$\lambda(e^5, X_0) = \lambda_1 + \lambda_2 + \lambda_3 + \lambda_4 + \lambda_5 = \text{div } F$$

A specific case is:

$$\varepsilon = (R-R_c)/R_c,\ R_c = 27\,\frac{\pi^4}{4}\,/(1+S+S/r),\ S = 4.8 \times 10^{-4},\ r = 10^{-2}$$

$$\varepsilon = 24,\ R = 15681,\ X_0 = (164.78, -1.13, -0.03, -1.19, 0.88),$$

where X_0 is slightly above the steady convective state. We have

$$\lambda_1 = 11,\ \lambda_2 = 0,\ \lambda_3 = -0.3,\ \lambda_4 = -0.3,\ \lambda_5 = -213.3$$

Then

$\lambda(e^5, X_0) = -202.86 \approx -202.82$ On the other hand $\lambda_+ = 11$, $\lambda_- = (-0.3-0.3-213.3)/3 = -71.3$. Thus the fractal dimension of the attractor is $D = 1 + 1 \{ 1 + 11/71.3 \} = 2.15$.

An illustration of this "strange attractor " is given in Figures 9(a) and 9(b). For R and S positive this attractor although embedded in a d = 5 dimensional phase space is very much like the strange attractor in the original Lorenz model.

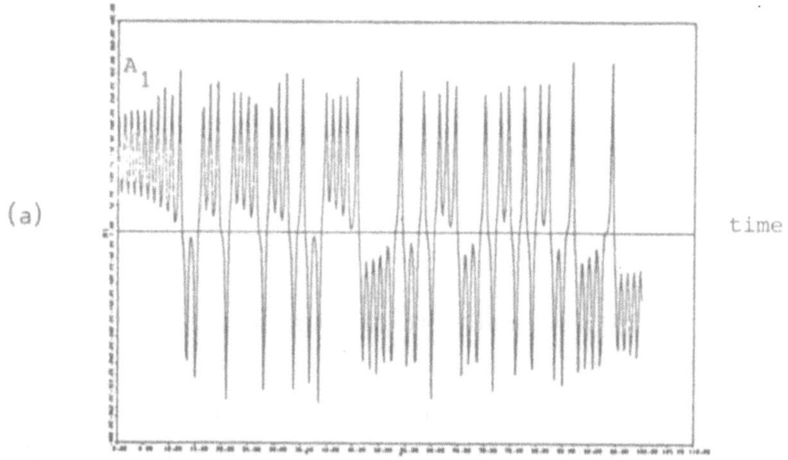

Fig.9. Strange attractor in the(two-component)Lorenz model (a)actual computer output for the flow velocity(A_1) *vs* time(arbitrary units)

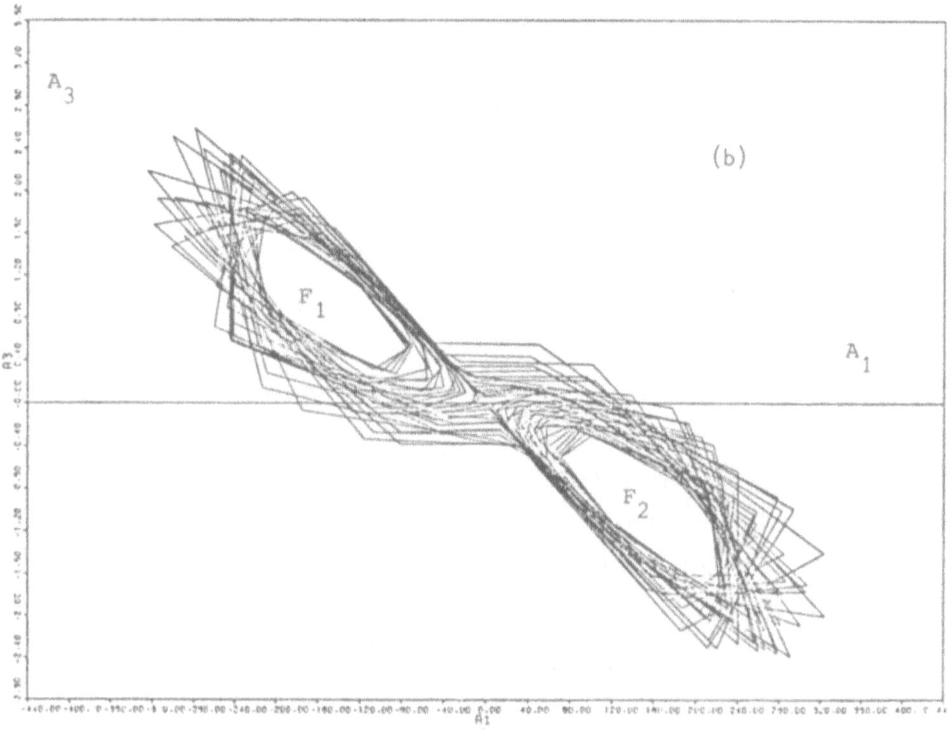

Fig.9. (b)phase space(restricted to temperature,A_3,and velocity,A_1) of the "flow" around the two unstable fixed points (F_1 and F_2)

4. DISCRETE MAPS: FEIGENBAUM'S CASCADE AND POMEAU'S INTERMITTENCIES

4.1 PERIOD-DOUBLING CASCADES /6,8-12,19,33/

4.1.1. Exercise : Myrberg map

Consider the non-invertible one-dimensional discrete map of the interval $-1 \leqslant x \leqslant 1$,

$$X_{n+1} = 1 - aX_n^2 \qquad (1)$$

For $-1/4 \leqslant a \leqslant 3/4$, sucessive iterations of (1) bring, with rapid convergence, any initial condition to a fixed point $X_F(a)$ (the largest root of $X_{n+1} = X_n = X_F$). Illustration of the iteration is given in Figure 1, where real time output of a programmable pocket calculator is displayed. Starting at $X_0 = 0.5$, it takes about one minute and some nine iterations to reach $X_F(a = 0.25) = 0.82843$. Other cases, chosen at random ($a = 1, a = 1.2505$, and $a = 1.4011552$) are also illustrated in Fig. 10.

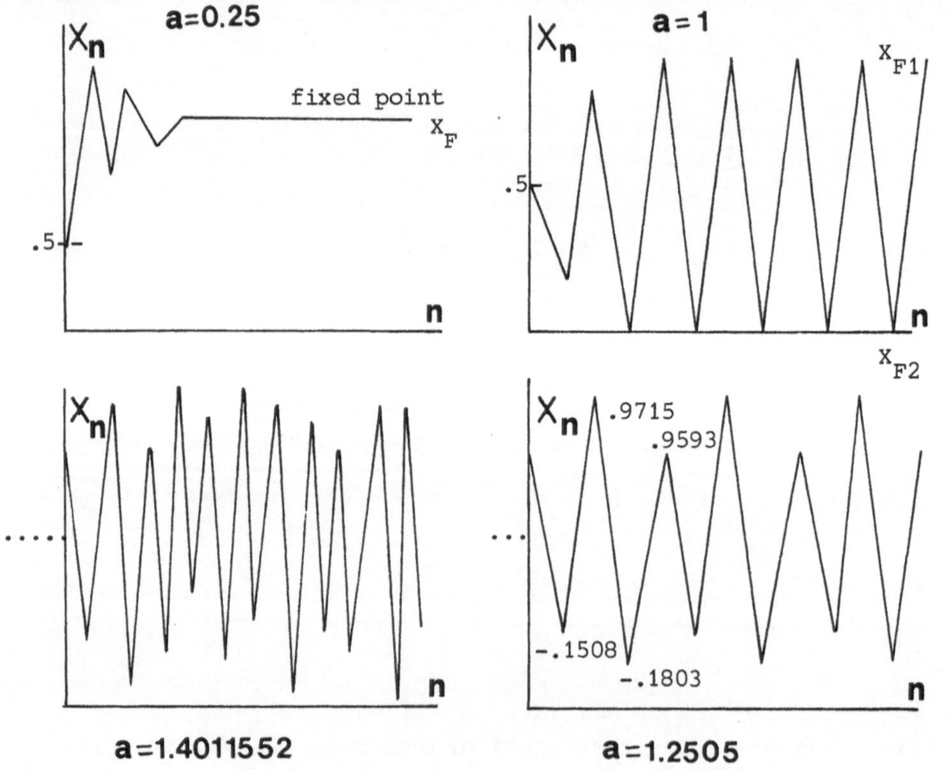

Fig. 10

The first striking feature of the map, for chosen range of values of a and x, is that an infinite sequence of period-doublings appears. Table 1 gives a systematic account of the sequence, up to n = 10. At n = 1 we reach a (fixed) point (period $2^0 = 1$), at n = 2 we reach two points, X_{F1}, X_{F2} (F1 → F2→F1..) ..., period $2^1 = 2$), and so on (see Figure 10). The value $a_* \approx 1.4011552$,,, is the accumulation point of the sequence, that defines the limit of $n = \infty$, $a_* = a_\infty$. Although this behavior and some other properties described below are known in the mathematics and engineering of nonlinear parametric excitation I believe that its rediscovery by Feigenbaum has opened a line of fertile approach to the study of the onset of turbulence.

The second striking property is that a_∞ is attained in geometric progression with ratio $\delta \approx 4.669$...Feigenbaum and others have shown that this constant is universal (see Table 1 and the comments in section 4.1.3 further below)

TABLE 1. PERIOD-DOUBLING CASCADE IN MAP $X_{m+1} = 1 - a\,X_m^2$

n (2^n cascade)	a_n	$a_n - a_{n-1}$	$\delta = (a_n - a_{n-1})/(a_{n+1} - a_n)$
0	.75		
		.5	
1	1.25		4.233738275
		.1180989394	
2	1.3680989394		4.551506949
		.0259472172	
3	1.3940461566		4.645807493
		.0055850823	
4	1.3996312389		4.663938185
		.0011975035	
5	1.4008287424		4.668103672
		.0002565289	
6	1.4010852713		4.668966942
		.000054943399	
7	1.401140214699		4.669147462
		.000011767330	
8	1.401151982029		4.669190003
		.000002520208	
9	1.401154502237		4.669196223
		.000000539752	
10	1.401155041989		
....			

accumulation point $a_* = a_\infty = 1.4011552\ldots$ $\delta = 4.669\ldots$

Thirdly, using the data given in Table 1, it can be easily checked that

$$n(a) \lg 2 \sim \nu \lg |a_* - a| \text{ with } \nu = -0.449807 \quad (2)$$

when n is taken continuously varying with a, ν is a universal exponent as shown by Feigenbaum and other authors. The analogy with critical phenomena is quite striking /29/.

Fourthly, Feingenbaum's cascade has a symmetric reflection for $a \geqslant a_*$, where the chaotic regime hides a highly *structured* background.

On the other hand, although comparison of results for a one-dimensional map cannot be established with data from experiment in flow instability it strikes indeed the similarity of the power spectrum of the Myrberg's map and a spectrum obtained by Libchaber and Maurer in experiments of Rayleigh-Bénard convection in Helium /24,25,30,31/. The two spectra are given in Figure 11.

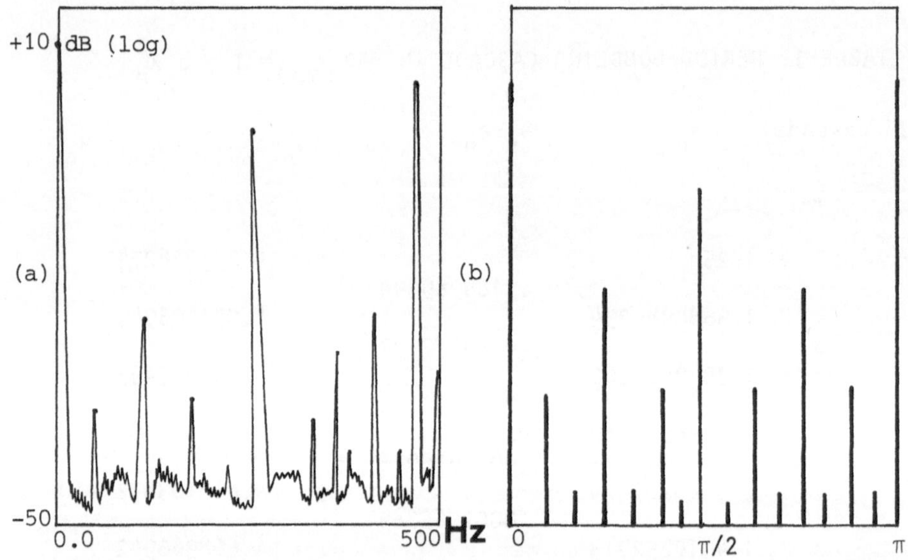

Fig.11 Power spectra (a) Bénard convection at R= 43 R_c, (b) period of length 64, case $1 - 1.401155....X^2$ /6/.

4.1.2. Exercise : Hénon's dissipative map /20/

Consider now the two-dimensional discrete map

$$X_{n+1} = Y_n + 1 - aX_n^2 \quad (1.a)$$

$$Y_{n+1} = bX_n \quad (1.b)$$

where a and b are parameters at our disposal. The mapping 1 invented by Hénon /20/ consists of the following sequence of steps (see Fig.12).

T' : X' = X, Y' = Y + 1 - aX^2, which presserves "area" although changes Y.

T'': X'' = b X', Y'' = Y', which contracts (b < 1) or expands (b > 1) area according to the value given to b.

T''': X''' = Y'', Y''' = X'', which presserves area while exchanging axes.

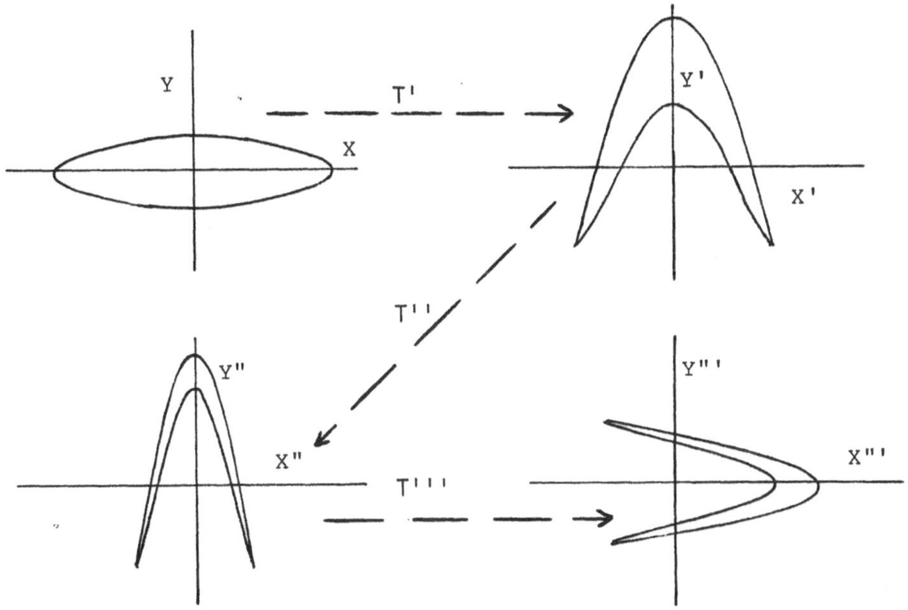

Fig. 12. Sequential description of Hénon's map.

For suitable values of the parameters, a and b, the Hénon flow (1) has negative divergence. The Jacobian $\partial(X_{n+1},Y_{n+1})/\partial(X_n,Y_n) = -b$ is negative for b positive, and lgb is negative for b < 1. Then Henon's mapping is *dissipative*. It is also invertible while Myrberg-Feigenbaum's is not. In fact, it is the most general quadratic map with constant Jacobian, under canonical form. As it is two-dimensional it can very well reproduce the Poincaré mapping of a real, three-dimensional flow problem. It has, however, the minor inconvenience that when starting far from the origin and due to the X^2 term points can escape to infinity.

Hénon aimed (a "reduccionist" approach he said) at finding a discrete map as simple as possible yet endowed with the same essential properties of the Lorenz model.

The map has a fixed point

$$X_{n+1} = X_n = X^* = \{-(1-b)+[\ (1-b)^2 + 4a\ ]\ \}^{1/2}/2a \qquad (2.a)$$

$$Y_{n+1} = Y_n = Y^* = b\ X^* \qquad (2.b)$$

with the stability region $(1-b)^2/4 < a < 3\ (1-b)^2/4$. Let us choose following Hénon /20/, $b = 0.3$. Then $a_1 = 3(1-b)^2/4 = 0.3675$ is the upper limit of stability of the fixed point. The following computations have been done with a pocket calculator.

1. $a = 0.3674 < a_1$. Here and in the following we take the initial condition $X_0 = Y_0 = 0$. After some three thousand iterations the fixed point is attained through an oscillatory approach . We have $X^* = 0.9524$, $Y^* = 0.2857$. From the initial condition we have reached *one* point. If we take Hénon's map as the Poincaré map of a three-dimensional problem then the original system is in a limit cycle.

2. $a = 0.3675...$ From the origin the iteration after some ten thousand steps brings us to the following *two-point* state

$(X_{F1} = .9347...\ \ Y_{F1} = .2909...), (X_{F2} = .9698..., Y_{F2} = .2804...)$. At this state further iteration shows a limit cycle behavior. We have $F_1 \to F_2 \to F_1 .....$

3. $a = 1.0$. From the origin we arrive at the following *four-point state*: $(X_{F1} = -.65..., Y_{F1} = .38...), (X_{F2} = .95.., Y_{F2} = -.19...)$, $(X_{F3} = -0.10..., Y_{F3} = .28...), (X_{F4} = 1.27..., Y_{F4} = -.03...)$. We have the limit cycle $F_1 \to F_2 \to F_3 \to F_4 \to F_1 ....$

For a slightly higher value of <u>a</u> we get the *eight-point state*.

4. $a = 1.055$, $X_0 = Y_0 = 0$ is still the initial condition, Henon's map brings us to the following *sixteen-point state*

$(X_{F1} = 1.2498546, Y_{F1} = -.0628636)$, $(X_{F2} = -.71, Y_{F2} = .37)$
$(X_{F3} = .84, Y_{F3} = -.21)$, $(X_{F4} = .03, Y_{F4} = .25)$
$(X_{F5} = 1.25, Y_{F5} = .01)$, $(X_{F6} = -.63, Y_{F6} = .37)$
$(X_{F7} = .94, Y_{F7} = -.19)$, $(X_{F8} = -.13, Y_{F8} = .28)$
$(X_{F9} = 1.26, Y_{F9} = -.03)$, $(X_{F10} = -.72, Y_{F10} = .37)$
$(X_{F11} = .82, Y_{F11} = -.21)$, $(X_{F12} = .07, Y_{F12} = .24)$
$(X_{F13} = 1.24, Y_{F13} = .02)$, $(X_{F14} = -.60, Y_{F14} = .37)$
$(X_{F15} = .98, Y_{F15} = -.18)$, $(X_{F16} = -.20, Y_{F16} = .29)$

Thus, we reach a 2^4- point state. The sequence follows indeed Feingebaum's 2^n period-doubling as we increase the values of a until we reach an accumulation point at $a = 1.4 = a^* = a_\infty$.

5. At $a = 1.4$ we reach a strange attractor. In fact it appears for $1.06 < a < 1.55$. Its one-dimensional Lyapunov exponents are $\lambda_1 = .42$ and $\lambda = -1.62$. We have $\lambda_1 + \lambda_2 = -1.20 = \lg 3 = \lg b$, as expected. There is one expanding direction and one contracting direction, and formally $m_0 = 0$ (See Section 3.2). Thus a straightforward application of Equation 3.12(15.b) yields the fractal dimension, D =1.26, a result earlier obtained by Feit / 13/.

4.1.3. Summary of some universal features of period-doubling cascades.

Although a more complete account of the universal features of Feigenbaum's cascades and their relation to critical phenomena can presumably be found in Prof. Martin's lecture notes, I here list some of them, just for the record:

i. Universality of the geometric progression to chaos: Near a_∞, period-doubling cascades proceed in geometric progression with the universal ratio $\delta \approx 4.466920160\ 91029909....$ In fact many other dynamical systems(Lorenz model, five-modal Navier-Stokes, Hénon's map, etc) lead to cascades with such universal constant .

ii. The inverse maximum Lyapunov exponent of the flow(in Myrberg's map there is only *one* one-dimensional exponent; for multidimensional flows see Section 3.2) which is an estimate of correlation time between nearby trajectories in phase space, diverges at a_∞ with a universal power-law of exponent $\nu \approx - 0.4498069...$

iii. The period-doubling cascade is "robust": if a one-parameter family shows subharmonic bifurcations to periods n=1,2 and 4, then there is a higher chance of finding the period n=8 at further parameter variation than of finding a period n=4 when only periods n=1 and 2 have appeared.

iv. Another universal constant exists when we follow two nearby points as the cascade proceeds *i.e.*, there is also universality in real space geometric features.

v. Moreover, there is symmetry with respect to a_∞ of the (subharmonic) period-doubling cascade. Further details can be found in the various references listed below / 6,9-12,22/.

4.2. POMEAU'S INTERMITTENCIES

4.2.1. A curious Poincaré map /2,3/

Consider a system in a limit cycle. Its Poincaré map shows a fixed point, a steady state as illustrated in Figure 13(a). To study the stability of the time-periodic orbit in real or phase space, it suffices to study the stability of the steady state, S, in the Poincaré map (Floquet theory). We have $A(t) = A(t + T)$ where T is the time-period in the limit cycle. Let us assume that A is a function of a parameter, say the Rayleigh number, R, in Bénard convection. A may be the amplitude of the flow velocity in real space. If the limit cycle is unstable at a given value of R, the Poincaré map will show a sequence of points S_1, S_2... See Fig.13(b).

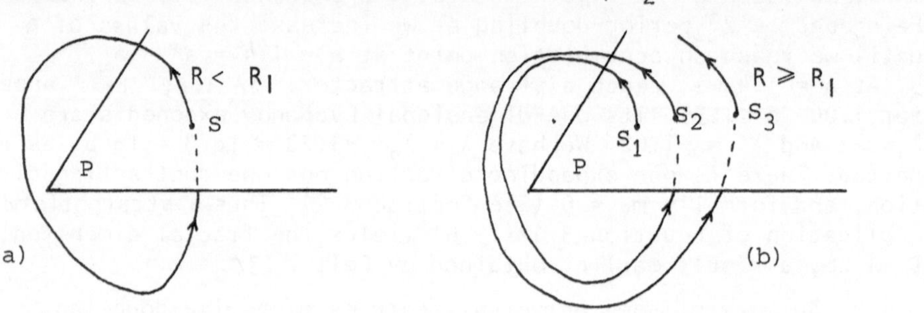

Fig.13. A Poincaré map (a) stable limit cycle, (b) unstable case.

If $\delta S_1 = S_1 - S$, $\delta S_2 = S_2 - S$,... to a first approximation we have $\delta S_2 = M\,\delta S_1$, where M is a matrix that accounts for the eventual expansion of δS. It is $M = M(R)$. Let R_I be the value of R at which the limit cycle starts being unstable. Then instability of the limit cycle corresponds to the existence of, at least, one eigenvalue of M, λ (M,R) of modulus greater than unity. The eigenvalues λ, are in general complex numbers. Thus, three possibilities exist and they are illustrated in Figure 14.

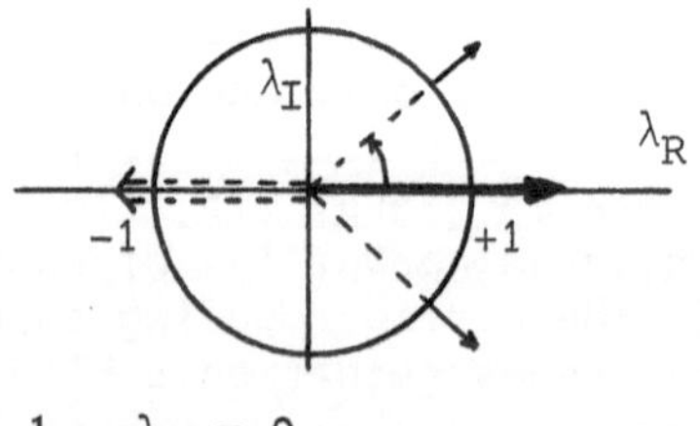

Fig.14. The unit circle in the λ-plane, $\lambda = \lambda_R + i\,\lambda_I$ illustrating three possibilities for instability.

i) λ crosses the unit circle at $\lambda_R = +1$, $\lambda_I = 0$

ii) λ crosses the unit circle at some angle θ, $\lambda_R(\Theta) \neq 0$, $\lambda_I(\Theta) \neq 0$

iii) λ crosses the unit circle at $\lambda_R = -1$, $\lambda_I = 0$

Here we restrict consideration to case i). Cases ii) and iii) lead to a torus (two frequencies) and Feigenbaum's cascade, respectively /3/. In case i) there is one unstable direction only. The Poincaré map $A(t) \rightarrow A(t+T) \rightarrow A(t+2T)$... corresponds to $S_o \rightarrow S_1 \rightarrow S_2$... /Fig.13(b)/. Thus we have the discrete map $X_n = f(X_{n-1}, R^o)$.

For convenience we introduce $\varepsilon = (R - R_I)/R_I$ and choose X such that the limit cycle is stable at $\varepsilon \leqslant 0$ and $X = 0$. To a first approximation the discrete mapping can be described by the continuous function $f(X,\varepsilon) = X + \varepsilon + aX^2$ + higher order terms: a curve is firstly a straight line and then, if necessary, a quadratic function to a first approximation ! a and ε measure the distance to the fixed point in P which corresponds to the T-periodic limit cycle in phase space. For simplicity we choose a = 1. At $\varepsilon < 0$, $f(X,\varepsilon) = X = X + \varepsilon + X^2$, has two solutions $X = \pm(-\varepsilon)^{1/2}$. These two solutions are the fixed points of the discrete mapping generated by $A(t) \to A(t+T) \to A(t+2T), \ldots$. It appears that one of the fixed points is stable and corresponds to the T-periodic stable limit cycle whereas the other fixed point is unstable (one may be node and the other saddle).

Let us assume that the quadratic approximation to $f(X,\varepsilon)$ is the parabola sketched in Figure 15.

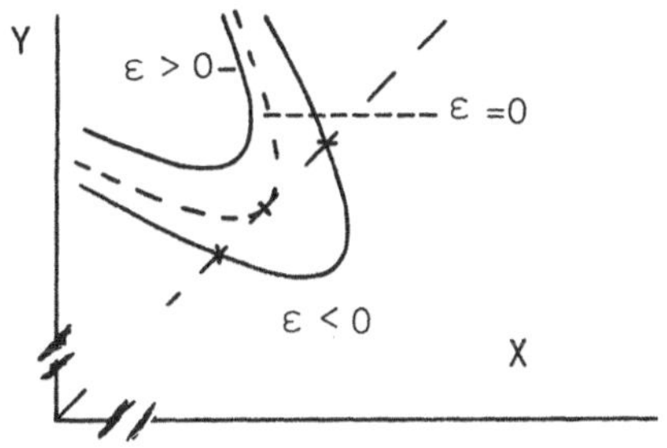

Fig.15 .Local behavior of the Poincaré map around the transition from stable limit cycle to intermittency.

The two fixed points F_i (i=1,2) of the discrete iteration are the intersections of $Y = X$ and $Y = f(X,\varepsilon)$ in Figure 15. At $\varepsilon = 0$ they collapse and the two lines are tangent to each other. There is no intersection at $\varepsilon > 0$. Then Figure 16 illustrates the discrete iteration when the limit cycle is unstable. It can be appreciated that the lower the value of $\varepsilon \gtrsim 0$ the longer is the time interval or number of iterations needed to cross the channel using the

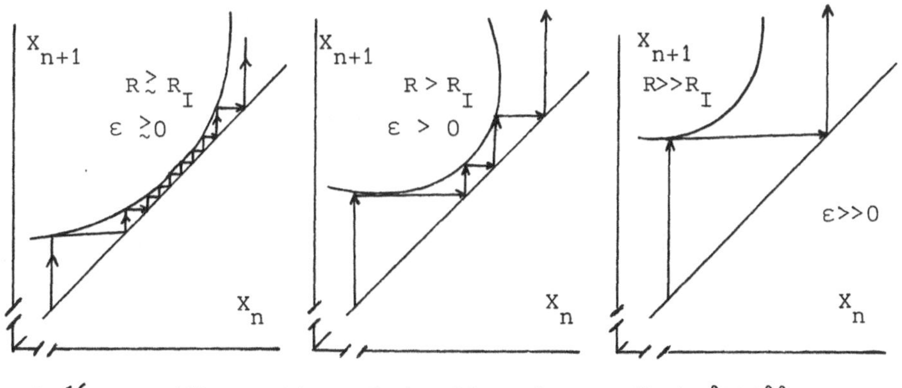

Fig.16 . An illustration of the discrete map that *locally* approximates the Poincaré map when the limit cycle becomes unstable and intermittency appears. See case i) main text.

map. In fact this time interval diverges like $\varepsilon^{-1/2}$ at $\varepsilon = 0$. Indeed the discrete map $X_{n+1} = X_n + \varepsilon + X_n^2$ at $\varepsilon \geqslant 0$ can be well described by the differential equation, $dX_n/dn = \varepsilon + X_n^2$ which leads to $n\,\varepsilon^{1/2} = \arctan(X_n\,\varepsilon^{-1/2})$. The divergence of X_n at the two outlets of the channel corresponds to the limits $n\,\varepsilon^{1/2} \to \pm\,\pi/2$. Thus in terms of iterations N_i we have $N_i \sim \varepsilon^{-1/2} \sim (R - R_I)^{-1/2}$ which is a rough, albeit reasonable estimate of a system possessing such instability mechanism. The limit cycle behavior is expected to show time intervals of (almost) T-periodic behavior followed by periods of wild and unpredictable behavior (turbulent bursts) as sketched in Figure 17.

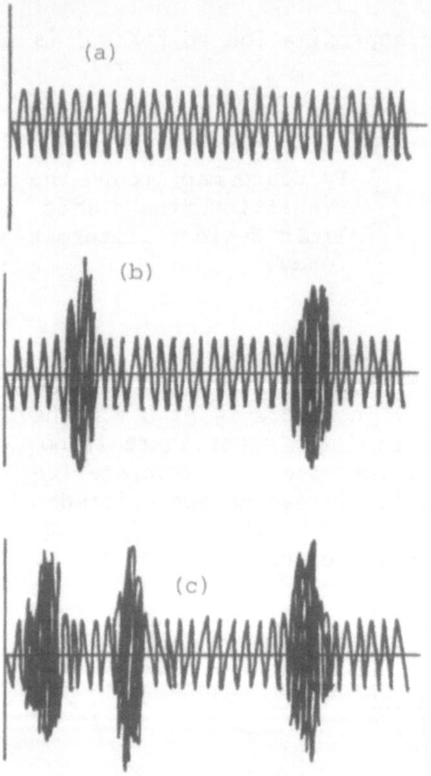

Fig. 17. Pomeau's intermittency (sketch)
(a) stable limit cycle below the onset of intermittency ($R < R_I$, $\varepsilon < 0$).
(b) the periodic cycle alternates with turbulent bursts at $R \geqslant R_I$, $\varepsilon \geqslant 0$.
(c) at higher values of $R > R_I$ the life-time of the cycle becomes shorter and shorter Note that at R_I this life-time diverges like the power $\varepsilon^{-\frac{1}{2}}$. With 'intermittency' one may expect 'selfsimilarity' on a smaller time scale. However, this is not a general rule, and does not happen here.

QUESTION : How would it be that the system's behavior alternates T-periodicity with turbulent bursts? A plausible answer is that when instability shows in ,say, one direction only there still remain parts in the T-periodic attractor that tend to keep the system in the limit cycle. The existence of an unstable direction, however, expells the system out of the basin of attraction of the limit cycle, and the system gets wild, *i.e.*, chaotic (turbulent). The further away the transition point, R_1, the higher are the chances to remain chaotic. In the wild intervalthe system explores a large enough part of the phase space and thus gets close enough to the basin of the T-periodic orbit. It is then trapped there for a while.

4.2.2. Lorenz model and a related exercise

The great discovery made by Pomeau/28,40/is that the *ad hoc* qualitative behavior sketched in the preceeding section does indeed happen in the Lorenz model for some values of the Rayleigh number $R \sim 166.07$ Rc, In the (X,Y,Z) notation mentionned in Section 3.3. $r = R/Rc = 166.07$, $\sigma = 10$, $b = 8/3$ whereas Rc denotes the onset for steady convection (first onset of steady convective instability) Actually this is what happens with the Poincaré map when using the Y-values at $X = 0, dX/dt > 0$ in the Lorenz model. For $148.4 < r < 166.07$ the Lorenz model operates in a stable limit cycle.

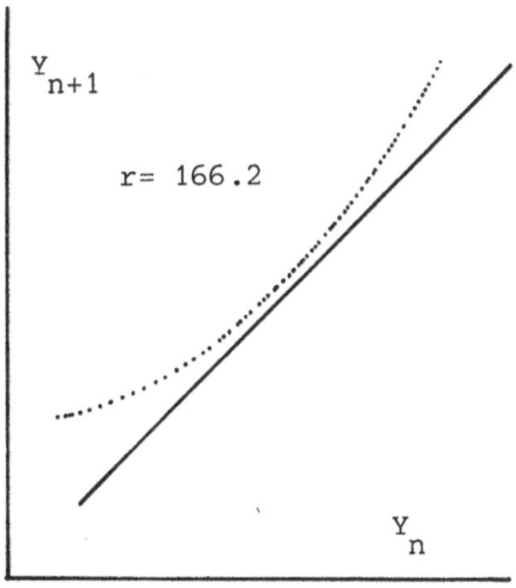

Fig.18.Local behavior of the Y-coordinate for the Poincaré map in the Lorenz model slightly above the onset of instability of limit cycle operation.The dotted line(computer output)can be nicely approximated by a parabola ! Taken from /40/.

Figure 18 shows the output of a computer experiment.It corresponds to Figure 5 of /40 /. The result is sharp: $N_i^2 \sim (r - r_I)^{-1}$, *i.e.* $N_i \sim \varepsilon^{-1/2}$ as predicted with the *ad hoc* construction sketched in the preceeding section. For those readers that would like more accessible exercise here is a map

$$\Theta \to 2\Theta + r \sin 2\pi\Theta + 0.1 \sin 4\pi\Theta \qquad (1)$$

$$(\text{mod } 1, 0 \leqslant \Theta \leqslant 1$$

that at $r \approx 0.24706$ has behavior similar to that described for Lorenz'model. This map applies $0 \leqslant \Theta \leqslant 1$ twice on itself and all numerical calculations can be performed with a desk-top calculator.

ACKNOWLEDGMENTS

My thanks to J.Carlos Antoranz for his help in the preparation of these Notes. He has taken the pain of computing everything I asked for. Calculations have been carried out at various places. The major computer work has been done at the University of Oslo, thanks to the hospitality of Prof. E. Palm, and at I.N.I.A. (Madrid). I also express my deepest gratitude to Prof. J.I. Fernández-Alonso, Dean of Sciences, Universidad Autónoma de Madrid, for letting me and my collaborators the free use of offices and other facilities at the Cantoblanco Campus (Madrid) where our research has been carried out. This research has been sponsored by the Stiftung Volkswagenwerk.

REFERENCES

1. ANTORANZ,J.C.and M.G.VELARDE,1981,Optics Commun.,to appear
2. ARNOLD,V.,1978,ORDINARY DIFFERENTIAL EQUATIONS,M.I.T.press, Cambridge,Mass.
3. ARNOLD,V.,1980,CHAPITRES SUPPLEMENTAIRES DE LA THEORIE DES EQUATIONS DIFFERENTIELLES ORDINAIRES,Mir,Moscow.
4. BOCCARA,N.,1976,SYMETRIES BRISEES,Hermann,Paris.
5. BENETTIN,G.L.GAGANI,A.GIORGILLI and J.M.STRELCYN,1978,C.R. Acad Sci.Paris,*286*,431
6. COLLET,P.and J.P.ECKMANN,1980,ITERATED MAPS ON THE INTERVAL AS DYNAMICAL SYSTEMS,Birkhäuser, Boston.
7. DEGIORGIO,V.,1979,Phys.Rev.*A20*,2193
8. DERRIDA,B.and Y.POMEAU,1980,Phys.Lett.*A80*,217
9. FEIGENBAUM,M.J.,1978,J.Stat.Phys.*19*,25
10. FEIGENBAUM,M.J.,1979,Phys.Lett.*A74*,375
11. FEIGENBAUM,M.J.,1979,J.Stat.Phys.*21*,669
12. FEIGENBAUM,M.J.,1980,Los Alamos Science,Summer issue.
13. FEIT,S.,1978,Commun.Math.Phys.*61*,249
14. GINZBURG,V.L.,A.P.LEVANYUK and A.A.SOBYANIN Phys.Rep.(Phvs.Lett. C.)*57*,152
15. GIGLIO,M.,1975,in FLUCTUATIONS,INSTABILITIES,AND PHASE TRANSI_ TIONS,Plenum Press,New York.
16. GOLLUB,J.P.and S.V.BENSON,1980,J.Fluid Mech.*100*,449
17. GUCKENHEIMER,J.,J.MOSER and S.E.NEWHOUSE,1980,DYNAMICAL SYSTEMS, Birkhäuser,Boston.
18. HAKEN,H.,1977,SYNERGETICS,Springer-Verlag,New York.
19. HELLEMAN,R.H.G.,1980,in FUNDAMENTAL PROBLEMS IN STATISTICAL ME_ CHANICS,vol 5,North Holland, Amsterdam.
20. HÉNON,M.,1976,Commun.Math.Phys.*50*,69
21. HIRSCH,M.W.and S.SMALE,1974,DIFFERENTIAL EQUATIONS,DYNAMICAL SYSTEMS AND LINEAR ALGEBRA,Academic Press,New York
22. HUBERMAN,B.A. and J.RUDNIK,1980,Phys.Rev.Lett.*45*,154
23. KINCAID,J.M.and E.G.D.COHEN,1975,Phys.Rep.(Phys.Lett.C)*22*,57
24. LIBCHABER,A.,this volume.
25. LIBCHABER,A,and J.MAURER,1978,J.Phys.(Paris)Lett.*39*,369
26. LORENZ,E.N.,1963,J.atmos.sci.*20*,130
27. MANDERBROT,B.,1977,FRACTALS,Freeman,San Francisco.

28.MANNEVILLE,P.and Y.POMEAU,1980,Physica,*1D*,219
29.MARTIN, P.C., this school, not published (editor's notes).
30.MAURER,J.and A.LIBCHABER,1979,J.Phys.(Paris)Lett.*40*,419
31.MAURER,J.and A.LIBCHABER,1980,J.Phys.(Paris)Lett.*41*,515
32.MAY,R.,1976,Nature,*261*,459
33.MAY,R.and G.F.OSTER,1980,Phys.Lett.*A78*,1
34.MORI,H.1980,Progress Theor.Phys.*63*,1044
35.MORI,H.1980,Progress Theor.Phys.*63*,1931
36.MORI,H.and H.FUJISAKA,1980, in SYSTEMS FAR FROM EQUILIBRIUM, Springer-Verlag,New York.
37.NICOLIS,G.and I.PRIGOGINE,1977,SELF-ORGANIZATION IN NONEQUILIBRIUM SYSTEMS,Wiley, New York.
38.NORMAND,C.,Y.POMEAU and M.G.VELARDE,1977,Rev.Mod.Phys.*49*,581
39.OSEDELEC,V.I.,1968,Trans.Moscow Math.Soc.*19*,197
40.POMEAU,Y. and P.MANNEVILLE,1980,Commun.math.phys.*74*,189
41.RABINOVICH,M.I.,1978,Soviet Phys.Usp.*21*,443
42.RISTE,T.,this volume, and references therein.
43.RUELLE,D. and F.TAKENS,1971,Commun.math.phys.*20*,167
44.SCHECHTER,R.S.,M.G.VELARDE and J.K.PLATTEN,1974,Adv.Chem.Phys. *26*,265
45.SHIMADA,I.and T.NAGASHIMA,1979,Progress theor.phys.*61*,1605
46.SHTEINBERG,V.A.,1970,J.appl.math.mech(PMM) *35*,335
47.SWINNEY,H.L. and J.P.GOLLUB,1978,Phys.today, August,41
48.SWINNEY,H.L. and J.P.GOLLUB,1981,(editors)HYDRODYNAMIC INSTABILITIES AND THE TRANSITION TO TURBULENCE,Springer-Verlag,New York.
49.VELARDE,M.G.,1976,in FLUID DYNAMICS,Gordon and Breach, N.Y.
50.VELARDE,M.G. and R.S.SCHECHTER,1972,Phys.Fluids,*15*,1707
51.VELARDE,M.G. and J.C.ANTORANZ,1979,Phys.Lett.*A72*,123
52.VELARDE,M.G. and I.ZUÑIGA,1979,J.Phys.(Paris)*40*.725
53.VELARDE,M.G. and C.NORMAND,1980,Sci.Amer.*243*,92
54.VELARDE,M.G. and J.C.ANTORANZ,1981,J.Stat.Phys.*24*,235
55.VERLARDE,M.G. and J.C.ANTORANZ,1981,Progress theor.phys. to appear.

THE PHYSICAL MECHANISM OF OSCILLATORY AND FINITE AMPLITUDE INSTABILITIES IN SYSTEMS WITH COMPETING EFFECTS

H.N.W. Lekkerkerker

Faculteit van de Wetenschappen
Vrije Universiteit Brussel
1050 Brussels, Belgium

I. INTRODUCTION

The onset of convection in a horizontal layer of fluid heated from below, the so-called Rayleigh-Bénard instability, has been extensively investigated for a long time [1,2]. In recent years there has been considerable interest in the new phenomena that occur in systems in which two competing stability influencing effects with different relaxation times are present. The first example of such systems that was studied in detail is the case of thermal convection in binary mixtures.[3,4] More recently it has been pointed out that thermal convection in nematic liquid crystals can also be advantageously considered from the same point of view [5]. It has been found that under certain conditions both binary mixtures [3,4] and nematic liquid crystals [6,7], when heated from below, show oscillatory and finite amplitude instabilities. In this contribution I will discuss the common physical mechanism underlying these phenomena from the point of view of the energy balance of convective disturbances.

II. OSCILLATORY INSTABILITY

According to Chandrasekhar [8] "The onset of thermal instability will be as overstable oscillations if it is possible (at a lower adverse temperature gradient than necessary for the onset of stationary convection) to balance in a synchronous manner the periodically varying amounts of kinetic and other forms of energy with similarly varying rates of dissipation and liberation of energy". Applied to binary mixtures the above principle implies that oscillatory convection is possible if the following balance can be maintained

$$\dot{F}^{v}_{kin} + \dot{F}^{g}_{kin} = \dot{F}^{o}_{kin} \tag{1}$$

Here $\dot{F}^{v}_{kin}$ is the rate of viscous dissipation of kinetic flow energy $\dot{F}^{g}_{kin}$ is the rate of production of kinetic flow energy by the buoyancy force and $\dot{F}^{o}_{kin}$ is the rate of change of kinetic energy due to the oscillatory nature of the velocity field. For the oscillatory convective disturbance

$$v_x(\vec{r}) = -2\frac{q_z}{q_x} v \cos\omega t \sin q_x x \cos q_z z$$

$$v_y(\vec{r}) = 0 \tag{2}$$

$$v_z(\vec{r}) = 2 v \cos\omega t \cos q_x x \sin q_z z$$

($q_z = \frac{\pi}{d}$, d : thickness layer)
with frequency ω such that

$$Dq^2 \ll \omega < \chi q^2$$

(D : mass diffusion constant, χ : heat diffusivity)
one obtains [9] (per unit volume)

$$\dot{F}^{v}_{kin} = -\eta \frac{q^4}{q_x^2} v^2 \cos^2\omega t \tag{3}$$

and

$$F^{g}_{kin} = g\rho\alpha\beta v^2 \left[\frac{1}{\chi q^2}\cos^2\omega t + \frac{\omega}{(\chi q^2)^2}\cos\omega t \sin\omega t + \frac{S}{\omega}\cos\omega t \sin\omega t \right] \tag{4}$$

(η : viscosity, ρ : density, α : thermal expansivity, $\beta = -\frac{\partial T}{\partial z}$)
The dimension less parameter S which is defined as

$$S = \frac{k_T}{T}\frac{\alpha'}{\alpha} \tag{5}$$

(k_T : thermal diffusion ratio, α' : solutal expansivity) is assumed to be negative, meaning that due to the Soret effect the more dense component moves to the warm lower boundary and thus has a stabilizing effect. The first two terms in square brackets on the righ hand side of Eq. (4) are due to the presence of the temperature gradient where as the third term is due to the presence of the Soret driven concentration gradient. It can be shown that $\dot{F}^{o}_{kin}$ is negligible compared to $\dot{F}^{v}_{kin}$ and $\dot{F}^{g}_{kin}$. Taking this into account it follows from Eqs (1), (3) and (4) that oscillatory convection is possible for $\beta > \beta^{o}_{c}$ where

$$\frac{g\alpha\beta^{o}_{c}d}{\chi\nu} = \frac{q^6 d^4}{q_x^2} \tag{6}$$

The oscillation frequency at the critical point is given by

$$\omega_c = \chi q^2 \sqrt{-S} \tag{7}$$

We see that the $\cos^2 \omega t$ term in $\dot{F}^{g}_{kin}$ serves to balance the viscous dissipation, whereas the required oscillation frequency is determined by the balancing of the $\cos \omega t \sin \omega t$ terms in $\dot{F}^{g}_{kin}$. Physically this can be interpreted that due to the oscillatory character of the convective disturbance the stabilizing effect of the concentration gradient is virtually eliminated whereas the destabilizing effect of the temperature gradient is retained. Note in this connection that the threshold given by Eq. (6) is the same as for the onset of stationary convection in an one-component liquid.

Homeotropic nematic liquid crystals heated from below [6,7] show similar characteristics as binary mixtures in which the Soret effect drives the more dense component to the warm lower boundary. Like in the case of binary mixtures the oscillatory instability in homeotropic nematic liquid crystals can be explained on the basis of the competition between a destabilizing effect and a stabilizing effect [10]. The destabilizing effect here is again the temperature gradient whereas the stabilizing effect this time is the heat focusing effect arising from a combination of the shear-director coupling and the anisotropic heat conductivity. Again applying the synchronous energy balance principle one obtains [10]

$$\frac{g\alpha\beta^{o}_{c} \; d^4}{\chi\nu} = \frac{q^6 d^4}{q_x^2} \tag{8}$$

and

$$\omega_c = \chi q^2 \sqrt{A} \tag{9}$$

The dimensionless parameter A is given by

$$A = \frac{C\chi_a}{\chi} \tag{10}$$

where C is the shear-director coupling constant, χ_a is the difference between the heat conductivity parallel and perpendicular to the director and χ is a suitable average of these quantities. Note that the threshold given by Eq. (8) is again the same as for the onset of stationary convection in an one-component liquid. Further note that the parameter A here plays the same role as the parameter S for binary mixtures.

III. FINITE AMPLITUDE INSTABILITY

Finite amplitude instabilities arise when the modification of the basic state (basic flow, basic temperature gradient) by a finite amplitude convective disturbance facilitates the onset of instability. In terms of an energy balance treatment it means that if one expands the rate of change of the energy of a (stationary) convective disturbance in terms of its amplitude

$$\dot{F} = av^2 + bv^4 + \ldots \tag{11}$$

that the coefficient b is positive.

That the coefficient b can be positive for a binary mixture, where the Soret effect drives the more dense component to the warm lower boundary, is due to the fact that convection modifies the temperature and concentration gradient to a different extent [11]. For the velocity field given by Eq. (2) with $\omega = 0$ one obtains for the mean steady-state temperature and concentration distribution

$$\frac{d\langle T\rangle}{dz} = -\beta \left[1 + \frac{v^2}{\chi^2 q^2} \cos 2q_z z \right] \tag{12}$$

$$\frac{d\langle c\rangle}{dz} = \frac{k_T}{T}\beta \left[1 - \frac{2v^2}{D^2 q^2} \sin^2 q_z z \right] \tag{13}$$

Since $D \ll \chi$ it follows from the above equations that convection has a much stronger influence on the concentration gradient than on the temperature gradient. The result is that finite amplitude convection leads to a strong reduction of the stabilizing concentration gradient while the destabilizing temperature gradient is hardly affected. Taking into account the modification of the temperature and concentration gradient one obtains for the energy balance

$$\dot{F} = \dot{F}^{v}_{kin} + \dot{F}^{g}_{kin}$$

$$= \frac{g\rho\alpha\beta}{\chi q^2} (1 + S \frac{\chi}{D}) v^2 - \frac{g\rho\alpha\beta}{2\chi^3 q^4} (1 + 3S \frac{\chi^3}{D^3}) v^4 \qquad (14)$$

Thus for

$$S < - \frac{1}{3} \frac{D^3}{\chi^3}$$

the coefficient of the v^4 term becomes positive and finite amplitude instability becomes possible.

So far no similar treatment for homeotropic liquid crystals has been worked out. The situation is complicated here by the fact that the steady state temperature distribution will not only be affected by convection but also by the effect of non-linearities on the orientation of the director [12].

IV. CONCLUDING REMARKS

The analysis of the foregoing sections allows us to state that the physical mechanism underlying both oscillatory instability and finite amplitude instability in thermal convection is the competition between a stabilizing effect with a long relaxation time and a destabilizing effect with a short relaxation time.

REFERENCES

1. S. Chandrasekhar, Hydrodynamic and Hydromagnetic Stability (Clarendon Press, Osford, 1961).
2. C. Normand, Y. Pomeau and M.G. Velarde, Rev. Mod. Phys. 49, 581 (1977).
3. J.S. Turner, Ann. Rev. Fluid Mech. 6, 37 (1974).
4. R.S. Schechter, M.G. Velarde and J.K. Platten, Adv. Chem. Phys. 26, 265 (1974).
5. H.N.W. Lekkerkerker, J. Phys. Lett. (Paris), 38, L-277 (1977).
6. E. Guyon, P. Pieranski and J. Salan, C.R. Acad. Sc. Paris, 287 Série B, 41 (1978).

7. E. Guyon, P. Pieranski and J. Salan, J. Fluid Mech. 93, 65 (1979).
8. S. Chandrasekhar E, Max Planck Festschrift 1958 (Veb. Deutscher Verlag der Wissenschaften, Berlin 1958) p. 103.
9. H.N.W. Lekkerkerker, Physica 93A, 307 (1978).
10. H.N.W. Lekkerkerker, J. Phys. (Paris) 40, C3-67 (1979).
11. H.N.W. Lekkerkerker, Light Scattering in Liquids and Macromolacular Solutions, V. Degiorgio, M. Corti and M. Giglio, Eds. (Plenum Publishing Corporation, 1980),p. 231.
12. E. Dubois-Violette and F. Rothen, J. Phys. (Paris) 40, 1013 (1979).

INVESTIGATION OF FLUCTUATIONS AND OSCILLATORY STATES IN RAYLEIGH-BÉNARD SYSTEMS BY NEUTRON SCATTERING

T. Riste and K. Otnes

Institute for Energy Technology

2007 Kjeller, Norway

1. INTRODUCTION

In order for the neutron scattering technique to be useful in the study of the Rayleigh-Bénard problem the sample must be a nematic liquid crystal. The elongated molecules have an anisotropic scattering power, i.e. the scattering depends on the orientation of the molecules relative to the (horizontal) scattering vector. There exists a well-known coupling between orientation and flow in nematics [1], and the onset and changes in the flow can be detected through the ensueing change in the molecular orientation.

There is an impressive literature [2] on instabilities in liquid crystals, including experiments by macroscopic methods. These studies have, however, had less emphasis on the phase-transition properties of the instabilities. Graham [3] has called attention to the possibility of using liquid crystal samples for studying the role of fluctuations at convective instabilities. The feasibility of using neutron scattering for such studies of fluctuating and oscillating states has been demonstrated earlier in [4] and in [5,6] , respectively. The rationale for such experiments is the following: At the microscopic level liquid crystals are known to exhibit strong orientational fluctuations [1]. Fluctuations at the long-wavelength cutoff, given by the dimension of the vessel, are most strongly excited. These inner fluctuations couple to hydrodynamic fluctuations [3] and are sufficiently slow that they may be detected in realtime neutron-intensity measurements. Using the whole convection roll as the scattering volume, the experiment is sensitive to the most relevant wavelength range.

2. OBSERVATIONS OF OSCILLATORY STATES IN A HOMEOTROPIC NEMATIC SAMPLE

Our previously reported experiments [4-8] were on a nematic sample in planar configuration, i.e. with molecular axes (the director) horizontally aligned by a magnetic field. For a homeotropically aligned sample (i.e. director parallel to the vertical temperature gradient) and heated from below, the first theoretical predictions and experimental results were for a quiescent, conductive state only[9]. Already six years ago we found experimentally [10] that this configuration led to an oscillatory, convective state. Stimulated by later theoretical [11,14] and experimental [12] work we have resumed these experiments. Here we give a preliminary report on the character of the oscillations.

The sample was, as in all of our previous experiments, fully deuterated para-azoxyanisole (PAA). It was contained in an aluminium vessel of parallel-epipedic shape with dimensions (3x30x30) mm^3. Of the long dimensions one is parallel to the horizontal scattering vector and the other to the vertical magnetic field. A vertical temperature difference ΔT was obtained by setting the difference of the power fed to the electrical heating elements at the top and bottom of the vessel. ΔT could be kept constant within $\pm$ .01° for hours. The convection threshold depends on the magnetic field, but is $\sim 1^{\circ}$ for H $\sim$ 100 Ørsted. The temperature at the midheight of the aluminium side wall was kept constant at 121°C.

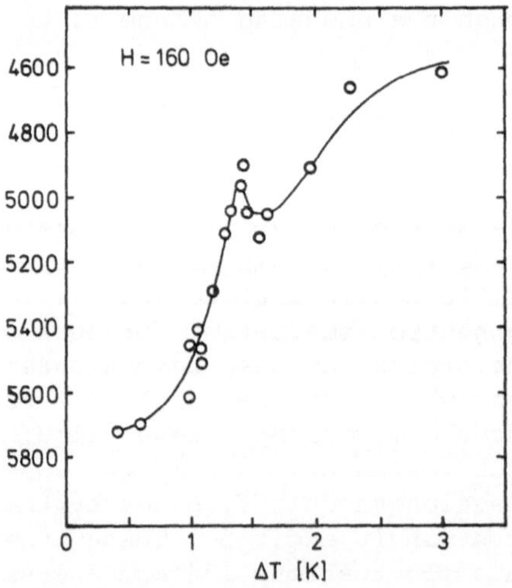

Fig. 1 Neutron intensity in arbitrary units, versus vertical temperature difference across the sample. The curve is a guide to the eye

In Fig. 1 we show the scattered neutron intensity as a function of the temperature difference ΔT. Each point is an average of several hours of counting. In this vertical-field con-

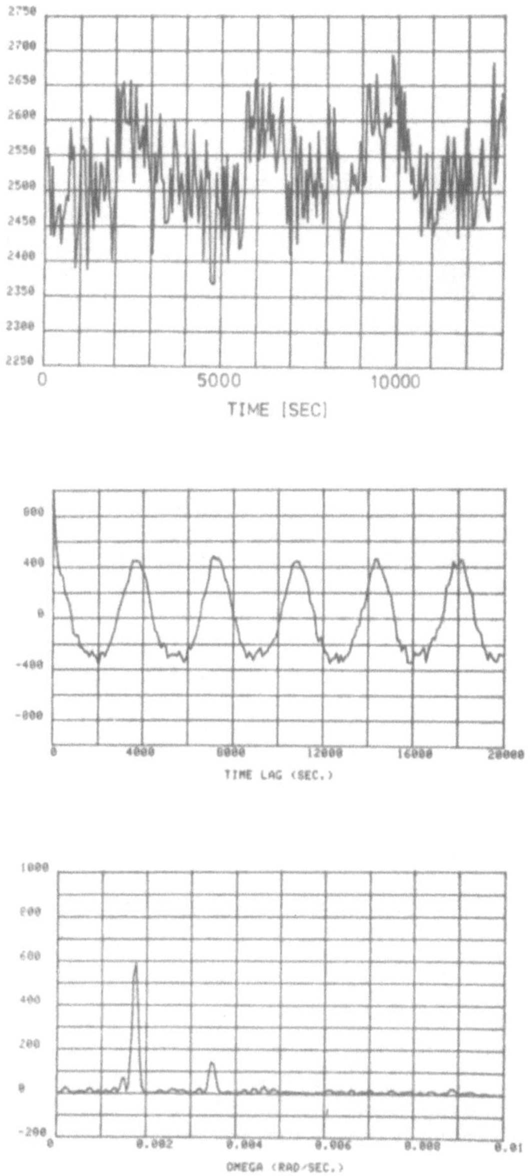

Fig. 2 Example of data in the oscillatory region at $\Delta T = 1.5^{\circ}$ (a) shows a portion of the raw data, (b) and (c) are the autocorrelation function and the power spectrum, respectively, of an extended set of the same data.

figuration the intensity is a maximum in the conductive state. By inverting the intensity scale the curve displays the phase-transition character of the instability. There is indication that the system enters the convective state through pretransitional fluctuations. Other data at lower applied fields presently being studied, reveal more clearly that the instability is an inverted bifurcation. We believe that the family of curves [15] can be explained from an evolution of the Landau potential, as in Fig. 3 of Velarde's paper [16]. The convective state is mostly found to be oscillatory (a limit cycle), but only within a narrow region of ΔT do the oscillations persist with low damping. The nonlinear character of the oscillations is shown in Fig. 2 a,b,c. The presence of the second harmonic (c) in the power spectrum reflects the anharmonic appearance of the raw data (a) and of the autocorrelations (b). A corresponding distortion of the limit cycle has been predicted by Velarde and Antoranz [14]. The period of the oscillations agrees quite well with the theoretical predictions [13].

REFERENCES

1. de Gennes, P.G., The Physics of Liquid Crystals (Clarendon Press, Oxford, England 1974).
2. For a review, see Dubois-Violette, E., Durand, G., Guyon, E., Manneville, E., and Pieranski, P., in Liquid Crystals, editor L. Liebert (Academic Press, N.Y. 1978) p. 147.
3. Graham, R., in Fluctuations, Instabilities and Phase Transitions, editor T. Riste (Plenum, N.Y., 1975) p.313.
4. Pedersen, A.M. and Riste, T., Z.Phys.B 37, 171 (1980)
5. Møller, H.B. and Riste, T., Phys.Rev.Lett. 34, 996 (1980)
6. Otnes, K. and Riste, T., Phys.Rev.Lett. 44, 1490 (1980)
7. Møller, H.B., Riste, T., and Otnes, K., in Fluctuations, Instabilities and Phase Transitions, editor T. Riste (Plenum, N.Y., 1975) p. 313.
8. Riste, T., Otnes, K., and Møller, H.B., in Proceedings of the International Conference on Neutron Inelastic Scattering, 1977 (International Atomic Energy Agency, Vienna, 1978) Vol I, p. 511
9. Pieranski, P., Dubois-Violette, E., and Guyon, E., Phys.Rev. Lett. 30, 736 (1973)
10. Otnes, K., and Riste, T., unpublished
11. Lekkerkerker, H.N.W., J.Physique Lett. 38, L-277 (1977) See also preceding paper this volume.
12. Guyon, E., Pieranski, P., and Salan, J., J. Fluid Mech. 93, 65 (1979).
13. Velarde, M.G., and Zussiga, I, J. Physique 40, 725 (1979)
14. Velarde, M.G. and Antoranz, J.C., Phys.Lett. A72, 123 (1979), Phys. Lett. A80, 220 (1980)
15. Riste, T., and Otnes, K., to be published
16. Velarde, M.G., this volume

A RAYLEIGH BÉNARD EXPERIMENT: HELIUM IN A SMALL BOX

A. Libchaber and J. Maurer

Ecole Normale Supérieure, Groupe de Physique des Solides

24 rue Lhomond, 75231 Paris Cedex 05, France

This is a limited excursion in the field of hydrodynamical instabilities, in itself an infinite domain of research. It is first restricted to a Rayleigh Benard experiment, and we will study the case of a small Prandtl number fluid ($0.4 < P < 1$). To simplify the problem some more we shall restrict ourselves to the geometry of a small rectangular box with two or three convective rolls present. This somewhat artificial case allows us to truncate the degrees of freedom of the system and thus to define some simple bifurcations to turbulence.

Thermal convection provides a simple experimental example of a non linear physical system. We are dealing with a fluid confined in a box as compared to experiments with unconfined flows where the ordered structure moves in space and time. The control parameter is the temperature difference between top and bottom plates which can be highly stabilized. Local thermal bolometers allow an easy measurement of the temperature profile and temperature oscillations.

A low Prandtl number fluid has the following simple property. Above convection and up to the onset of the first time dependent instability, the so-called oscillatory instability, the convective structure has a two dimensional shape. For Prandtl numbers larger than one the convective structure acquires a three dimensional character before any bifurcation to a time dependent state (cross rolls, zig zag, skewed varicose).

Let us finally try to explain why our experiment is performed in a small confined geometry. In a large cell, G. Ahlers[1] showed that immediately above convection, a low frequency noise is present. We believe that in large cells with many convective rolls the wavenumber selection[2] leads to a perpetual motion of the convective structure which never achieves a state with a stable pattern. In our first experiment[3] with a cylindrical cell of aspect ratio $\Gamma = R/d = 6$ (R : radius, d : height) two distinct time dependent phenomena were observed and are shown on figure 1. As one increases the Rayleigh number above the critical Rayleigh number for convection (R_c), a low frequency noise starts at about twice the convection threshold ($R/R_c \simeq 2$). For $R/R_c \sim 3$ a distinct higher frequency mode appears, the oscillatory instability.

We are thus faced with two distinct phenomena. Using a two rolls convective cell we can freeze out the effect of the wave-number selection and have a stable structure up to the onset of the oscillatory instability.

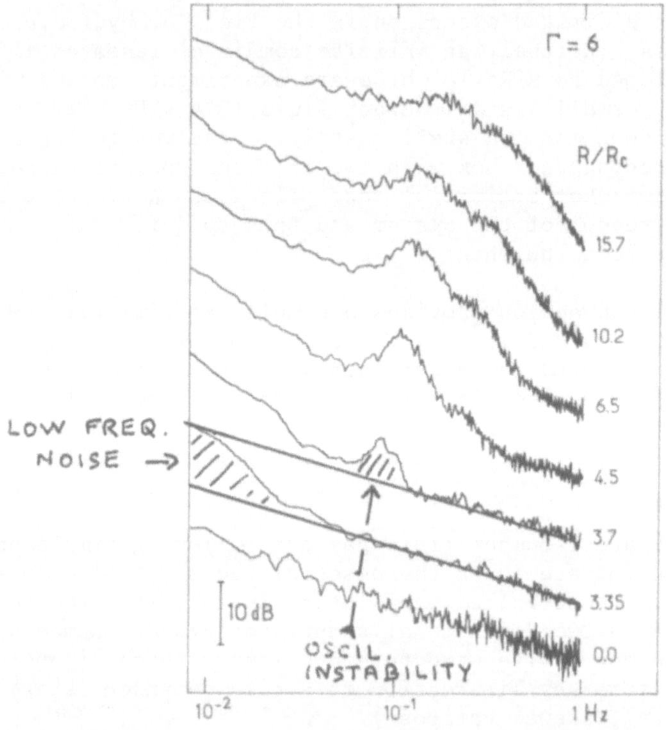

Fig. 1. Power spectra as a function of frequency. Both coordinates are in logarithmic scale. The curves have been translated along the vertical axis for clarity. The lower curve shows the 1/f character of the bolometer.

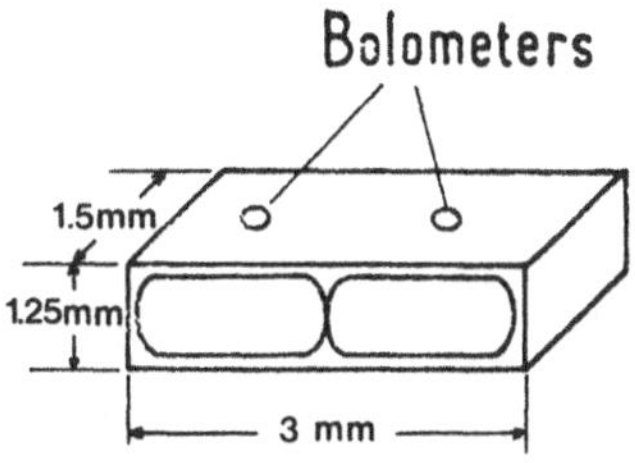

Fig. 2. Schematic of the experimental cell with the position of the local bolometers.

I. THE EXPERIMENT

This is just an outline of the experimental set-up and we refer to previous publications for a more detailed presentation. For our confined cell we use a rectangular geometry instead of a circular one to lift the degeneracy of the convective roll pattern. We thus get a non ambiguous pattern with rolls perpendicular to the largest lateral side, as shown by Stork and Müller[4]. A typical sample shape and dimension is presented on Fig. 2.

The fluid used is liquid Helium (^{4}He). We change the Prandtl number by varying the temperature in a range from 2.5 to 4.5°K and the pressure from 1 to 5 atmospheres. We refer to the NBS chart[5] for the values of the Prandtl number and all the relevant physical constants of the fluid. The Prandtl number can be easily changed from 0.48 to 1.

Given our typical cell height, $d \cong 1$ mm, the onset of convection corresponds to a ΔT in the millikelvin range. The thermal time constant for heat diffusion from bottom to top plate is around twenty seconds.

$$\tau \cong \frac{d^2}{\kappa} \cong 20 \text{ s} \qquad\qquad \kappa \sim 5\ 10^{-4} \text{ cm}^2/\text{sec}$$

An important experimental character of our set-up is that the thermal conductivity of the fluid is very close to the thermal conductivity of the lateral boundaries. In contrast the top and bottom plates have a thermal conductivity extremely high compared to the fluid. This is one of the advantages of low temperature.

Table of thermal conductivities

Bottom plate : high purity copper	∿ 100 Watt/cm °K
Top plate : sapphire crystal	∿ 10 Watt/cm °K
Helium	∿ 1 to 5 10^{-4} Watt/cm °K
Lateral side walls : araldite, nylon, teflon	∿ 1 to 5 10^{-4} Watt/cm °K

The overall set-up is shown on Fig. 3.

We use two bolometers to measure the local temperature as shown on Fig. 2. The bolometers are inserted in the top sapphire plate. They are cut and machined from an Allen-Bradley resistance (the

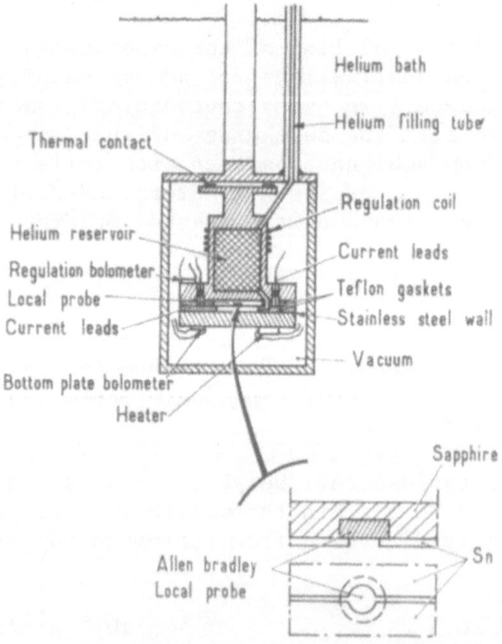

Fig. 3 The upper part is the detail of the experimental system. The lower part is an enlarged view of the local probe seen both in a transverse cut and from the bottom. Dimensions are not conserved.

formula for the variation of the resistance with temperature is

$$\mathrm{Log}\ R + \frac{C}{\mathrm{Log}\ R} = A + \frac{B}{T}\ .)$$

The effective dimension of a bolometer is less than 0.1 mm. The noise factor of the bolometers has a 1/f overall variation in the frequency range where the experiment is performed (refer to Fig. 1 for $R/R_c = 0.0$).

We use a balanced bridge at helium temperature and a lock-in detection to record the temperature signal from each bolometer.

The last important point concerning the experiment is the thermal regulation and the way the temperature constraint ΔT is applied. In our set-up the top plate only is highly regulated, within 10^{-6} K. The whole system is under very high cryogenic vacuum. We heat the bottom plate with a regulated power supply. The sample geometry is such that the thermal conductivity through the helium liquid is very small compared to the thermal conductivity through the lateral boundary. If we use an electronic image let us say the sample impedance is larger, by two orders of magnitude, than the parallel impedance of the lateral walls. We are thus applying a "constant voltage source" which means that we impose a stable temperature difference ΔT. This point if often overlooked in many experiments and it is quite a crucial one.

II. BASICS OF CONVECTION : BUSSE[6] THEORY

The Oberbeck-Boussinesq approximation is used, this means that all fluid properties are assumed constant, with the exception of the temperature dependence of the density, taken into account in the gravity term. To get dimensionless numbers one uses d as a length scale, d^2/κ as a time scale, $(T_2 - T_1)/R$ as a temperature scale.

The governing equations become then :

$$P^{-1}\left(\frac{\partial}{\partial t}\vec{v} + (v.\nabla)\vec{v}\right) = -\nabla p + \theta\vec{k} + \nabla^2\vec{v} \qquad (1)$$

$$\nabla.\vec{v} = 0 \qquad (2)$$

$$\frac{\partial}{\partial t}\theta + (v.\nabla)\theta = R\ \vec{k}.\vec{v} + \nabla^2\theta \qquad (3)$$

where $\vec{k}$ is a vertical unit vector opposed to gravity, θ the deviation from the static temperature distribution, ∇p all terms that can be written in gradient form, and where R and P are the Rayleigh number and the Prandtl number.

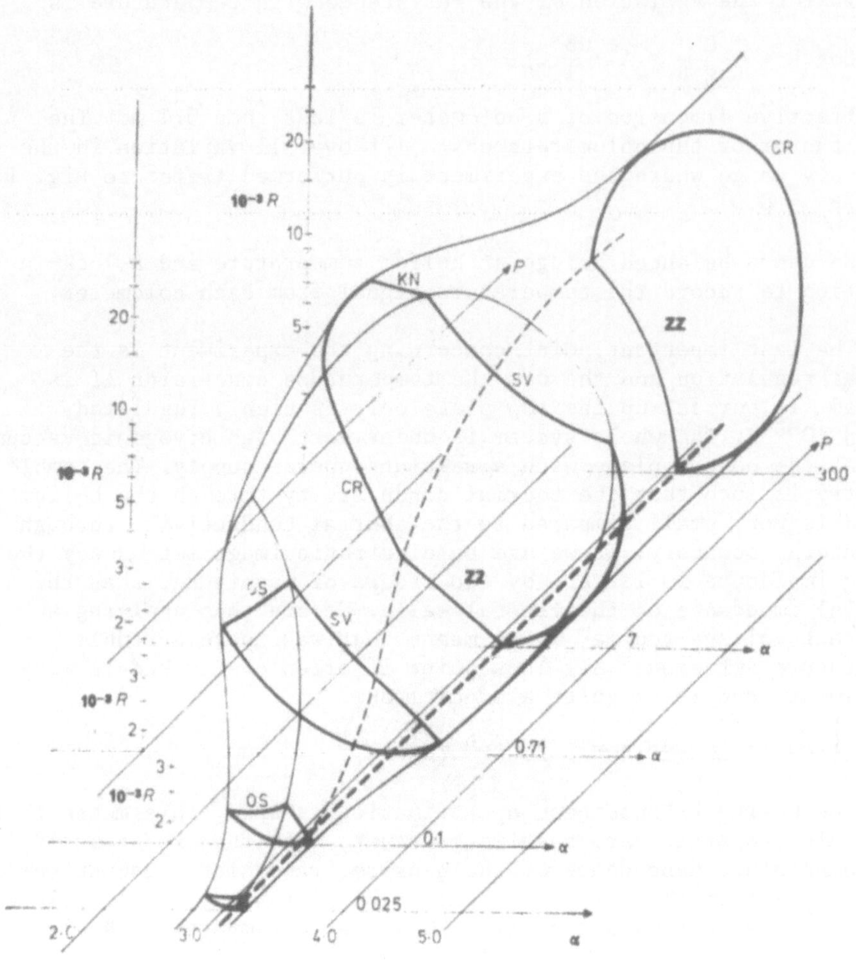

Fig. 4 Region of stable convection rolls in the R.P.α space. OS is the oscillatory instability, SV the skewed varicose, CR the cross roll, KN the knot and ZZ the zigzag one (from ref. 6).

For a good physical interpretation of the onset of convection and the meaning of the Rayleigh number see Velarde and Normand[7] article in Scientific American.

The onset of convection defines a critical Rayleigh number and a critical wavenumber for a fluid layer of infinite horizontal extent

$$R_c = 1707 \qquad \alpha_c = 3.117$$

The wavenumber is defined as $\alpha = \frac{2\pi d}{\lambda}$ where λ is the horizontal

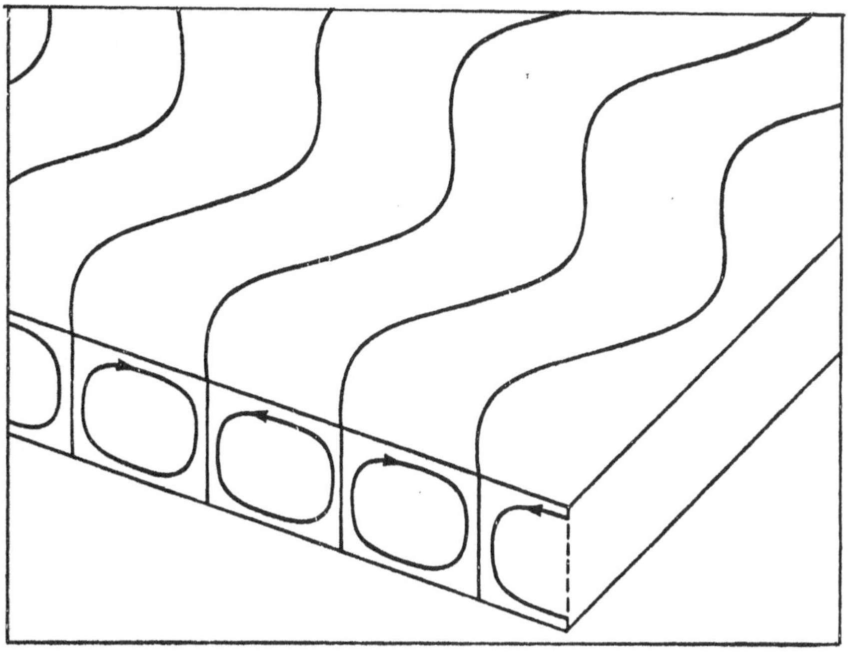

Fig. 5 Sketch of the oscillatory instability (from reference 6).

dimension of the cross section of two adjacent rolls (they define a unit cell as they rotate in opposite directions).

The fact that $\alpha_c \sim \Pi$ as $\lambda_c \sim 2d$ indicates that convective cells with a nearly square cross section are preferred at threshold. The physical reason for this choice is as follows[6] : for $\alpha < \alpha_c$ the potential energy released by the vertical motion is too small compared to viscous dissipation in the horizontal motion. For $\alpha > \alpha_c$ heat conduction between up and down fluid diminishes the buoyancy force.

Whereas for the onset of convection the Rayleigh number is the only relevant parameter, the Prandtl number enters insofar as the nonlinear properties of convection are concerned. Thus, depending on the Prandtl number, various instabilities are observed above the convection threshold. A beautiful analysis of the various possible convective structures has been performed by Busse[6], and is called the "Busse balloon". It is shown on Fig. 4.

In the experimental work reported here ($P < 0.7$) two instabilities will be of concern : the oscillatory instability and the skewed varicose.

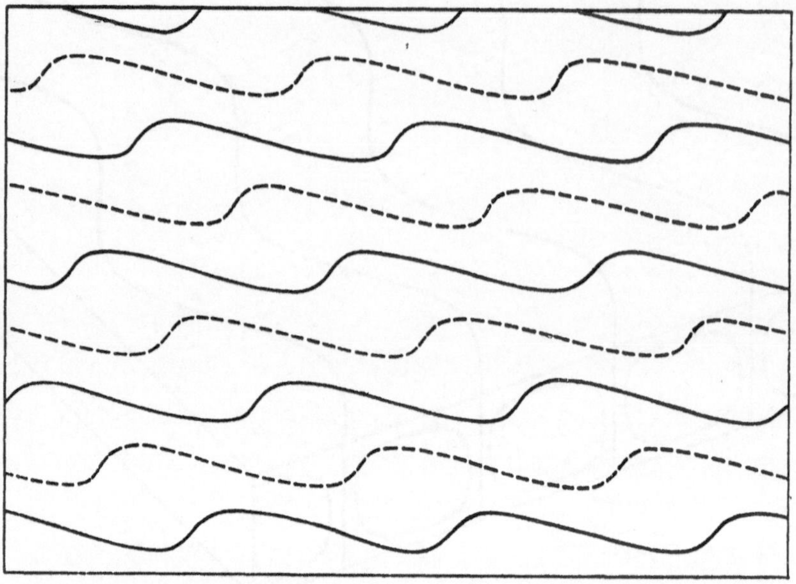

Fig. 6 Distortion of a horizontal pattern of convection rolls by the skewed varicose instability. Full and broken line represent the up and down going motion of the fluid (ref. 6).

For a very small P, the relevant nonlinear term is the momentum advection one, the $(v.\nabla)v$ term of equation (1). Thus starting from bidimensional rolls at R_c, the first instability which appears is a time dependent one, called the oscillatory instability. It consists of a transverse, time dependent, oscillation of the convective rolls, which propagates along the roll axis (Fig. 5).

As one increases P the nonlinear advection term in the heat equation (3), the $(v.\nabla)\theta$ term, can no longer be neglected. Thermal boundary layers at the top and bottom plates, perturb the perfect bidimensional up and down motion of the fluid. This leads to the skewed varicose instability which consists of a static periodic thickening and thinning of the convective rolls (Fig. 6).

III. THE WAVENUMBER SELECTION : EFFECT OF SIDE WALLS AND THE DYNAMICS OF A 3 ROLLS TO 2 ROLLS TRANSITION

Above the onset of convection a band of possible wavenumbers is allowed for the convective structure, in contrast with usual situations in physics where a well defined mode is selected. This effect may be relevant to Ahlers[1] observation that in large aspect

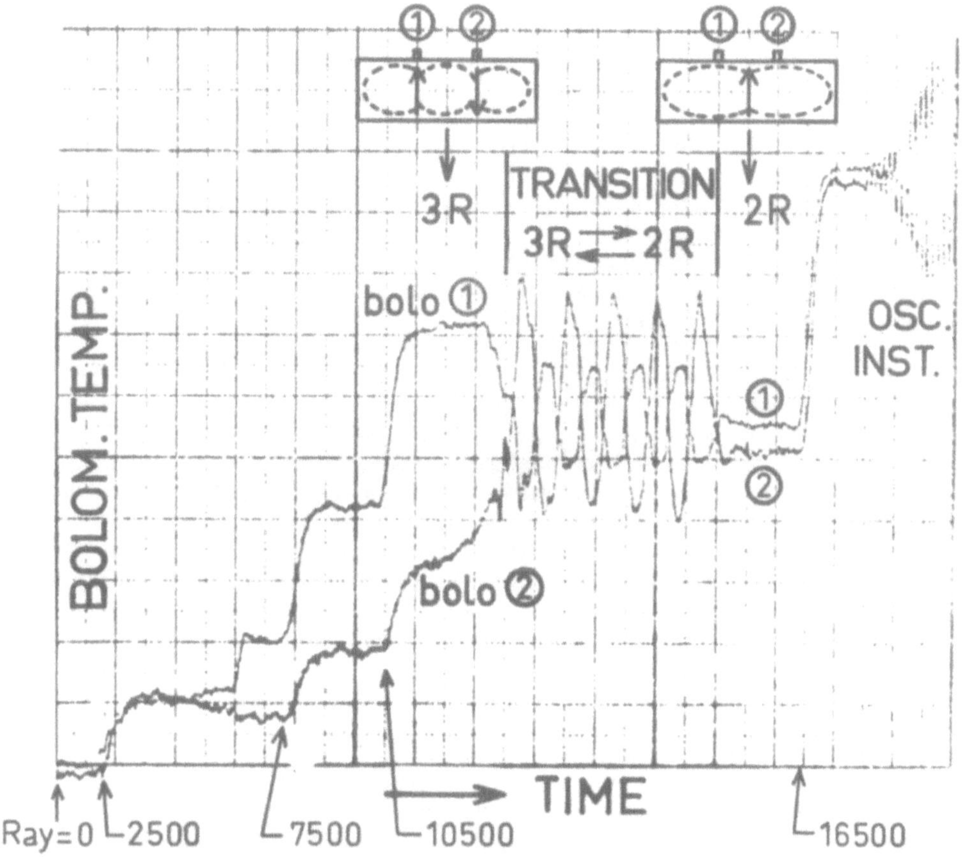

Fig. 7 Recordings of the local temperature seen by the bolometers as one varies the Rayleigh number stepwise. The regions labelled 3R and 2R correspond respectively to a three rolls and two rolls pattern. The fast oscillation is the oscillatory instability. The slow oscillation corresponds to the sloshing of the rolls pattern.

ratio cells, a low frequency noise appears above the onset of convection. In this context the experimental observation by Koschmieder[8], that beyond the onset of convection, there is a general tendency for a wavelength increase of the roll pattern, is important. From a theoretical point of view this problem has been addressed recently by Cross et al[9], Pomeau and Zaleski[10].

In a long and careful paper Cross et al have tried to answer the important question of the effect of sidewalls on the wavelength selection of the convective structure. They consider 2D motion in a large rectangular cell. The main result is that the presence of sidewalls, no matter how distant, severely restricts the possible wavevectors that can occur. Specifically the band of available q about q_0 is reduced from a size $|q| \sim ((R - R_0)/R_0)^{1/2}$ to $|q| \sim (R - R_0)/R_0$. The result is strongly dependent on the Prandtl number P, and also on the sidewall conductivity as compared to the liquid one. For a small Prandtl number they do find a wavelength increase of the roll pattern.

Going back to the experiment with a small cell and a small P, as the Rayleigh number increases, one reaches a transition[11] from 3 rolls to 2 rolls.

Fig. 7 warrants some explanation, and is inserted here for pedagogical reasons. It is a direct laboratory recording of the local temperature seen by the bolometers as a function of time for five different values of the Rayleigh number, applied stepwise. In a three rolls pattern one bolometer is at a higher temperature than the other one as it experiences the arrival of hot fluid, the second one being at a location where cold fluid is going down. In a two rolls pattern they experience the same D.C. temperature. Thus a direct observation of the recording shows that one starts with a three rolls state at low Rayleigh number and that for $R_a = 10500$ the bolometers see a sloshing motion of the rolls before the system transits to a two rolls state. Then for a higher Rayleigh number the oscillatory instability sets in.

The conclusion is thus twofold. First as one increases the Rayleigh number we do observe a wavelength increase of the roll pattern, which in our quantized size cell leads to a transition from three or two rolls. Second this transition is a dynamic one with an instability region where the rolls move back and forth without finding a stable position.

As far as this instability is concerned we have found various possible trajectories. We show in Fig. 8 one of the typical recordings by one bolometer ; the second one sees the same temperature oscillations but in phase opposition as in Fig. 7. The main oscillation period is of about 100 sec (10 mHz) which scales with the diffusion time along the sample $\tau = (d^2/\kappa)(L/d)^2$. It starts for R/R_c close to three. The limit cycle can be qualitatively analyzed as follows.

3 rolls → one damped oscillation to 2 rolls → 3 reversed rolls

It looks as if when the three rolls pattern is destabilized a

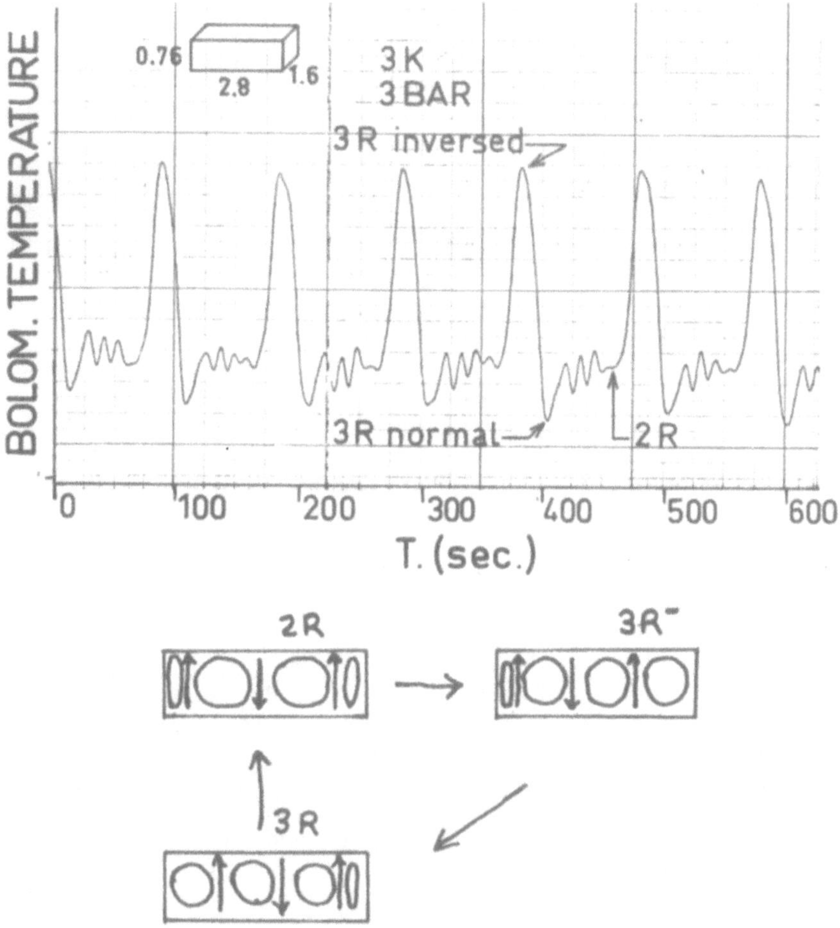

Fig. 8 Recordings of the sloshing motion of the rolls and schematic of the rolls motion in the cell with the limit cycle (3R and 2R mean three and two main rolls).

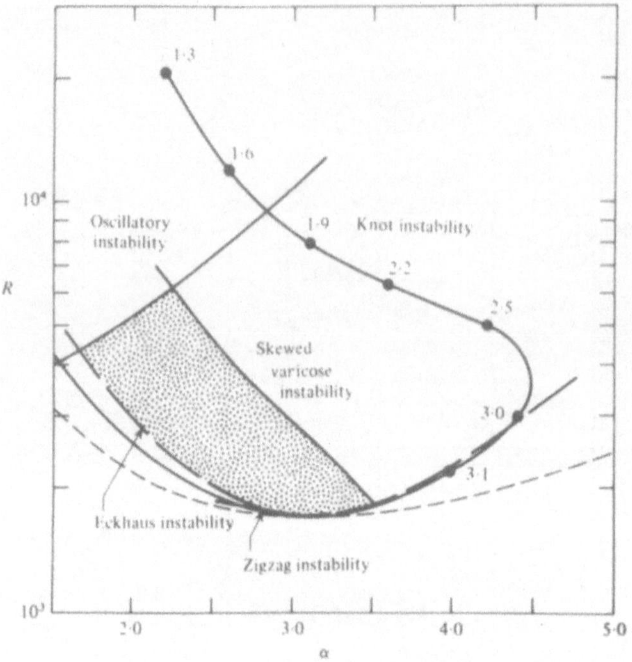

Fig. 9 Stability boundaries of convection rolls for a wavenumber $\alpha = 3.117$ (solid line) and $\alpha = 2.2$ (dashed line) (ref. 12).

fourth roll germinates at one lateral boundary, and this leads to the following consecutive pattern.

In the next chapter we will start to study the routes to chaos for restricted geometries with two convective rolls present. We will see that the bifurcations to turbulence can be simple analyzed in a model with two or three oscillators present. But one should keep in mind that for larger aspect ratio cells, and even with three rolls present, the routes are more complex and should always include the wavelength selection problem which leads to very low frequencies oscillations or noise.

IV. THE OSCILLATORY INSTABILITY

From now on we shall reduce our study to the case of a confined geometry with two convective rolls present. A typical experimental recording of the oscillatory instability was shown on Fig. 7 for $R_a = 16500$. It represents the first bifurcation to a time dependent state in a two rolls geometry.

The theory for the stability boundary of the oscillatory instability as a function of the Rayleigh number and Prandtl number is due to Busse and Clever[12]. Fig. 9 is taken from their paper.

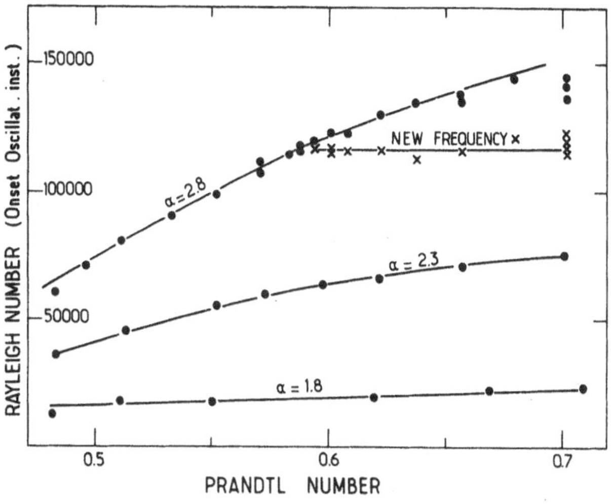

Fig. 10 Onset of the oscillatory instability (black dots). The crosses correspond to the onset of a new time dependent oscillation which we associate to the presence of the skewed varicose instability.

Whereas the onset of convection is related to a critical Rayleigh number, the onset of the oscillatory instability is related to a critical Reynolds number of the convective flow

$$\frac{vL}{\nu} = R_{e\ cr}$$

Taking for the velocity the vertical convective velocity, for L the rolls wavelength λ one gets

$$\frac{\kappa}{d}\,(R - R_c)^{1/2}\,\frac{\lambda}{\nu} = R_{e\ cr}$$

$$\boxed{\frac{(R - R_c)^{1/2}}{\alpha\, P_r} = R_{e\ cr}}$$

Thus the critical Rayleigh number for the onset of the oscillatory instability increases with the wavenumber and the Prandtl number as shown on Fig. 9.

The experimental results[13] for a cell with two rolls are shown on Fig. 10. In the experiment we define somewhat arbitrarily the wavenumber by $\alpha = 2\Pi d/L$ where L is the largest lateral side. We used three different samples. Busse analysis is relevant to a sample of infinite lateral dimensions whereas our experiment is in a confined geometry. Nevertheless the agreement is qualitatively good. The

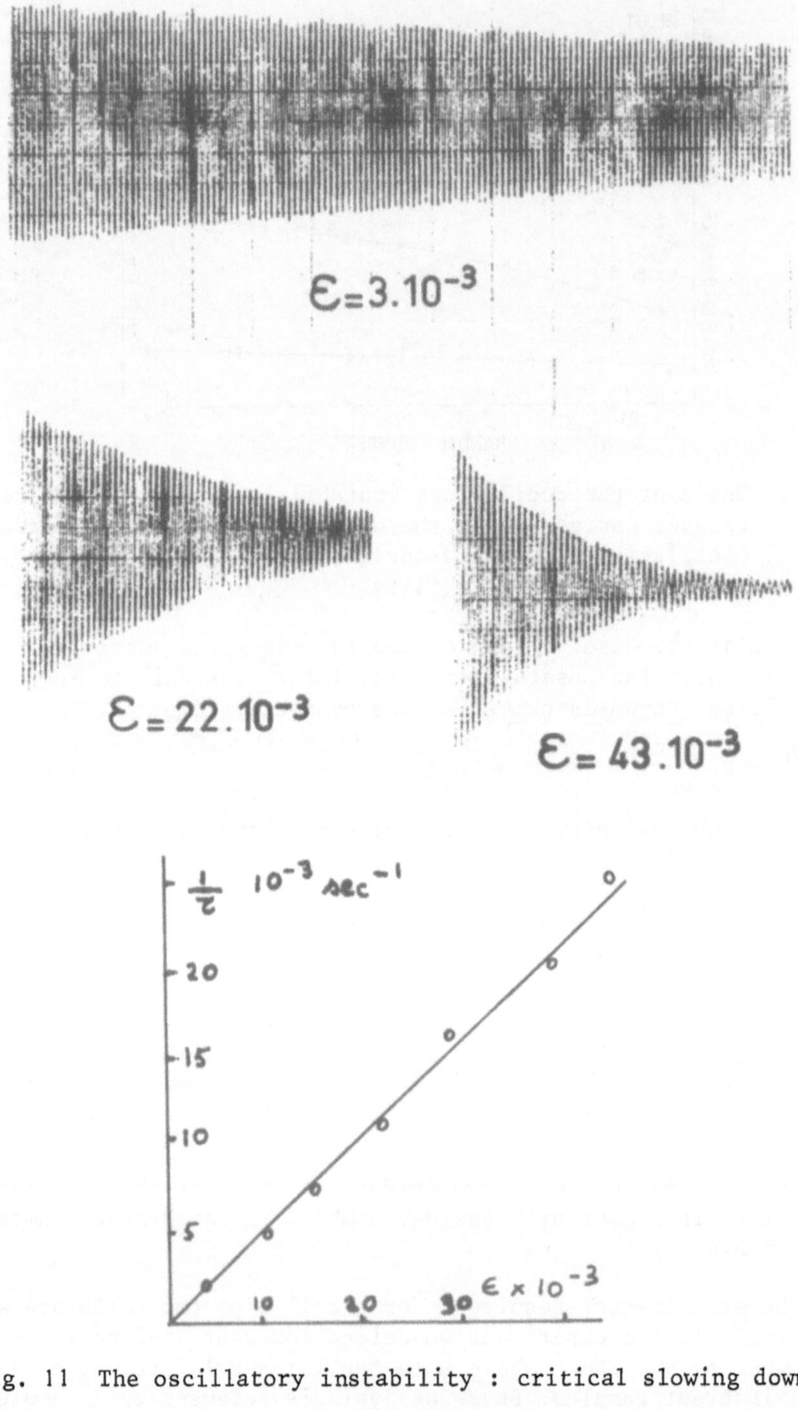

Fig. 11 The oscillatory instability : critical slowing down

main difference introduced by the lateral walls seems to be an increase by about an order of magnitude of the value of the Rayleigh number at onset of the oscillatory instability.

Before the onset of the oscillatory instability the vorticity field is horizontal. A vertical component of the vorticity is introduced by this instability. Some general considerations developed by

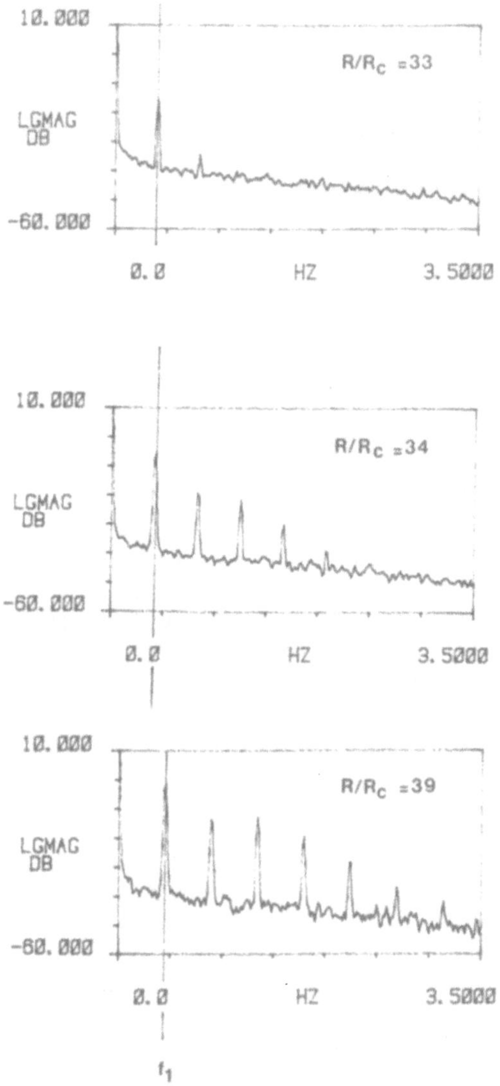

Fig. 12 Evolution of the frequency spectrum of the oscillatory instability as a function of R/R_c, for the sample corresponding to Fig. 2.

Siggia and Zippelius[14] are relevant now. The presence of a vertical vorticity term, whether it comes from the presence of a defect in the structure, a curvature of the rolls caused by the lateral boundaries, or the onset of an oscillating mode, may lead to a time dependent motion of the convective structure. In fact large scale changes in pattern require a vertical vorticity.

The onset of the oscillatory instability shows the characteristic slowing down of the Landau Hopf bifurcation. We test it as follows.

Starting from a Rayleigh number just above the onset of the oscillatory instability, R_0, we decrease the Rayleigh number by various amounts. We then get exponential decreases of the amplitude of the oscillation. Plotting the inverse of the relaxation time, inferred from this exponential decrease, as a function of ε ($\varepsilon = 1 - R/R_0$) we get a linear dependence.

This is shown on Fig. 11 for a sample of $\alpha = 2.75$ and $P_r = 0.48$.

V. TWO OSCILLATORS : ENTRAINMENT AND LOCKING

Starting from the onset of the oscillatory instability, the harmonic content of the oscillator (called from now on f_1) increases with Rayleigh number[15] as shown on Fig. 12.

For a further increase of the control parameter a new bifurcation appears associated to a second oscillator of frequency f_2. There is no theory as for the physical origin of this new mode. Its frequency is experimentally related to the largest lateral size, decreasing as L/d increases. This oscillator may be related to an instability of the boundary layers at the top and bottom plate.

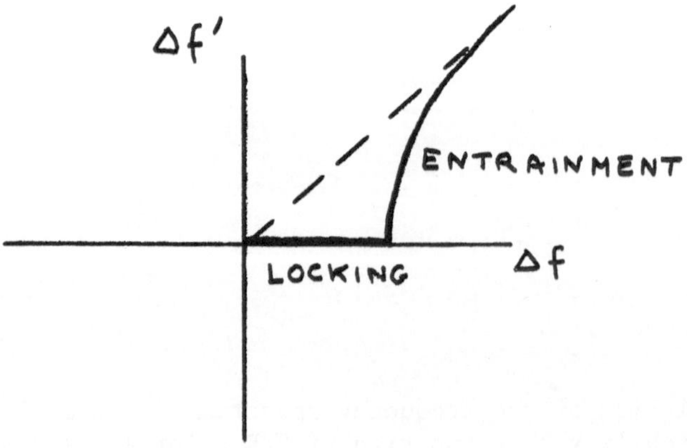

Fig. 13 Theoretical curve of entrainment and lock-in[16]

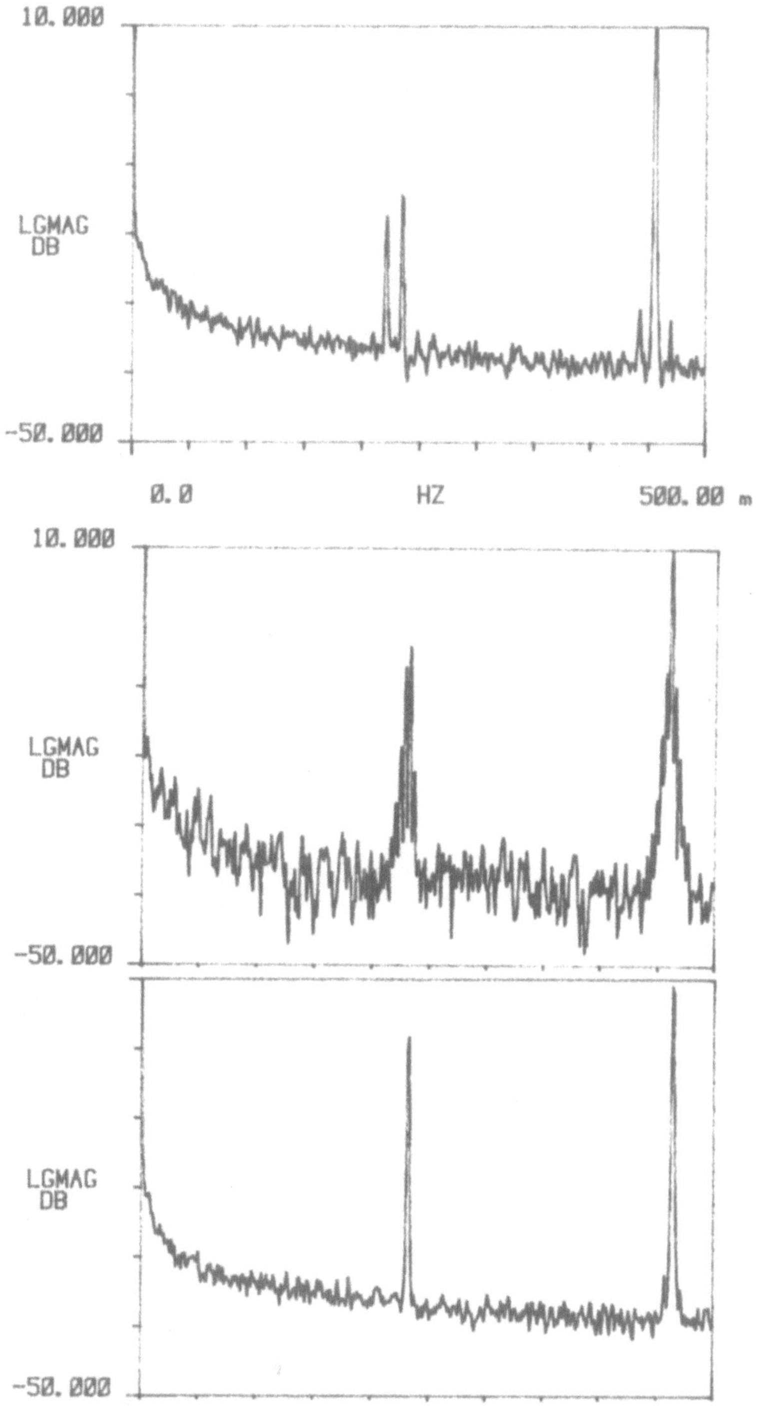

Fig. 14 Typical scenario : two noncommensurate frequencies, entrainment, locking.

First f_2 is noncommensurate with f_1, defining an invariant torus.

There is then a critical Rayleigh number where f_2 interacts with f_1, leading to an entrainment followed by a locking state[16].

A typical scenario is shown on Fig. 14 (for the sample of Fig. 2). The locking state observed shows no hysteresis and may be compared to the relevant theories in dynamical systems[17]. For a general approach to the theory of locking see Stratonovich et al[18].

In this scenario only one locking state was observed, but in various experiments with cells with largest aspect ratio (lower wavenumber for the rolls) a cascade of locking states exists. Flaherty and Hoppenstead have discussed extensively such regimes for the equation[19]

$$\ddot{y} + \frac{1}{\varepsilon}(y^2 - 1)\dot{y} + y = \frac{B}{\varepsilon}\cos \omega t$$

See also for a beautiful study of coupled relaxation oscillators the work of Gollub et al[20], and for a Rayleigh Benard experiment in water, Gollub and Benson study[21]. The results of one of our experiments[11] is shown on Fig. 15 with two clear lock-in for a frequency ratio 6.5 and 7 showing hysteresis and many other lock-in with no hysteresis

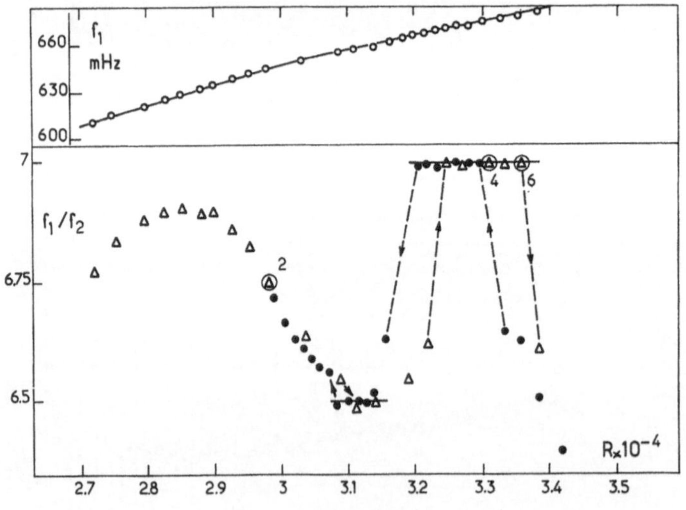

Fig. 15 Top curve : evolution of the frequency f_1. Bottom curve : the various lock-in states.

VI. THE ROUTES TO TURBULENT CONVECTION

In small boxes many routes to turbulent convection have been observed : Gollub and Benson[21] in water ($2.5 < P_r < 5$), Bergé and Dubois[22] in silicone oil ($P_r \sim 130$), Ahlers and Behringer[23] in small cylindrical cells of Helium. In all the studies the general consensus is that the presence of two oscillators is enough to reach a turbulent state. This is in some contradiction with the old picture of turbulence by Landau. It is closer to Lorenz, Ruelle and Takens[24] picture where chaos may occur when only few modes are present.

In our experiments the routes to turbulence are strongly dependent on the Prandtl number. In fact the important factor is the presence or not of static instabilities before we reach the two oscillators regime. More specifically it depends on whether the convective structure, before any time dependent phenomena, has a 2D or 3D character. Summing up our results :

Low P_r (2D character) $f_1 \to f_1, f_2 \to$ locking $\to$ period multiplication

High P_r (3D character) $\begin{cases} f_1 \to f_1, f_2 \to \text{intermittency to chaos} \\ f_1 \to f_1, f_2 \to f_1, f_2, f_3 + \text{chaos} \end{cases}$

We will now detail the two main routes to chaos, i.e. the period doubling cascade and intermittency.

VI.A. The Period Doubling Bifurcation to Chaos, Feigenbaum[25] Scheme

A good general approach to this scenario is given by Collet and Eckmann book[26]. A more physical approach is to present it as a cascade of parametric amplification. In this respect one should note the pionnering work of Rayleigh[27] on parametric amplification.

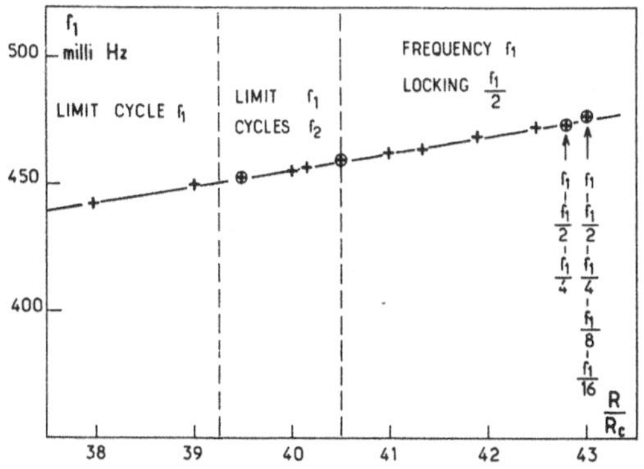

Fig. 16 Evolution of the period doubling cascade

Finally in this topic two textbooks are quite relevant : Landau mechanics and Bender and Orzag[28] on advanced mathematical methods.

Let us come back to our experimental results and pursue the scenario shown on Fig. 14. If we increase the Rayleigh number the subharmonic f/4 appears and then a cascade of subharmonics up to f/16 with a low frequency noise superposed. The evolution is shown on Fig. 16.

The Fourier spectrum of the last two steps of the cascade are shown on Fig. 17. The experimental observations are as follows :

- There is an accumulation point (called R_∞) in Rayleigh number

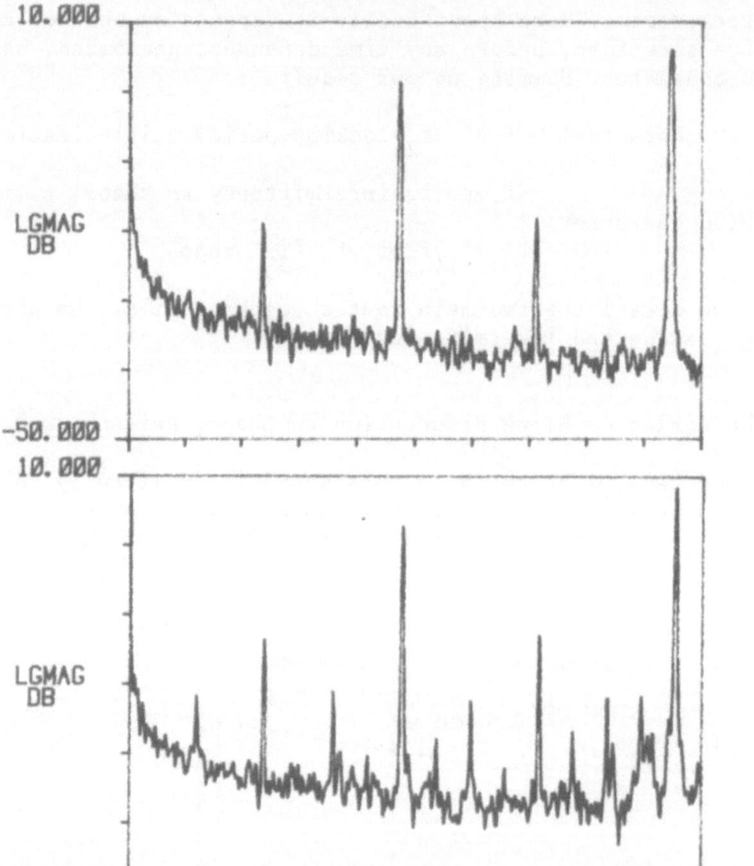

Fig. 17 The last two steps of the cascade

for the cascade but we were unable to observe subharmonics below f/16.

- There is a decrease of the power spectrum of every subharmonic by about 10 dB.

- The low frequency noise has a finite bandwidth and starts at the end of the cascade only.

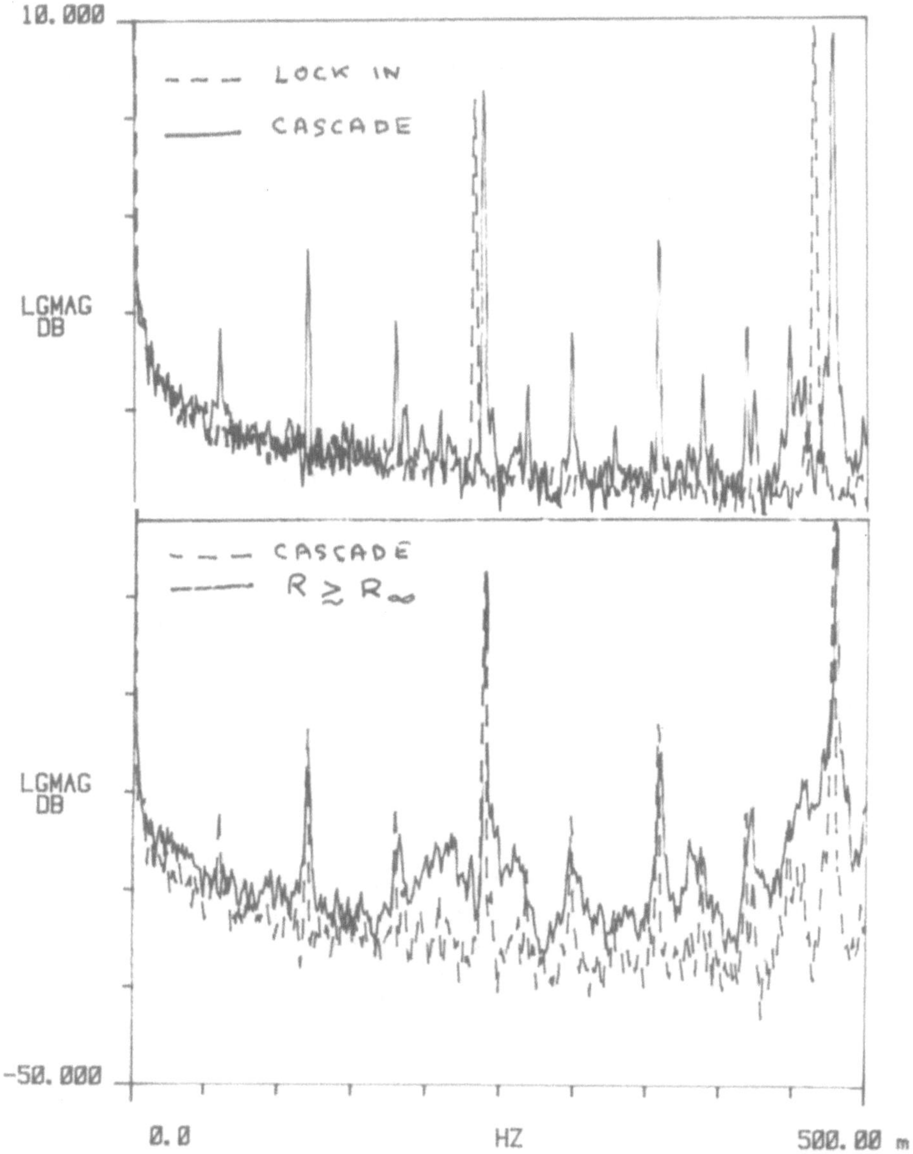

Fig. 18 The abrupt increase of the noise power above R_∞

To show clearly this effect we present two recordings in Fig. 18. In the first one we have superposed two spectra, one for the locking state f_1 and $f_{1/2}$, the other one being the final cascade. One can see that the base line is unaffected and that the noise is just barely modulating the various peaks. In the second one we have superposed the cascade and a recording for a very small increase of the Rayleigh number beyond it. There is a very large increase in the noise power.

There are two essential predictions of the theory.

- The Rayleigh number for the onset of each subharmonic should follow a geometric progression

$$\frac{R_n - R_{n+1}}{R_{n+1} - R_{n+2}} \xrightarrow[n\to\infty]{} \delta = 4.6692$$

- The peaks of the successive subharmonics should decrease by a constant amount in dB as $n \to \infty$.

The experimental precision on δ is quite bad. We get

$\delta = 3.5 \pm 1.5$

Two calculations[29] have proposed a number for the ratio of the successive subharmonics. If we define ϕ_n/ϕ_{n+1} as the ratio of the subharmonic power one gets

	Feigenbaum	Nauenberg Rudnick	This experiment
$\frac{\phi_n}{\phi_{n+1}}$	43	21	7 to 12

There is quite a discrepancy but let us recall that the numbers are asymptotic values for large n, whereas in the experiment we measure only the three first bifurcations.

Let us close this part by pointing out that recently Giglio et al[30] in a similar experiment in water have also measured the period doubling cascade with a better precision.

Another prediction of the theory is that beyond the accumulation point, a mirror image of the cascade should exist[31,26]. In other words the cascade is observed starting from the quiescent state of the fluid and increasing the heating. But the same one should be observed starting from the turbulent state and decreasing the heating. We show on the four next power spectra this phenomena. (see pages 24 and 25).

If we plot, within our precision, the results as a function of the frequency f_1, which is proportional to the Rayleigh number one gets

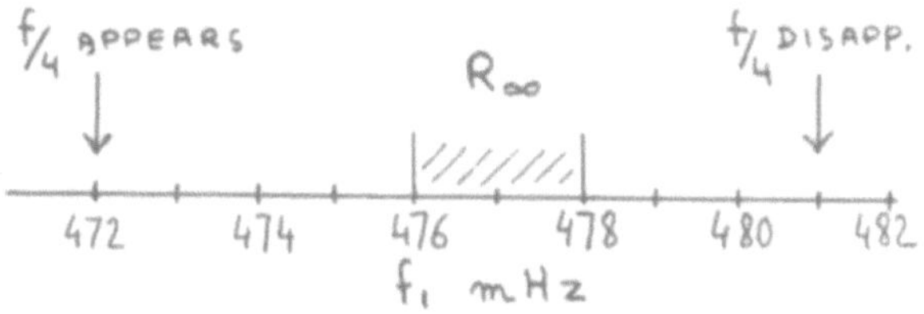

To conclude let us present an experimental recording of the cascade by plotting on an XY recorder the signals from one bolometer as a function of the signal from the other bolometer (let us recall that in our experiment we have a bolometer on top of each convecting roll). In those recordings we have just filtered the high harmonic contents of the frequency f_1.

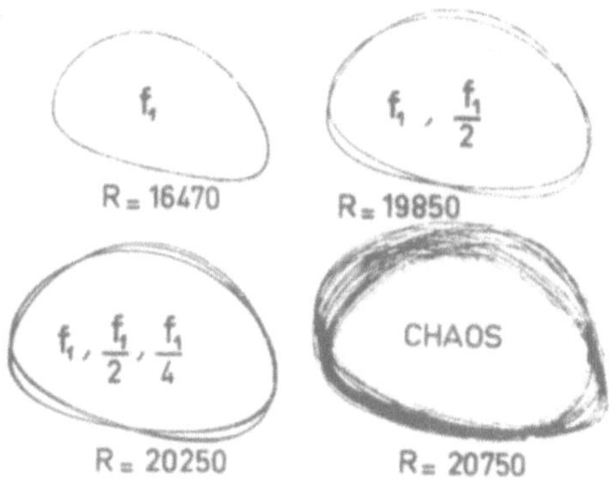

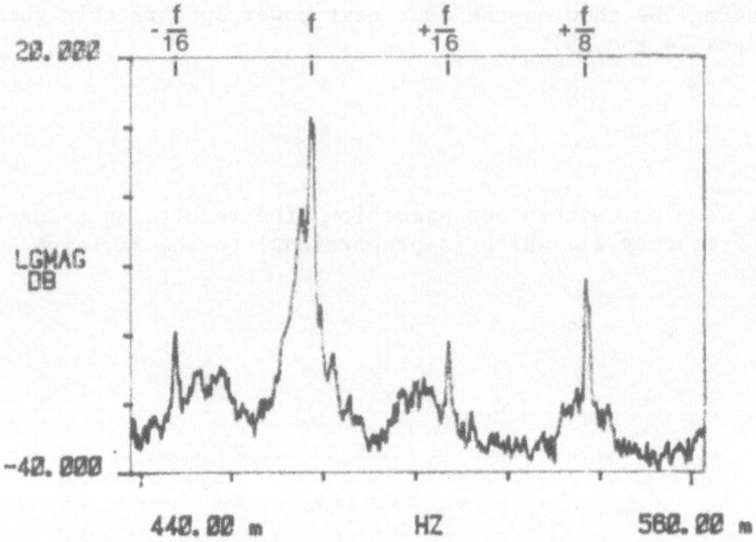

A : above the accumulation point. All the subharmonics are present plus noise.

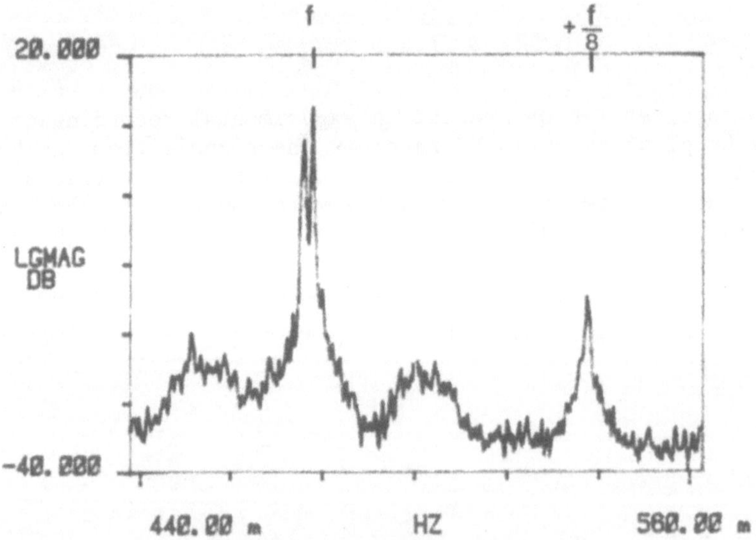

B : a slight increase in Rayleigh number $\frac{f}{16}$ disappears.

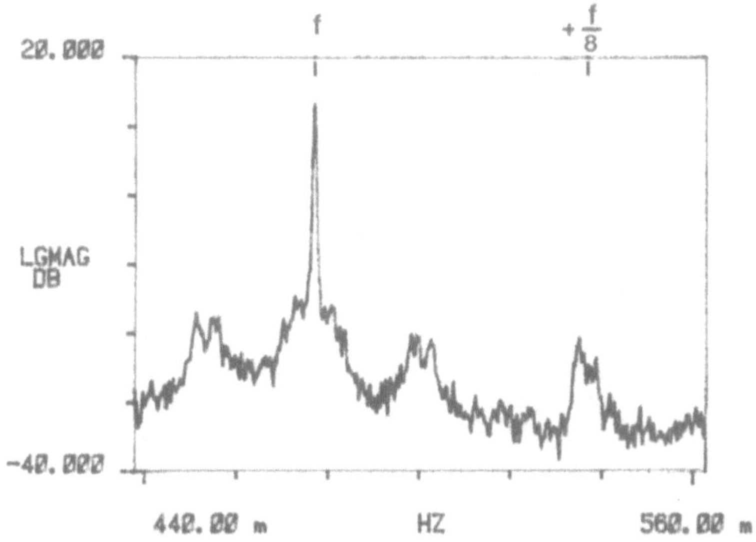

C : new increase in R_a. $\frac{f}{8}$ disappears.

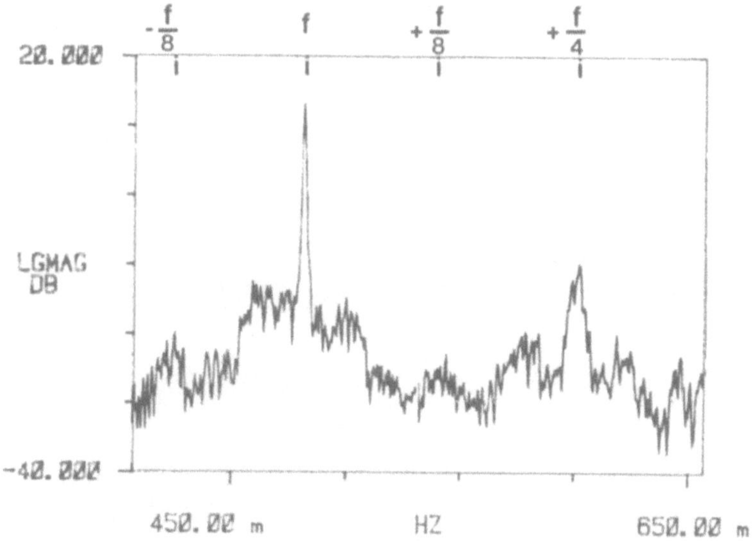

D : larger increase in R_a. $\frac{f}{4}$ disappears. Note that the bandwidth has been changed.

Conclusion

- The qualitative picture proposed by Feigenbaum seems to be correct. Quantitatively there are discrepancies, may be associated to the fact that we observe only the very first bifurcations.

- One of the main difficulties of this experiment is that one often bifurcates to other frequency divisions like the division by 3 or 5. We have observed clearly the following sequences

$$f \to \frac{f}{2} \to \frac{f}{6}$$

$$f \to \frac{f}{2} \to \frac{f}{4} \to \frac{f}{20}$$

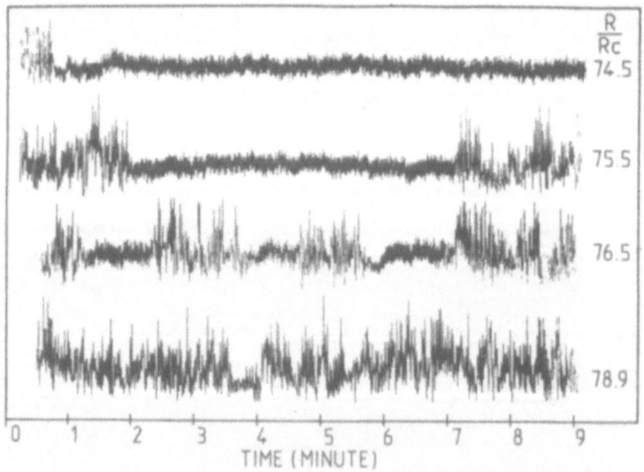

Fig. 19 Intermittent transition to turbulence for a cell $\alpha = 2.7$ $P_r = 0.62$.

VI.B. Intermittency as a Route to Turbulence

Manneville and Pomeau[32] have shown that an intermittent transition to chaos exists for the Lorenz model. Experiments by Gollub[21] in water and Bergé[33] in silicon oil have shown intermittent transitions to chaos.

In our experiments in helium an intermittent transition to chaos is measured[13] at high P and high wavenumber. We first reach a state with two oscillators with uncommensurate frequencies. The transition to turbulence is intermittent for a range of Rayleigh numbers $74.5 < R/R_c < 79$.

Defining the onset of noise bursts at $R/R_c = 74.4$ we find that the laminar time intervals without noise diverge like $\tau \simeq e^{-\beta}$, $1 < \beta < 1.5$. The data are shown on Fig. 19.

REFERENCES

1. G. Ahlers, R.W. Walden, Phys. Rev. Lett., 44:445 (1980).
2. J. Wesfreid, V. Croquette, Phys. Rev. Lett., 45:634 (1980).
3. A. Libchaber, J. Maurer, Journal Phys. Lettres, 39:369 (1978).
4. K. Stork, U. Müller, J. Fluid Mech., 71:231 (1975), 54:599 (1972).
5. R.D. Mc Carthy, Thermophysical Properties of ^{4}He, Bur. Stand. Tech. Note n° 631 (1972).
6. F.H. Busse, Report Progr. Physics 41:1929 (1978).
7. M. Velarde, C. Normand, Scientific American, 243:78 (July 1980).
8. E.L. Koschmieder, Adv. Chemical Physics, 26:177 (1974).
9. M.E. Cross, P.G. Daniels, P.C. Hohenberg, E. Siggia, Phys. Rev. Lett., 45:898 (1980).
10. Y. Pomeau, S. Zaleski, C. R. Acad. Sc. Paris, 290 (série B):505 (1980).
11. J. Maurer, A. Libchaber, J. Physique Lettres, 40:419 (1979).
12. B.H. Busse, R.M. Clever, J. Fluid Mech., 91:319 (1979).
13. J. Maurer, A. Libchaber, J. Phys. Lettres, 41:515 (1980).
14. E. Siggia, A. Zippelius, "Dynamics of Defects in Rayleigh-Benard Convection", preprint.
15. A. Libchaber, J. Maurer, J. de Physique, Coll. C3, 41:51 (1980).
16. A. Adler, Proc. I.R.E., 34:351 (1946).
17. G. Iooss, Math. Studies, 36, New York (1979).
    G. Iooss, W.F. Langford, Annals of the New York Academy of Sciences, Vol. 327 (1980).
18. R.L. Stratonovich, Topics in the Theory of Random Noise, Gordon and Breach (1967).
    W.E. Lamb Jr., Phys. Rev., 134:429 (1964).
    B. Van der Pol, Phil. Mag., 3:65 (1927).
19. J.E. Flaherty, F.C. Hoppensteadt, Study Appl. Math., 58:5 (1978).
20. J.P. Gollub, E.J. Romer, J.E. Socolar, Journ. Stat. Phys., 23:321 (1980).

21. J.P. Gollub, S.V. Benson, J. Fluid Mech., 100:449 (1980).
22. M. Dubois, Colloque Pierre Curie, Paris (1980).
M. Dubois, P. Bergé, J. Physique, 42:167 (1981).
23. G. Ahlers, R.L. Behringer, Phys. Rev. Lett., 40:712 (1978).
24. L. Landau, E. Lifshitz, Fluid Mechanics, Chapt. 3, Pergamon, Oxford (1959).
D. Ruelle, F. Takens, Comm. Math. Phys., 20:167 (1971).
E.N. Lorenz, J. Atmos. Sci., 20:130 (1978).
25. M.J. Feigenbaum, Phys. Lett., 74A:375 (1979) ; Comm. Math. Phys., 77:65 (1980).
P. Coullet, C. Tresser, A. Arneodo, Phys. Lett., 72A:268 (1979).
26. J. Collet, J.P. Eckmann, "Iterated Maps of the Interval as Dynamical Systems", Birkhaüser (1980).
27. Lord Rayleigh, "The Theory of Sound", Vol. I, Chapt. 3, Dover (1945).
28. C. Bender, S. Orzag, "Advanced Math. Methods for Scientists", McGraw Hill (1978).
29. M. Nauenberg, J. Rüdnick, "University and the Power Spectrum at the Onset of Chaos", preprint.
30. M. Giglio, S. Musazzi, U. Perini, "Transition to Chaos via a Well Defined Ordered Sequence of Period Doubling", preprint.
31. S. Grossmann, S. Thomae, Z. Naturforsch., 32a:1353 (1977).
S. Thomae, S. Grossman, "Correlations and Spectra of Periodic Chaos Generated by the Logistic Parabola", preprint.
A. Wolf, J. Swift, "Universal Power Spectra for the Reverse Bifurcation Sequence", preprint.
B. Huberman, A. Zisook, Phys. Rev. Lett., 46:626 (1981).
32. P. Manneville, Y. Pomeau, Phys. Lett., 75A:1 (1979).
Y. Pomeau, P. Manneville, Physica, D1:219 (1980).
33. P. Bergé, M. Dubois, P. Manneville, Y. Pomeau, J. Phys. Lettres, 41:341 (1980).

PERIOD DOUBLING BIFURCATION ROUTE TO CHAOS

Marzio Giglio, Sergio Musazzi, Umberto Perini

CISE S.p.A.

P.O.B. 12081, 20100 Milano, Italy

A theory recently formulated by Feigenbaum[1,2] predicts that the transition to chaotic behaviour via a sequence of period doubling bifurcations has a universal character. Although at this stage the extent at which the theory is applicable is not entirely clear, it is generally believed that it should hold for a large class of nonlinear systems, provided that phase trajectories remain confined in a phase region of adequately low dimension.

The theory applies to the following situation. By gradual increase of an appropriate stress parameter λ, a nonlinear system is brought into chaotic regime through a sequence λ_n of period doubling bifurcations (chaos resides at the accumulation point of this sequence). The theory introduces two universal numbers which characterize in a precise way this route to chaos.

The first number refers to the way the λ_n sequence is constructed. Indeed as n is increased, the ratio $\delta_n = (\lambda_{n+1}-\lambda_n)/(\lambda_{n+2}-\lambda_{n+1})$ should approach the number $\delta = 4.6692...$ The second number is instead related to some regular features in the spectrum characterizing the dynamic behaviour of the system. If one calls S(i) the amplitude of the spectral component of frequency $f_{1/2^i}$, the ratio S(i)/S(i+1) should approach the value $\mu = 6.5740....$

In this contribution we will describe some experimental results on the transition to chaos of a Rayleigh Benard type instability. Previous work in this area has been presented by Libchaber and Maurer[3,4] and Gollub[5]. The reader is also invited to look at the lecture contribution from Libchaber[6] in this volume, where more recent results obtained in a low Prandtl number system are reported. The system we have investigated is a low aspect ratio Rayleigh Benard cell filled with water. The inner volume of the cell is 25.0 mm wide, 15.0 mm long and 7.9 mm high. The lateral boundaries are made of four 5 mm thick, 7.9 mm high glass plates of optical quality. Top and bottom plates are made out of aluminium alloy, and the cell is assembled by glueing together all the components. This type of assemblage guarantees good uniformity of the vertical temperature gradient on the boundaries. The temperature difference between the plates is controlled by a two stage temperature controller utilizing two peltier heat pumps, symmetrically located one above the top plate, one below the bottom plate. Temperature differences can be established with a short time constant (less than a minute), and keeping the temperature of the horizontal midplane constant. The stability of the temperature difference is better than 2 mdeg, and this should be compared with ΔT_{chaos} which is close to 7°C. All the data have been obtained with an average temperature T = 35.950.

The measurements performed are measurements of the vertical and horizontal temperature gradients. The technique employed is a laser beam deflection technique, and so the quantities actually measured are average gradients along the line traversed by the beam inside the cell. The actual position of the beam for all the measurements was horizontal, parallel to the short side, at midheight and close to the lateral boundary ($\sim$ 3 mm). Beam deflections along the vertical and horizontal axis have been measured with a solid state beam position sensor. The accuracy is of the order of a few microradians, while angular oscillations are typically of the order of one milliradian.

The outputs of the sensor have been used in two ways. Since both outputs are simultaneously available, we can plot one signal against the other on an XY recording system (digital oscilloscope or pen recorder). In this way we record the "orbits" described by the temperature gradient. These orbits are the visualization of the actual motion of the beam centroid. Alternatively each signal can be fed to a Fast Fourier Transform analyzer for spectral analysis.

Before we describe the actual set of data, we must discuss an important point related to the way the system is prepared before a set of measurementsis taken. If the temperature difference is increased in stepwise manner from zero, and measurements are taken at each fixed ΔT, we noticed that the system behaves in a not easily controlled manner. Orbits have a very complicated appearence, and the route to chaos is not uniquely defined. This is probably due to the fact that above the convection threshold the system can assume either a two roll or a three roll planform. Some erraticity in the wavelength selection mechanism may possibly explain the observed behaviour. We discovered however that if we suddenly apply a large temperature difference (larger than ΔT_{chaos}) and then rapidly come back to a smaller value, then the system starts oscillating very regularly. This operation has a permanent effect. If afterwards the temperature difference is changed in a stepwise manner, the system can be brought in a very reproducible manner through a sequence of well defined dynamical states. We have performed some crude shadowgraph observations, in order to understand the planform of the instability. The system retains symmetry with respect to a vertical plane passing through the center and parallel to the short side of the cell. It seems that the oscillations are caused by propagating features passing under the beam at regular time intervals. We believe we are not dealing with oscillations of a simple spatial structure like in the case of the oscillatory instability studied by Libchaber and Maurer[3,4,6]. We will present the results of three series of measurements. Each series has been initiated with one quenching operation, followed by a sequence of temperature difference changes. Measurements have been taken at each fixed temperature difference.

In Fig.1 we report a sequence of orbits at different values of R/R_c. Sequential splitting of the orbits leading to $f_1/8$ can be observed in the figure. We point out that signals are very stable. Indeed, all the orbits in Fig.1 have been retraced at least fifteen times. Orbits at the same R/R_c and taken at different days are virtually superimposable.

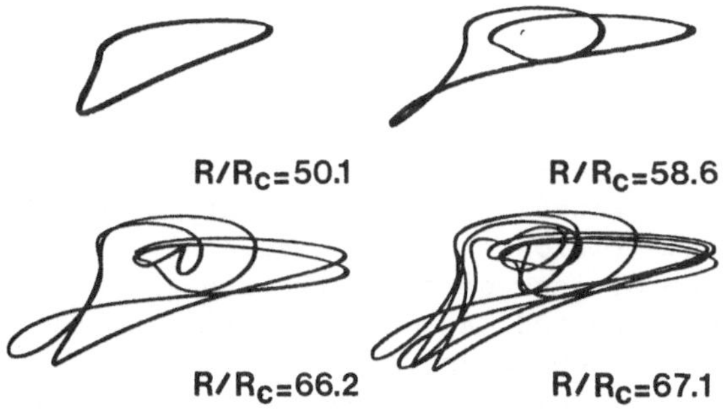

Fig.1. Temperature gradients orbits for different values of R/R_c.

Another point we must make is that orbits occasionally show up in a different fashion. After the $f_1/8$ and $f_1/16$ bifurcation the orbits exhibit a very regular oscillatory motion. More precisely, immediately after the bifurcations the newly split limbs of the orbit come close together, eventually superimpose, swing away in the opposite direction, and then come back again, and the cycle repeats itself. An orbit in this mode is shown in Fig.2. As one can notice, the orbit has a nicely combed appearence, but it never retraces itself, thus indicating that the oscillatory motion occurs at an incommensurate frequency. Notice that after the R/R_c value in Fig.2, the orbit assumes again a perfectly stable form indicated at the end of the sequence in Fig.1. Above the $f_1/16$ bifurcation, however, we have never been able to observe orbits in a non oscillatory motion.

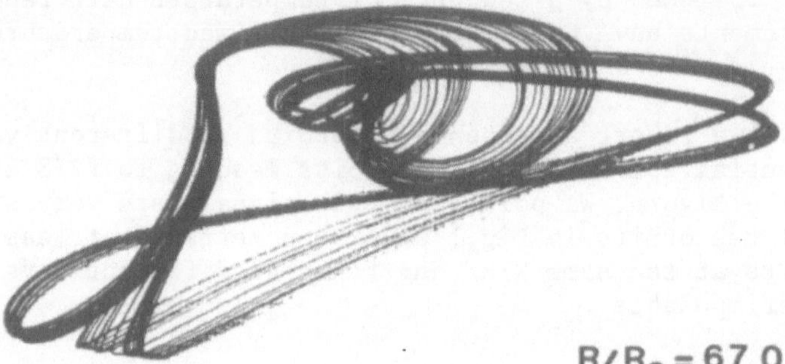

Fig.2. Example of orbit in the oscillatory mode. Spectral analysis reveals that $f_1/8$ components are already present. Compare this figure with the last one in Fig.1.

What we have just described is essential in order to understand some features of the spectra we observe. In Fig.3 we report three spectra. The first is above the $f_1/8$ bifurcation, while the second and third are extremely close to and immediately above the $f_1/16$ bifurcation. Because of lack of space the spectra at the previous transitions are not reported. As one can notice in the first spectrum, the emerging $f_1/8$ component and all its odd multiples are in the form of finely split doublets. This is a consequence of the slow oscillatory motion of the newly split orbit limbs which introduces an almost hundred per cent amplitude modulation on these emerging components. According to the picture, all the other peaks are sharp, while the doublet splitting is constant on all the $f_1/8$ components.

Incidentally, this splitting is very close to $f_1/38$. As we said, further increase in R/R_c makes the orbits become stable, and in the second spectrum the $f_1/8$ components are now sharp. In this spectrum one can barely observe the appearence of the $f_1/16$ components. A small increase in R/R_c makes the emerging doublets appear unanbigously as shown in the third spectrum. Orbit oscillations now occur with a smaller period, and the splitting is very close to $f_1/19$ (twice the splitting at the previous bifurcation).

At this stage, we can attempt to estimate both the μ and δ Feigenbaum numbers.

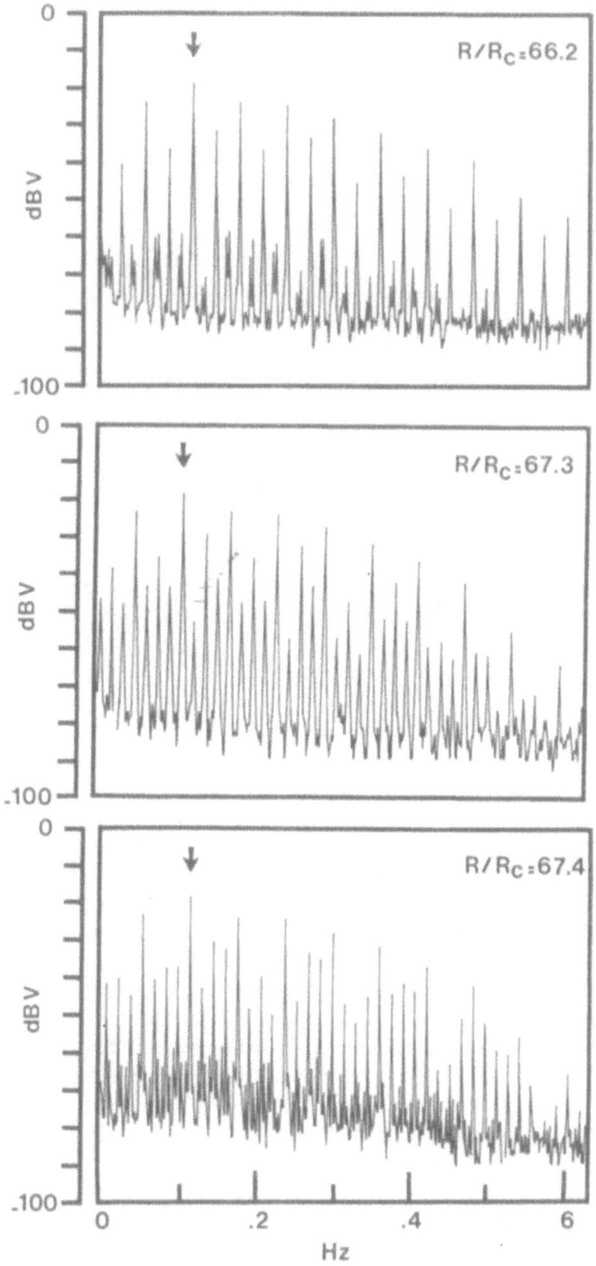

Fig.3. Spectra of the horizontal temperature gradient for different values of R/R_c. At $R/R_c = 66.2$ the system is just above the $f_1/8$ bifurcation. At $R/R_c = 67.3$ the system is extremely close to the $f_1/16$ bifurcation and slightly above it at $R/R_c = 67.4$. The arrow indicates the position of f_1.

We recall that S(i) is the amplitude spectrum value for the $f_1/2^i$ subharmonic, and we define $\mu_{n,i}$ the ratio S(i)/S(i+1) evaluated close to the λ_n bifurcation. The spectra we have used to calculate the μ_n,i values are those close to the $f_1/8$ and $f_1/16$ transitions, that is the first two spectra shown in Fig.3. Since the spectra are very rich, and since there are no clear cut theoretical rules on how to calculate the $\mu_{n,i}$ values on such spectra, we have taken for S(i) the value averaged over all odd multiples. If the averages are calculated up to $2f_1$, we obtain $\mu_{3,1}=4.1$, $\mu_{4,1}=3.8$, $\mu_{4,3} = 3.8$. Averages up to $4f_1$ yield $\mu_{3,1} = 4.0$,$\mu_{4,1} = 4.2$, $\mu_{4,2} = 3.6$. The values for $\mu_{2,0}$, $\mu_{3,0}$, $\mu_{4,0}$ calculated in both cases are however rather small and close to unity. These numbers have to be compared with the asymptotic $\mu = 6.57...$ calculated by Feigenbaum, and the more recent $\mu = 4.58$ .... calculated by Nauenberg and Rudnick[7].

The location of the bifurcations is given by the following sequence λ_n (calling $\lambda = R/R_c$) : $\lambda_o = 34.34$ $\lambda_1 = 50.52$ $\lambda_2 = 62.47$ $\lambda_3 = 66.25$ $\lambda_n = 67.32$. From this sequence we can calculate the first three terms of the δ_n sequence. We obtain $\delta_o = 1.3$, $\delta_1=3.2$, $\delta_2= 3.5$. These numbers have to be compared with the asymptotic Feigenbaum value $\delta = 4.66$ .... Values for δ_n however can also be evaluated in a different way. From the known position of the first bifurcations we can estimate the location of λ_∞ . We can then define a sequence $\delta_n = (\lambda_\infty - \lambda_n) / (\lambda_\infty - \lambda_{n+1})$ which has the same accumulation point as before. Four values of δ_n can be generated in this way. They are $\delta_o = 2.0$, $\delta_1 = 3.3$, $\delta_2 = 3.6$, $\delta_3 = 4.3$. Typical error bars on these values, as well as on the previous one, can be estimated to be around 15%.

Finally, a comment on what happens above the $f_1/16$ bifurcation. We have fairly strong hints that our system leaves the Feigenbaum route after this bifurcation. First, orbits never go back to the stable mode. Second, we never observe the appearence of $f_1/32$ components, although the resolution and stability in R/R_c was adequate. Third, while up to $R/R_c = 67.3$ the μ estimates calculated on any spectrum maintain fairly constant values around 4, above the $f_1/16$ transition they deviate substantially from this value. This is the case for example of the μ values calculated on the last spectrum in Fig.3. We cannot refrain from noticing that the premature termination of the Feigenbaum sequence is associated with the cross over between the last subharmonic frequency associated to the spectrum and the orbit oscillations frequency.

We are greatly indebted to A.Libchaber for illuminating discussions. Thanks are also due to F.Busse, V.Degiorgio and to M.Corti for the loan of the analyzer.

This work has been partially supported by CNR/CISE Contract n.80.00016.02.

REFERENCES

1. M.J.Feigenbaum, Phys.Lett. 74A, 375 (1979).
2. M.J.Feigenbaum, Commun.Math.Phys. 77, 65 (1980).
3. J.Maurer and A.Libchaber, J.Phys.(Paris) Lett. 41, L515 (1980).
4. A.Libchaber and J.Maurer, J.Phys (Paris) Coll. C 3 41, C 3 51 (1980).
5. J.P.Gollub and S.V.Benson, J.Fluid Mech. 100, 449 (1980).
6. A.Libchaber, Lecture Notes in this volume.
7. M.Nauenberg and J.Rudnick, to be published.

SPACE-TIME SYMMETRY IN DOUBLY PERIODIC CIRCULAR COUETTE FLOW

M. Gorman and Harry L. Swinney

Department of Physics
University of Houston
Houston, Texas 77004
and
Department of Physics
The University of Texas
Austin, Texas 78712

INTRODUCTION

Our recent experiments(1-3) on the circular Couette system (two concentric cylinders with the outer cylinder fixed and the inner cylinder rotating) have determined the spatial and temporal characteristics of the doubly periodic flow regime. In many cases doubly periodic flow is the final preturbulent regime, so a thorough understanding of this regime is important for understanding the onset of turbulence in such cases.

Using ideas from dynamical systems and symmetry arguments based on group theory, Rand has correctly predicted some of these experimental results(4). Although the mathematical language of this theory is unfamiliar to most physicists, the central ideas of his argument are relatively simple.

We will attempt to present a heuristic explanation of Rand's results, eliminating the mathematical details as much as possible. Instead we will concentrate on the central ideas and the results; the proofs themselves can be found elsewhere(4). After a review of the pertinent experimental results, we will discuss the basic ideas of Rand's theory and their relationship to our experiments.

REVIEW OF EXPERIMENTAL RESULTS

The flow in this system is characterized by the following parameters: the Reynolds number, $R=\Omega r_i(r_o-r_i)/\nu$, where Ω is the rotation frequency of the inner cylinder, r_o and r_i are the outer and inner radii, respectively, and ν is the kinematic viscosity; the radius ratio, $\eta=r_o/r_i$; and the aspect ratio, $A=h/(r_o-r_i)$, where h is the height of the fluid. In our experiments $\eta=0.88$ and $A=20$. The working fluid was water at room temperature and the flow was made visible by placing anisotripic platelets in the flow. When the platelets are oriented parallel to the line of sight one sees through to the blackened inner cylinder; when they are perpendicular to the line of sight, one sees reflected light. Other experimental details, such as the techniques used to obtain power spectra and the determination of the modulation patterns are described elsewhere[2].

If the Reynolds number is increased from zero, a critical value of the Reynolds number, R is reached where the purely azimuthal laminar flow becomes unstable and Taylor vortex flow (TVF) appears. In TVF toroidal vortices encircle the inner cylinder and are stacked in the axial direction; see Fig. 1 (a). The number of vortices is labeled N. The velocity now has nonzero radial and azimuthal components, but is still time-independent. At $R/R_c=1.2$ these vortices become unstable and wavy vortex flow (WVF) appears in which traveling azimuthal waves are superimposed on the vortices; see Fig. 1 (b). The number of waves is labeled m. The velocity power spectrum for WVF is characterized by a single fundamental frequency, f_1, and its harmonics. At $R/R_c=10$ (this value depends on m and N) this periodic flow becomes unstable and a new flow, modulated wavy vortex flow (MWVF), appears in which amplitude of the waves is modulated as they travel around the annulus. The effect of the modulation can be seen in Fig. 2 (a) and (b). Figure 2 (a) shows a state at an instant in which all the vortex outflow boundaries are simultaneously flattened; Fig. 2 (b) shows a state in which only one of the waves is completely flattened at a given time.

Our measurements show that, within the experimental uncertainty, the modulation can be characterized by an integer k, where $\Delta\phi=k(2\pi/m)$ is the phase difference between the modulation of successive waves in the direction of rotation of the inner cylinder[1,2]. The velocity power spectrum for MWVF contains two fundamental frequencies, f_1 and f_2, where f_2 corresponds to the frequency of the modulation. There are multiple stable states with different values of f_2 for a given value of f_1. From the relationship between k and $\Delta\phi$ it can be seen that there are only m distinct values of k (and hence m distinct modulation patterns) for a given m.

(a)

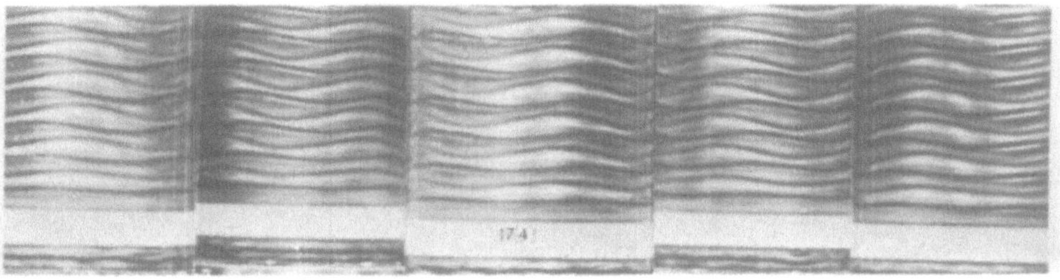

(b)

Fig. 1 (a) Taylor vortex flow (TVF); (b) Wavy vortex flow (WVF)

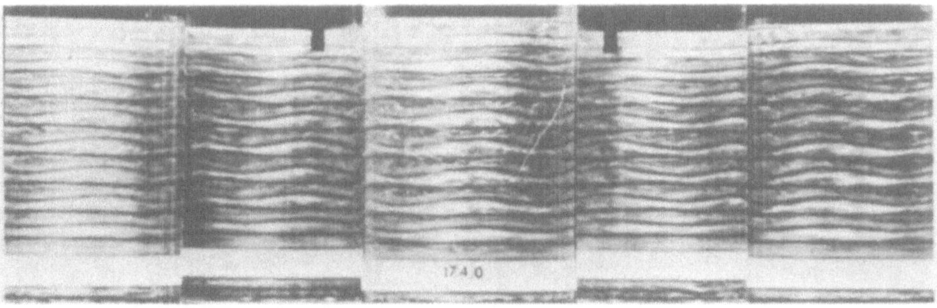

(a)

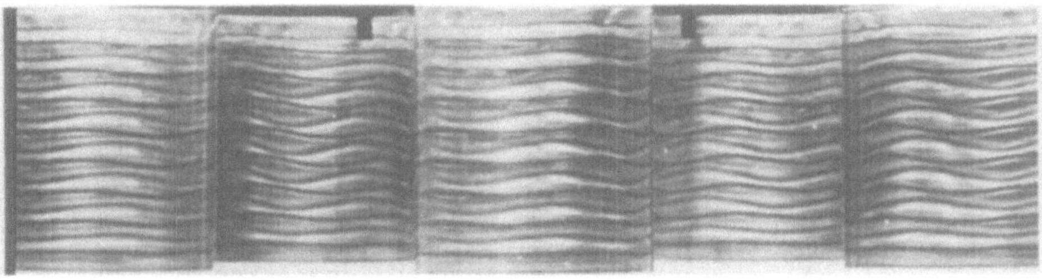

(b)

Fig. 2 (a) Modulated wavy vortex flow (MWVF)-4/0 state
(b) 4/1 state

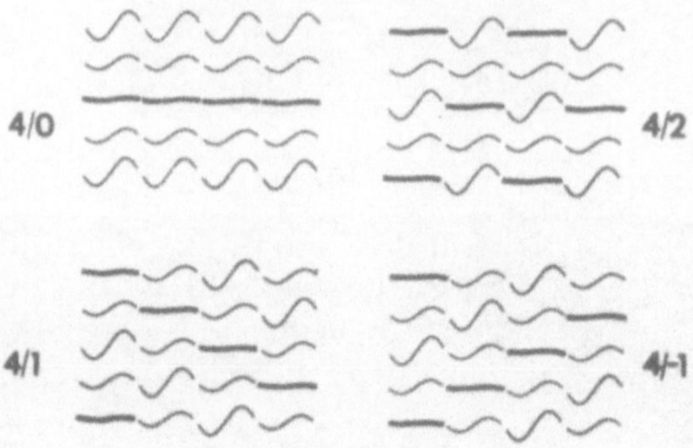

Figure 3. Schematic representations of the evolution in time of the modulation patterns in a reference frame rotating with the waves. The diagrams show the evolution of a single vortex outflow boundary with time. Time increases downward and the azimuthal angle increases rightward. The heavy black lines show waves that have the maximum modulation. (greatest flattening)

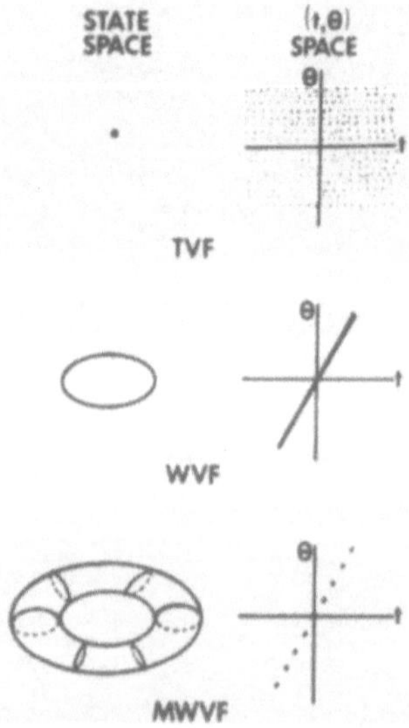

Figure 4. A schematic diagram of (a) the orbits of v_t in state space and (b) the elements of $Z(v)$ in (t,θ) space for TVF, WVF, and MWVF.

Figure 3 shows schematic diagrams of the time evolution of the outflow boundary for each of the possible k values for m=4. The waves are viewed in a reference frame rotating with the waves, and the waves are displayed as if they were unwrapped from around the annulus. The direction of propagation is to the right. Because k is modulo m we can choose k to range from $-(m/2)-1$ to $m/2$ for even m and from $-(m-1)/2$ to $(m-1)/2$ for odd m.

Each MWVF state can be labeled with the integers m/k; for m=4 we have observed the 4/0, 4/1, and 4/2 states. Although the 4/-1 state has not yet been observed the 5/-1 and 6/-1 states have been found. Using different initial conditions, we have observed twelve different MWVF states, with m ranging from 3 to 7. It is highly likely that other stable m/k states can be reached with other initial conditions. A MWVF state is typically stable from 10R to $20R_c$.

THEORY

The theory of Rand analyzes the motion in state space of a point that represents the fluid velocity vector field, v_t. The evolution of the fluid in time is described by the orbit in state space. Rand assumes that the motion in state space is a periodic attractor (limit cycle) for WVF and an attracting torus for MWVF. Because the boundary conditions for the flow are circularly symmetric, the equations of motion must be invariant under axial rotations; hence following a solution for a time t and then rotating that solution by an angle θ is equivalent to rotating the initial conditions by θ and then following the solution defined by this rotation for a time t. This can be written mathematically as

$$\phi(t,\ R_\theta(v)) = R_\theta\phi(t,v)$$

The theory examines those translations (t, θ) for which the field v_t is invariant. The space-time symmetry group, Z(v), is defined to be the set of translations (t, θ) such that $v_t=R_\theta(v)$. Rand proves that: for TVF, Z(v) contains the entire (t, θ) plane; for WVF, Z(v) contains the line $\theta=\omega\tau \bmod 2\pi/m$, where ω is the angular velocity of the waves; and for MWVF, Z(v) is the discrete set of points $p(\tau,\ \omega\tau+n(2\pi/m)$, where τ is the basic time in which v_t repeats, p is any integer, and n is an integer given by the theory. Thus in a frame rotating with the angular velocity ω, the pattern will repeat after time , but shifted in angle by $\Theta=n(2\pi/m)$ where Θ is measured with respect to the rotating frame.

Thus there are three spaces of interest: state space; (t, θ) space in which the elements of the group Z(v) exist; and real space, in which the wave patterns exist. The orbits in state space and the elements of Z(v) in (t, θ) space are summarized graphically in Fig. 4 for each of the flow regimes. From Eq. (1) and the assumptions described above for WVF and MWVF, Rand derives the properties of the space-time symmetry group Z(v). The results for Z(v) imply selection rules for the types of doubly periodic flows that can occur in circularly symmetric systems.

For pedagogical purposes we will now examine the symmetry properties of the wave patterns. From Fig. 1 (a) one can see that TVF is invariant under any combination of axial rotations or translations in time. Hence the space-time symmetry symmetry group contains the entire (t, θ) space. From Fig. 1 (b) one can see that the wave pattern for WVF is invariant under rotations of θ only for certain times t such that $\theta=\omega t \bmod 2\pi/m$ where ω is the phase velocity of the waves. Thus the symmetry operations of this WVF pattern lie on a line in (t, θ) space, $\theta=\omega t$. From Fig. 2 (b) or (c) one can see that the wave pattern for MWVF will repeat only after a specific time $\tau=(\Omega_2)^{-1}$ where Ω_2 is the modulation frequency measured in the rotating frame. When the pattern repeats it is shifted in angle. Our empirical result for the temporal modulation phase angle, $\Delta\phi=k(2\pi/m)$ is equivalent to Rand's result for the spatial phase angle, $\Theta=n(2\pi/m)$ (n and k are different integers; see ref. 2).

The results of the previous paragraph suggest that both the orbit of v_t in state space and the symmetries of the wave patterns in real space have the same group representation, Z(v). We do not know whether this result is true in general but it is not surprising since the wave pattern provides a representation of v_t.

Rand's proofs concern the vector field v_t in state space; however, the symmetries of the wave patterns in real space graphically illustrate the group theoretical characteristics of the theory. This agreement between theory and experiment represents a significant step forward in understanding the dynamics of ciruclar Couette flow as well as the dynamics of other circularly symmetric doubly periodic systems.

This work was supported by National Science Foundation Grant CME 79-09585 to the University of Texas.

REFERENCES

1. M. Gorman, H. L. Swinney, and D. Rand, submitted to Phys. Rev. Lett.
2. M. Gorman and H. L. Swinney, submitted to Jour. Fluid Mech.
3. M. Gorman and H. L. Swinney, Phys. Rev. Lett. 43 1871 (1979).
4. D. Rand, Arch. Rat. Mech. Anal., to appear.

THE STRUCTURE AND DYNAMICS OF NON-STATIONARY TAYLOR-VORTEX FLOW

A. Brandstäter, U. Gerdts, A. Lorenzen and G. Pfister

Universität Kiel
D-2300 Kiel, W-Germany

In going from laminar to turbulent flow, the circular Couette flow passes through a sequence of space and time periodic flow patterns [1]. Increasing the Reynolds number, the first of the nonstationary flows may either be (i) the "wavy mode", in which azimuthal waves are superimposed upon the stationary Taylor vortices, or (ii) the "jet instability", where the fast outward flow oscillates in axial direction. For the wide gap geometry of our apparatus ($R_1/R_2 = 0.5$) both modes have wave number $n = 1$. Either of these modes may occur separately, depending on the boundary conditions and the wavelength of the stationary Taylor cells, so that neither one is a condition for the occurence of the other. We have studied both flow modes using conventional Laser-Doppler-Velocimetry and improved rate cross-correlation techniques.

The order parameter concept allows a parametrization and theoretical description of hydrodynamic instabilities [2, 3]. For the case of stationary Taylor vortices Donnelly and Schwarz [4] and more recently Gollub and Freilich [5] showed that the order parameter dependences on Reynolds number and time follow the theoretical predictions. They did not take into account any spacial dependence of the order parameter as Wesfreid et al [6] did for the Bénard instability and Pfister and Rehberg [7] for the Taylor vortices.

The evolution of the order parameter V is governed by the time dependent Ginsburg-Landau equation [8]:

$$\tau_o \cdot \frac{\partial V}{\partial t} = \varepsilon V - \frac{V^3}{V_o^2} + \xi_o^2 \frac{\partial^2 V}{\partial x^2} \qquad (1)$$

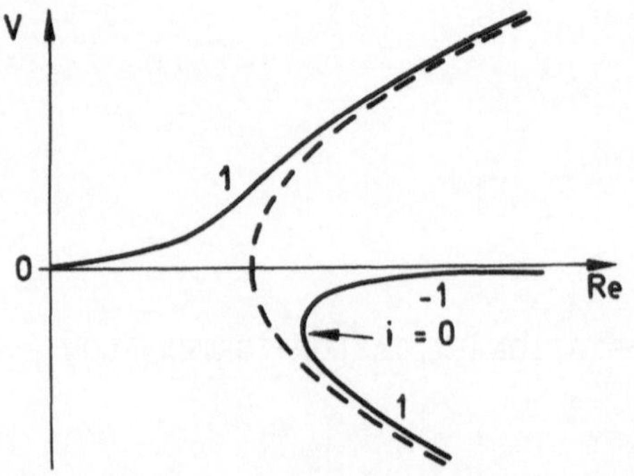

Fig. 1. Decoupling of supercritical bifurcation in the case of stationary Taylor vortex flow with fixed end plates. The branches denoted with 1 are stable, those, with -1 are unstable. The upper stable one is the normal state, the lower stable one is the state where all vortices rotate the wrong way.

with $\varepsilon = (Re - Re_c)/Re_c$, where Re is the Reynolds number: $Re = r_1 (r_2 - r_1) \Omega_1/\nu$, $\Omega_1 = 2\pi \cdot$ rotation rate of inner cylinder, ν = kinematic viscosity, Re_c the critical Reynolds number for infinitely long cylinders, τ_o = linear amplification rate, V_o = normalization constant such that $\sqrt{\varepsilon} \cdot V_o = V$ for cylinders of infinite length and ξ_o = length unit $\approx$ cell dimension.

This model holds for the most stable wavelength λ of Taylor vortices only. Changing the wavelength by 5 % we found changes in the critical Reynolds number of the onset of wavy modes up to 15 %. Similar results were published by Benjamin and Mullin [9].

The finite cylinder length in any experiment causes the onset of stationary Taylor vortices to be a continous process, hence the transition is no instability. This is qualitatively shown in figure 1. The transitions to the time dependent flow patterns remain sharp and give definite critical Reynolds numbers which are shifted to higher values when the cylinder length is decreased in steps of the cell length. The time evolution of the order parameter is then given by [10]:

$$V(t) = \left(\frac{V_e \exp(2t/\tau)}{\exp(2t/\tau) + (V_e/V_i)^2 - 1} \right)^{1/2} \qquad (2)$$

where $\tau = \tau_o/\varepsilon$ and V_i is the inital, V_e the equilibrium oscillation amplitude, $V_e = V_o \sqrt{\varepsilon}$ following the well known Landau law. For the linear growth rate of the instability near $\varepsilon = 0$ we obtain $\tau = \tau_o/\varepsilon$, i. e. a critical exponent of -1. This is the typical slowing down known from continous phase transitions.

Our measurements were done in a wide gap geometry ($R_1/R_2 = 0.5$) using siliconoil as working fluid. Experimental details can be found in [7, 11, 12, 13]. In those experiments where we induced the wavy vortex state, the top end plate was replaced by a plate tilted by an angle of 10 degrees to the plane perpendicular to the cylinder axis, that rotated independently of the inner cylinder.

(i) Wavy mode

Visualization experiments and rate cross correlation measurements [14, 15] have shown that the realized vortex state is a circumferential travelling wave with wave number n = 1 superimposed upon the stationary Taylor vortex flow and a frequency of about $0.08 \cdot \omega_i$ (ω_i the frequency of inner cylinder).

The measurements of the time dependence of the oscillation amplitude were done in the inward flow regime where the spatial gradient of the velocity component is rather linear and the amplitude is fairly high. Measurements of the oscillation amplitude at other locations in the gap resulted in the same time constant and Reynolds number dependence. We measured the onset and decay of the wavy vortex oscillation after sudden increases or decreases to supercritical and subcritical Reynolds numbers respectively. The envelopes of the oscillation were fitted with equ. 2 giving the time constants shown in fig. 2. The full lines represent fits with

$$\tau = 4.215 \cdot \varepsilon^{-1} \quad \text{for} \quad \varepsilon > 0$$

$$\text{and } \tau = 3.86 \cdot \varepsilon^{-1} \quad \text{for} \quad \varepsilon < 0 \,,$$

while the critical rotation rate is $f_c = 3.884$ Hz.

For the equilibrium amplitude we found the square root law $V_e = V_o \sqrt{\varepsilon}$ valid up to $\varepsilon = 0.1$, where $V_o = 50.9$ mm/s and $f_c = 3.886$.

In the subcritical region we induced waves with wave number n = 1 by replacing the fixed top end plate with a tilted rotating plate. Fig. 3 shows the oscillation amplitude of the third vortex against the excitation frequency at three different subcritical rotation frequencies of the inner cylinder. The critical frequency

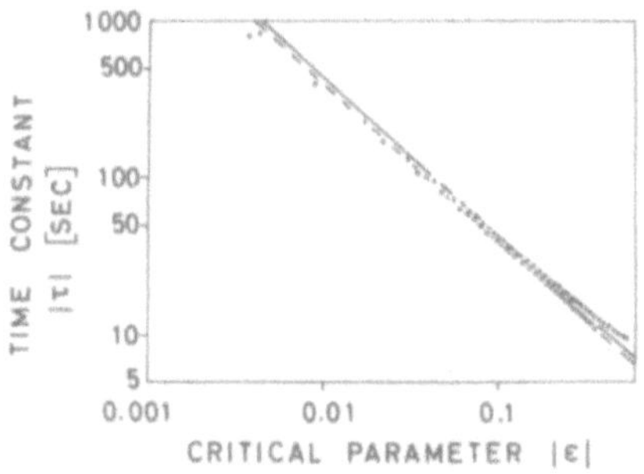

Fig. 2. Double logarithmic plot of time constant versus absolute critical parameter ε. o measurementpoints for ε < 0 and + for ε > 0.

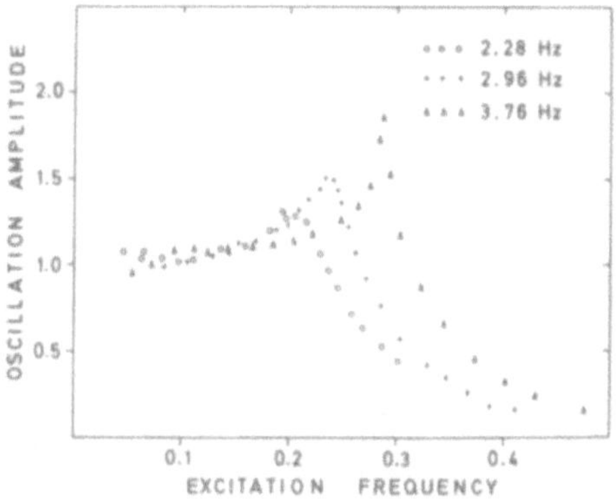

Fig. 3. Stimulated oscillation amplitude of the wavy mode versus excitation frequency at three subcritical Reynolds numbers.

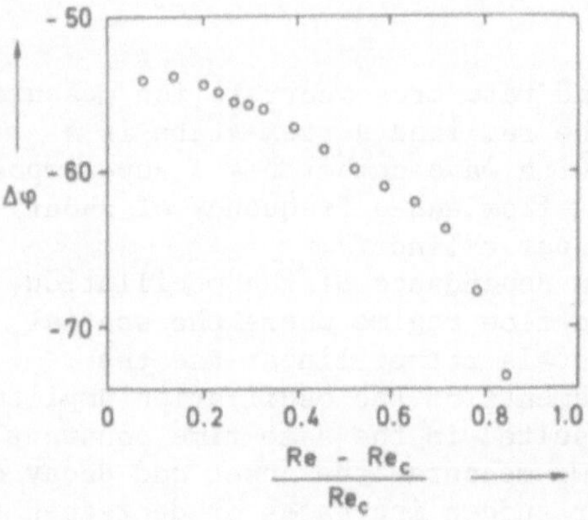

Fig. 4. Phasedifference between oscillating amplitude of outward and inward flow versus critical parameter ε. For $\varepsilon \approx 1$ and $\Delta\varphi > 80^{\circ}$ this mode becomes unstable.

for the onset of the wavy state was 3.90 Hz in this experiment. The three curves show a resonance like behaviour of the system due to a frequency dependent perturbation. The resonance peak sharpens with decreasing difference to the critical Reynolds number.

Measurements of the velocity cross-correlation function showed that adjacent vortex pairs oscillate at nearly the same phase. There is a phase difference, however, between the oscillating amplitudes of outward and inward flow, which increases with increasing Reynolds number as shown in fig. 4.

These measurements show that the wavy mode with wave number n = 1 is a collective movement of all vortices in phase. The oscillation amplitude must vanish at top and bottom of the gap. The consequences of these boundary conditions are considered elsewhere [7].

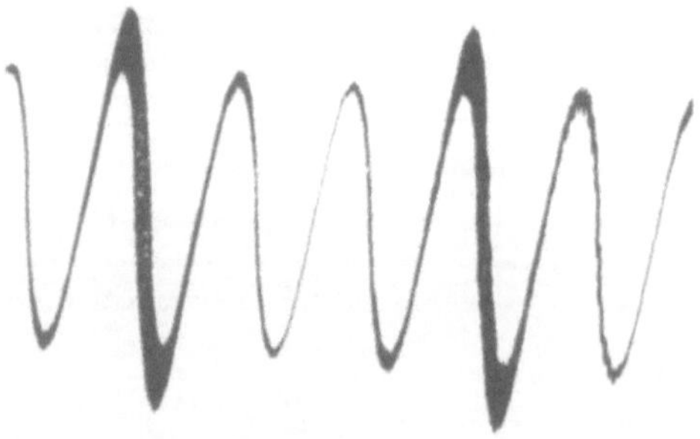

Fig. 5. Recording of axial velocity versus axial coordinate showing jet instability oscillations associated with two particular vortices only.

Fig. 6. Recording of axial velocity versus axial coordinate showing jet instability oscillations of identical amplitude associated with all vortices.

(ii) Jet Instability

In contrast to the wavy mode, the jet instability may be present at isolated vortex pairs as shown in fig. 5. The graph shows an axial profile of vertical velocity. One clearly recognizes two isolated areas of oscillation associated primarily with the outward jet. This suggests that the coupling between adjacent vortex pairs via the region of inward flow is rather weak. Experience has shown that such isolated jet instabilities can only be obtained by rapid rotational acceleration of the inner cylinder. By slowly varying the rotation rate one obtains jet instabilities of the same amplitude in all vortices like shown in fig. 6. The oscillation frequency is about 0.4 of the frequency of the inner cylinder. Cross-correlation measurements have shown that adjacent vortices oscillate with nearly opposite phase, hence next to nearest neighbours at nearly the same phase. The equilibrium amplitude of the jet instability again follows the Landau square root law $V_e = V_o \sqrt{\varepsilon}$ up to $\varepsilon \approx 0.1$.

We thank K. Schätzel for reading the manuscript.

References

[1] R. C. Di Prima and H. L Swinney; in Topics in Appl. Physics, Vol. 45: Hydrodynamic Instabilities and the Transition to Turbulence, Edt. H. L. Swinney and J. P. Gollub, Springer-Verlag 1980, p. 139 and papers cited therein.
[2] P. C. Martin; J. Physique 37, C1-57 (1976)
[3] P. Bergé; in Dynamical Critical Phenomena and Related Topics, Edt.: C. P. Enz, Springer-Verlag (1979), p. 288
[4] R. J. Donnelly and K. W. Schwarz; Proc.Roy.Soc. Lond. A 283, 531-556 (1965)
[5] J. P. Gollub, M. H. Freilich; Phys. Fluids 19, 618-625 (1976)
[6] J. Wesfreid, Y. Pomeau, M. Dubois, C. Normand and P. Bergé; J. Physique Lett. 39, 725-731 (1978)
[7] G. Pfister and I. Rehberg; Phys. Letters 1981 (in print)
[8] H. Haken; Synergetics, p. 186, Springer-Verlag (1977)
[9] T. Mullin and T. B. Benjamin; Nature 288, p. 567 (1980)
T. B. Benjamin, T. Mullin; Phil.Trans.Roy.Soc.A 1981 (i.print)
[10] A. Davey; J. Fluid Mech. 14, 336-368 (1962)
[11] R. Vehrenkamp, K. Schätzel, G. Pfister, B. Fedders and E. O. Schulz-DuBois; Physica Scripta 19, 379-382 (1979)
[12] K. Schätzel; Optica Acta 27, 45-52 (1980)
[13] U. Gerdts; Diplomarbeit University of Kiel (1980)
[14] G. Pfister; Proc. from the 4th Internat. Conf. on Photon Correlation Techniques in Fluid Mech., Stanford, CA (1980)
[15] G. Pfister and U. Gerdts; Phys. Letters (1981) (in print)

PATTERN FORMATION DURING CRYSTAL GROWTH: THEORY

J. S. Langer

Physics Department and
Center for the Joining of Materials
Carnegie-Mellon University
Pittsburgh, PA 15213 USA

1. INTRODUCTION

These lectures will focus on two different versions of the problem of pattern formation during crystal growth. We shall look first at the free growth of dendrites and, next, at directional solidification of cellular structures and eutectics. In the first case, a study of the stability of the dendrite tip and related sidebranching deformations leads naturally to an hypothesis that the dendrite operates at or near a point of marginal instability. This hypothesis seems to be in agreement with experimental observations. The second case is a direct analog of the Bénard problem in hydrodynamics. Again, a marginal instability seems to play some role in pattern selection; but the situation is unclear at the moment. We shall see that there are possibilities for a variety of kinds of interesting behavior in these systems.

Much of what I shall say here, especially regarding the dendrite problem, is contained in a recently published review[1]; therefore I shall only outline the general features of the dendrite theory before mentioning some recent developments and outstanding questions. The eutectic stability theory is newer; so here I shall present more of the background and describe some less developed ideas.

In all of these lectures, I should like to present solidification both as a specific, scientifically and technologically interesting area of investigation and as a prototype for a possibly broader class of pattern forming processes. All of the characteristic ingredients are here: instabilities, nonlinear dynamics, sequences of increasingly complex configurations and, ultimately, chaotic behavior. Whether or not the apparent similarity between these phenomena and others discussed at this Institute is truly deep and important remains to be seen.

2. DENDRITES

A stability theory of dendritic solidification has been described in a series of recent publications by Müller-Krumbhaar and myself.[2,3] One convenient way to summarize this theory is to start with a short exercise in dimensional analysis. Consider a solid forming in an undercooled melt, for example, ice freezing in a sample of pure water which has been cooled uniformly to a degree or so below the melting point. The rate of solidification is governed by the rate at which the latent heat generated at the liquid-solid interface can be conducted away from the growing solid. To describe this process, we must solve a heat-flow problem in which the moving interface is the heat source and in which the temperature of this interface differs from the melting temperature by the Gibbs-Thomson term, that is, the product of the interfacial curvature and a capillary constant. Thus, the material parameters of primary interest are: $\alpha(\alpha')$, the thermal diffusion constant for the liquid (solid); $c_\rho(c'_\rho)$, the specific heat per unit volume; L, the latent heat per unit volume; and γ, the solid-liquid surface tension. We shall neglect, at least for the moment, crystalline anisotropies, molecular attachment effects, and any effects having to do with convection in the fluid. The only control parameter at our disposal is the initial undercooling of the melt, $T_M - T_\infty$, where T_M is the melting temperature and T_∞ is the temperature in the liquid infinitely far from the region where solidification is taking place.

When a dendritic, i.e. treelike, solid structure grows freely in such a system, it turns out that the velocity of the frontmost tip, the sharpness or radius of curvature of this tip, and the spacing of the sidebranches which emerge behind the tip all are accurately reproducible functions of the undercooling. The appropriate unit in which to measure temperature is

$$\Delta T \equiv L/c_p \tag{2.1}$$

so that the dimensionless undercooling is

$$\Delta \equiv \frac{T_M - T_\infty}{\Delta T} . \tag{2.2}$$

If we restrict ourselves to the (unprimed) parameters characteristic of the liquid phase, then the only material-dependent length in the problem is the capillary length

$$d_o = \frac{T_M \gamma}{(\Delta T) L}, \qquad (2.3)$$

which generally turns out to be a microscopic quanitity, of order angstroms. (Justification for the factor $T_M/\Delta T$ in d_o requires a more detailed examination of the equations of motion than is necessary here.) The only quantity with the dimensions of a velocity is α/d_o. Thus, the growth velocity v and the tip radius ρ can be written in the forms

$$v = \frac{2\alpha}{d_o} V(\Delta) \qquad (2.4)$$

and

$$\rho = d_o \tilde{\rho}(\Delta) \qquad (2.5)$$

where V and $\tilde{p}$ are dimensionless functions of Δ which may depend, in addition, upon ratios like α'/α and may possibly depend upon crystalline anisotropies.

The dimensionless growth rate $V(\Delta)$ is shown in Fig. 1 for the two most carefully studied dendritic systems: ice (open circles) and succinonitrile (dark triangles). The velocity measurements for ice were performed by Fujioka[4,5], and those for succinonitrile by Glicksman, Schaefer and Ayers[6]. Both sets of data were obtained by essentially the same technique: the seeding of a crystal through a capillary tube at the center of a sample of undercooled fluid, and the subsequent photographic study of the dendritic structure which grows out from this seed. It is surprising, and possibly accidental, that the data for these two dissimilar substances appear to lie on a single smooth curve. However, there is reason to believe that the function $V(\Delta)$ for various pure substances should differ only by a weakly material-dependent multiplicative constant, so that their logarithmic plots in Fig. 1 should be parallel and fairly close to one another.

The solid curve in the figure is the function $V(\Delta)$ as determined by the stability theory. There are two basic elements of this theory. First, it is assumed that the time-averaged, steady-state motion of the tip of the dendrite can be approximated by the Ivantsov[7,8] solution for an isothermal, cylindrically symmetric, paraboloidal 'needle-crystal'. That is, we assume that the tip radius ρ is sufficiently large compared to d_o that we can neglect capillary corrections in estimating the steady-state growth velocity. We are then left with a version of the Stefan problem

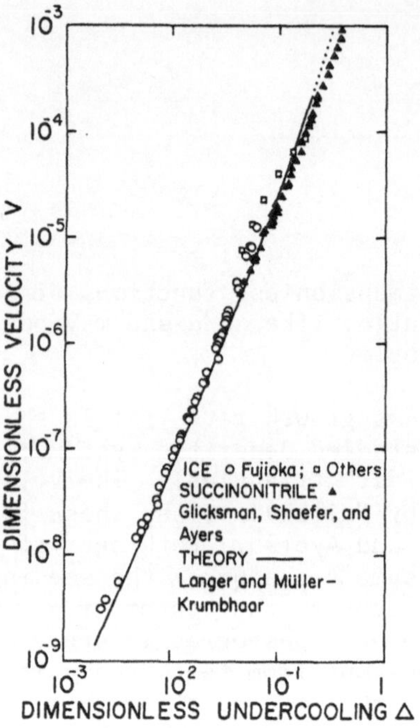

Figure 1. Comparison between theory and experiment for dendritic growth rate as a function of undercooling.

which is exactly soluble. The result of interest here is:

$$\Delta = pe^{p}E_1(p); \quad E_1(p) = \int_p^\infty e^{-y}\,\frac{dy}{y}\,; \tag{2.6}$$

where p is the thermal 'Péclet number'

$$p = \frac{\rho V}{2\alpha} = \tilde{\rho}V \tag{2.7}$$

and E_1 is the standard exponential integral.

Like all steady-state theories, even those including capillary forces, equation (2.6) provides us with only a single relationship between Δ, V, and $\tilde{\rho}$, instead of the two separate functions (2.4) and (2.5) which are required for comparison with experiment. The second, stability-related, ingredient of the new theory is a simple relationship of the form

$$\left(\frac{\lambda_s}{2\pi\rho}\right)^2 = \sigma^* \tag{2.8}$$

where

$$\lambda_s = 2\pi\left(\frac{2\alpha d_o}{V}\right)^{1/2} \tag{2.9}$$

and σ^* is a constant whose value is roughly 0.02 for succinonitrile. Equation (2.8) can be rewritten in the form

$$V = \sigma^* p^2 \tag{2.10}$$

which, in combination with (2.6) produces the V(Δ) shown in Fig. 1.

The arguments leading to the stability criterion (2.8) are given in detail in Ref. 2. This sequence of papers consists primarily of a linear stability analysis for the steady-state, paraboloidal, solidification front described above. The linear equation of motion for general deformations of the paraboloidal surface is quite complicated; and most of our information about solutions of this equation has had to be obtained by numerical means. One striking result of the analysis is the appearance of sidebranching instabilities of the kind shown in Fig. 2. This picture shows the theoretical dendrite tip in five successive positions as it grows up the page. Superimposed upon the steady-state paraboloid is a special eigenmode of the stability operator,

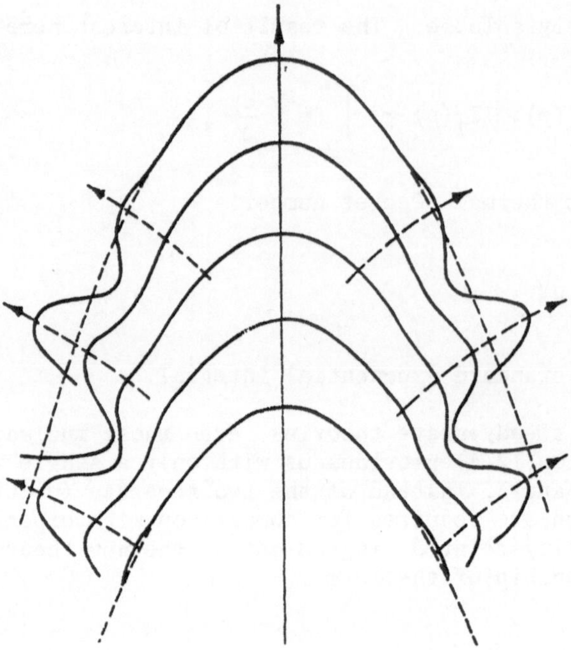

Figure 2. Sidebranching instability.

computed here for the particular value of the ratio λ_s/ρ which we believe identifies the natural operating point of the dendrite. As can be seen, the oscillatory deformations grow in amplitude as they move down the dendrite, and their velocity is such as to keep them nearly stationary in the laboratory frame of reference. This sidebranching instability occurs at all values of v and ρ; thus, the sidebranches appear to be a very general feature of the solidification mechanism. It is also encouraging to note that Fig. 2 is very similar to the experimental photographs of succinonitrile dendrites[6].

A further result of the numerical analysis is that, when the dendrite is flat enough, that is, when ρ is large, the tip becomes unstable against sharpening or splitting deformations. This is the paraboloidal version of the Mullins-Sekerka instability[9,10]. Because the Mullins-Sekerka instability underlies much of what I shall talk about in all of these lectures, it will be useful to say a little more about it at this point. A necessary, but not sufficient, requirement for this class of instabilities is that the solidification front be moving into a region of supercooled or supersaturated fluid. That is, cooling or solute-rejection must take place through the liquid rather than the solid. In such a situation, a forward bulge in the solidification front will increase the local rate at which latent heat or rejected solute diffuses away into the fluid; thus, the bulge may grow rather than decay. This unstable growth is limited by capillary forces to deformations of sufficiently large extent. For example, a planar solidification front, advancing at velocity v into an undercooled melt, is unstable against all sinusoidal perturbations of wavelength greater than the λ_s defined in (2.9).

From dimensional considerations alone, one can guess that if a stability criterion for the dendrite exists at all, it will have the form of (2.8) so that the natural tip radius is proportional to the stability length λ_s. The problem is to understand the stability mechanism in enough detail to determine the constant σ^*. Müller-Krumbhaar and I have argued as follows. The dendrite tip is stable for a continuous range of steady-state situations in which the radius ρ is smaller than, and v is bigger than, the critical value determined by (2.8). This kind of stability means that, if a bump forms near the tip, it may generate sidebranches, but the tip eventually will return to its original size and shape. As ρ increases toward its critical value, however, tip deformations persist for longer and longer times until, at supercritical values of ρ, an instability occurs. Now, an advancing dendrite tip is always subject to thermal fluctuations. According to the above argument, a fluctuation which causes ρ to increase will persist for a longer time than one which tends in the other direction; thus the dendrite should drift toward its point of marginal instability. Note that this drift toward instability is an intrinsically nonlinear process. Presumably, once the system reaches the point of marginal instability, its actual steady-state behavior will be governed by a fully nonlinear equation of motion. So far, we have not been able to carry out any of this analysis explicitly; but I believe that it may be possible to do so in the near future. Equation (2.8) amounts simply to a statement that the system operates at (or near) its point of marginal instability; and the value $\sigma^* \cong 0.02$ is our best estimate based on linear stability analysis.

Having completed this relatively brief introduction to the dendrite problem, I should like to describe - even more briefly - several recent developments.

First, Müller-Krumbhaar and I have performed a detailed numerical study of a two-dimensional model of the sidebranching instability[3]. Our purpose here was to investigate the importance of the supposedly small, capillarity-induced, steady-state shape correction which had been omitted in earlier work and also to test the accuracy of our numerical methods. By working in two dimensions, a geometry suitable for the thin-film experiments that I shall describe later but not yet used for free-dendrite experiments, we were able to solve some parts of the problem analytically before resorting to the computer. Our conclusion was that the shape correction is, indeed, relatively unimportant and that our estimate of σ^* remains roughly consistent with experiment but is quantitatively less reliable than we had originally supposed[2]. The most surprising result, however, was that the spectrum of sidebranching modes does not seem to determine the sidebranch spacing in the way we had expected. Instead of having a well-defined spatial periodicity comparable to the tip radius as in our earlier calculations, the marginally unstable modes of deformation at σ^* seem to occur at very long wavelengths. We suspect that the appearance of long-wavelength instabilities has to do with our neglect of crystalline anisotropy. All of the models that we have considered so far lack a preferred growth direction which might be provided by anisotropy in surface tension or attachment kinetics. It seems likely that such an anisotropy will induce nonuniform growth rates along the curved surface of the dendrite, and that this nonuniformity will select sidebranching modes with compatible wavelengths. We now are investigating this possibility, as well as the possibility that anisotropy plays an important role in determining the nonlinear behavior near the point of marginal stability.

The second development that I want to mention also involves crystalline anisotropy. The direct calculation of σ^* for general crystalline systems is very difficult, and we have tried to circumvent this problem in several ways. The most successful procedure seems to be a spherical approximation which, although known to be highly inaccurate for the steady-state part of the theory and also very poor in its two-dimensional version[3], may still be useful for the special purpose of determining how the point of marginal stability depends on various system parameters. The idea is to approximate the tip of the dendrite by a solid sphere of radius ρ growing at radial velocity v into its undercooled melt. The analysis for the growing sphere has been described in the original papers by Mullins and Sekerka[9]. They find that a deformation proportional to the spherical harmonic of order j will grow with an amplification rate ω_j, where

$$\omega_j = \frac{v}{\rho}(j-1)\left\{1 - \frac{2\alpha d_o}{\rho^2 v}\left[\frac{(j+1)(j+2)}{2}\right]\left[1 + \frac{j\alpha' C'_p}{(j+1)\alpha C_p}\right]\right\}. \quad (2.11)$$

Note that the dimensionless group of parameters, $2\alpha d_o/\rho^2 v$, is precisely what appears on the left-hand side of (2.8). Thus, the marginal-stability condition for the j'th mode ($\omega_j = 0$) can be used to write

$$\frac{1}{\sigma^*} = \frac{1}{2}\,(j+1)(j+2)\left[1 + \frac{j\alpha' C'_p}{(j+1)\alpha C_p}\right]. \qquad (2.12)$$

This formula contains explicitly the ratio of thermal conductivities in solid and liquid phases plus the harmonic index j which we may suppose is determined by the crystalline symmetry. That is, j might be the index of the lowest spherical harmonic which is consistent with the symmetry of the crystallographically preferred deformations of the dendrite tip. I strongly doubt that this prescription can be taken very seriously, but am willing to consider the possibility that (2.12) does describe the α-dependence of σ^* with j being a phenomenological constant determined only by crystal symmetry.

During the last year or so, Huang and Glicksman[11] have repeated and extended the succinonitrile experiments, taking care to account for convective effects which turn out to be increasingly important at smaller undercoolings. They find excellent agreement with the theory for $\sigma^* = 0.0195$. Because $\alpha' C'_p \cong \alpha C_p$ for succinonitrile, this value of σ^* agrees almost exactly with (2.12) for $j = 6$. Succinonitrile is a cubic crystal, and its dendrites exhibit four-fold symmetry about their growth axes. It is tempting to observe that this symmetry is consistent with $j = 6$ if we think of the tip as sitting at one of the six corners of a regular octahedron, its four neighboring corners being the embryonic sidebranches. Ice dendrites exhibit stronger crystalline effects and even develop facets on the basal plane. For ice, the ratio $\alpha' C'_p/\alpha C_p$ is about four; and $\sigma^* = .025$ for $j = 3$, which seems to be consistent with the experimental data. Is $j = 3$ determined by the hexagonal symmetry of ice? Or does the whole analysis break down for such a strongly anisotropic solid?

The last of the new points that I want to mention has to do with the effect of impurities, i.e. dilute solutes, on dendritic growth rates and morphologies. As a rule, the solubility of an impurity will be much smaller in the solid than in the liquid, thus the solidification rate will be limited by the rate at which the rejected impurities diffuse away from the region in which solidification is taking place. Because chemical diffusivities ordinarily are orders of magnitude smaller than thermal diffusion constants, one might guess that the presence of impurities would retard dendritic growth rates. There is evidence, however, that just the opposite can happen-that dendritic velocities may be markedly enhanced by the addition of small amounts of impurity[12].

Many different kinds of effects could be important here. For example, there could be specific chemical interactions that cause solutes to bind to the solid surface; and the surface tension or molecular attachment rate may be strongly sensitive to the chemical state of the surface. Much more theoretical and experimental work will be needed before we shall be able to sort out these processes and understand which are relevant to various physical situations. My point is that, under some circumstances, the velocity enhancement may be indicative of a stability effect and might be used as a sharp test of the stability theory.

The effect of interest here can be understood qualitatively by noting that the stability length λ_s, defined in (2.9) is the geometric mean of d_0 and $2\alpha/v$, the latter being roughly the thickness of the layer of warm fluid ahead of a solidification front which is advancing at speed v. If this front is rejecting impurities, then there also will be an impurity-rich layer of thickness 2D/v, where D is the chemical diffusion constant. This impurity layer may destabilize the interface on a length scale which is smaller than λ_s by the ratio $\sqrt{D/\alpha}$, thus leading to sharper and faster dendritic structures. A quantitative version of this argument, based on the spherical approximation described above, has been published recently[13]. I understand that Glicksman is now doing some of the relevant experiments.

3. DIRECTIONAL SOLIDIFICATION

For the rest of these lectures, I shall turn to a different version of this class of problems, specifically, directional solidification. Here the dominant kinetic effect will be chemical rather than thermal diffusion; that is, the motion of the solid-liquid interface will be governed by the diffusion of rejected solute - rather than heat - through the liquid. Thermal effects will enter via an imposed temperature gradient which controls the orientation and velocity of the solidification front. With this extra degree of control, the system can be made to undergo one or more of the standard, Bénard-like, pattern-forming transitions.

The basic features of the system are shown in Fig. 3. The sample consists of a long rod or a long thin strip of the working material which is drawn, at a predetermined velocity v, through a fixed temperature gradient established by stationary hot and cold contacts at A and B, respectively. The temperatures at A and B are chosen so that the sample is molten at A and solid at B, and so that the interface is visible in between.

Strictly speaking, this system is one in which both the thermal and chemical diffusion fields are nontrivially coupled to one another. As long as the thermal diffusion constant is much larger than the chemical diffusivity, however, and as long as the latent heat is not too big, we may ignore the heat generated at the interface. Then, without much additional loss of generality, we may assume that the thermal conductivities are about the same in both

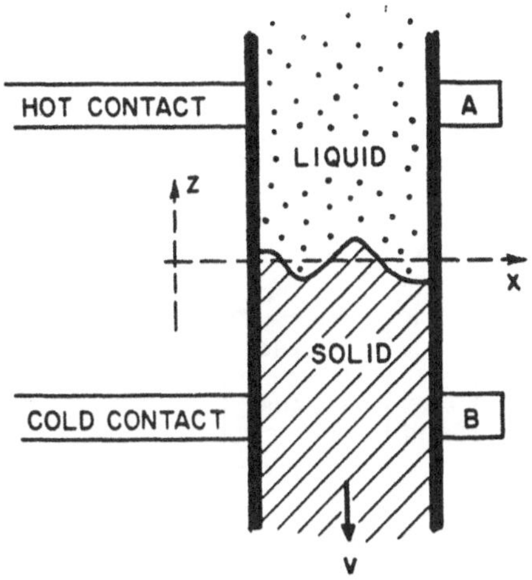

Figure 3. Schematic arrangement of a directional solidification experiment.

liquid and solid phases, and write simply that the temperature throughout the region AB is

$$T = T_0 + Gz. \tag{3.1}$$

Here, z is the position coordinate shown in Fig. 5; T_0 is a reference temperature, and G is the impressed temperature gradient.

We can understand much of what happens in this system without doing elaborate calculations. Let us suppose that the sample is long enough that we can achieve steady-state motion with the given v and G. Now consider the phase diagram in Fig. 4. Because of Eq. (3.1), the T axis in this figure is effectively a z axis, and we can draw the composition profile C(z) directly on the diagram. This is the heavy dashed line, shown here for the case of a flat interface at temperature T_0, that is, at z = 0. Note that the steady-state condition requires that all of the solute in the liquid be absorbed by the solid; thus $C = C_0$ at infinity on both sides of the interface. The intersection of the vertical line, $C = C_0$, with the solidus determines the temperature T_0 and, therefore, the position of the interface.

The dashed line in Fig. 4 also illustrates the "spike" in solute concentration which must build up in front of the advancing interface. The concentration gradient in this region must be just large enough to drive the rejected solute forward into the

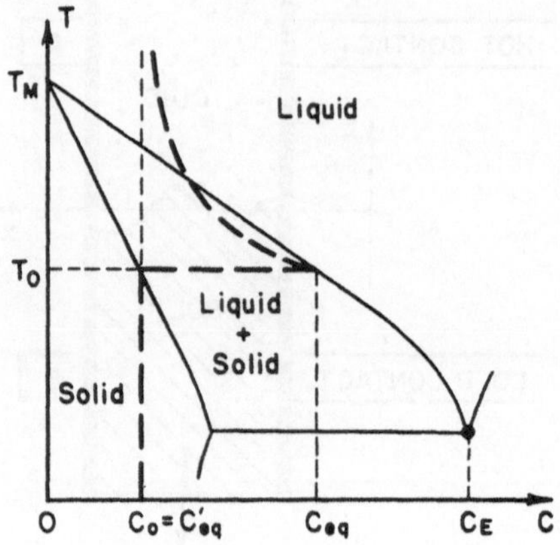

Figure 4. Portion of the phase diagram for a binary solution.

liquid at the velocity v. As v increases, this gradient increases, and, as is shown in the figure, the concentration profile enters the two-phase region. Thus, with increasing v (or decreasing G), we encounter a situation known as "constitutional supercooling"[14]. The fluid immediately in front of the solidification front is supersaturated; and the front may become unstable via the chemical version of the Mullins-Sekerka mechanism.

This instability is particularly interesting because, under some conditions, the interface restabilizes into a periodic cellular pattern reminiscent of Bénard convection. Looking down on the interface (through the liquid), one sometimes sees parallel arrays of grooves in the advancing solidification front. Under other circumstances, these grooves form hexagonal patterns. One particularly elegant set of observations was made some years ago by Jackson[15], who looked at a thin sample of $C\,Br_4$ contained between two microscope slides and passed at constant velocity across a pair of hot and cold plates as shown in Fig. 3. Stability was most conveniently controlled in this case by successively adding small amounts of impurity rather than by increasing the velocity. The system was seen to undergo two transitions, first from a flat interface to a cellular one, and, second, from cellular to dendritic.

Rather than going into the details of the Mullins-Sekerka analysis and its nonlinear generalizations, I should like next to present a qualitative overview of the theoretical situation, emphasizing the analogy to the hydrodynamic problem. What follows is, I hope, a fair summary and generalization of an as-yet incomplete set of calculations performed primarily by Wollkind and Segel[16] and by myself and various collaborators[1,17,18].

Let the z coordinate of the interface in Figure 3 be described by the variable $\zeta(\vec{x},t)$, a function of position $\vec{x}$ and time t. To perform the linear analysis, we can write

$$\zeta(\vec{x},t) \propto \exp(i\vec{k}\cdot\vec{x} + \omega t) \tag{3.2}$$

where $\vec{k}$ is a two-dimensional wave vector and ω is the amplification rate. The diffusion equations with appropriate boundary conditions at the interface ζ, linearized in ζ, produce a spectrum of stability eigenvalues $\omega(k)$. (Strictly speaking, this works exactly as stated only in a quasi-stationary approximation, that is, only when the diffusion fields respond so quickly to the motion of the interface that memory effects may be neglected. This approximation is usually accurate for experimentally interesting situations.) Schematic graphs of $\omega(k)$ are shown in parts (a) of Figures 5 and 6. Note that both of the cases shown in these figures involve instabilities; that is, ω is positive for some range of values of k.

The control parameter most nearly analogous to the Rayleigh or Reynolds numbers for this class of systems is the dimensionless group:

$$\nu \equiv \frac{v(\Delta C)}{GD\left|\frac{dC_{eq}}{dT}\right|} , \tag{3.3}$$

where v is the growth velocity, $\Delta C = C_{eq} - C'_{eq}$ (see Figure 4) is the miscibility gap, G is the temperature gradient defined in Eq. (3.1), D is the chemical diffusion constant for the liquid, and $|dC_{eq}/dT|$ is the slope of the liquidus. There are a number of other dimensionless groups of parameters which occur in this problem. For example, the ratio of the capillary length d_0 (the chemical version of (2.3)) to the diffusion length 2D/v involves the surface tension and plays an important role in the analysis; but, for present purposes it will be convenient to think of this ratio as being fixed. Similarly, the ratio of the solute diffusivity in the solid to that in the liquid or the ratio of the slope of the solidus to that of the liquidus are needed in order to construct accurate theories, but will not be considered explicitly here.

Note that ν increases with velocity v and solute concentration C_0 (ΔC increases with C_0 according to Figure 4), and decreases with thermal gradient G. Thus any one of these parameters may be used

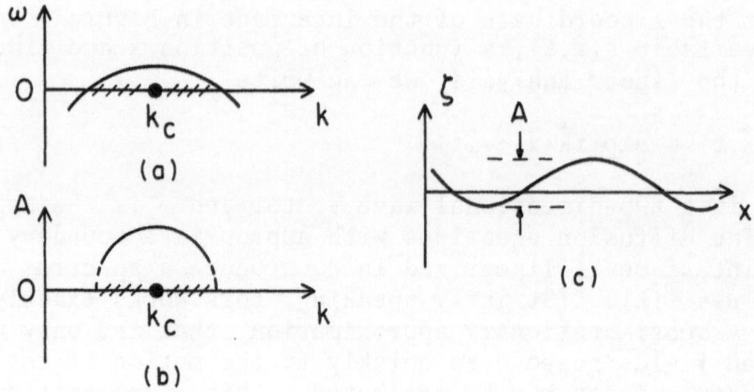

Figure 5. Interfacial deformations for weak instability.

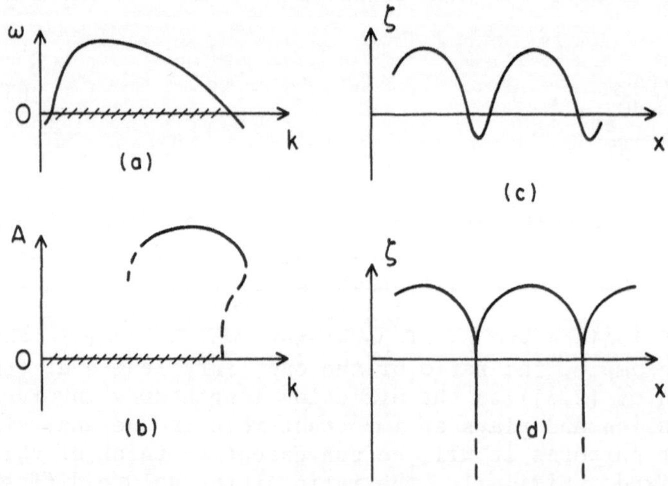

Figure 6. Interfacial deformations for strong instability.

experimentally to control stability. At small values of ν, ω is negative at all k and the interface is stable against deformations at all wavelengths. At a critical value ν_c (generally of order unity), the interface becomes marginally unstable at wavenumber $k = k_c$; and for $\nu > \nu_c$ there is a finite range of wavenumbers around k_c describing deformations against which the interface is unstable. Figures 5a and 6a show schematically the corresponding ranges of positive ω for ν just slightly larger than ν_c and appreciably larger than ν_c respectively. The solid curve in Figure 7 indicates, again schematically, the range of unstable k-values as a function of ν.

To make contact with the dendrite theory and also with a special aspect of the eutectic theory to be discussed later, it is useful to note that the interface may be stabilized by geometrical constraints even in the absence of a temperature gradient. In the limit of vanishing G, that is, infinite ν, the largest wave number k_s at which the interface is unstable turns out to be proportional to $(v/2Dd_0)^{\frac{1}{2}}$. In other words, the interface will undergo an instability only if it extends over a region larger than the chemical Mullins-Sekerka stability length, once again the geometric mean of the capillary length d_0 and the thickness of the boundary layer, 2D/v.

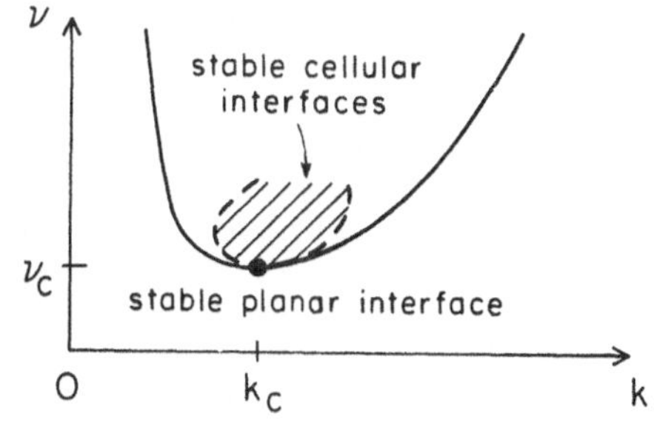

Figure 7. Schematic regions of stability in the ν-k plane.

The analysis so far is purely linear and is the solidification-theoretic analog of the calculation of the critical Rayleigh number for the onset of convection in a Bénard system. To go further than this, as in the hydrodynamic problem, one must develop a nonlinear theory. This can be done by perturbative methods, specifically, by working at values of ν and k close to ν_c and k_c and expanding in powers of the amplitude of the deformation ζ. We have supplemented this procedure, in a few cases, by nonperturbative numerical methods valid at somewhat larger values of ν. Some of these results are summarized in Figures 5 and 6, parts b and c. These diagrams pertain to the restabilized cellular structures which may occur after the onset of instability - the analog of Bénard rolls. Figure 5c shows a small-amplitude cellular structure, the displacement ζ as a function of a one-dimensional position x, and identifies the amplitude A and the period $2\pi/k$. The variation of A with k is shown in Figure 5b. Note that, at this small value of ν, there is a stationary cellular structure for each wavenumber at which the initial planar interface was unstable; but not all of these cellular interfaces are stable themselves. It would have been satisfying had only one of these structures been fully stable because that would have solved the pattern-selection problem. As it turns out, however, stable cellular interfaces occur over the finite range of periodicities k indicated by the solid curve in Figure 5b and by the shaded region in the ν-k plane in Figure 7. Just as in the Bénard problem, even the fully nonlinear equations of motion for the solidification problem seem incapable of determining a unique solidification pattern at a given set of growth conditions.

The same sequence of schematic illustrations is repeated in Figure 6, this time for a value of ν large enough that cellular steady-state solutions to the nonlinear equations no longer exist at all the periodicities at which the initial instability occurred. As ν increases, these cellular structures develop flatter fronts and sharper grooves until, at some upper critical value of ν, the completely smooth solutions of the kind shown in Figure 6c seem to disappear altogether. The experimental pictures and a few of our own exploratory (unpublished) calculations indicate that the steady-state solutions break down in the groove roots where the concentrated solute is absorbed into the solid. Most cellular structures seen experimentally appear to have sharp grooves of the kind shown in Figure 6d; and the actual behavior in these groove roots may be quite complicated. Here we are approaching the situation, to be considered in the last part of these lectures, in which two distinct solid phases are formed from a single-phase fluid. So far as I know, the stability of the large-amplitude cellular structures, with smooth or sharp grooves, has not been studied theoretically in any systematic way. The solid portion of the A versus k curve shown in Figure 6b is a qualititive guess based on incomplete calculations.

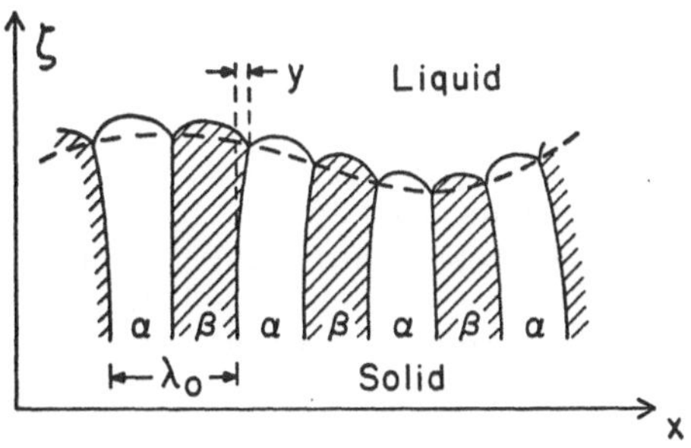

Figure 8. Schematic illustration of a lamellar eutectic growing up the page with a deformed solidification front.

4. EUTECTICS[19]

When the concentration of the melt is large, say, in the neighborhood of the eutectic composition C_E shown in Figure 4, then two distinct solid phases must be formed during directional solidification. Generally such phases emerge in the form of parallel plates or "lamellae" or, alternatively, rods of one phase in a matrix of the other. The lamellar version is illustrated in Figure 8 where a periodic array of solid phases α and β is shown growing upward into the fluid. These lamellae may be visualized as semi-infinite plates perpendicular to the plane of the paper. In what follows, however, it will be convenient to consider primarily the thin-film geometry in which the thickness of the sample is small compared to the lamellar spacing.

At first glance, it seems that this problem must be very much more complicated than the one involving cellular interfaces with only a single solid phase. Indeed, a complete solution of the eutectic problem including a calculation of the detailed shape of the interface is a formidable task. But it turns out that, because of certain geometrical constraints associated with the lamellar structure, it is possible to derive a relatively simple set of equations of motion for the positions of the lamellae at the solidification front. These equations are not exact and, as we shall see, are limited in an important way. With these equations, however, we enter the eutectic problem at a level analogous to that of the nonlinear theory of Bénard rolls[20] or cellular solidification fronts. In effect, we jump over the problems of formation and structure of the cells or rolls and go directly to the question of pattern selection. Thus the eutectic model may be a broadly useful theoretical device.

The starting point for this discussion is the steady-state theory of eutectic growth as worked out by Jackson and Hunt[21]. These authors consider (among other cases) a lamellar eutectic growing at constant velocity with a fixed, uniform spacing λ. In their calculation they introduce several of the crucial approximations referred to above. Specifically, they compute the composition field in the liquid by assuming that the lamellar fronts are locally flat. Capillary effects are introduced by saying that the average curvature of a lamellar front - the forward bulge shown in Figure 8 - produces an average undercooling of the front via the Gibbs-Thomson condition. These two assumptions are sufficient for a calculation of the undercooling of the liquid-solid interface as a function of λ. The result is:

$$\Delta T(\lambda,v) = T_E - T_{interface} = \tfrac{1}{2}\,\Delta T_{min}\left(\frac{\lambda}{\lambda_{min}} + \frac{\lambda_{min}}{\lambda}\right) \qquad (4.1)$$

where T_E is the eutectic temperature and λ_{min} is a stability length of the form of λ_s in (2.9), ($\lambda_{min}^2 \propto Dd_o/v$.) Both λ_{min} and the temperature ΔT_{min} are system-dependent quantities which depend on the composition of the melt. As defined here, ΔT_{min} is proportional to the dimensionless group $(d_o v/D)^{\frac{1}{2}}$.

The two parts of the right-hand side of (4.1) can be understood qualitatively as follows. The first, proportional to $\lambda v/D$, is a diffusion effect. Each advancing α region rejects β molecules and vice versa; thus the fluid ahead of each solid region is supersaturated and the interfacial temperature is correspondingly depressed. The larger λ or v became, the larger must be the concentration of, say, β molecules in front of each α region in order to drive the required diffusion flux. Therefore this term in (4.1) increases with λ. The second term in (4.1), proportional to d_o/λ, is the capillary term mentioned above. The forward bulges shown in Figure 8 are necessary in order that the capillary forces balance at the α-β-liquid triple points; in fact, the opening angles are completely determined by the force-balance conditions. Thus the average curvature of a lamellar front depends only on its width.

Equation (4.1) is as much as one can say about the eutectic growth mode on the basis of steady-state theory alone. The situation is similar to that of the dendrite where steady-state theory yields a single relation (2.6) between velocity v, tip radius ρ, and undercooling Δ, but where nature seems to select only one mode from this continuum of possible states. These situations have been known and related to one another in the metallurgical literature for many years. If one departs from directional solidification and simply quenches the eutectic fluid to some temperature below T_E, lamellar arrays will form and grow with apparently well defined λ and v. Now, in direct analogy to the dendrite problem as posed above, it is ΔT which is fixed in (4.1); and we are left with only a single relation between λ and v which is insufficient to determine each of these quantities separately.

The conventional way out of this dilemma has been to notice that (4.1) predicts that v passes through a maximum as a function of λ at fixed ΔT, and to hypothesize that the natural mode of operation of the system is at the point of maximum velocity. This hypothesis has been tested very carefully for the case of the dendrite by Glicksman and collaborators[6]. The succinonitrile experiments described previously were performed first as a test of a steady-state theory in which (2.6) was generalized to include capillary effects in analogy to (4.1). The actual calculation by Nash and Glicksman[22] was far more accurate and technically difficult than (4.1), but was performed in such a way as to be relevant only in the context of the maximum-velocity hypothesis. The remarkable conclusion of these efforts was that the maximum-velocity theory disagrees with experiment by almost an order of magnitude and, in view of more recent work, appears to be qualitatively incorrect. The marginal stability theory described above produces distinctly different results for the dendrite and, so far, seems to be in satisfactory agreement with the data.

The equivalent of the maximum velocity hypothesis for directional solidification of eutectics at fixed v is a condition of minimum undercooling. Equation (4.1) has been written in a form which emphasizes the fact that ΔT passes through a minimum as a function of λ at $\lambda = \lambda_{min}$. It has generally been assumed in metallurgical work that λ_{min} is, in fact, the correct lamellar spacing; and experiments - at least those using non-facetted eutectic solids - are generally consistent with the prediction that $\lambda^2 v$ is a constant[23]. To my knowledge, however, the actual value of that constant has never been checked to see whether it really agrees with the theory. The difficulty here is that the theoretical formula involves various thermodynamic parameters and, especially, surface tensions which are not easy to obtain.

An interesting difference between the eutectic and the dendrite is that, in the eutectic model, λ_{min} actually locates a point of marginal instability. The fact that growth with λ less than λ_{min} should be unstable was pointed out on physical grounds by Jackson and Hunt[21], who attribute the argument to Cahn. The mathematical derivation of this instability[24] provides a useful description of modes of deformation of the eutectic growth pattern.

Suppose that the solidification front is slowly deformed on a length scale much greater than λ. Let this deformation be described by the function $\zeta(x,t)$, the dashed curve in Figure 8, which measures the forward displacement of the interface away from its undeformed position in the frame of reference moving with velocity v. As in part 3 above, we consider the temperature gradient G to be fixed in the moving frame so that

$$G\zeta = -\Delta T(\lambda) \tag{4.2}$$

Next, define $y(x,t)$ to be the sidewise displacement of lamellae originally at position x, as shown in the figure. That is, the local lamellar spacing is

$$\lambda(x,y) \simeq \lambda_0(1 + \partial y/\partial x) \tag{4.3}$$

where λ_0 is the original spacing of the undeformed system. The crucial new assumption - the one which permits the stability problem to be solved without a much more elaborate calculation - is that ζ and y are coupled by the condition that each lamella must grow in a direction which is locally perpendicular to the solidification front. Thus

$$\partial y/\partial t \simeq -v\,\partial\zeta/\partial x \quad . \tag{4.4}$$

Taking two derivatives with respect to x on both sides of (4.2), then, we obtain nonlinear partial differential equation for $\lambda(x,t)$:

$$\frac{\partial\lambda}{\partial t} \simeq \frac{v\lambda_0}{G}\,\frac{\partial^2}{\partial x^2}\,\Delta T(\lambda) \quad . \tag{4.5}$$

Equation (4.5) may conveniently be written in the form of a diffusion equation:

$$\frac{\partial\lambda}{\partial t} = \frac{\partial}{\partial x}\mathcal{D}(\lambda)\,\frac{\partial\lambda}{\partial x} \tag{4.6}$$

where $\mathcal{D}$ is a λ-dependent diffusion constant

$$\mathcal{D}(\lambda) = \frac{v\lambda_0}{G}\,\frac{\partial}{\partial\lambda}\,\Delta T(\lambda) \propto 1 - (\lambda_{min}/\lambda)^2 \tag{4.7}$$

The stability properties described by Jackson and Hunt are immediately apparent in Eqs. (4.6) and (4.7). Constant -λ solutions of (4.6) are differentially stable if $\mathcal{D}$ is positive, that is, if $\lambda > \lambda_{min}$. Nothing in this equation of motion by itself, however, tells us which member of this continuous set of stable solutions is selected in nature. The marginal stability mechanism, as described in the case of the dendrite, tells us that externally driven fluctuations should cause the system to drift toward $\lambda = \lambda_{min}$. If that intrinsically nonlinear mechanism is dominant, then our theory is nicely consistent with the conventional minimum-undercooling hypothesis for eutectics; but, as we shall see, that is not the whole story.

Before going further into the nonlinear and statistical aspects of the eutectic problem, I should like to mention that Datye[25] and I recently have used the physical assumptions described here - especially the one requiring lamellae to grow in directions normal to the local orientation of the solidification front - to make a complete linear stability analysis of the Jackson-Hunt model. Our analysis is complete in the sense that it deals with discrete lamellae rather than being a continuum approximation of the kind just described, it includes displacements of the lamellar fronts both parallel and perpendicular to the direction of growth, and it permits local variations of the volume fraction - the local ratio of the widths of α to β lamellae. We find that the above argument which predicts marginal instability at λ_{min} remains exactly correct at all values of melt composition and even in the limit of vanishing temperature gradient. At off-eutectic compositions, a boundary layer or solute "spike" like that shown in Figure 4 forms in front of the interface; and it might be supposed that this layer would induce a Mullins-Sekerka instability at wavelengths much larger than the lamellar spacing. No such effect seems to occur, however. Apparently the extra degree of freedom associated with local variations of the volume fraction stabilizes the eutectic in a way which does not happen with single-phase solids. The one really new phenomenon that we have found in this analysis is a short-wavelength, oscillatory instability which occurs at sufficiently off-eutectic melt compositions. Oscillating modes of this kind have been seen experimentally[26], and they may also have something to do with the transition from eutectic to cellular or dendritic growth modes which is observed in off-eutectic melts.

I should like to conclude these lectures with some further discussion of the nonlinear aspects of the eutectic problem. One of the lessons learned from the linear analysis just described is that the nonlinear equation (4.6) contains most of the interesting dynamics of the lamellar growth process. That is, the basic instability and its dependence on λ are correctly described, and there do not seem to be any other features within the framework of the Jackson-Hunt model which might be essential to the mechanism of pattern selection. There are, however, some important limitations in what we have done so far. Specifically, (4.6) does not contain enough detail about the shape of the liquid-solid interface to describe the appearance of new lamellae or the termination of existing ones.

Equation (4.6) does give us a partial picture of the termination mechanism[24]. Suppose that, as illustrated by the solid curve in Figure 9(a), the system has reached a point where $\Lambda \equiv \lambda/\lambda_{min}$ is greater than unity almost everywhere but a fluctuation has caused λ to drift to slightly subcritical values in some finite region. At the minimum of this curve, $\partial\lambda/\partial x = 0$, $\partial^2\lambda/\partial x^2 > 0$, and $\frac{\partial\lambda}{\partial t} < 0$; thus λ decreases. Another useful point to consider is x_1, defined by $\lambda(x_1,t) = \lambda_{min}$. Using (4.6), we find

$$\frac{dx_1}{dt} = -\frac{\partial\lambda/\partial t}{\partial\lambda/\partial x} \propto -\left(\frac{\partial\lambda}{\partial x}\right)_{x=x_1} \quad . \tag{4.8}$$

The sign of the right-hand side of (4.8) implies that the pair of points labelled x_1 are approaching each other. The resulting behavior is indicated by the arrows and the dashed curve in Figure 9(a) and by the schematic illustration of the actual event in Figure 9(b). What is happening here is that all of the intensity of an initially diffuse and shallow fluctuation is being concentrated at a point. When λ touches zero, the lamella at that point disappears. For such rapid variation of λ, however, the equation of motion has lost its validity. The physical system presumably reverts to a state with fewer lamellae and larger average λ, from which configuration the process may start all over again; but Eq. (4.6) does not tell us how this recovery takes place.

The phenomenon which is completely missing in our model so far is the creation of new lamellae. If no such mechanism existed, then the occasional termination of lamellae would cause the system to drift away from rather than toward λ_{min}. In fact, in thin-film eutectic experiments, new lamellae seem to be created easily at boundaries rough enough to cause large fluctuations. One also sees the splitting of interior lamellae under circumstances where the system has been perturbed by some defect or where the growth rate has been changed abruptly. Note that λ_{min}, defined following (4.1), is a stability length of precisely the same form as the Mullins-Sekerka length described previously - the geometric mean of the capillary length and the diffusion length. If the temperature gradient is not too large, then there will exist a maximum λ, say, λ_{max}, determined by morphological stability, which need not be appreciably different from λ_{min}. The front of a lamella which is wider than λ_{max} will be unstable against deformations which, one may assume, will lead to splitting. Jackson and Hunt[21] have attempted a different approach to the splitting problem in which they compute a λ_{max} at which the lamellar front develops an infinitely steep-sided depression at its center. They presume that the steady-state solution breaks down at such a point. Their results are qualitatively similar to what one obtains from the preceding

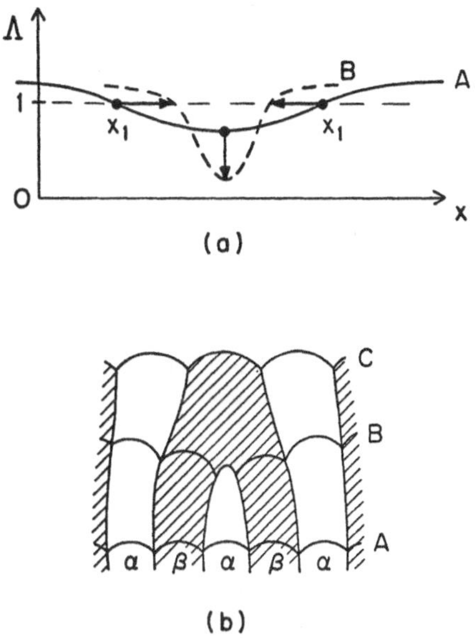

Figure 9. Schematic illustration of an unstable fluctuation which terminates a lamella. Part (a) shows the function Λ at two successive instants A and B. The corresponding sequence of solidification fronts is shown in Part (b). Front C occurs sometime later when the system has restabilized at a larger average value of Λ.

crude stability argument; but neither approach has yet been worked out in adequate detail.

These considerations have led us to examine the following caricature of the eutectic problem. Let Λ_i denote a set of lamellar widths measured in units of λ_{min}, and suppose that these Λ_i obey a set of equations of motion of the form

$$\frac{d\Lambda_i}{d\tau} = (\Lambda_{i+1} - 1)^2 - 2(\Lambda_i - 1)^2 + (\Lambda_{i-1} - 1)^2 + \eta_i - \eta_{i-1}, \quad (4.9)$$

where τ is a time-like variable and η_i is a noise source with correlation

$$\langle \eta_i(t)\eta_j(t') \rangle = \theta\, \delta_{ij}\, \delta(t-t') \quad (4.10)$$

Equation (4.9) is a spatially discrete version of (4.6) except that we have replaced the form $\Lambda + \Lambda^{-1}$, which we did not take seriously near $\Lambda = 0$ anyway, by $(\Lambda-1)^2$ which has the same minimum at $\Lambda = 1$. Its continuum version is sometimes called the "porous medium equation". The noise source in (4.9), and also the boundary conditions near i=1 and i=N, are chosen so that the total width of the system remains fixed:

$$\frac{d}{d\tau} \sum_{i=1}^{N} \Lambda_i \equiv \frac{dL}{d\tau} = 0 \quad . \quad (4.11)$$

The solutions of (4.9) become meaninglessly singular if the Λ_i are permitted to be less than unity; but the onset of this singularity is just the onset of the termination of a lamella. We therefore supplement the equations of motion (4.9) with the prescription that, if Λ_i goes to zero, then Λ_{i+1} is combined with Λ_{i-1} and the total number of lamellae N decreases by two. We also introduce a splitting mechanism at $\Lambda_{max} = \lambda_{max}/\lambda_{min}$ in a similar way; that is, we supplement (4.9) with the prescription that, if Λ_i is greater than Λ_{max}, then it splits into three new lamellae of width $\Lambda_i/3$. In the latter case, N increases by two. The total width L remains unchanged in both processes. Note that Λ_{max} has become the principal control parameter.

At the time that these notes are being written, we have not yet made a systematic study of this model; but several interesting features have emerged from preliminary considerations. The simplest approach to the problem is to start with a naive mean-field approximation. Let $\bar{\Lambda}$ be the average lamellar width, and consider a single lamella of width Λ not necessarily equal to $\bar{\Lambda}$. If, in a first approximation, we assume that the lamellae on either side of Λ have widths $\bar{\Lambda}$, then (4.9) without fluctuations may be written in the form

$$\frac{d\Lambda}{d\tau} \simeq 2(\bar{\Lambda}-1)^2 - 2(\Lambda-1)^2 \equiv -\frac{\partial\psi}{\partial\Lambda} \qquad (4.12)$$

Here, for reasons to appear immediately, we have introduced a velocity-potential in Λ-space, $\psi(\Lambda)$. To include the fluctuations, we can use (4.12) to write an equation of motion for the distribution function $n(\Lambda,\tau)$:

$$\frac{\partial n}{\partial \tau} = -\frac{\partial J}{\partial \Lambda} + \text{source terms}, \qquad (4.13)$$

where the Λ-flux J is

$$J = -\frac{\partial\psi}{\partial\Lambda} n - 2\theta\frac{\partial n}{\partial\Lambda} \qquad (4.14)$$

and the "source terms" are quantities of the form $\delta(\Lambda-\Lambda_s)$ which account for example, for the insertion of lamellae at $\Lambda_s=\Lambda_{max}/3$ because of splittings at Λ_{max}. The latter terms will not be needed in any detail for this discussion.

Note that (4.12) has a stable fixed point at $\Lambda=\bar{\Lambda}$ and an unstable one at $\Lambda=2-\bar{\Lambda}$. As long as $\bar{\Lambda}$ is large enough and the noise level θ is small, we expect the distribution $n(\Lambda)$ to be sharply peaked at $\bar{\Lambda}$. The question is: what is $\bar{\Lambda}$? In a steady-state situation, $\bar{\Lambda}$ must lie somewhere between the unstable fixed point of (4.12) and Λ_{max}, because lamellae which wander beyond these limits either terminate or split. The peak of the distribution must find its steady-state position at a point near the center of the allowed range of Λ's where these two kinds of events happen at the same rates. Within the allowed range, the Λ dependence of n will be dominated by a factor $\exp(-\psi/2\theta)$, which is the solution of (4.14) for vanishing J, and which obviously exhibits the expected sharp peak at $\bar{\Lambda}$. Equality of the termination and splitting fluxes then requires that $\psi(2-\bar{\Lambda}) \simeq \psi(\Lambda_{max})$, which immediately yields the result $\bar{\Lambda} \simeq (1+\Lambda_{max})/2$. A more detailed solution of the rate equations indicates that the fluctuations displace $\bar{\Lambda}$ slightly toward $\Lambda=1$ in accord with the marginal-stability argument. But, in contradiction to the marginal-stability or minimum undercooling theories, Λ_{max} also plays a crucial role in setting the scale of the solidification pattern.

The validity of the above analysis requires that $\Lambda_{max}/3$, the width of the lamellae which are formed upon splitting at Λ_{max}, be within the region of attraction of the stable fixed point of (4.12). That is, $\Lambda_{max}/3 > 2-\bar{\Lambda}$ which, in turn, implies $\Lambda_{max} \gtrsim 1.8$. This condition arises mathematically as a requirement for the existence of a self-consistent, steady-state $n(\Lambda)$ in the mean-field approximation, but it is easy to see what is happening physically. When

Λ_{max} is so small that the lamellae which are formed in a splitting event are automatically unstable against termination, then the distribution n itself must be unstable. This will happen even though there exists a finite range of Λ between 1 and Λ_{max} in which the uniform system is dynamically stable.

The eutectic caricature described by Eq. (4.9) and the supplementary rules is well suited for computer simulation. It was, in fact, designed for that purpose. Preliminary results, obtained very recently in collaboration with R. Mathur, seem to confirm the general features of the mean-field theory. For Λ_{max} greater than 1.8, the system settles into something close to the predicted spacing and is quite stable there, even in the presence of appreciable noise. At smaller values of Λ_{max}, fluctuations may excite localized oscillating modes in which one lamella repeatedly terminates and reappears. At yet lower Λ_{max}, these modes seem to interact with one another and spread throughout the system. We suspect but do not know for sure, that there exists a well defined value of Λ_{max} below which the system remains intrinsically chaotic even when the external noise vanishes. In any case, the external noise is strongly amplified at small Λ_{max}.

Figure 10 is a picture of this mathematical system in action, drawn here for the case $\Lambda_{max} = 1.6$ and no noise, $\theta = 0$. The system was started at $\tau = 0$ at the bottom of the diagram with a random distribution of spacings Λ between 1 and 2; and its progress was plotted moving up the page to $\tau = 10$. This particular run settled down eventually to a state containing just one local oscillation which seemed to persist indefinitely. The overall effect of this picture is similar to metallurgical micrographs showing longitudinal sections of certain directionally solidified rod-like eutectics in which the rods seem to disappear and reappear in an irregular manner. So far as I know, no similar effects have been seen in thin-film lamellar eutectics; but I doubt that anyone has yet looked for such effects in a systematic way.

At the moment, it is too early to say whether this mathematical system or something like it can really provide a vehicle for answering important questions about pattern formation. My reason for presenting it so prematurely is that it points out very clearly that the problem is richer in possibilities than had been expected on the basis of the minimum-undercooling or marginal-stability hypothesis. Remember that these hypotheses had predicted simply that $\bar{\Lambda} = 1$. The elementary considerations of the last few paragraphs tell us, however, that Λ_{max} plays a major role - not just in fixing $\bar{\Lambda}$ - but also in determining how the system responds to noise and, possibly, whether its intrinsic behavior is regular or chaotic. There are also two other dimensionless parameters, the noise strength θ and the total width L, which may shift the steady-state averages, modify the response to external perturbations, and otherwise have

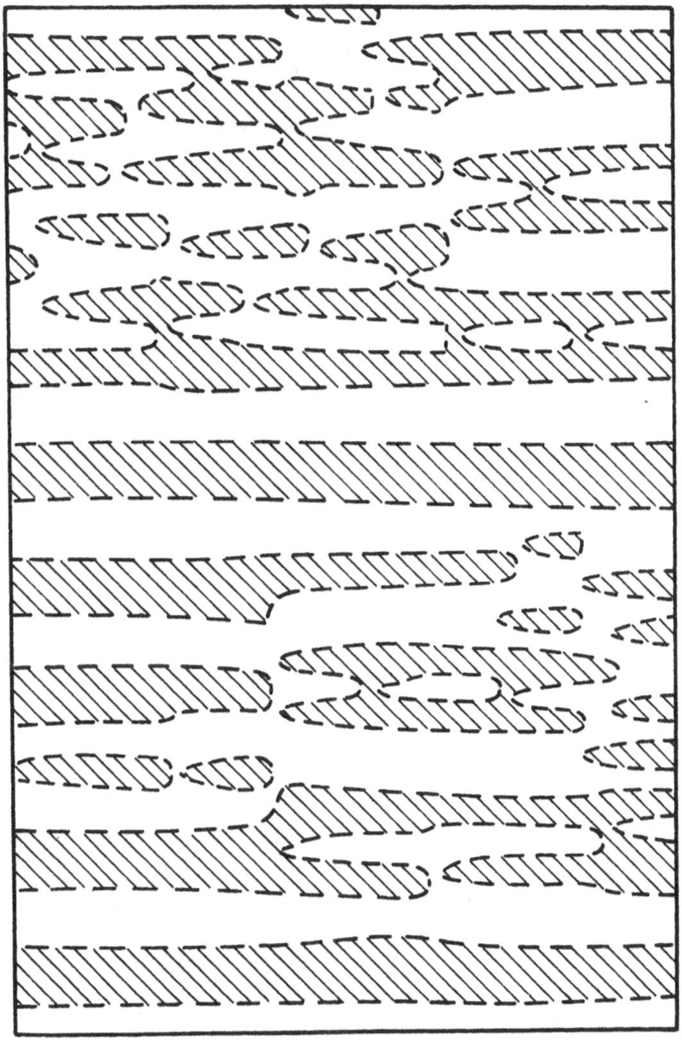

Figure 10. Eutectic caricature for $\Lambda_{max} = 1.6$, $\theta = 0$, and random initial spacings at $\tau = 0$ (bottom of figure).

qualitative effects on the behavior of the system. A detailed analysis of these effects is surely possible within the framework of this eutectic caricature, and the results of such analyses may be interesting.

REFERENCES

1. J. S. Langer, Rev. Mod. Phys. 52, 1 (1980).
2. J. S. Langer and H. Müller-Krumbhaar, Acta Metallurgica 26, 1681; 1689; 1697 (1978).
3. H. Müller-Krumbhaar and J. S. Langer, Acta Metallurgica.
4. T. Fujioka, Ph.D. Thesis, Carnegie-Mellon Univ. (1978).
5. J. S. Langer, R. F. Sekerka, and T. Fujioka, J. Crystal Growth 44, 414 (1978).
6. M. E. Glicksman, R. J. Shaefer, and J. D. Ayers, Metall. Trans. A7, 1747 (1976).
7. G. P. Ivantsov, Dokl. Akad. Nauk. SSSR 58, 567 (1947).
8. G. Horvay and J. W. Cahn, Acta Metallurgica 9, 695 (1961).
9. W. W. Mullins and R. F. Sekerka, J. Appl. Phys. 34, 323 (1963).
10. W. W. Mullins and R. F. Sekerka, J. Appl. Phys. 35, 444 (1964).
11. S. C. Huang and M. E. Glicksman, General Electric Technical Reports No. 80CRD207 and 80CRD208, September 1980.
12. C. Lindenmeyer, Ph.D. Thesis, Harvard University (1959).
13. J. S. Langer, Physicochemical Hydrodynamics 1, 41 (1980).
14. W. A. Tiller, K. A. Jackson, J. W. Rutter, and B. Chalmers, Acta Metallurgica 1, 428 (1953).
15. K. Jackson, "Defect Formation, Microsegregation, and Crystal Growth Morphology", in Solidification, American Society for Metals, Metals Park, Ohio, (1971).
16. D. Wollkind and L. Segel, Phil. Trans. Roy. Soc. (London) 268, 351 (1970).
17. J. S. Langer and L. A. Turski, Acta Metallurgica 25, 1113 (1977).
18. J. S. Langer, Acta Metallurgica 25, 1121 (1977).
19. There is a very large metallurgical literature on the subject of eutectic solidification. Useful reviews include L. M. Hogan, R. W. Kraft, and F. D. Leakey, "Eutectic Grains", Advances in Materials Research 5, H. Herman ed., Wiley-Interscience (1971); G. Lesoult Ann. Chim. Fr. 5, 154 (1980).
20. A. C. Newell and J. A. Whitehead, J. Fluid Mech. 38, 279 (1969).
21. K. A. Jackson and J. D. Hunt, Trans. Met. Soc. of AIME 236, 1129 (1966).
22. G. E. Nash and M. E. Glicksman, Acta Metallurgica 22, 1129 (1974).
23. R. M. Jordan and J. D. Hunt, Met. Trans. 2, 1385 (1971).
24. J. S. Langer, Phys. Rev. Lett. 44, 1023 (1980).
25. V. Datye and J. S. Langer, to be published.
26. K. A. Jackson, private communication.

ELECTROTHERMAL INSTABILITIES AT MAGNETIC CRITICAL POINTS

M. Ausloos

Université de Liège

Institut de Physique B5

4000 Sart Tilman/Liège 1, Belgium

ABSTRACT

Characteristic oscillations in the vicinity of magnetic ordering temperatures are explained in terms of the *continuously forced ballast resistor model*. Because of the highly non linear (I-V) characteristics in the vicinity of the critical temperature, the sample splits into various temperature domains. The "thermal grain boundary" velocity is a sensitive function of the temperature derivative of the Seebeck coefficient (which diverges near the critical point).

ELECTRONIC TRANSPORT COEFFICIENTS AT T_c

In the vicinity of a magnetic ordering temperature of metallic systems (Gd, Ni, Cr, alloys and compounds), the temperature derivative of electronic transport coefficients presents some singular behavior assimilated to a λ anomaly. In fact, in the case of previously called second order phase transition the behavior is supposed to be like that of the magnetic specific heat, c_m i.e. defining ε as $|T-T_c|/T_c$,

$$\frac{d\rho}{dT} \simeq \varepsilon^{-\lambda} ; \ \frac{dS}{dT} \simeq \varepsilon^{-\mu} ; \ c_m \simeq \varepsilon^{-\alpha} \qquad (1)$$

where ρ and S are the electrical resistivity and thermoelectric power respectively. Theoretical arguments have indicated |1-2| that $\alpha \simeq \lambda \simeq \mu$.

As exemplified by the lectures (and lecturers) of this School, systems under conditions "<u>far away from</u>

equilibrium" are receiving much attention. In fact, measurements of the electrical resistivity near an ordering temperature under "quasi static" conditions are typical examples of measurements on systems continuously forced out of equilibrium.

Let us consider a temperature range $|\varepsilon|$ close to a critical temperature T_c such that $|\varepsilon| \simeq 10^{-1}, 10^{-3}|$. It is well known that the system takes a very long time to relax into an equilibrium state, -such a time increasing to infinity when approaching T_c. Furthermore a zero rate r of heating or cooling is impossible to reach. In the best conditions $r \simeq 10$ mK/min |3|. Therefore the system is never in equilibrium, and is naturally forced away from a non equilibrium state toward another one as time proceeds. In particular this is due to the constant supply of energy to the system through the maintenance of a finite electrical current. Such conditions on $d\rho/dT$ near T_c can be understood following recent theoretical work on the ballast resistor |4-8|

DIFFERENTIAL NEGATIVE RESISTIVITY CONDITIONS NEAR T_c

The ballast resistor is a one dimensional wire of resistance R in a tube of gas. When a voltage V is applied to the wire a current $I=V/R$ flows into the wire, and some heat loss $A=RI^2$ is dissipated. The steady state conditions can be rewritten as $I=\sqrt{A/R}$, $V=\sqrt{AR}$. Both A, R and hence also V are monotonous (not necessarily increasing) functions of temperature T.

As pointed out by Ross and Litster|4|, an electrothermal instability arises if a negative differential resistivity is present, i.e. if dI/dV or dV/dI becomes negative for some temperature range. Obviously,

$$\frac{1}{I}\frac{dI}{dV} = \frac{1}{2}\left(\frac{d\ell n A}{dT} - \frac{d\ell n R}{dT}\right)\frac{dT}{dV} \tag{2}$$

$$\frac{1}{V}\frac{dV}{dI} = \frac{1}{2}\left(\frac{d\ell n A}{dT} + \frac{d\ell n R}{dT}\right)\frac{dT}{dI} \tag{3}$$

Defining the thermoelectric power or Seebeck coefficient as

$$S = -e(\nabla V/\nabla T) \tag{4}$$

Eq.(2) implies that dI/dV is negative if, e.g.,

$$\frac{d\ell n A}{dT} < \frac{d\ell n R}{dT} \quad \text{and} \quad -IS > 0 \qquad \text{or conversely.} \tag{5}$$

Condition (5) is in principle realizable near T_C for any sign of the thermoelectric power S by appropriately choosing the current direction. On the other hand, Eq.(3) implies that dV/dI is negative if, e.g., the term in parentheses is negative and $V(dI/dT)^{-1}<0$. Such a condition can be realized near an antiferromagnetic transition.

If we assume that $A=q(T-T_B^\circ)$ where q relates the energy flow into the gas to the temperature difference between the wire and the heat bath starting temperature, the instability conditions become

$$(1/R)\,|dR/dT|>(T-T_B^\circ)^{-1} \tag{6}$$

This gives the range of temperature (T-, T+) in which the instability occurs. Let us point out that the system breaks either into high and low electric field (voltage drop) domains, or into strong and weak current filaments, depending on the shape (N or S like) of the characteristic I-V curve, and on the above instability conditions.

CONTINUOUSLY FORCED BALLAST RESISTOR

In order to describe the microscopic effects due to the instability conditions, one writes the conservation of internal energy in the system, and use the phenomenological equations for the electrical current and the heat flux along the wire. Under the condition of electroneutrality in the wire, this equation reads |6|

$$c_V \frac{\partial T}{\partial t} = \frac{\partial}{\partial x}\lambda\frac{\partial T}{\partial x} + \tau I\frac{\partial T}{\partial x} - q(T-T_B^\circ) + RI^2 \tag{7}$$

where $T\equiv T(x,t)$ describes the time and space temperature distribution in the wire, while the Thomson coefficient is defined as $\tau=T(dS/dT)$, and λ is the thermal conductivity of the wire. The $T(x,o)=T(x)$ stationary states can be easily identified in particular when $\tau=0$ and λ is a constant. In such a case, an instability occurs if and only if

$$I\geqq\hat{I}^2=(A'/R')=q/R'_c \quad ; \quad T_B\leqq\hat{T}_B=T_c-R_c\hat{I}^2/q=T_c-(R_c/R'_c) \tag{8}$$

where the index c denotes quantities measured at the "usual" (magnetic) critical point, $(R''_c=0)$. In the case of an S type instability, the definition of the critical parameters is in the (V^2,T_B) plane.

An inhomogeneous distribution of T leads to define thermal grains (TG) and TG boundary widths $d=(\lambda/\sqrt{R'})$ in which the temperature rapidly varies.|7| To let τ finite (it is "theoretically infinite" at T_C) implies a "finite velocity" v for the TG boundary|8|,

$$v = (\tau/c)I \qquad (9)$$

The proportionality factor remains finite, but may not be a smooth function away from T_C, hence different damping regimes may appear when v is greater or smaller than $\hat{v}=(\tau/c)\hat{I}$. Furthermore, if we introduce the fact that the rate of heat exchange r is small, then one can approximate the left hand side of Eq.(7) by $c_v r$. This does not change the structure of the steady state equation except that the bath temperature is redefined as $\tilde{T}^{o}_{B}=T^{o}_{B}+c\,|r|/q$.

The temperature fluctuations and their correlation $\langle T_i T_j\rangle$ for an homogeneous field T_s influence $d\rho/dT$ in a metal like N_i near T_C. The first derivative of ρ has a λ shape anomaly on which is superimposed oscillations due to $\langle T_i T_j\rangle$, with a characteristic frequency Ω such that (see fig.1)

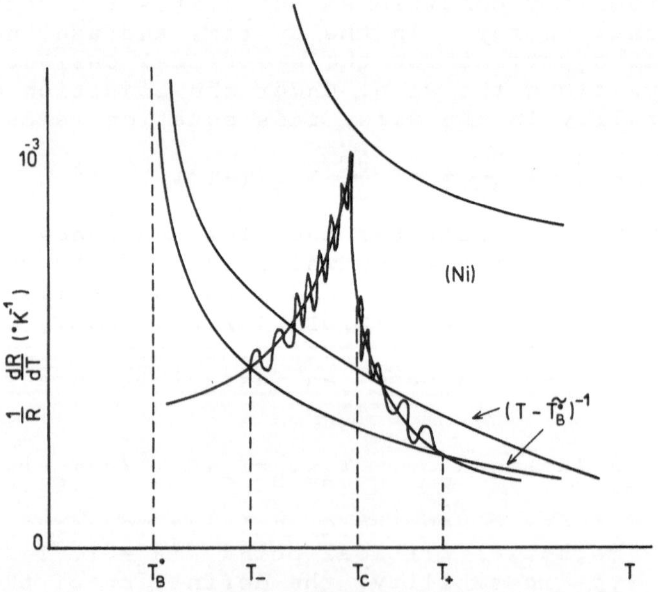

Fig. 1. Sketch of (1/R)(dR/dT) near the Curie point of a metal like Nickel, showing the instability regimes, and the predicted oscillations due to the finite thermal grain boundary velocity.

$$\Omega T_c \simeq \omega_o \tau_{k\simeq o} = v\ \tau_{k\simeq o}/L \tag{10}$$

where $\tau_k \equiv |\lambda(k^2+\tilde{d}^{-2})|^{-1}$ c and $\tilde{d}$ is the TG boundary width modified to take into account the Thomson coefficient. L is the length of the sample.

Acknowledgments : Part of this work has been supported by NATO (grant 1481). Discussions with J.B.Sousa and his colleagues in Porto have been quite stimulating.

REFERENCES

1. M.Ausloos, J.Phys.A 11 (1978) 1621.
2. M.Ausloos and K.Durczewski, Phys.Rev. B 22 (1980) 2439, for an exhaustive list of references.
3. J.B.Sousa, M.M.Amado, R.P.Pinto, J.M.Moreira, M.E. Braga, M.Ausloos, J.P.Leburton, J.C.Van Hay, P. Clippe, J.P.Vigneron and P.Morin, J.Phys.F 10 (1980) 933.
4. B.Ross and J.D.Litster, Phys.Rev.A 15 (1977) 1246.
5. R.Landauer, Phys.Rev. A 15 (1977) 2117.
6. D.Bedeaux, P.Mazur and R.A.Pasmanter, Physica 86A (1977) 355.
7. R.A.Pasmanter, D.Bedeaux and P.Mazur, Physica 90A (1978) 355
8. M.Büttiker and R.Landauer, this volume.

INSTABILITIES DURING CRYSTAL GROWTH (EXPERIMENTS)

J.H. Bilgram

Laboratorium für Festkörperphysik
Eidgenössische Technische Hochschule
8093 Zürich, Switzerland

ABSTRACT

In this paper mainly instabilities during crystal growth from the melt (during freezing) are discussed. The solid-liquid transition shows several asymmetries with respect to freezing and melting. The non faceted - faceted transition and the sharp interface - diffuse interface transition occur during the freezing process only. There are dissimilarities in the dynamics of freezing and melting at conditions far from thermodynamical equilibrium. They might origin from the diffuse interface layer which first has been detected at growing ice single crystals by quasi elastic light scattering.

1. INTRODUCTION

'Instabilities during crystal growth' is a contradiction in itself, because the aim of crystal growth is the preparation of homogeneous single crystals with a high volume to surface ratio. Typical examples for the successful application of the art of the crystal growers are the Czochralski growth of silicon crystals and the hydrothermal growth of quartz crystals. The most famous example for crystaline material which has a very small volume to surface ratio are snow flakes. Their forms give information on the growth conditions. The great variety of natural snow flakes is documented in the book of Bentley and Humphreys /1/.

Crystal growth occurs always at non equilibrium conditions and there are two steps necessary:

i) the rearrangement of molecular configurations

ii) the transport of the heat ot crystallization and the rejection of particles away from the interface.

Both processes are kinetic processes and the origin of instabilities must therefore result from the kinetics of crystal growth. Crystals can be grown from the vapour phase, from solution, from the melt or from another solid. In all these techniques instabilities can be observed. A discussion of all these techniques would be beyond the extend of this paper. Crystal growth from the melt has been studied by the author because it is a means to study the freezing process. The solid-liquid transition shows several asymmetries with respect to freezing and melting:

1. The solid-liquid transition is a first order transition with a hysteresis and a metastable region. This region is not symmetrical to the thermodynamical melting point. Liquids can be supercooled by tens of degrees or more but it is very difficult or perhaps impossible to superheat crystals /2/.

2. Crystals in contact with their pure melt do not facet during melting.

3. A diffuse interface layer is formed at a solid-liquid interface at high freezing rates.

4. Segregation of impurities and their accumulation at the solid-liquid interface occurs during the freezing process only.

These anisotropies are typical for the freezing process. In this paper we will discuss instabilities which occur at the solid-liquid interface during crystal growth and which are not symmetrical with respect to freezing and melting. With increasing driving force (heat removal from the solid and supercooling of the liquid) these instabilities can be observed as a sequence of changes in growth dynamics and interface morphology. Very close to equilibrium conditions crystals in contact with their pure melt grow or melt normal to their interface. In a first transition which will be discussed in chapter 2 normal growth is replaced by lateral growth (corresponding to asymmetry 2). In chapter 3 results from light scattering experiments are presented which give evidence for a transition from a sharp to a diffuse interface (asymmetry 3). Growth of dendrites into supercooled melt is discussed in chapter 4 and compared with the dynamics of dendritic melting of superheated crystals in chapter 5 (asymmetries 1 and 3).

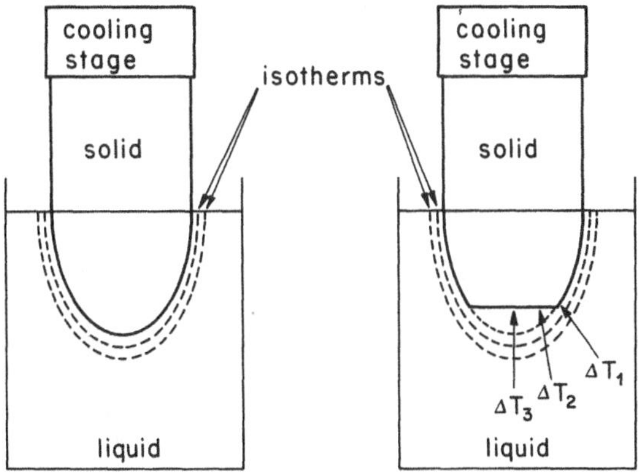

Fig. 1. Equilibrium shape and shape of a growing ice crystal.

2. NORMAL GROWTH AND LATERAL GROWTH

Two types of crystal surfaces can be observed: Facets and rough interfaces. Some of the properties of these crystal surfaces can be illustrated easily in an experiment. As shown in fig. 1 an ice single crystal is growing into its melt. The crucible with pure water is held at a temperature between 0^oC and 4^oC. To prevent convections the temperature should be below the temperature of maximum density. The c-axis of this crystal is oriented parallel to the growth direction. The crystal can be cooled. At equilibrium conditions or if the crystal is melting, the surface will follow exactly the shape of an isotherm. The interface is "rough". During growth a facet normal to the c-axis is formed. The water does not have the same undercooling ΔT at the whole solid liquid interface $\Delta T_1<\Delta T_2<\Delta T_3$. Such a facet at a perfect single crystal is assumed to be "flat" in molecular scale. A measure for the degree of roughness is the ratio N_A/N of N_A molecules which are added to a perfect surface with N possible sites. $N_A/N = 0.5$ means a rough interface and $N_A/N = 1$ or 0 is facet. Whether an interface is rough or faceted is predicted by the so called Jackson α factor /3//4/.

$$\alpha = \xi \cdot \frac{L}{k_B T_M}$$

ξ is a geometrical factor $Z_\ell/Z<1$. Z_ℓ is the number of nearest neighbours in the plane parallel to the crystal surface considered and Z is the total coordination number. (L/k_BT_M) is the ratio of the binding energy at the interface to the thermal energy. From a calculation of the free energy change per atom added to the surface it has been found that for $0 \leq \alpha \leq 2$ surfaces are rough and for $2 \leq \alpha \leq \infty$ the surface is smooth.

Some Examples:

1. Copper. The melting entropy of copper is

$$\Delta S_f \text{ (Cu)} = 10 \frac{\text{J}}{\text{mol K}}$$

and we obtain $\alpha = \xi \cdot 1.2$. This is a typical value for metals. That means: no faceting for all orientations of the crystal. This is in agreement with the observation, that metals do not facet in contact with their pure melts. The use of bulk properties for the calculation of the binding energy L can lead to a negative interfacial energy for materials with low melting entropy /5/ /6/.

2. Ice

$$\Delta S_f \text{ (H}_2\text{O)} = 21{,}99 \frac{\text{J}}{\text{mol K}}, \quad \alpha = \xi \cdot 2{,}64$$

ξ depends on crystal orientation. Each water molecule is linked by 4 H-bonds to the lattice. The c-axis is the axis of highest symmetry in the lattice. Each water molecule has one bond parallel to it whereas three bonds are close to a plane normal to the c-axis. Hence ξ is $\frac{3}{4}$ for the basal plane.

$$\alpha_{\text{basal plane}} = \xi \cdot 2{,}64 \approx 2$$

For all other directions α is less than 2. Ice growing into water facets only at the basal plane. This is in agreement with the observations during crystal growth.

3. Salol. The next example has high melting entropy

$$\Delta S_f \text{ (Salol)} = 62.8 \frac{\text{J}}{\text{mol K}}; \quad \alpha = \xi \cdot 7{,}5$$

There are many orientations of Salol which facet during growth of the crystal /7/. A study of the growth morphology of biphenyl has been published recently /8/.

If we neglect diffusion processes then growth rate v normal to a facet of a dislocation free crystal is determined by two dimensional nucleation. The growth rate v has the following form

$$v = K_1 \exp\left(- \frac{K_2}{k_B T}\right).$$

The peculiarity of this mechanism is that growth velocity is significant only after achieving a threshold supercooling. Once such a nucleus has been formed atoms can be adsorbed at the step formed by it and the facet plane. The crystal grows layer by layer. If the crystal is not perfect but contains screw dislocations the growth law has a parabolic form

$$v = K_3 \, (\Delta T)^2 .$$

In the case of a rough interface no threshold has to be overcome and the molecules can be adsorbed at any place of the interface. No surface diffusion is necessary to find a step to be bound permanently. For a rough surface the crystal grows normal to its interface and growth rate is proportional to the supercooling

$$v = K_4 \, \Delta T .$$

These models are only valid for small supercooling.

The Nonfaceted - Faceted Transition

It has been tried some 30 years ago to describe the rough and faceted state of the interface as an analogy to a two dimensional ferromagnet /9/. In this case a second order transition is expected from the faceted to a rough interface. These early calculations led to a critical transition temperature T_R which was above the melting temperature of the crystal. For the solid-vapor interface new calculations of the so-called roughening transitions led to a different picture with a transition temperature below the melting point. For a review see /10/, /11/. These calculations have stimulated several new experiments.

For an experimental verification of the surface roughening transition a system is necessary where it is possible to change α continuously. That means L/k_BT_M has to be varied. For a solid in contact with its pure melt this seems to be impossible because only very high pressures can influence the melting point significantly and crystals usually do not facet at all in contact with their pure melt under equilibrium conditions. They facet only at growth conditions. That means a second order roughening transition in such systems can not be expected. These restrictions do not exist for crystal growth from the vapor or from solutions.

The Solid-Vapor Interface

The first experimental evidence for a roughening transition has been reported by Jackson and Miller /12/. They used a crystal-vapor system which is close to the model system based on the Ising model. Two substances have been chosen: C_2Cl_6 and NH_4Cl. Both have a high vapor pressure at the melting point (Table 1). By means of three different techniques a gradient of the chemical potential of the vapor molecules has been established at the interface

a) vapor transport in a closed ampule at variable temperature,

b) migrating vapor bubble in a temperature gradient,

c) a thermal gradient over a crystal which is in contact with the vapor.

For C_2Cl_6 an abrupt and reversible change of growth morphology from faceted crystals to rounded dendritic growth has been observed at temperatures close to $T_R = 100^oC$. Similar results have been obtained with NH_4Cl ($T_R = 365^oC$). In both cases a transition has been observed only at non-equilibrium conditions using crystal growth as a means of detection.

An experiment which can be interpreted in terms of a one dimensional roughening transition at a solid vapor interface has been performed by Lampert and Reichelt /13/. They studied the evaporation of rock salt. In evaporation and growth of crystals at low supersaturation spirals play a dominant role. Anisotropy in interface free energy and transport coefficients are typical for faceted interfaces. These anisotropies are reflected in an anisotropy of the growth spirals. With increasing distance from the center the growth spiral becomes more rounded. A decrease of anisotropy is also found with increasing temperature for all distances from the center and in a linear extrapolation the anisotropy of the evaporation spirals vanishes at about 650°C. It would be interesting to see in an

Table 1

	C_2Cl_6	NH_4Cl
Sublimation temperature (°C)	185	335
Melting point (°C)	186	525
Vapor pressure at MP (atm)	1	41.4
T_R (°C)	100	365
Vapor pressure at T_R	40 Torr	3 atm
k_BT_R/L_v	1/16	1/16
T_R/T_M	0.81	0.80

analogous experiment under growth conditions whether this one dimensional roughening transition shows a hysteresis effect.

The Solid-Solution Interface

Metals do not facet in contact with their pure melts but they form facets in contact with other liquid metals and alloys. Interface properties influence faceting. This has been demonstrated by Passerone et al. /14/. Samples of Zn-Bi-In alloys have been prepared and annealed at 336°C for 250 hours. Zn does facet in contact with Bi, it does not facet in contact with pure In. Neither Bi nor In are incorporated in Zn or adsorbed at grain boundaries of Zn. Therefore the grain boundary surface tension γ_{GB} does not change with alloy composition and has always the value of pure Zn. In the samples at 336°C solid Zn is in contact with liquid Bi-In alloy and the parameter that can be varied is the concentration ratio of In and Bi. After annealing the samples are quenched. The solid liquid interface tension $\gamma_{s\ell}$ has been determined by measuring the dihedral angle θ formed at the intersections between a grain boundary and a solid liquid interface. The faceting of Zn in contact with the melt has been observed by means of scanning electron microscopy. The experimental results are: $\gamma_{s\ell}$ decreases with increasing In concentration. There is an isothermal faceted to non faceted transition. The transition occurs in the composition range where important changes in $\gamma_{s\ell}$ take place. A linear dependence of the factor on $\gamma_{s\ell}$ can be deduced for systems where chemical adsorption at the interface can be neglected.

The Solid-Melt Interface

By changing pressure, the melting temperature of Helium can be varied over a wide range on a relative scale. At temperatures below 1 K the melting entropy tends very quickly to zero.

$$\Delta S_f/R\ (1.2\ K) = 0.0482 \qquad \Delta S_f/R\ (0.4\ K) = 0.000048$$

Hence the Jackson α-factor can be varied over a wide range and the interface morphology can be studied as a function of melting temperature. Landau et al./15/ have observed that during growth at temperatures below 1 K the hcp crystals are always faceted when growth rate exceeds about 1 mm/s. Similar to normal crystals no facets can be observed during melting. All facets correspond to principal axes. No faceting has been observed at the bcc phase. The area of a facet has been measured as a function of temperature from 0.5 to 1.1 K /16/. It has been observed that He forms facets at low temperatures which vanish with increasing temperatures. Independent whether the area of facets has been measured for increasing or decreasing

Table 2

	T_M(°C)	T_{max}(°C)	T_t(°C)
O-terphenyl	55	35	20 to 25
Salol	41.6	22	13 to 17
Thymol	51.5	0	-10 to -15

T_M = melting point,

T_{max} = temperature of maximum growth rate,

T_t = temperature of morphology transition

temperature, the same values have been found and a roughening temperature of 1.1 K has been extrapolated.

MC calculations predict the roughening transition of a solid vapor interface at high temperatures corresponding to high vapor pressure and high driving force for crystal growth. Interface morphology has been studied during crystal growth from highly supercooled melt by Miller /17/. The experiments have been done with o-therphenyl, Salol and Thymol.These are organic substances with high melting entropy. The substances have been carefully cleaned by zone refining and sealed in a thin cuvette formed by two microscope cover glasses (20 x 20 mm) with an interior spacing of 20 ... 25 μm. Starting with a poly crystalline seed the interface is faceted and the individual crystals maintain a position which deviates from a mean interface depending on the crystal orientation. When the system was cooled below a critical temperature of about 15°C for Salol the interface changes from a bizarr shape to an optical smooth appearance and all crystallites have the same growth rate at a given undercooling. To come back to the anisotropic regime it is necessary to heat the system by about 10°C (Table 2). This transition takes place at temperatures below the temperature of maximum growth rate where growth rate decreases with increasing supercooling. A comparison of these experiments with the dynamic roughening transition has been criticised /18/ because viscosity of the melt plays an important role in the growth of crystals from highly supercooled liquids.

3. SHARP INTERFACE-DIFFUSE INTERFACE

In 1960 Cahn has discussed the possibility of a diffuse interface which would enable a normal growth with low driving force /19/. This is probably the first paper where a diffuse interface in front of a growing crystal is discussed. The concept has been applied earlier by Cahn and Hilliard /20/ to critical systems. There are also some recent computer simulations which deal with surface layers on crystals. These calculations as well as the model of Cahn lead to surface regions only a few lattice constants thick.

There is evidence from light scattering experiments that there exists an interface layer with a thickness of a few μm in front of a crystal growing into its melt /21/. This layer is nucleated at conditions far from thermodynamical equilibrium. In this chapter some of these light scattering experiments /22//23/ will be summarized together with recent measurements of the thickness of the layer /24/.

Light Scattering at the Solid Liquid Interface

For the investigation of a diffuse interface and its dynamics up to now light scattering seems to be the method which combines best time resolution and the ability to probe small sample volumes. The spectrum of the light scattered in a pure liquid consists of a central component, two Brillouin components and Raman lines. The relative intensities of the central line I_c and the Brillouin components I_B are given by the Landau-Placzek intensity ratio

$$\frac{I_c}{2I_B} = \frac{(C_p - C_v)}{C_v}$$

Where C_v and C_p are the specific heats at constant volume and constant pressure respectively. If there are structural changes in the liquid close to the interface they will give rise to a contribution to the central line. They will also influence the other spectral regions,but the quasielastic scattering gives the most direct evidence for the existence of fluctuations at the interface. Most interesting would be to study the solidification of simple liquids because the structure of simple liquids e.g. rare gases and metals is known. There are two conditions which limit the application of light scattering technique:

1. The material must not absorb light.
2. The intensity level of any light scattered from interfaces or inhomogeneities in the bulk of the melt or of the crystal has to be low, because the separation of several sources of scattered light might be very difficult.

Condition one excludes metals and condition two leads to experimental difficulties:

a) the melt has to be free from dust
b) the solid has to be a single crystal.

It is an unsolved problem to grow large single crystals from rare gases Ar, Kr, Xe. One of the reasons is the high thermal expansion coefficient of rare gas solids which is about 10^{-3} and a factor of 100 bigger than the thermal expansion coefficient of any container in which the crystal might grow. Hence the conditions 1 and 2 made it up to now impossible to do a Rayleigh light scattering experiment at the solid-liquid interface of substances that form simple liquids.

Assume that it is possible to grow single crystals of sufficient size to neglect straylight from surfaces. Assume that it is possible to prepare a dust-free melt. The intensity of the background straylight is then determined by the scattering on thermal fluctuations in the liquid. There is one liquid with a vanishing intensity of the central-line at temperatures close to the melting point: water. Due to the density maximum the Rayleigh intensity vanishes at 4°C /25/. Hence the water-ice interface is unique for an experiment which tries to detect fluctuations at the solid liquid interface. Another very important advantage of water is that ice crystals are available in our laboratory and we have great experience in zone refining of water /26/. This is the technique we use for the production of pure water.

Quasi Elastic Light Scattering Technique

In this technique light of a laser beam is scattered by inhomogeneities. The light scattered into a direction defined by a scattering vector k is detected by a photomultiplier. Temporal fluctuations in the spatial distribution of these inhomogeneities give rise to changes in the interference pattern of the scattered light and thus to fluctuations in the photocurrent of the detector. The correlation function of the photocurrent gives information on the type of the fluctuations (chaotic or periodic) and their time scale (decay time or periodicity). For a review on the technique see the book by B. Chu /27/.

Quasi elastic light scattering from small noninteracting spheres suspended in water is treated as an example for the introduction of some assumptions and data used in this experimental technique. Due to random walk the concentration of the spheres is not homogeneous through the whole volume. Assuming linear superposition of concentrations, a Fourier analysis can be done of the spatial concentration distribution. The intensity of the light scattered into a direction defined by a scattering vector k is proportional to the square of the amplitude of a Fourier component c(t,k). k corresponds to a spatial wavelength $\Lambda = \frac{2\pi}{k}$. The amplitude decays according to a diffusion law

$$\frac{dc}{dt} = D \frac{d^2c}{dx^2}$$

leading to a decay time

$$\tau = \frac{1}{k^2 D}$$

which is the decay constant of the intensity correlation function. The spatial wave lengths Λ which are accessible vary with the scattering angle θ according to Bragg's law from Λ = 1835 Å for $\theta = 180^o$ to Λ = 22015 Å in the case of foreward scattering ($\theta=10^o$). In the experiment τ is measured. $1/\tau$ is proportional to k^2 and the proportionality is the translational diffusion constant. Using Stokes-Einstein relation

$$D = \frac{k_B T}{6\pi\eta a}$$

the radius a of the spheres can be calculated. If a diffusion constant of $D = \Gamma/k^2 = 2\cdot10^{-8}$ cm^2/s has been measured in an experiment where spheres are diffusing in water at 0^oC ($\eta = 1{,}787\cdot10^{-2}$ poise) one obtains for the hydrodynamic radius of these spheres $a = 5.8\cdot10^{-6}$ cm.

Light Scattering at the Ice-Water Interface

For all light scattering experiments it is indispensable to exclude any dust from the scattering volume. Therefore the light scattering experiment at the solid liquid interface is done in situ during zone refining. The zone refining apparatus as shown in fig. 2 is placed in a cold room at -18^oC. The two heating tubes provide a positive temperature gradient in front of the interface (fig. 3). Impurity concentrations in the liquid zone are below ppb level. The interface is always stable relative to dendritic instabilities. A Mullins-Sekerka instability has never been observed in these pure crystals. During the experiment the ice-crystal is lowered into the cooling bath exactly at the growth rate, so that the ice-water interface does not move relative to the optical system. ψ_0 is the angle of incidence of the laser beam. The photomultiplier can be rotated around the growth tube. It detects light scattered parallel to the interface plane. θ is the angle between the projection of the

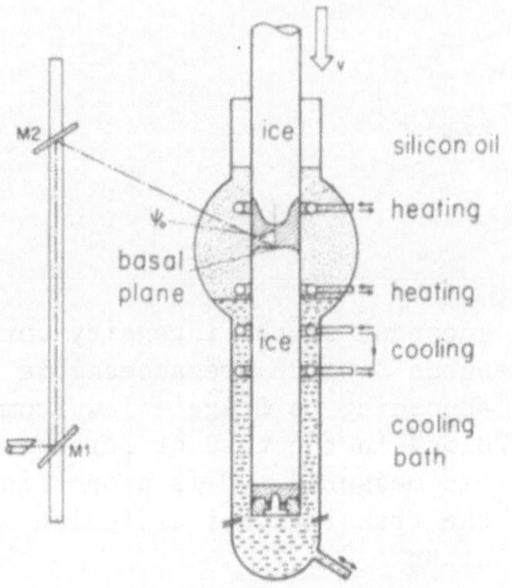

Fig. 2. Zone refining apparatus

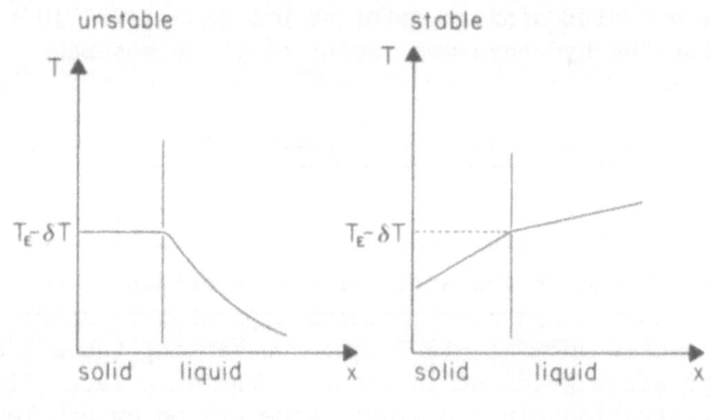

Fig. 3. Temperature gradients at the solid-liquid interface which stabilize and destabilize the interface

incident laser beam on the interface plane and the direction of the observed scattering. For $\psi_0 = 90^o$, θ is the scattering angle. $\psi_{crit.}$ is the critical angle of total reflection.

Intensity Hysteresis and Dynamics

Starting with a crystal at equilibrium conditions no light is scattered quasielastically from the interface. Only unmodulated straylight (background) and Brillouin scattered light is detected. When the growth velocity exceeds 1.5 µm/s the onset of quasielastic light scattering occurs. The intensity is proportional to the growth rate as long as v is positive (fig. 4). The maximum growth rate achieved is 5 µm/s. After reverting to equilibrium conditions quasi-elastic scattering vanishes and can be initiated again by exceeding the growth rate of 1.5 µm/s. This behavior is observed at the facet as well as at rough interfaces. Dynamical scattering occurs in a thin interface region only. In the crystals the intensity of the Raman scattering light is large compared to the intensity of the Rayleigh and Brillouin components together.

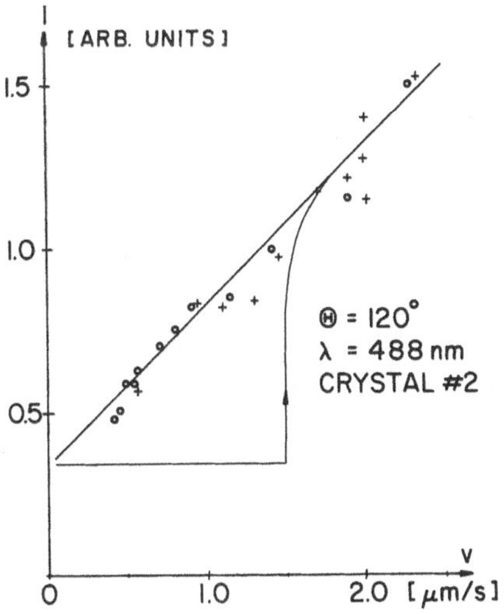

Fig. 4. Hysteresis behavior of the scattered intensity as a function of growth rate.

From the angular dependence of the intensity of the scattered light one concludes that the upper limit of the diameter of the scattering inhomogeneities is below the wavelength of light. Measurements of the autocorrelation function of the photocounts can be fitted by one single exponential. There is no indication for a distribution of decay times. For experiments with the scattering vector in the interface plane the line width increases with the square of the scattering vector (fig. 5). The line width does not depend on growth rate. These measurements have been done some time after a steady state of the growth process has been reached and

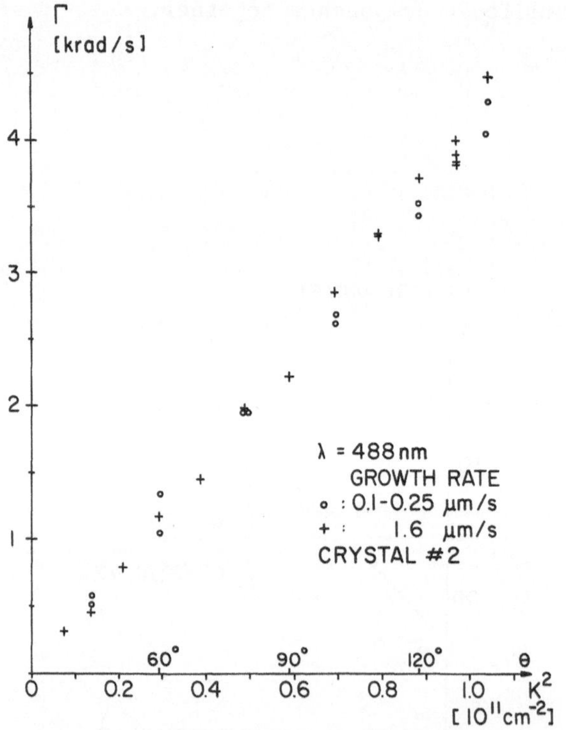

Fig. 5. Line width of the scattered light versus the square of the scattering vector. Measurements with scattering vector in the interface plane are plotted.

with the scattering vector oriented parallel to the solid-liquid interface plane. For different crystals and growth conditions the slope Γ/k^2 can change, but in all experiments the measured Γ is proportional to k^2.

Three models will be discussed to interpret the data plotted in fig. 5. Only the assumption of a diffuse interface (model 3) does not lead to contradictions with experiments. In the first model we interpret these measurements in terms of spheres diffusing in water. The measured diffusion coefficient corresponds to spheres with a hydrodynamic radius of $3.9 \cdot 10^{-6}$ cm. There are three arguments which rule out this interpretation:

1. Measurements of the scattered intensity for an angle of incidence $\psi_0 > \psi_{crit}$ relative to the intensity scattered at $\psi_0 < \psi_{crit}$ led to a ratio of 4 instead of 2 as expected for this model /21/
2. Spheres of only one diameter are observed. The single exponential fit compares very well with fits obtained in light scattering experiments on mono disperse latex spheres.
3. There is no thermodynamical reasoning to explain the nucleation of such spheres at a supercooling of about 0.1^oC only /22/

In a second model a corrugated interface is assumed. The measured decay times τ of the fluctuations of the ice-water interface are interpreted as the decay times of disturbances of a sharp solid-liquid interface /22//23/. Assuming a linear dependence of growth velocity v on supercooling ΔT

$$v = \mu \cdot \Delta T$$

The line width $\Gamma = 1/\tau$ can be calculated by using Gibbs-Thomson equation:

$$\Gamma = \frac{1}{\tau} = \frac{\mu \gamma_{s\ell} k^2}{\Delta S_f}$$

Using $\mu = 10^{-4}$ m/sK /28/, $\gamma_{s\ell} = 29 \cdot 10^{-3}$ J/m^2 /29/ we obtain for the slope of the fitted line in fig. 5 $\Gamma/k^2 \approx 3 \cdot 10^{-8}$ cm^2/s. The agreement of this simple model with the experimental data is astonishing. It has to be kept in mind that the dynamics of the system are hidden in the quantity μ which is not known very precisely. The interfacial energy $\gamma_{s\ell}$ has been measured at equilibrium conditions. The meaning of this quantity at conditions far from thermodynamical equilibrium is not clear.

In an experiment where the scattering plane coincides with the solid-liquid interface it is not possible to decide whether the interface is sharp and corrugated or three dimensional and diffuse. Therefore measurements of the scattered intensity and line width have been done using experimental configurations where the scattering vector is perpendicular to the interface. Such measurements give information on the existence of a fourier component perpendicular to the solid-liquid interface. The corresponding spatial wavelength gives an estimate for the thickness of the layer. In fig. 6 measurements with scattering vector parallel (circles) and perpendicular (triangles) to the solid liquid interface are compared. No significant difference in line width can be found. This rules out the

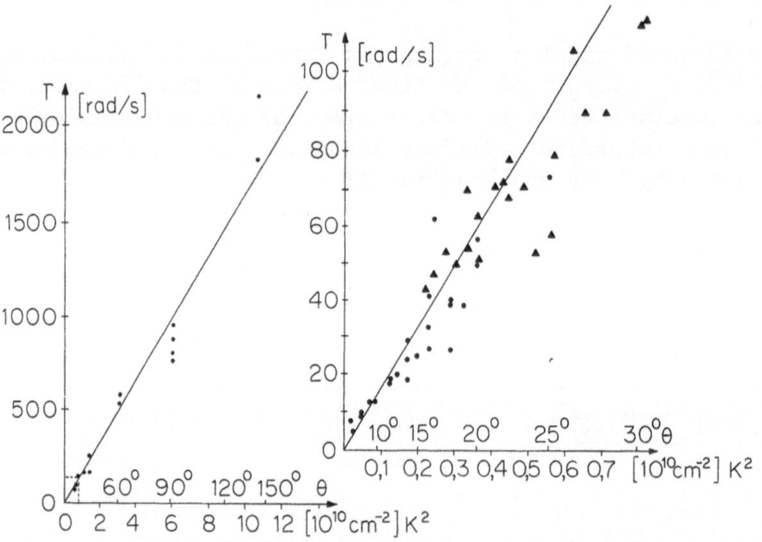

Fig. 6. The line widths of light scattered with k parallel (circles) and k perpendicular (triangles) to the solid-liquid interface are compared. The insert is expanded by a factor of 20. The data points at high scattering angles have been taken in the transient regime when the steady state has not yet been reached.

model of the corrugated interface. In this model only the projection of the scattering vector on the interface plane should be effective. Therefore a decrease of the line width is expected by tilting the scattering vector out ot the plane of the ice surface. The isotropy of the decay rate of the fluctuations is observed for all orientations of the scattering vector. The model of the corrugated interface is also not compatible with intensity measurements at variable ψ_o /21/. In fig. 6 the smallest k-vectors which have been chosen correspond to a spatial wavelength of 1.7 µm.This is a lower limit for the thickness of the interface layer /24/. These measurments have been done at a growth rate of 1.5 µm/s. In earlier measurements where the coherence properties of the scattered light have been determined, an upper limit for the thickness of 6 µm has been found /22/.

Transient Phenomena

The line width of the Rayleigh line of water corresponds to a diffusion constant of the thermal diffusivity of $D_{th} \approx 10^{-3}$ cm^2/s. During the transient from the equilibrium state, where no quasi-dynamical scattering at the interface takes place, to the scattering steady state the intensity of the Rayleigh line increases drastically. At the same time linewidth decreases to a value corresponding to a diffusion constant of about 0.4 10^{-7} cm^2/s (fig. 7). The Rayleigh line width has been measured at the ice-water interface just after the onset of the scattering, as long as the scattered intensity has not reached the steady state value. It has been found that Rayleigh line width decreases during the build up of the scattering power. There also occur fluctuations in the line width during the measurements. To check whether these fluctuations are a property of the interface layer, Γ has been measured simultaneously at two k vectors. It has been found that the two measured line widths vary synchronously and that Γ is always proportional to k^2.

Light Scattering at the Solid-Liquid Interfaces of D_2O and Salol

H_2O has many peculiar properties which are determined by the H-bond network. If the interface layer is one of those properties of water, then it should be influenced by an exchange of the protons by deuterons. /30/. Light scattering experiments at the solid-liquid interface of D_2O led to the same dynamical behavior which has already been found for H_2O. The intensity hysteresis is more pronounced in the D_2O system, the scattering process does not start before the crystal has reached a growth velocity of 2.2 µm/s /31/.

Salol is a substance with high melting entropy. That is one of the reasons why it has been chosen for this experiment. By

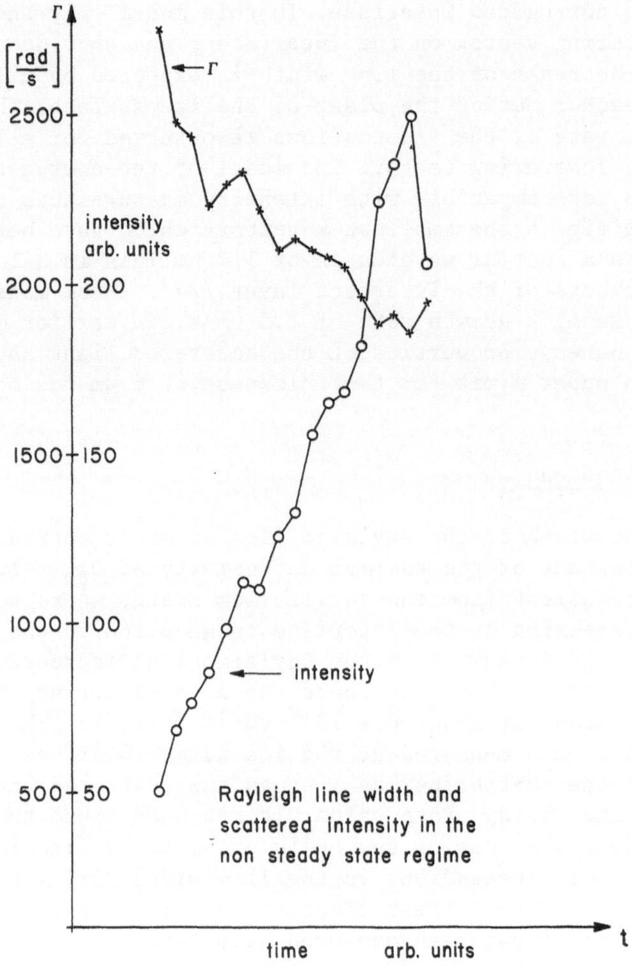

Fig. 7. Transient phenomena during the onset of scattering

illuminating the solid-liquid interface from the solid side it is possible to observe two instabilities which succeed one another. First at low growth rates a transition to periodic fluctuations is initiated. The measured frequency increases linearly with the scattering vector k. At high growth rates after a second step the scattering behavior changes to a diffusive mode similar to the one already known from the ice-water interface. For comparable geometry the decay time measured in the Salol system is by a factor 20 longer than the one in the H_2O experiment /32/.

The properties of supercooled water show at temperatures around 225 K a temperature dependence which can be fitted by critical exponents /33//34/. A similar behavior has been observed for other liquids e.g. Salol /35/. It might be that the diffuse interface layer is a typical property of the liquid-solid transition far from equilibrium and that the properties of the interface layer can be compared with the properties of highly supercooled liquids. In this case the thickness of the interface layer can be regarded as a correlation length and the kinetics of crystal growth are determined by the kinetics in the interface layer.

4. GROWTH INTO SUPERCOOLED MELT

In the previous chapter there has been a positive temperature gradient in the liquid. Now we change to configurations where crystals grow into supercooled melt (fig. 3). Whether a crystal surface is stable at such conditions is treated in the theory of Mullins and Sekerka. It is a macroscopic theory and has no direct connection to the properties of the interface on an atomic scale. Originally it has been developed for isotropic crystals with rough interfaces. Experiments and theory have been reviewed recently /36//37/.

Growth of Dendrites

The growth rate of dendrites has been measured very carefully by Glicksman et al. /38/. Two methods have been used:

i) Growth in a capillary, this is the widely used method. It has the advantage that the supercooling at the solid liquid interface can be determined from the temperature of the surrounding bath using some corrections for the heat flow. The great disadvantage of this method is that the wall of the capillary does influence the growth kinetics.

ii) Free growth into a supercooled melt. Data obtained with this method will be discussed here only. Any disturbances from walls are eliminated, but now convection in the liquid can lead to erroneous data.

Succinonitrile has been studied because it is a model substance for the solidification of materials with low melting entropy and it is transparent. Ultrahigh purity material has been prepared by mulitple destillation and zone refining. To insure that the dendritic morphology and growth kinetics remain free of any external influence a special apparatus has been designed. It consists of a 4 cm diameter spherical growth chamber. In the center of it there ends a capillary tube. A seed crystal can grow through this tube. Upon its exit from the capillary, the dendritic front can expand freely into the spherical volume of supercooled liquid. A typical experiment consisted of melting the entire specimen and immersing the apparatus in the thermostat for at least 25 min. a time long enough to insure adequate uniformity of temperature throughout the spherical portions of the specimen. Growth velocities for supercoolings from 1° to

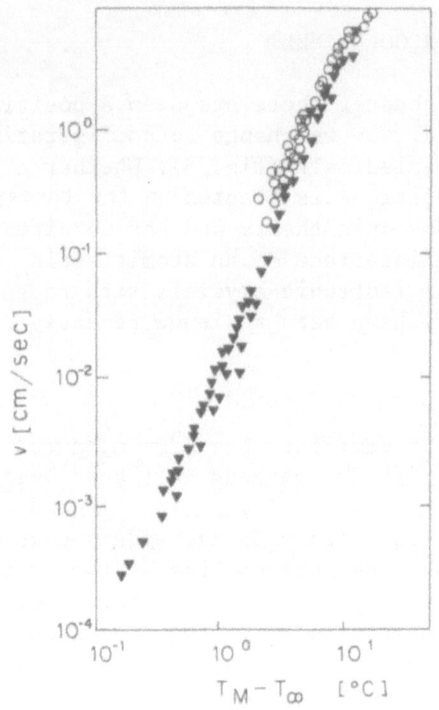

Fig. 8. Free dendritic growth of ice into supercooled water

about 10° have been reported. At these growth rates disturbances from convective heat transport can be neglected. In new measurements the temperature range has been extended to supercoolings as low as 0.04° /39/. Very precise measurements of the free dendritic growth of succinonitrile have been done by Lappe /40/. He found an increase of growth velocities if impurities are present in the melt with a mole fraction between $5 \cdot 10^{-5}$ and 10^{-3}.

An other substance where free dendritic growth has been studied is ice. Data from various authors have been collected by Sekerka /41/ (fig. 8). Growth rates range from $3 \cdot 10^{-4}$ cm/s up to about 10 cm/s at initial supercoolings between 0.1° and 10°C.

Langeret al. /42/ have shown that the measured growth rates of ice and succinonitrile fall on a single smooth curve (fig. 9) when the data are plotted in the form of a dimensionless growth rate

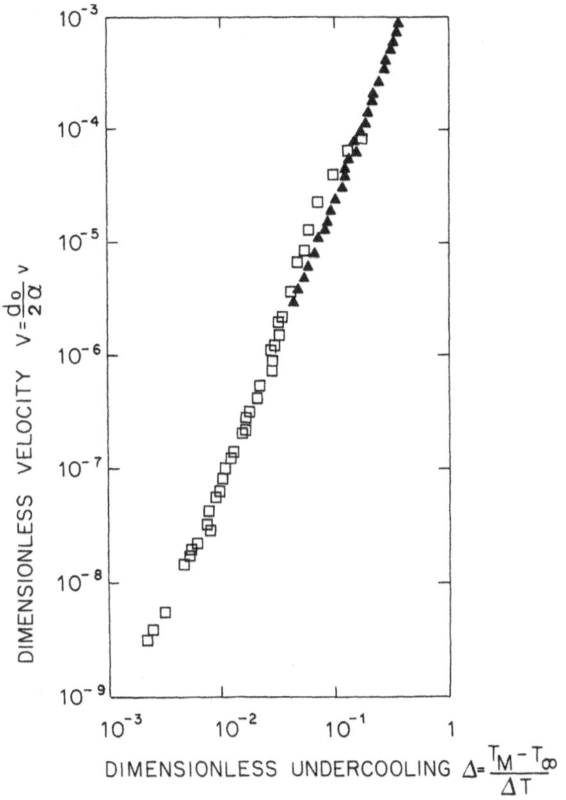

Fig. 9. Dendritic growth rates of ice (squares) and succinonitrile (triangles) are plotted on scales of dimensionless variables.

$$V = \frac{v\,d_o}{2\alpha}$$

versus the normalized supercooling

$$\Delta = \frac{T_m - T_\infty}{(\Delta T)}$$

where α is the thermal diffusivity of the melt and d_o the capillarity length given by

$$d_o = \frac{T_m\,\gamma_{s\ell}}{(\Delta T)L_v}$$

T_m and T_∞ are the equilibrium melting temperature and the temperature of the liquid well away from the region in which solidification is taking place. (ΔT) is the unit undercooling L_v/C_p. A theory /43/ /44/ has been developed for materials with isotropic growth dynamics. For the experiment of Glicksman succinonitrile has been selected because of its high degree of isotropy. Ice is not a material which has this feature. It does facet and the dendrites grow predominantly in the basal plane. But it is a fact that the universal law $V(\Delta)$ describes the growth rate of succinonitrile and ice.

5. THE DYNAMICS OF FREEZING AND MELTING

Although the process which leads to the diffuse interface layer is not understood it can be assumed, that this layer influences the growth dynamics of the crystals. If this is correct an asymmetry has to be found regarding high freezing rates and high melting rates. In front of the growing crystal there exists a diffuse layer with slow fluctuations in the supercooled "liquid", whereas close to the melting crystal there is no supercooling and no interface layer. Therefore the dynamic parameter for melting is expected to be higher than the one for the freezing process.

For low growth rates Miksch /45/ has looked for such an asymmetry. He found the same dynamic parameter during freezing and melting. This experiment has been done close to equilibrium conditions at growth rates up to 0.3 μm/s and at temperatures close to the triple point temperature. At such conditions no diffuse interface layer is expected.

High freezing rates are best studied during dendritic growth. In the case of ice and water it is possible to compare these freezing rates with growth rates of water dendrites into a superheated crystal. This analogon to dendritic crystal growth has been observed 1856 by J. Tyndall in the Swiss Alps /46/. Usually these figures are called Tyndall flowers. A sketch prepared by Tyndall /47/ is reproduced in fig. 10. Ice can be brought to a temperature above $0^{o}C$ by internal heating e.g. absorption of infrared or microwave radiation. At a temperature of about $0.3^{o}C$ water figures start to develop in the ice, which have the shape of dendrites and contain always a vapor buble. Käss and Magun /48//49/ have done quantitative experiments where the growth rates of water dendrites have been measured. Constraining effects of crystal surfaces and proximity effects of neighboring dendrites could not be avoided. Therefore these growth rates have to be regarded as a lower limit. On the other hand there are great advantages in comparison to the dendritic crystal growth experiments: complications introduced by convection and solute

Fig. 10. J. Tyndall's sketch of "Tyndall flowers"

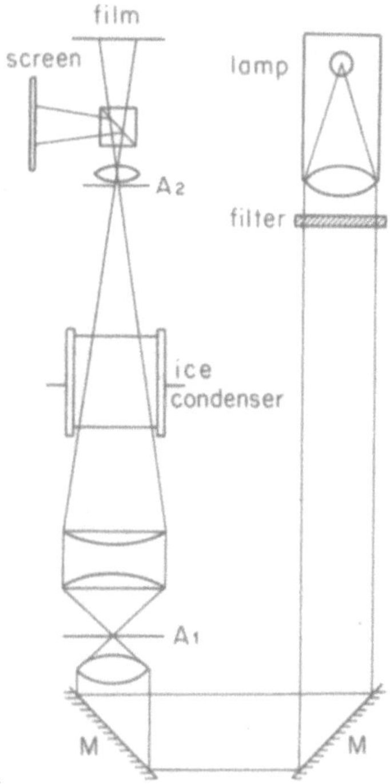

Fig. 11. Apparatus for the measurement of the growth rates of Tyndall flowers

redistribution do not exist. In fig. 11 the setup used in the experiments is shown. After the ice crystal has been brought to $0^{o}C$ it is heated in a 10 MHz field with a heating rate of $0.14^{o}C/s$. The crystal can be observed on a screen and simultaneously photographed. The diameters of a few Tyndall flowers nucleated in one piece of ice are plotted in fig. 12 as a function of time. The ice has been heated for 6.6 s. The growth continues after the heating has been stopped. The growth rate is about 2 mm/s. The superheating of the ice is below $0.9^{o}C$. $0.9^{o}C$ would be the temperature of the ice if no heat would have been consumed by the existing stars and by the melting surface of the ice. The growth rate is high if there are only few Tyndall flowers in the sample. If we compare these melting rates with the growth rates of ice dendrites growing into water with a comparable supercooling,the melting rates of the liquid dendritic figures are about two orders of magnitude higher than the growth rates of ice dendrites plotted in fig. 8. The thermal diffusivity of ice is by a factor of 6 higher than the one of water /50/. The resulting dimensionless growth velocity of water dendrites are more than one order of magnitude higher than the corresponding freezing velocities plotted in fig. 9.

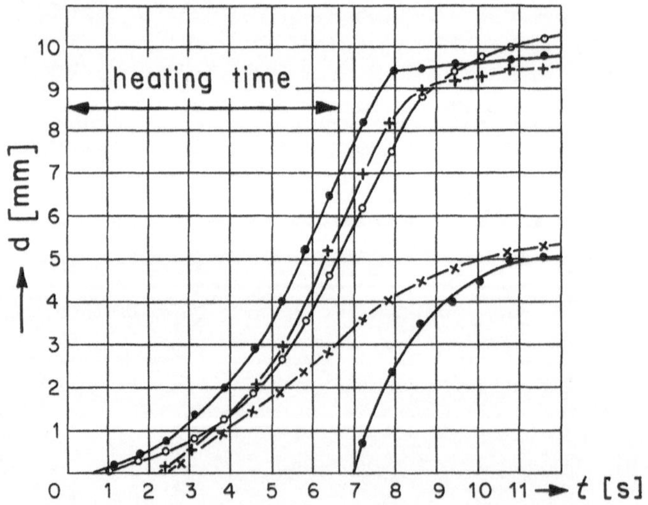

Fig. 12. The diameter of Tyndall flowers as a function of time. The heating time was 6.6 s.

The asymmetry of growth rates and melting rates which is found only at high growth rates supports the assumption that the diffuse interface layer influences the freezing dynamics at high growth rates (high undercooling). In this case the universal law of dendritic growth V(Δ) for faceting and non faceting crystals is a consequence of the existence of the diffuse interface layer.

ACKNOWLEDGEMENTS

The author is indebted to W. Känzig for his interest in this work. Thanks are due to colleagues who participate in the light scattering experiments: P. Böni and U. Dürig.

This work is supported by the Swiss National Science Foundation.

REFERENCES

1. W. A. Bentley and W. J. Humphreys, "Snow Crystals", Dover, New York (1962)
2. A. R. Ubbelohde, "The molten state of matter", Wiley, Chichester (1978)
3. K. A. Jackson, Mechanism of Growth, in: "Liquid Metals and Solidification", ASM Cleveland (1958)
4. K.A. Jackson, Theory of Crystal Growth, in: "Treatise on Solid State Chemistry, Vol. 5, Changes of State," H.B. Hannay, ed., Plenum, New York (1975)
5. D. Nason and W. A. Tiller, On the Lattice-Liquid Model for Interface Roughening, J. Cryst. Growth 10:117 (1971)
6. L. Coudurier, N. Eustathopoulos et P. Desre, Rugosité atomique et adsorption chimique aux interfaces solid-liquide des systèmes metalliques binaires, Acta Metall.26:465 (1978)
7. A. Neuhaus und G. Nitschmann, Zur Ausdeutung der Wachstumsergebnisse nach dem Nacken-Kyropoulos-Verfahren, Z. Elektrochemie 56:483 (1952)
8. H. J. Human, J. P. van der Eerden, L. A. M. J. Jetten and J. G. M. Odekerken, On the Roughening Transition of Biphenyl: Transition of Faceted to Non-Faceted Growth of Biphenyl for Growth from different Solvents and the Melt, J. Crystal Growth 51:589 (1981)
9. W. K. Burton and N. Cabrera, Crystal Growth and Surface Structure, Disc. Faraday Soc. No. 5:33 (1949)
10. J. D. Weeks and G. H. Gilmer, Dynamics of Crystal Growth, in: 'Advances in Chemical Physics' XL:157 (1979)
11. H. Müller-Krumbhaar, Surface Dynamics of Growing Crystals, in: "Festkörperprobleme XIX", Vieweg, Braunschweig (1979)

12. K. A. Jackson and C. E. Miller, Experimental Observation of the Surface Roughening Transition in Vapor Phase Growth, J. Crystal Growth 40:169 (1977)
13. B. Lampert and K. Reichelt, Anisotropy of Round Evaporation Spirals on Rocksalt Surfaces, J. Crystal Growth 47:77 (1979)
14. A. Passerone, R. Sangiorgi and N. Eustathopoulos, Isothermal Faceted to Non-Faceted Equilibrium Transition of Solid-Liquid Interfaces in Zn-Bi-In Alloys, Scr. Metall. 14:1089 (1980)
15. J. Landau, S. G. Lipson, L. M. Määttänen, L. S. Balfour and D. U. Edwards, Interface between Superfluid and Solid ^{4}He, Phys. Rev. Letters 45:31 (1980)
16. J. E. Avron, L. S. Balfour, C. G. Kuper, J. Landau, S. G. Lipson and L. S. Schulman, Roughening Transition in the ^{4}He Solid-Superfluid Interface, Phys. Rev. Letters 45:814 (1980)
17. C. E. Miller, Faceting Transition in Melt-Grown Crystals, J. Crystal Growth 42:357 (1977)
18. W. T. Griffith, On the Transition from Faceted to Non-Faceted Growth in Melt-Grown Crystals, J. Crystal Growth 47:473 (1979)
19. J. W. Cahn, Theory of Crystal Growth and Interface Motion in Crystalline Materials, Acta Metall. 8:554 (1960)
20. J. W. Cahn and J. W. Hilliard, Free Energy of a Nonuniform System. I. Interfacial Free Energy, J. Chem. Phys. 28:258 (1958)
21. J. H. Bilgram and P. Böni, Interface Fluctuations of Growing Ice Crystals, in: "Light Scattering in Liquids and Macromolecular Solutions", V. Degiorgio, M. Corti and M. Giglio, ed., Plenum Publ. Corp. New York (1980)
22. H. Güttinger, J.H. Bilgram, W. Känzig, Dynamic Light Scattering at the Ice Water Interface during Freezing, J. Phys. Chem. Solids 40:55 (1979)
23. J. H. Bilgram, H. Güttinger and W. Känzig, Fluctuations of the Ice-Water Interface during Solidification, Phys. Rev. Letters 40:1394 (1978)
24. P. Böni und J. H. Bilgram, Fluktuationen in der Grenzschicht zwischen Eis und Wasser (H_2O), Helv. Phys. Acta, to be published (1981)
25. C. L. O'Connor and J. P. Schlupf, Brillouin Scattering in Water: The Landau-Placzek Ratio, J. Chem. Phys. 47:31 (1967)
26. J. Bilgram, H. Wenzl and G. Mair, Perfection of Zone Refined Ice Single Crystals, J. Crystal Growth 20:319 (1973)
27. B. Chu, "Laser Light Scattering", Academic Press, New York (1974)
28. W. B. Hillig, The Kinetics of Freezing of Ice in the Direction Perpendicular to the Basal Plane, in: "Growth and Perfection of Crystals" R. H. Doremus, B. W. Roberts and D. Turnbull, ed., John Wiley, New York (1958)
29. S. C. Hardy, A Grain Boundary Groove Measurement of the Surface Tension between Ice and Water, Philos. Mag. 35:471 (1977)

30. H. E. Stanley, A polychromatic correlated-site percolation problem with possible relevance to the unusual behavior of supercooled H_2O and D_2O, J. Phys. A: Math. Gen. 12:L 329 (1979)
31. B. Zysset, P. Böni und J. H. Bilgram, Lichtstreuung an der Phasengrenze fest/flüssig von D_2O, Helv. Phys. Acta to be published (1981)
32. U. Dürig and J. H. Bilgram, Periodic Fluctuations at the Solid-Liquid Interface of Salol, this volume
33. J. Teixeira and J. Leblond, Brillouin Scattering from Supercooled Water, J. Physique 39:L-83 (1978)
34. C. A. Angell, Supercooled Water, in: "Water a comprehensive Treaise, Vol. 7", F. Franks, ed., Plenum, New York (1981)
35. A. N. Hunter, Measurements of the Velocity and Absorption of High-Frequency Ultrasonic Waves in Supercooled Liquids, Proc. Phys. Soc. 69:965 (1956)
36. R. F. Sekerka, Morphological Stability, in: "Crystal Growth, An Introduction" P. Hartman, ed., North Holland, Amsterdam (1973)
37. R. T. Delves, Theory of Interface Stability, in: "Crystal Growth" B.R. Pamplin, ed., Pergamon Press, Oxford (1975)
38. M. E. Glicksman, R. J. Schaefer and J.D. Ayers, Dendritic Growth - A Test of Theory, Metall. Trans. 7 A:1747 (1976)
39. S. C. Huang and M. E. Glicksman, Fundamentals of Dendritic Solidification - I. Steady-State Tip Growth, Acta Metall. 29: 701 (1981)
40. U. Lappe, Experimentelle Untersuchung des dendritischen Wachstums von Kristallen in unterkühlten Schmelzen, Berichte der Kernforschungsanlage Jülich - Nr. 1671, Jülich (1980)
41. R. F. Sekerka, On the Modeling of Solid-Fluid Interface Dynamics, in: "Proceedings of the Darken Conference" p. 301, United States Steel Res. Lab. Monroeville, Pa. (1976)
42. J. S. Langer, R. F. Sekerka and T. Fujioka, Evidence for a universal law of dendritic growth rates, J. Crystal Growth 44:414 (1978)
43. J. S. Langer, Instabilities and pattern formation in crystal growth, Rev. Modern Phys. 52:1 (1980)
44. J. S. Langer, this volume
45. E. S. Miksch, Equilibration of the Ice-Water Temperature Standard, Rev. Sci. Instr. 36:797 (1965)
46. J. Tyndall, "The glaciers of the alps", London (1860)
47. J. Tyndall, "Die Wärme" Braunschweig (1871) transl. of "Heat considered as a mode of motion" London (1863)
48. M. Käss und S. Magun, Zur Ueberhitzung am Phasenübergang fest-flüssig, Z.Kristallogr. 116:354 (1961)
49. G. J. Krüger und S. Magun, "Negative" Schneekristalle, Photographie und Forschung, Heft 8 (1955)

50. D. W. James, The Thermal Diffusivity of Ice and Water between -40 and +60°C, J. Mater. Sci. 3:540 (1968)

PERIODIC FLUCTUATIONS AT THE SOLID LIQUID INTERFACE OF SALOL

U. Dürig and J.H. Bilgram

Laboratorium für Festkörperphysik
Eidgenössische Technische Hochschule
CH-8093 Zürich, Switzerland

ABSTRACT

The freezing process is studied by quasi elastic light scattering at the solid liquid interface of growing Salol crystals. Depending upon the experimental condition one observes two different dynamical states of the interface. One manifests itself in an oscillating autocorrelation function the other in an exponentially decaying one. Measurements of the frequency of the oscillations as a function of the scattering angle and the laser frequency suggest that the scattered light is Doppler shifted in the first case. In the second case the decay rate (line width) increases with the square of the scattering vector. The results of these measurements are compared with those obtained in a similar experiment performed on ice.

INTRODUCTION

There is little experimental information available about the dynamics of crystal growth on a microscopic scale. In our quasi elastic light scattering experiments we study the solidification process at the solid liquid interface far from thermal equilibrium. Experiments done on the ice-water system showed that fluctuations at the solid liquid interface occur once a critical growth rate has been exceeded. The measurements of the line width of the scattered light also indicate that far from thermal equilibrium a phase boundary layer is nucleated (1,2). Since water has some unique properties the studies were extended to a different substance. Salol (Benzoic acid 2-hydroxy phenylester) was chosen because large single crystals can be grown and it is transparent to the Argon laser light. The space group of the crystalline phase is Pbca (3).

EXPERIMENTAL

Single crystals growing along the 001 or 010 axis are used. In order to exclude any dust the measurements are performed in situ during zone refining. The setup is shown in figure 1. During the experiment the ampoule containing the crystal is lowered into the cooling bath exactly at the growth rate of the crystal so that the solid liquid interface does not move relative to the optical system. Figure 2 shows the scattering geometry. The interface is illuminated from the crystal side under condition of total reflection by a focused Argon laser beam of not more than 200 mW power at 488 nm. The evanescent wave is travelling along the liquid side of the interface. The scattered light is then observed from the liquid side under an angle of 5° with respect to the facet plane. θ is the angle between the projection of the incident laser beam and the projection of the direction of observation on the interface plane. The scattered

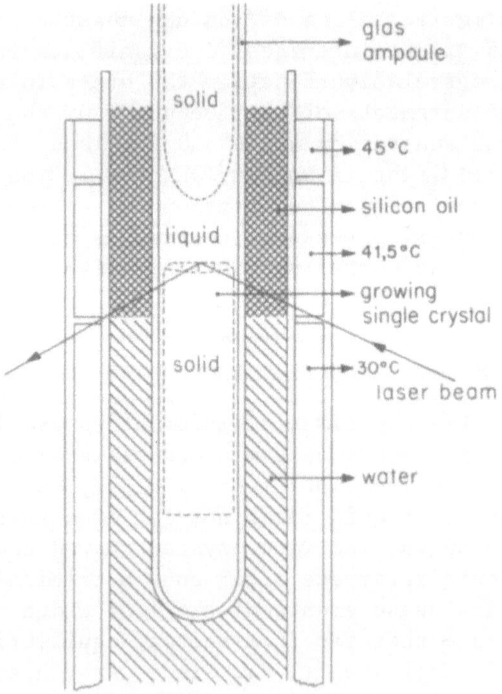

Fig. 1. Zone melting apparatus

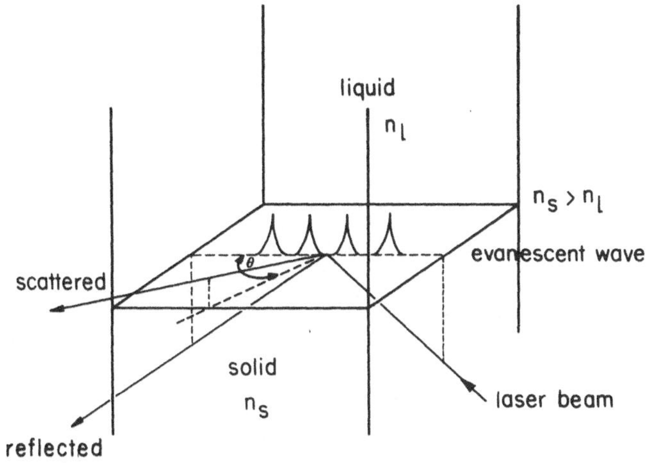

Fig. 2. Scattering geometry.
The incident laser beam is totally reflected at the solid liquid interface giving rise to an evanescent wave in the liquid.
n_ℓ and n_s designate the index of refraction of the liquid and the solid respectively.

light is detected by a photomultiplier which can be rotated around the growth tube. The photon pulses are fed into a digital correlator which is computing the autocorrelation function of the scattered intensity.

OBSERVATIONS

During crystal growth a facet develops. No scattering from the interface can be detected at growth velocities below 0.2 μm/sec. After the growth rate has been increased strongly anisotropic light scattering sets in at different areas on the facet. An oscillating autocorrelation function is observed at this stage. The amplitude of the oscillations does not decrease over more than 50 periods. After increasing the growth velocity above $v_{crit} = 0.4 \pm 0.1$ μm/sec one still observes the oscillatory scattering in the beginning. However after a delay of 1 to 2 hours a second type of scattering process sets in which is similar to the one observed in the ice-water experiment. During the smooth transition (15-10 min) the oscillations in the autocorrelation function become damped and

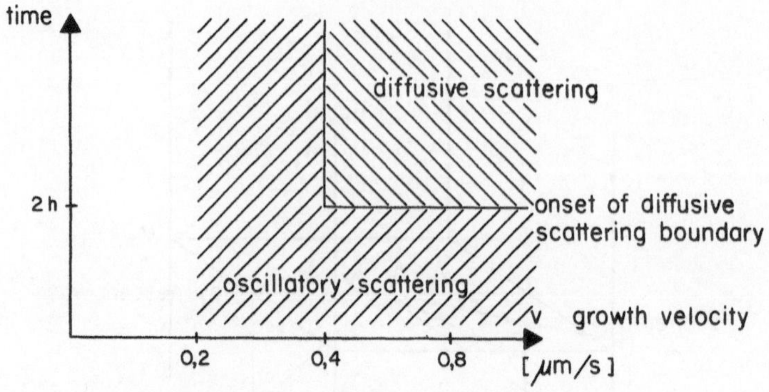

Fig. 3. Limits for onset of oscillatory and diffusive scattering.
The transition from oscillatory to diffusive scattering is not reversible. Once the diffusive state is reached the system does not revert to the oscillatory one.

disappear finally while an exponentially decaying one emerges. At the same time the scattering of the light becomes more and more isotropic. We call this regime diffusive scattering. Once the diffusive scattering is reached the system does not revert to the oscillatory state any more (figure 3). The growth velocities at which the two scattering processes set in vary considerably from sample to sample.

OSCILLATORY SCATTERING

In the case of the oscillatory scattering the light is only seen if the scattering vector points along a fixed direction in the facet plane. The frequency of the oscillations of the auto-correlation function depends linearly on the sin $^{\theta}/2$ (figure 4(a)). The proportionality factor can vary over a factor of two between different crystals. The frequency also increases as the inverse of the wavelength λ_L of the incident light (figure 4(b)). So far this has only been verified for the three intense Argon lines. These results can be interpreted in terms of a Doppler shift. Supposing that the scatterer moves with a velocity that has a component v_k parallel to the scattering vector the frequency shift of the scattered light is given by

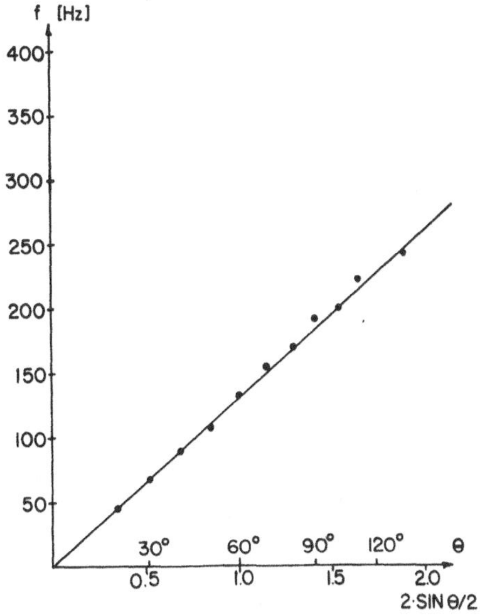

Fig. 4(a)

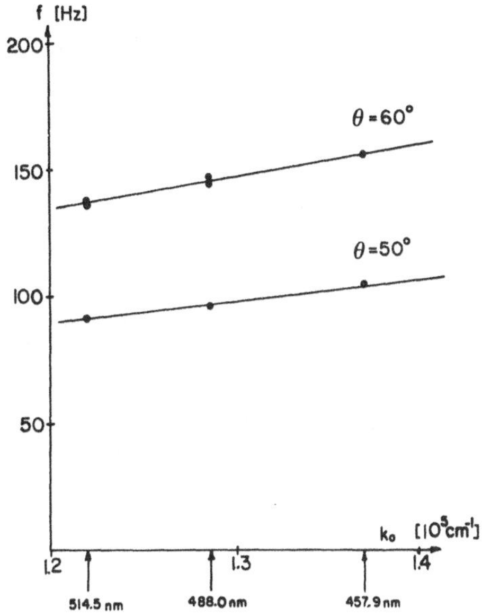

Fig. 4(b)

Fig. 4. Frequency of the oscillations of the autocorrelation function for a crystal growing parallel to the 001 axis at a velocity of 0.454 μm/sec
(a) as a function of $2 \cdot \sin \theta/2$
(b) as a function of the wavevector of the incident laser light

$$\Omega = v_k \frac{2\pi}{\lambda_L} \cdot 2 \cdot \sin \theta/2$$

Using this expression v_k can be determined from independent sets of experimental data, namely from the measurement of Ω versus $\sin \theta/2$ and from the measurement of Ω versus $^1/\lambda_L$. The data seem to support the Doppler hypothesis:

from Ω versus $2 \cdot \sin \theta/2$: $v_k = 63\ \mu m\ s^{-1}$

from Ω versus $^1/\lambda_L$: $v_k = 56\ \mu m\ s^{-1}$ $(\theta = 50^o)$

$v_k = 71\ \mu m\ s^{-1}$ $(\theta = 60^o)$

This velocity is larger than the growth velocity by a factor of 150.

DIFFUSIVE SCATTERING

The transition from the oscillatory to the diffusive scattering is not reversible. A similar hysteresis has been found in the ice water experiments. The measured line width Γ is proportional to the square of the scattering vector for growth along the 010 axis; for growth along the 001 axis a similar functional dependence is observed but the scatter in the datapoints is considerably larger (figures 5(a), (b)). We could not find a dependence of Γ on growth velocity. The slopes $\Gamma/\sin^2\theta/2$ are between 100 and 200 rad/sec depending on the crystal. These line widths are by a factor of 10-20 smaller than those measured in the ice-water system. For a corrugated interface the decay time of a fourier component with the wavevector k was calculated based on the Gibbs-Thomson equation as in Ref.(1)

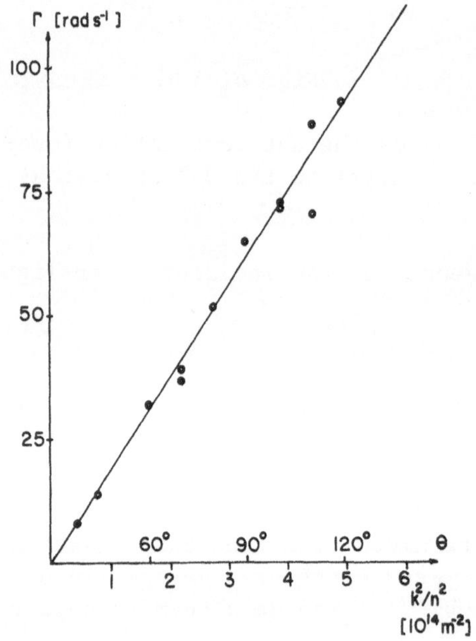

Fig. 5(a)

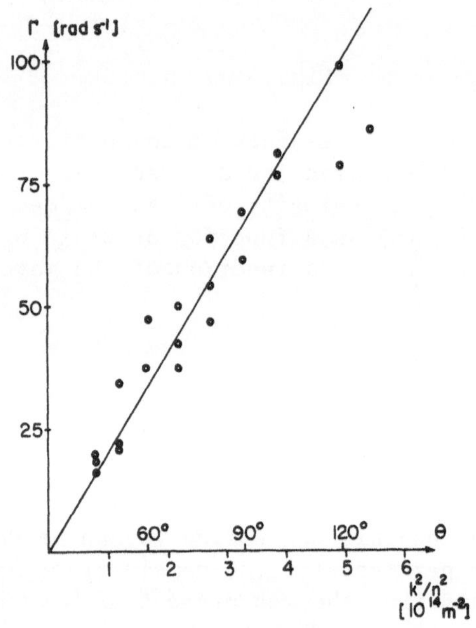

Fig. 5(b)

Fig. 5. Line width of the scattered light as a function of the square of the scattering vector for a crystal growing parallel to
(a) the 010 axis at a velocity of 0.45 μm/sec
(b) the 001 axis at a velocity of 0.78 μm/sec
n is the index of refraction in the scattering volume

$$\frac{1}{\tau} = \Gamma = \frac{\mu\, \gamma_{s\ell}\, k^2}{\Delta S_f}$$

$\gamma_{s\ell}$ designates the solid liquid interfacial energy and ΔS_f the melting entropy. μ is a dynamic growth factor based on the assumption that the growth velocity v of the crystal is proportional to the deviation of surface temperature from its equilibrium value

$$v = \mu \Delta T$$

Since the values for μ in the literature vary considerably we measured the temperature profile across the solid liquid interface with an array of thermocouples. From these measurements we obtained

$$\mu = 10^{-6}\ ms^{-1}K^{-1}$$

at a growth velocity v = 0.78 µm/sec. Table 1 shows the measured and calculated values of $\Gamma/\sin^2\theta/2$ for ice and Salol. The agreement between the measurements and the calculation is surprising considering the fact that the model is two dimensional and does not take into account the existence of an interface layer as it was observed in the ice experiments.

Table 1

	ICE	SALOL
γ_{sl}	$2{,}9 \cdot 10^{-2}\ Jm^{-2}$ (4)	$2 \cdot 10^{-2}\ Jm^{-2}$ (estimated)
ΔS_f	$1{,}117 \cdot 10^{6}\ Jm^{-3}K^{-1}$ (4)	$3{,}24 \cdot 10^{5}\ Jm^{-3}K^{-1}$ (5)
μ	$10^{-4}\ ms^{-1}K^{-1}$ (1)	$10^{-6}\ ms^{-1}K^{-1}$
$\Gamma/\sin^2\theta/2$ cal.	1600 rad s^{-1}	100 rad s^{-1}
$\Gamma/\sin^2\theta/2$ meas.	1000-2000 rad s^{-1}	100-200 rad s^{-1}

ACKNOWLEDGEMENTS

The authors thank W. Känzig for his steady interest and many fruitful discussions and critical reading of the manuscript and M. Wächter for careful preparation of the crystals.

This work is supported by the Swiss National Science Foundation.

REFERENCES

1. J. H. Bilgram, H. Güttinger, W. Känzig, Dynamic Light Scattering at the Ice Water Interface during Freezing, J.Phys.Chem.Solids 40:55 (1979).
2. J. H. Bilgram, P. Böni, Interface Fluctuations of Growing Ice Crystals, in: "Light Scattering in Liquids and Macromolecular Solutions", V. Degiorgio, M. Corti and M. Giglio, ed., Plenum, New York (1980).
3. A. Baptista, Estudo Radiocristalografico de Substancias Organicas I-Salicilato de Fenila, An.da Acad.Brasileira de Ciencias, 38:415 (1966).
4. S.C. Hardy, A Grain Boundary Groove Measurement of the Surface Tension between Ice and Water, Philos.Mag. 35:471 (1977).
5. H. Pollatschek, Die Bestimmung der an der Grenze fest/flüssig während der Kristallisation unterkühlter Schmelzen herrschenden Temperatur, Z.phys.Chem. A, 142:289 (1929).

ON THE DYNAMICS OF EPITAXIAL PHASE TRANSFORMATIONS*

Uzi Landman, C. L. Cleveland, C. S. Brown
and R. N. Barnett

School of Physics
Georgia Institute of Technology
Atlanta, Georgia 30332

The epitaxial growth of an oriented crystalline substance from a melt (or solution) onto a substrate (Liquid Phase Epitaxy-LPE) is a process of high scientific and technological interest.[1] The reproducibility, structural quality and purity of crystals obtained via LPE depend critically on a number of growth parameters such as the temperatures of the solid and melt, temperature gradients (in the direction of growth and laterally), the rates of heat dissipation and the compositions. The growth parameters, coupled with specific material characteristics such as thermal conductivities, latent heat, segregation and diffusion coefficients and crystalline anisotropies of material properties govern the kinetics and dynamics of the solidification process and the characteristics of the eventual product.

The two main processes occuring in solidification are molecular or density rearrangement and heat transport. The evolution of long-range, crystalline order in the system is coupled to a decrease in entropy and production of latent heat, which the system attempts to remove (through convection, diffusion and radiative transfer). These coupled processes and their rates which occur, in general, under non-equilibrium conditions, together with the imposed boundary conditions cause the system to pass through certain instabilities and dictate the mode of crystallization (see lectures by Langer[2] and Bilgram[3]) and the properties of the fluid-solid interface.

The purpose of our investigations is to provide information about the microscopic dynamics and kinetics of epitaxial crystallization processes. For this purpose we have developed a Surface Molecular Dynamics method[4] which allows one to perform detailed studies of the temporal evolution of interfacial phenomena. In the following sections we outline the method, present results for the epitaxial crystallization dynamics of liquids and provide a new picture of the process for the systems which we investigated.

Surface Molecular Dynamics (SMD)

The molecular dynamics (MD) technique is the epitome of a theorist's experiment. In MD the classical equations of motion of an interacting collection of particles are integrated and the recorded phase-space trajectories ($\vec{r}(t)$, $\vec{v}(t)$) are then analyzed.[5] In most MD studies periodic boundary conditions (pbc) are used. 3D pbc's are appropriate for the description of bulk properties. A different situation is presented by a system which contains a surface. While for a two-dimensional system the question of pbc's is simple, a semi-infinite system is a much harder problem. A schematic description of the system is shown in Fig. 1. The "bulk block" (B) and "surface block" (S), consist each of 500 particles, interacting via a 6-12 Lennard-Jones potential. The B-system possesses 3D pbc's and the S-system possesses only 2D pbc's and is free in the z direction. Consequently, while the dynamics of particles in the S-system is influenced by that of particles in the B-system, the reverse statement does not hold, (in the spirit of a bulk being an infinite reservoir whose dynamics and properties should not be influenced by the surface). A region of the S-system at the S-B interface is used as a "coupling region" on which we impose (time step by time step) a scaling of velocities such that the kinetic energies (kinetic temperature) in these layers are equal to the corresponding bulk ones. The thickness of this region (in these calculations, three layers) is chosen to be larger than the pair interaction range to assure that the rest of the S-system is not directly influenced by the bulk. The integration of the equations of motions is performed using a predictor-corrector method[5] with a time step $\Delta t^* = 0.0075$ and the evolution of the S and B systems is synchronized at each time-step.[6] The S-B system has been equilibrated as an fcc crystal (the density of the B system is adjusted to yield a

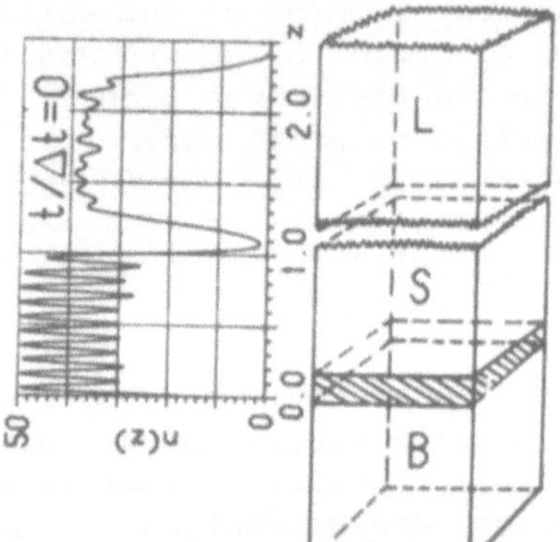

Figure 1. Schematic description of the bulk (B)-surface (S) liquid (L) system. The coupling region is hatched. Also included is the n(z) profile for the SCL system at t = 0. Distance in units of 7.94 σ.

vanishing average equilibrium pressure) exposing the (001) face, at a temperature $T^* = 0.4$ (Ar melts at $T^* = 0.7$).

The above SMD technique has an advantage over slab configurations or calculations in which a static or a random matrix representation of the bulk are used. It allows the investigation of equilibrium and non-equilibrium surface and interface phenomena in which the dynamics of the bulk reservoir is included. Using our method we currently investigate surface melting, defects, surface alloy dynamics, and transport phenomena.

Surface-Liquid Systems

Following the equilibration of a solid surface at $T^* = 0.4$, described above, we prepared two equilibrated liquid (L) samples: (a) a supercooled L-J liquid (SCL) film (2D pbc's) equilibrated at $T^* = 0.4$ and (b) a L-J liquid film equilibrated at $T^* = 0.737$, (HL). Subsequently, we positioned each of the liquid systems at a distance from the surface, chosen such that the smallest separation between a surface and a liquid particle is equal to 1.12 σ (the minimum of the L-J 6-12 potential). At this time, $t = 0$, we allow the liquid-solid surface systems to interact. Note that at $t = 0$ both the SCL and HL - surface combined systems are not in equilibrium. Analysis of the recorded particle trajectories versus time allows us to follow in detail the evolution of the systems and to investigate the energetics, kinetics and dynamics of density reorganization, ordering and structural changes as the systems develop towards equilibrium. The sample particle-number versus z profiles at selected time steps shown in Fig. 2 provide a visualization of certain aspects of the ordering process. These profiles are momentary snap-shots of the systems' configurations, and contain both permanent and transient features. Thus, for example, while the density profiles of solid layers are permanent features, certain of the fluctuations in the liquid regions are transient. Here, and in the following, layers are defined as regions whose thickness in the z (001) direction are that of a layer in the solid surface. Layers are numbered increasingly from the surface-bulk (S-B) interface with layer 10 being the top-most layer of the solid surface. A detailed study of the density profiles and a number of order parameters (see below) reveals that the first stage of the transformation involves a reorganization of particle density in a region of the liquid of thickness equivalent to 4-5 layers into a persistent stratified (layered) structure in the z (001) direction which lacks intralayer crystalline order. Layers in this region further removed from the solid surface are more diffuse than the closer ones. The process in which a 3D layered interface region is established is more pronounced in the SCL system and occurs at about the 1000th and 1800th time steps for the SCL and HL systems, respectively. This ordering event is accompanied by the production of latent heat which is expressed as

Figure 2. Particle number vs. z profiles for the SCL (top) and HL (bottom) systems, respectively, at selected time steps.

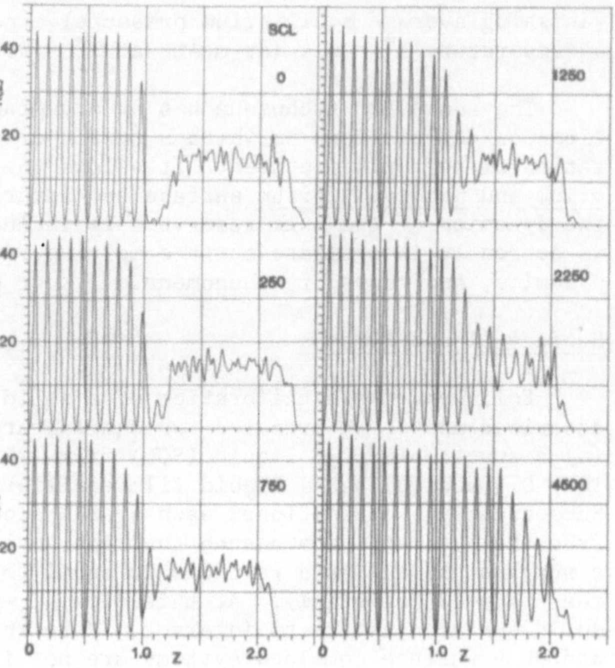

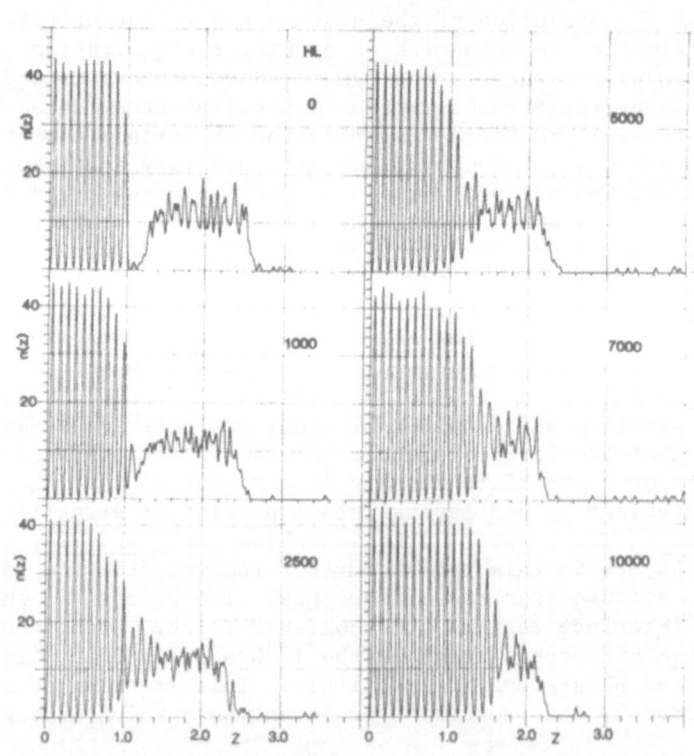

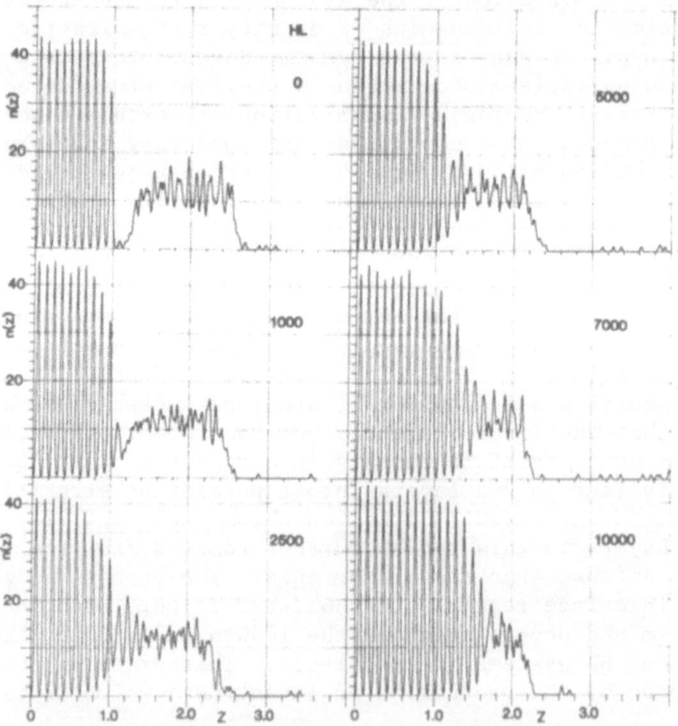

marked increases in the kinetic temperatures of the corresponding layers, shown in Fig. 3. The stability acquired by the "layering process" is further demonstrated by a drop in the layers' potential energies occurring simultaneously for an extended region of the fluid, as shown in Fig. 4.

Subsequent to the establishment of the 3D layered interface region, intra-layer ordering starts to develop. This stage of the transformation which is separated in time from the first stage occurs in a sequential (but coupled) layer-by-layer manner. A measure of intralayer ordering is provided by the layers projected pair correlation functions (pcf), $g_L(R,t)$, which are defined as

$$g_L(R,t) = \langle n_L(R,t)/2\pi R\Delta R \rangle$$

where L is the layer number, R is the projection of the position vector connection a central atom with any other atom in the layer onto the (001) plane, $n_L(R,t)$ is the number of atoms around the central one found in an annulus of radius R and of area $2\pi R\Delta R$ and $\langle \ldots\ldots \rangle$ denotes average over the central atoms in layer L at time t. Pair correlation functions and their time development for various layers for the SCL and HL systems are shown in Fig. 5. Typical pcf's for the solid and supercooled liquids are those given as $g_8(R,0)$ and $g_{14}(R,0)$ of the lhs panel and for the hot liquid as $g_{14}(R,0)$ in the rhs panel (note the differences between the $t = 0$ pcf's for the two liquids). The irregular fluctuations near $t = 0$ for $L = 12$ are due to poor statistics in this region of space at the beginning of the process. It is of interest to note the partial disordering of layer 10 in the SCL system at the initial stages of the transformation followed by a rather fast recovery. This effect is even more pronounced in the HL system, where the disordering extends further into the solid surface (about 3 atomic layers).

Following the development of the pcf's in a region of space which at $t = 0$ is occupied by a liquid, and crystallizing in the course of time (for example, $g_{14}(R,t)$), illustrates the intralayer reorganization of particle density and the fact that during this non-equilibrium evolution the fluid passes through states of intermediate order. The sequential character of the intralayer ordering is seen by comparison of pcf's for the sequence of layers.

The evolution of the system and the degree of ordering can be further analyzed in terms of translational and orientational order parameters.[4] For the purpose of illustration we discuss one of the orientational order parameters, (see Fig. 6) defined as

$$O_n^L(t) = |N_L^{-1}(t) \sum_{I,J\varepsilon L} N_{I,nn}^{-1}(t) \exp(in\theta_{IJ}) \textcircled{H}(R_{nn}-(r_I-r_J))|^2$$

where $N_L(t)$ is the number of particles in layer L at time t, $\textcircled{H}$ is

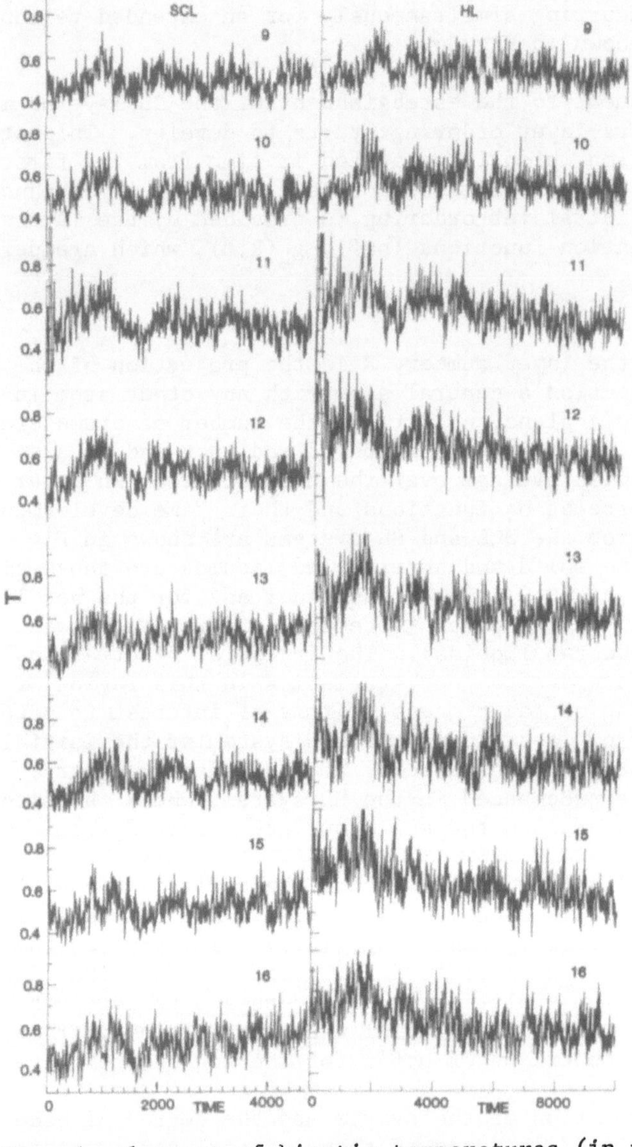

Figure 3. Time development of kinetic temperatures (in units of ε/k_B) of layers 9 through 16, for the SCL (left) and HL (right) systems respectively. Note the simultaneous rise at time steps 1000 and 1800 for the SCL and HL, respectively, due to layering in the fluid.

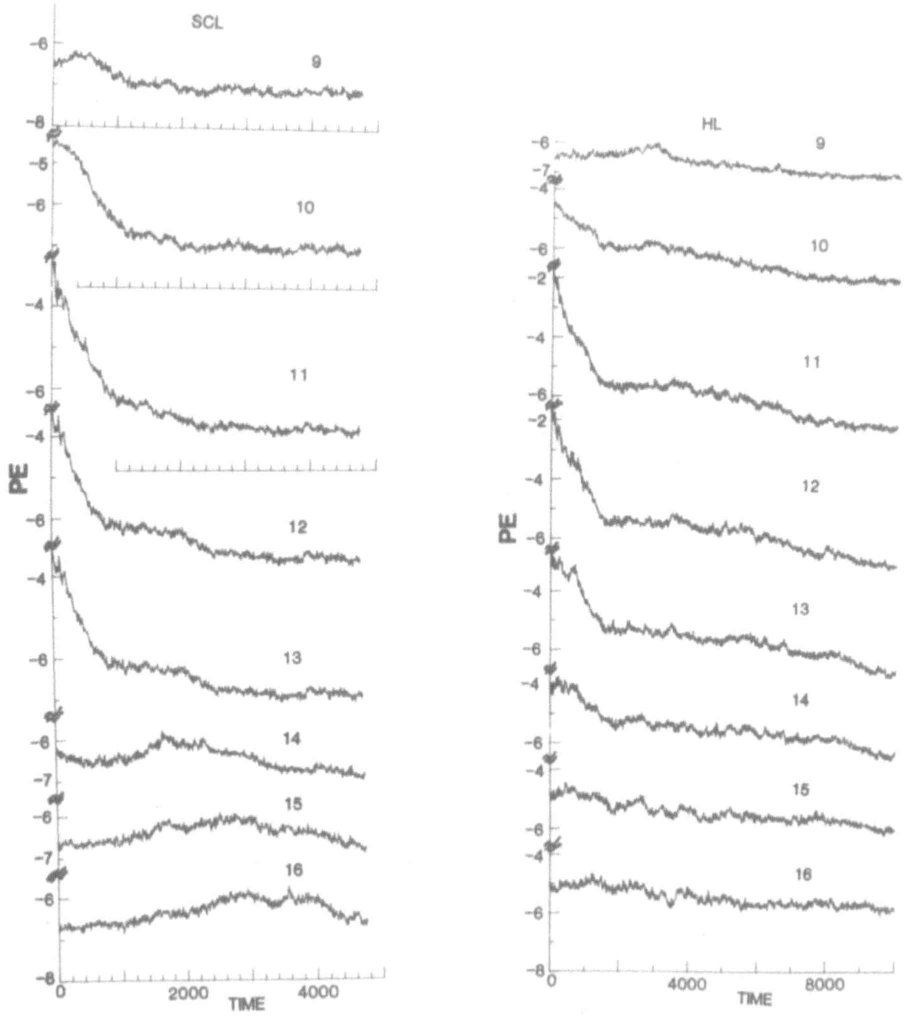

Figure 4. Time development of per particle potential energies (in units of ε) of layers 9-16 for the SCL (left) and HL (right) systems, respectively.

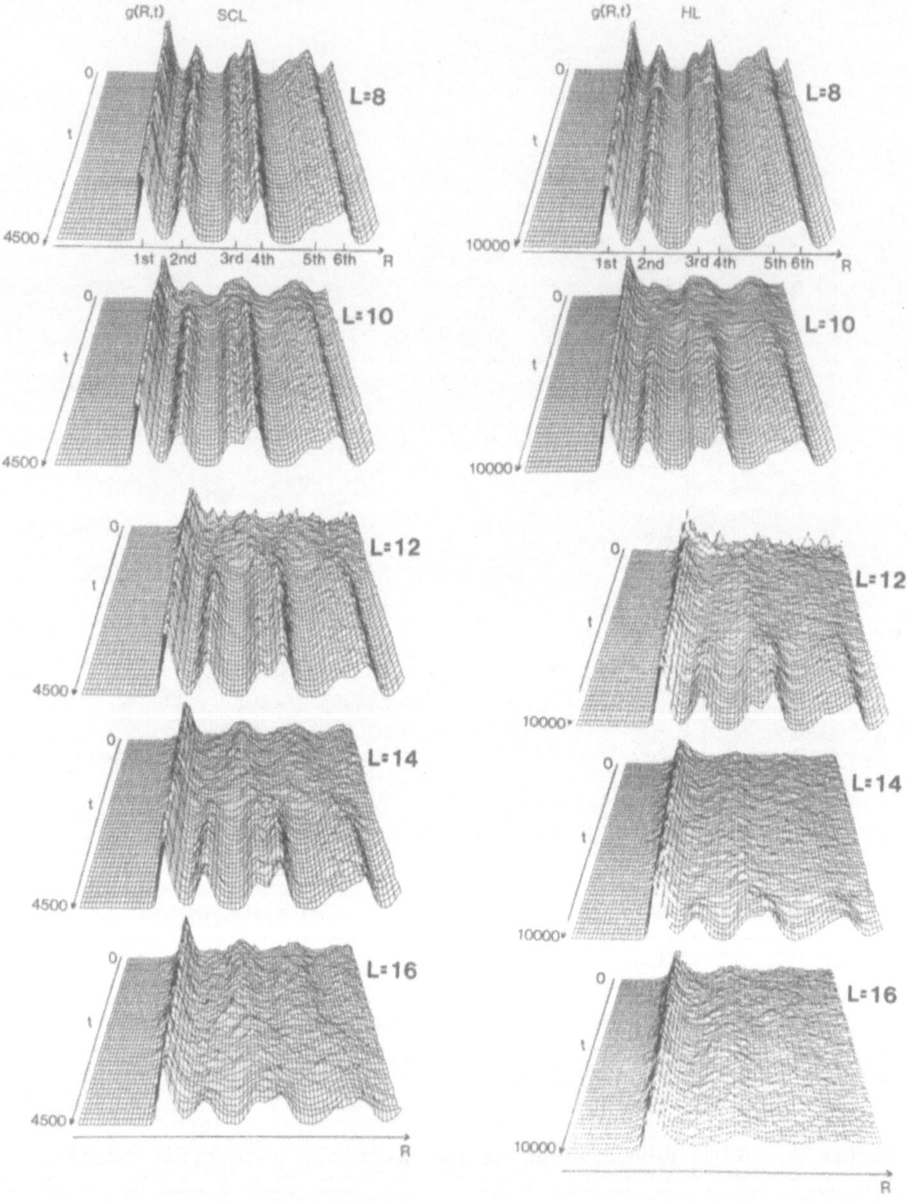

Figure 5. Time development of projected pair correlation functions, $g_L(R,t)$, for selected layers of the SCL and HL systems. The positions of the first 6 nearest neighbors in the solid surface are marked for L = 8.

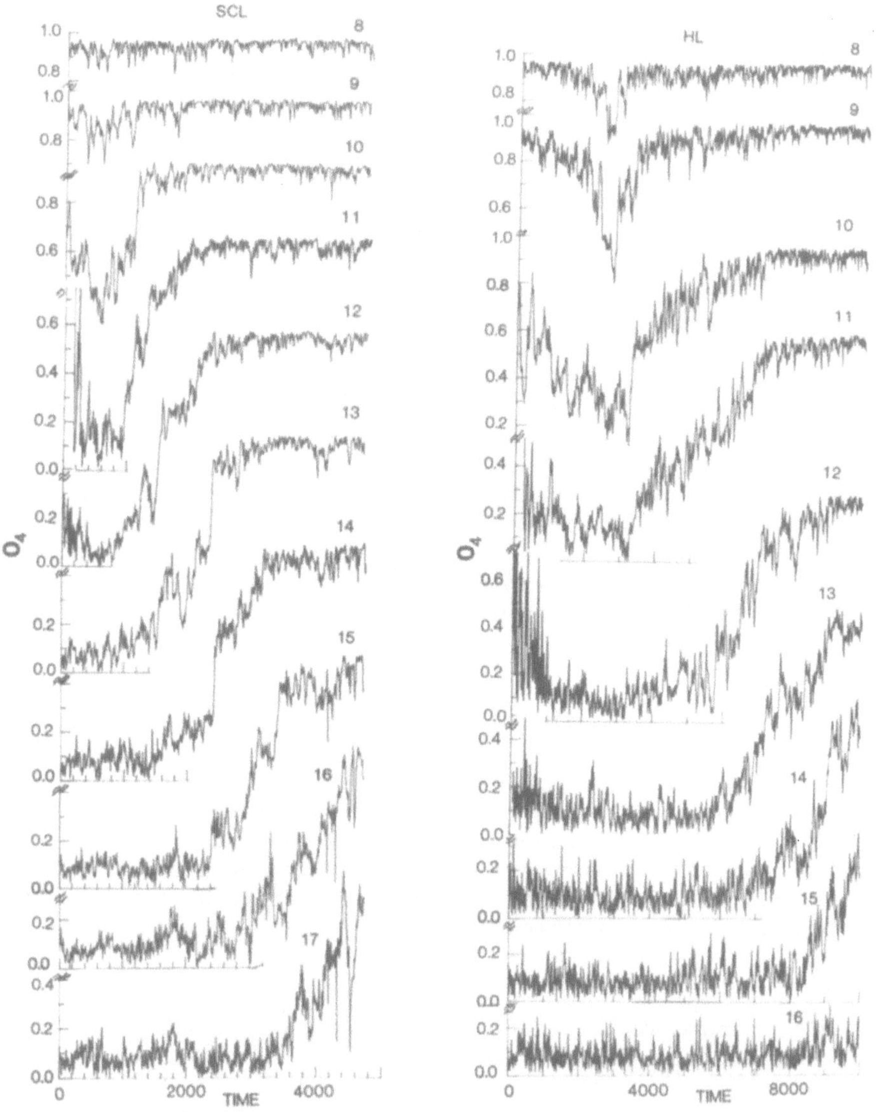

Figure 6. Time development of orientational order, O_4 in the layers, for the SCL and HL systems. Note the pronounced coupled order-disorder transformations in the layers, and the sequential nature of the process.

the Heavyside step function, R_{nn} is the neighbor distance, $N_{L,nn}$ is the number of neighbors to particle I within a sphere of radius R_{nn} and θ_{IJ} is the "bond" angle between I and J with reference to an arbitrary direction. Note that for n = 4, O_4^L takes the value 1 for a perfect cubic crystal and a small value in a liquid. The orientation order parameters show a disordering of solid layers (more extensive and extending deeper for the HL system) at the initial stages followed by a sudden ordering. Similar disordering occurs in particle positions along the (001) direction. This process of "roughening" of the substrate, is due mainly to the dissipation of the thermal energy released in the "liquid layering" process and it participates in the dynamics of the transformation. In the SCL system we observe a trend where the intralayer ordering of a certain layer, triggers ordering in the layer above it which in turn accelerates ordering in the lower layer, releasing heat which partially disorders the upper one. Eventually this layer recrystallizes followed by the crystallization of the layer above it, and so on. The sequential and strongly coupled nature of the layer crystallization processes are also observed in the HL system, though with less sharpness. The time separation between the eventual rises in O_4 (and in the translational order parameters) from layer to layer allows us to estimate the rate of crystallization as 100 m/sec and 5 m/sec for the SCL and HL systems, respectively. We also observe a gradual slowing of the intralayer ordering, due to the heating of the fluid during the crystallization of the first few layers. It is of interest to note that in the HL system crystallization of a liquid region is delayed until sufficient thermal energy has been dissipated to bring about a slight super-cooling of that region. In addition we observe that the kinetic temperature in fluid regions removed from the current crystallizing region is higher than that of the corresponding region in the solid substrate, thus dynamically establishing a time-dependent temperature gradient which favors the growth of planar structures. The mismatch in sound velocity at the crystallization interface and enhanced convective heat transport in the fluid may be responsible for this effect.

In the above we have outlined a new procedure for the investigation of surface and interface phenomena using the molecular dynamics technique. In contrast with previous MD studies of the equilibrium properties of crystal surface-fluid systems in coexistence,[7] we study the temporal evolution and dynamics of an initially non-equilibrium system. The wealth of detailed information thus obtained and the new picture of epitaxial phase transformations which emerges, provide the impetus for further study and prompt a revision of current theoretical models of these processes.

---

*Work supported by U.S. DOE Contract No. EG-S-05-5489.

References

1. J. C. Brice, "The Growth of Crystals from Liquids", (North-Holland, Amsterdam, 1973); "Crystal Growth", Ed. B. R. Pamplin, (Pergamon Press, Oxford, England, 1980).
2. J. S. Langer (this volume).
3. J. H. Bilgram (this volume).
4. U. Landman, C. L. Cleveland and C. S. Brown, Phys. Rev. Letts. 45, 2032 (1980).
5. A. Rahman, Phys. Rev. 136, A405 (1964).
6. Reduced units are used throughout the paper. The reduced time unit is $(\mu\sigma^2/\varepsilon)^{1/2}$ where μ is mass, σ and ε are the Lennard-Jones (L-J) 6-12 parameters. For Argon $\sigma = 3.4$A, $\varepsilon/k = 120^{\circ}$K and the time unit is 1.65×10^{-12} sec. Energy is in the unit of ε. The potential cutoff is $2.5\ \sigma$. In the figures, length is given in units of $7.94\ \sigma$.
7. J. Q. Broughton and F. F. Abraham, Chem. Phys. Letts. 71, 456 (1980); J. N. Cape and L. V. Woodcock, J. Chem. Phys. 73, 2420 (1980).

THE PINNING OF A DOMAIN WALL BY WEAKENED BONDS IN TWO DIMENSIONS

J.T. Chalker

Department of Theoretical Physics
University of Oxford
1 Keble Road
Oxford OX1 3NP, England

INTRODUCTION

The effect of thermal fluctuations on an interface between two co-existing phases can be understood as an example of the Hohenberg, Mermin and Wagner theorem [1]. The interface between a liquid and its vapour, in the absence of a stabilising force such as gravity, provides an instance of a two-dimensional system with a continuous symmetry, under translations 'vertically'. There is therefore no long-range order in the interfacial position; the interface is roughened by capillary waves, which are the relevant Goldstone modes. These considerations do not apply to a domain wall in the three-dimensional Ising model since the presence of a lattice renders the translational symmetry discrete. In this case it has been proven [2] that below the critical temperature of a corresponding two-dimensional Ising system the interface is localised near the ground-state position, whilst a range of evidence (low temperature series expansion [3], computer simulation [4] and mappings to other models [5]) suggests that just above this temperature, and well below the bulk critical point, the interface has a roughening transition to a diffuse phase dominated by capillary waves.

The effects of fluctuations on phase separation in two dimensions are intrinsically greater: it is well known that the domain wall in a homogeneous, planar Ising model is rough at all finite temperatures, despite the lattice [6]. Recently Abraham [7] has shown with an exact solution that a more drastic breaking of the continuous symmetry - a row of weakened exchange bonds next to the edge of the system - is sufficient to pin a domain wall at low temperatures. At higher temperatures the free energy of the wall is dominated by the entropy gain from wandering rather than the pinning energy, and there is a transition (which is similar in some but not all aspects to the usual roughening transition) to a diffuse phase.

There is a valuable simplification, the solid-on-solid model [8] which can be used to study the interface in Ising systems. This rests on the idea that, at temperatures well below the bulk critical point, the magnetization saturates and only fluctuations of the interface itself are important. The bond strength perpendicular to the domain wall is taken infinite (with boundary conditions which, however, guarantee two co-existing phases) so that all excitations of the bulk, together with configurations in which the wall overhangs itself, are suppressed. Here the solid-on-solid model is used to study the pinning of a domain wall in two dimensions [9]. Related work has been done by several other authors [10].

SOLUTION OF THE MODEL

In the solid-on-solid model corresponding to an NxN site square Ising model below its critical point (to which periodic boundary conditions have been applied between the 'left' and 'right' edges whilst the spins at the 'top' edge are held up and those at the bottom edge down), a configuration of the interface is described by N variables h_i ($1 \leq i \leq N$, $h_i = 1,2, .., N-1$) which give the number of down spins in each column of the Ising system. The correspondence between the models is one to one in the absence of bulk excitations and overhangs of the wall. The energy of a configuration $\{h_i\}$ is

$$E = \sum_{i=1}^{N} 2\{J|h_i - h_{i+1}| - \Delta\, \delta(h_i, 1)\} \qquad (1)$$

where periodic boundary conditions are taken: $h_{N+1} \equiv h_1$. Here J is the bond strength parallel to the wall in the original Ising model whilst Δ gives the degree of bond weakening at one edge. $\delta(h,1)$ is the Kronecker delta. This model can be solved by finding the eigenvalues and eigenvectors of the transfer matrix. In the thermodynamic limit the eigenvalues fill a band: at low temperatures there is also one larger eigenvalue with an associated, localised eigenvector, but at the roughening temperature, T_R, this merges into the top of the band. Explicit results [9] are:

(i) T_R is given by

$$\exp(-2\Delta/kT_R) + \exp(-2J/kT_R) = 1,$$

(ii) the magnetization, m(z), as a function of distance, z, from the weakened bonds is

$$m(z) = 2e^{-\kappa z} - 1$$

where $\kappa = \ell n(\exp(2\Delta/kT)-1) + 2(J-\Delta)/kT$,

(iii) the pair correlation function (defined as the probability that $h_{x+1} = z_2$, given $h_1=z_1$) is, asymptotically for large x,

$$\frac{\langle\delta(h_1,z_1)\delta(h_{x+1},z_2)\rangle - \delta(h_1,z_1)\rangle\langle\delta(h_{x+1},z_2)\rangle}{\langle\delta(h_1,z_1)\rangle} \equiv G(x;z_1,z_2))$$

$$G(x;z_1,z_2) \sim \left[\frac{\langle\delta(h_{x+1}),z_2)\rangle}{\langle\delta\ h_1,z_1)\rangle}\right]^{\frac{1}{2}}_{x} \begin{cases} f_1(x) & T>T_R \\ f_2(x) & T=T_R \\ e^{-x/\xi}\ f_1(x) & T<T_R \end{cases}$$

where
$$f_1(x) = x^{-\frac{1}{2}} \frac{1}{\sqrt{\pi a}} [\exp(-\frac{a(z_1-z_2)^2}{x}) - \exp(-\frac{a(z_1+z_2)^2}{x})]$$

$$f_2(x) = x^{-\frac{1}{2}} \frac{1}{\sqrt{\pi a}} \exp(-\frac{a(z_1-z_2)^2}{x}) + \exp(-\frac{a(z_1+z_2)^2}{x})] \quad (5a)$$

$$\xi^{-1} = \ell n[\frac{t^2-2t\cosh\kappa+1}{(t-1)^2}],\quad t = \exp(-2J/kT)$$

$$a = (1-t)^2/2t$$

and (iv) the specific heat of the domain wall has a simple discontinuity at T_R.

DISCUSSION

Power law decay of the pair correlation function at all temperatures above the transition is also found in the rough phase of an interface in three dimensions: in both cases the entire high temperature phase is critical, in the sense that there is a susceptibility which is infinite. The critical point itself, however, is quite different for the transition in three dimensions, which is related by duality to the Kosterlitz-Thouless transition and so has an essential singularity in the specific heat.

In order to understand the special aspects of the transition in two dimensions, it is useful to examine two variations on the model defined by (1). First, allow the column heights $\{h_i\}$ to vary continuously in (0,N) instead of restricting them to integer values. In the solution, eigenvalues and eigenvectors of a transfer matrix are then replaced by eigenvalues and eigenfunctions of a transfer integral. For a solid-on-solid model of an interface in three dimensions this modification restores the continuous translational symmetry of the interface which was broken by the lattice, and consequently the interface is rough at all finite temperatures. However, for the model studied here, translational symmetry remains broken by the weakened bonds and the behaviour of the continuous model is qualitatively the same as in the discrete case. Secondly, the effects of a pinning force acting in the bulk of a system, rather than near the edge, can be studied by allowing the column heights to take on negative as well as positive values: $h_i=0, \pm 1, \ldots \pm N$. The largest eigenvalue of the transfer matrix for this case is isolated at all temperatures, corresponding to the fact that the domain wall is now always bound close to the weakened bonds, and not rough at any finite temperature.

This difference can be understood using the equivalence between d-dimensional statistical mechanics and quantum field theory in d-1 space dimensions. A continuum approximation to (1), in which $\{h_i\}$ becomes a one-dimensional field, h(x), is

$$E\{h\} = \int\{J(\frac{dh}{dx})^2 + V(h)\}\, dx$$

where $V(h) = \begin{matrix} +\infty & h<0 \\ -\Delta & 0<h<1 \\ 0 & 1<h \end{matrix}$ for a pinning force at the edge of the system

$V(h) = \begin{matrix} 0 & h<0 \\ -\Delta & 0<h<1 \\ 0 & 1<h \end{matrix}$ for a pinning force in the bulk.

The probability that $h(x_o)=h_o$, given $h(0) = 0$, is proportional to $\int \mathcal{D}(h)e^{-\beta e\{h\}}$, where the functional integral is over all paths h(x) which satisfy the boundary conditions at $x=0$, x_o. This quantity is also, within the Feynman formulation of quantum mechanics, the probability that a (relativistic) particle of mass 2J moving in a one-dimensional potential V(h) will start from the origin and have position h_o at time x_o. Planck's constant in quantum mechanics plays the same role as temperature in statistical mechanics. It is well known [11] that the potential corresponding to weakened bonds in the bulk always has a bound state, whilst that representing weakened bonds near an edge gas a bound state only for sufficiently large values of $J\Delta/\hbar^2$.

ACKNOWLEDGEMENTS

I should like to thank Dr. G.A. Gehring for guidance and encouragement in this work. I am grateful to Dr. D.B. Abraham for introducing me to the subject and for many valuable discussions.

REFERENCES

[1] N.D. Mermin and H. Wagner, Phys. Rev. Lett. 17, 1133 (1966)
P.C. Hohenberg, Phys. Rev. 158, 383 (1967)

[2] R.L. Dobrushin, Theory. Prob. Appl. 17, 582 (1972)

[3] J.D. Weeks, G.H. Gilmer and H.L. Leamy, Phys. Rev. Lett. 31, 549 (1973)

[4] J.D. Weeks and G.H. Wilmer, Adv. Chem. Phys. 40, 157 (1979)

[5] S.T. Chui and J.D. Weeks, Phys. Rev. B14, 4978 (1976)

[6] G. Galavotti, Commum. Math. Phys. 27, 103 (1972)

[7] D.B. Abraham, Phys. Rev. Lett. 44, 1165 (1980)

[8] G.H. Gilmer and P. Bennema, J. Appl. Phys. 43, 1347 (1972)
J.D. Weeks, 'Ordering in StronglyFluctuating Condensed Matter Systems', Ed. T. Riste, (New York, 1980)

[9] J.T. Chalker, J. Phys. A. to be published

[10] S.T. Chui and J.D. Weeks, Phys. Rev. B. to be published
J.M.J. van Leeuwen and H.J. Hilhorst, Physica A to be published
T.W. Burkardt, J. Phys. A. 14, L63 (1981)

[11] L.D. Landau and E.M. Lifschitz, 'Quantum Mechanics', p.57 (London, 1958).

MELTING IN TWO DIMENSIONS

J.M. Kosterlitz

Department of Mathematical Physics
Birmingham University
Birmingham B15 2TT, England

I INTRODUCTION

In these lectures I will try to review the present state of our understanding of the various phases which can occur in thin films which crystallise at low temperatures. Most experimental investigations are on systems of gases adsorbed on graphite in which the substrate plays an important role but often obscures the theoretically important effects. There is some recent work on freely suspended films of liquid crystal in which the problem of substrate is minimised but other complications arise. Perhaps the system closest to the theoretical ideal is electrons trapped on the surface of helium but the experimental probes of this are rather restricted.

The main interest in such systems was stimulated by the realisation that a melting transition could be continuous in two dimensions in contrast to bulk systems where it is invariably first order. The proposed mechanism for this is the unbinding of topological defects or dislocations for which techniques have been developed that give almost quantitative predictions for many experimentally accessible (in principle) quantities. These techniques have been successfully used to explain the superfluid transition in thin helium films and is widely accepted as being correct for this. However, the dislocation theory of melting is still very controversial and there is a maze of conflicting evidence.

In these lectures, I will concentrate on the theoretical aspects of the problem and restrict myself to a brief discussion of the experimental and computer simulation situation. Most of the material in these lectures can be found, in more detail, in a series of papers by Halperin, Nelson, Ostlund and Zippelius[1-5]. The original application of dislocation mediated melting to thin films is due to Kosterlitz and Thouless[6] and there are several recent reviews on melting and related problems[7].

Section I is theoretical background. Section II reviews the dislocation theory of melting of a system of point-like atoms on a smooth substrate and summarizes the main predictions. The effects of a substrate with a periodic binding potential are briefly discussed. Section III discusses the fluid phases predicted by the dislocation theory and recent computer simulations are reviewed. In section IV the complications occurring in liquid crystal layers are discussed and in section V the modifications due to dislocations to the hydrodynamics of two dimensional systems near their melting temperature are introduced.

Theoretical Background

Ever since Bloch, Peierls and Landau[8] pointed out in the 1930's that a two-dimensional system with a continuous symmetry cannot have long-range order at any finite temperature, theorists have been intrigued by these. In the 1960's Hohenberg, Mermin and Wagner[9] put this on a rigorous basis but at the same time made it clear that lack of long-range order did not exclude a transition between two states without long-range order. At the same time, high temperature series calculations by Stanley and coworkers[10] seemed to indicate there is a transition. This apparent discrepancy was cleared up in the early 1970's when it was shown that a transition to a low temperature phase without long-range order is possible[6,11].

In a two dimensional solid on a smooth substrate a quantity measuring the degree of crystalline order is[1,2]

$$C_{\vec{G}}(\vec{r}) = \langle \rho^*_{\vec{G}}(\vec{r})\, \rho_{\vec{G}}(o) \rangle \tag{1.1}$$

where $\rho_{\vec{G}}(\vec{r})$ is the local Fourier component of mass density at a reciprocal lattice vector $\vec{G}$. The correlation function of eq.(1.1) is just the Debye-Waller factor whose Fourier transform gives the X-ray structure factor in the vicinity of $\vec{G}$. The long-range crystalline order means that a continuous translational symmetry has been broken when

$$\lim_{r\to\infty} C_{\vec{G}}(\vec{r}) = \text{constant} \neq 0 \tag{1.2}$$

which is equivalent to $\langle \rho_{\vec{G}}(\vec{r}) \rangle \neq 0$. Experimentally, this will show up as delta function Bragg peaks at reciprocal lattice vectors.

In two dimensions, however, this is expected to fall off with a power of r at low temperatures. This strange behaviour is illustrated by the standard continuum elasticity theory[13] of a collection of point particles. The equilibrium positions of the particles will generally be a triangular lattice with sites denoted by $\vec{r}$. The density at a point $\vec{x}$ is

$$\rho(\vec{x}) = \sum_{\vec{r}} \delta(\vec{x} - \vec{r} - \vec{u}(\vec{r})) \; ; \qquad \rho_{\vec{G}}(\vec{r}) = \exp i\vec{G}.\vec{u}(\vec{r}) \tag{1.3}$$

where $\vec{u}(\vec{r})$ is the displacement of an atom from $\vec{r}$. The X-ray structure factor is

$$S(\vec{q}) = <|\rho(\vec{q})|^2> = \sum_{\vec{r}} e^{i\vec{q}.\vec{r}} <e^{i\vec{q}.\left[\vec{u}(\vec{r}) - \vec{u}(o)\right]}> \tag{1.4}$$

In the vicinity of a reciprocal lattice vector $\vec{G}$, we immediately see that this is just the Fourier transform of $C_{\vec{G}}(\vec{r})$.

At low temperatures we expect that the displacements from equilibrium $\vec{u}(\vec{r})$ will be small so, expanding the free energy to second order, the statistical averaging in eqs. 1.1, 1.4 is with the free energy functional

$$F[\vec{u}] = \frac{1}{2} \int d^2\vec{r} \; (2\mu \, u^2_{ij}(\vec{r}) + \lambda u^2_{kk}(\vec{r})) \tag{1.5}$$

where u_{ij} is the symmetric strain tensor

$$u_{ij} = \frac{1}{2} \left(\frac{\partial u_i}{\partial r_j} + \frac{\partial u_j}{\partial r_i}\right) \tag{1.6}$$

The free energy functional of eq. 1.5 is assumed to describe fluctuations from equilibrium of wavelength large compared to some minimum cut off. The Lame coefficients λ and μ are the only ones allowed to quadratic order for isotropic systems and triangular lattices which have six-fold symmetry. There is a lot of (unknown) physics buried in λ and μ because of the implicit averaging over very short wavelength fluctuations, quantum effects etc., so they must be regarded as phenomenological parameters depending on pressure and temperature and obeying stability requirements

$$\mu > 0 \qquad \lambda + \mu > 0 \tag{1.7}$$

in the temperature range of interest. At this order, F cannot depend on the anti-symmetric part of $\partial_i u_j$

$$\theta = \frac{1}{2} (\partial_x u_y - \partial_y u_x) \tag{1.8}$$

because of rotational invariance for a system on a smooth substrate. Derivatives of θ are allowed in eq.(1.5) but are irrelevant for melting. In the presence of a substrate with a periodic binding potential this invariance is broken and F contains terms in θ^2.

Assuming for the moment that the displacement field $\vec{u}(\vec{r})$ is a single-valued function of position (i.e. ignoring dislocations and disclinations[14]), we can easily calculate the Debye-Waller factor for large $\vec{r}$:

$$C_{\vec{G}}(\vec{r}) \sim r^{-\eta_{\vec{G}}(T)} \tag{1.9}$$

where

$$\eta_{\vec{G}}(T) = k_B T|\vec{G}|^2(3\mu + \lambda)/4\pi\mu(2\mu + \lambda) \tag{1.10}$$

For $\vec{q}$ close to a reciprocal lattice vector $\vec{G}$, this leads to a power law singularity in the X-ray structure factor instead of the delta function Bragg peaks of bulk crystals

$$S(\vec{q}) \sim |\vec{q} - \vec{G}|^{-2+\eta_{\vec{G}}(T)} \tag{1.11}$$

which diverges at $\vec{q} = \vec{G}$ when $\eta_{\vec{G}}(T) < 2$. We thus expect that for the first few reciprocal lattice vectors at sufficiently low temperatures there are infinite peaks in $S(\vec{q})$. A sketch of the powder (angular) average of $S(\vec{q})$ is shown in fig. 1.1. Although this is rather different from the three dimensional structure factor it is also quite different from a liquid one since then one expects an exponential decay of $C_{\vec{G}}(\vec{r})$. The power law peaks require only non-zero elastic constants which we take as the definition of a two-dimensional solid although the positional order parameter $\langle\rho_{\vec{G}}(\vec{r})\rangle = 0$.

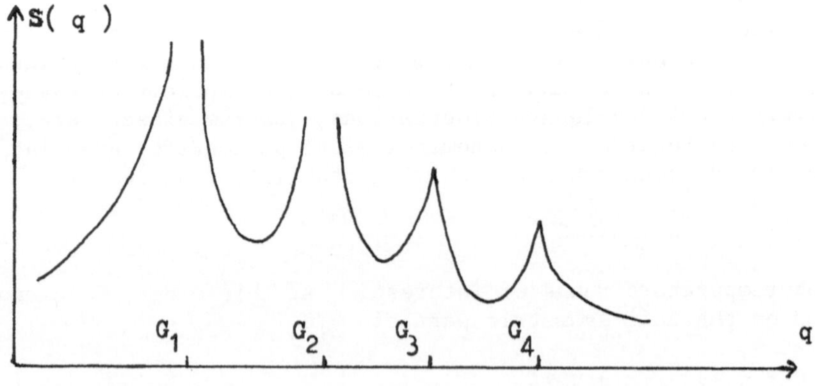

Fig.1.1 Powder average of the x-ray structure factor S(q).

It is almost impossible to measure this power law experimentally because of instrumental resolution, misaligned crystallites in a substrate and the lack of intensity because there is very few atoms in a monolayer. Moreover, slight deviations from the ideal theoretical model can alter dramatically the character of the system. The structure factor can be interpreted as the linear response function of the density to perturbations of wavenumber $\vec{q}$. Since the local density of the adsorbate couples linearly to the binding potential of the substrate, a substrate which is commensurate, or almost so, with the adsorbate will have an infinite response[2]. The adsorbate will be locked to the substrate and a lattice gas description may be more appropriate. An incommensurate substrate potential does not cause such a large response and will permit a floating incommensurate solid[2]. There is a gigantic literature on the properties of gases adsorbed on substrates like graphite[15] and both commensurate and incommensurate structures have been observed[16].

A system in which this complication does not occur is electrons pinned to the surface of helium studied by Grimes and Adams[17] and analysed by Fisher et al[18]. Longitudinal modes of the combined electron ripplon system were excited and sharp resonant absorption of the modulating rf. field were observed below about 0.5°K. The theoretical analysis was carried out on the assumption that a Wigner crystallisation of the electrons had taken place and excellent agreement with the observed resonant frequencies obtained. This is strong evidence for the existence of a two dimensional solid.

Freely suspended liquid crystal films[19-23] are good realisations of isolated liquid and solid phases in which to study melting by light[19,20] and X-ray[21] scattering. There have also been recent attempts to measure directly the shear modulus[22,23] by similar torsional oscillator techniques to those used to study the superfluid transition in thin helium films[24]. Lipid monolayers floating on water provide another two-dimensional system in which to investigate melting[25] but there are mechanisms at work other than those discussed here.

II DISLOCATION MEDIATED MELTING

The harmonic approximation for a two-dimensional crystal gives no long-range positional order at any finite temperature (eq.1.9) but no fluid phase in which $C_{\vec{G}}(\vec{r})$ should decay exponentially. In analogy to the planar rotor magnet and superfluids[6] we ask about the effect of singularities in continuum elasticity theory. The relevant singularities for solids are dislocations and disclinations[14] shown for a square lattice in fig. 2.1. A dislocation is characterised by the amount a reference contour which closes in a perfect crystal fails to close

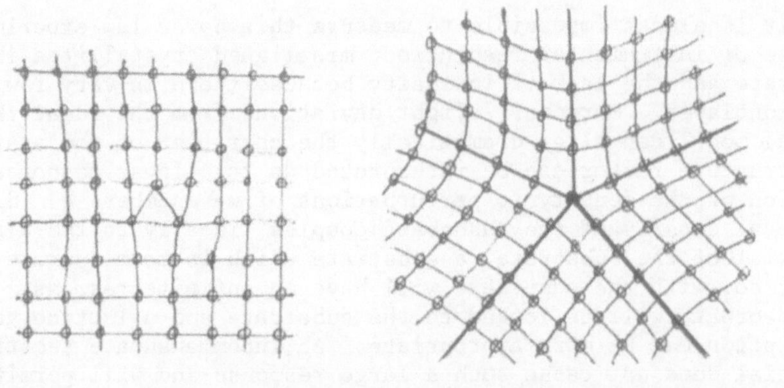

Fig.2.1 Dislocation (a) and disclination (b) in a square lattice. The Burger's vector of the dislocation is the amount by which the contour in (a) fails to close.

in the dislocated crystal. This is known as a Burger's vector $a_0\vec{b}$ and in continuum elasticity theory is defined by

$$\oint_C w_{ij}\, dr_j = a_0 b_i \qquad (2.1)$$

where $w_{ij} = \partial u_i/\partial r_j$, and a_0 is the lattice spacing. Since the positional order parameter $\rho_{\vec{G}}(\vec{r})$ is a single valued function of position, $a_0\vec{b}$ must be a vector of the underlying lattice

$$\exp ia_0\vec{b}.\vec{G} = 1 \qquad (2.2)$$

An elementary dislocation which has $|\vec{b}| = 1$ is illustrated for the triangular lattice in fig. 2.2. Note that the core of the dislocation where the distortion of the lattice is large can be regarded as being made up of two points of 5 and 7-fold rotational symmetry which are embryonic disclinations. Although these are irrelevant for the present discussion they will become significant later. The conditions 2.1 and 2.2 mean that the dimensionless Burger's vector

$$\vec{b} = n_1\vec{e}_1 + n_2\vec{e}_2 \qquad (2.3)$$

where $\vec{e}_1$ and $\vec{e}_2$ are unit vectors spanning the lattice and n_1. n_2 are integers. The six elementary dislocations of the triangular lattice which is the interesting case here have Burger's vectors $\pm(1,0)$, $\pm(\frac{1}{2}, \frac{\sqrt{3}}{2})$, $\pm(\frac{1}{2}, -\frac{\sqrt{3}}{2})$.

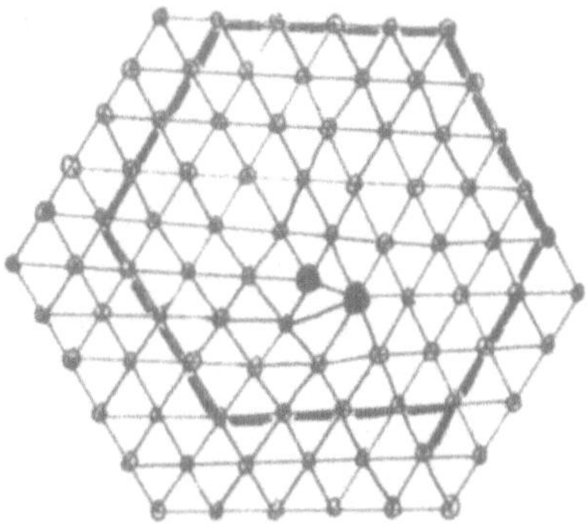

Fig.2.2 Dislocation in a triangular lattice with a Burger's circuit. The core is made up of two embryonic disclinations indicated by heavy dots one lattice spacing apart.

In the presence of dislocations, the displacement field is multivalued (eq.2.1) and, according to standard texts on elasticity theory, the interaction energy of a pair of dislocations of equal and opposite Burger's vector $\vec{b}$ separated by $\vec{R}$ is

$$E(\vec{R})/k_BT = \frac{K}{4\pi}\{b^2 \ell n\left|\frac{\vec{R}}{a}\right| - \frac{(\vec{b}.\vec{R})^2}{\vec{R}^2}\} + \frac{2E_c}{kT}b^2 \quad (2.4)$$

where

$$K = 4\bar{\mu}(\bar{\mu} + \bar{\lambda})/2\bar{\mu} + \bar{\lambda}) \quad \text{and} \quad \bar{\mu} = \mu a_o^2/kT$$

and a is a dislocation core size. The definition of the core size a is not unique but we take it as the size of the region near the centre of the dislocation where distortions are so large that linear elasticity theory fails. The energy of such regions is assumed to be represented by a constant "core energy" E_c. Note that in general the core size will bear little or no relation to the lattice spacing.

In the presence of an applied external stress field σ_{ij} there will be a force on a dislocation given by

$$f_i = \varepsilon_{ji}\,\sigma_{j\ell}\,b_\ell \quad (2.5)$$

which will tend to cause the dislocation to move in a direction such as to relax the stress. Bound dislocation pairs will be polarised by an applied stress and the resulting polarisation will lead to a reduction in the elastic constants λ and μ. If free dislocations are present, on the other hand, they will continue to flow and the polarisation will increase without limit.

The system can no longer sustain shear and melting has occured.

The mechanical stability of the system is thus determined by the absence of free dislocations in thermal equilibrium. This is in exact analogy to the stability or otherwise of a flow in a superfluid film determined by the absence of free vortices in thermal equilibrium . A criterion for stability can be deduced by considering a system with only two dislocations $\pm\vec{b}$. The mean square separation in thermal equilibrium from eq.2.4 is

$$\langle \vec{R}^2 \rangle \propto 1/(K - 16\pi) \tag{2.6}$$

which means that for $K > 16\pi$ dislocations are present in bound pairs and triplets only. The upper limit of stability against free dislocation formation is given by

$$\mu(\lambda + \mu)/(2\mu + \lambda) = 4a_o^2\, kT \tag{2.7}$$

As we shall see later, this criterion remains valid in a system of interacting dislocations provided μ and λ are interpreted as their observed or renormalised values as determined by the response to an infinitesimal stress. Note that eq.2.7 is an upper limit with respect to free dislocation formation. It does not exclude some other mechanism such as an explosive increase in the number of isolated defects such as vacancies and interstitials below T_m which probably leads to a first order transition[27]. This could be caused by large anharmonic terms in the elastic free energy but in what follows we shall ignore this possibility. However, such a first order transition has been seen in a related but much simpler model of superfluidity in He^3-He^4 mixtures in which the He^3 impurities play the role of vacancies and interstitials[28].

Dislocation Mediated Melting at a Smooth Substrate

In this section we will review the theory of a two dimensional crystal on a smooth substrate whose only role is to keep the crystal in a plane but has no effect on motions parallel to the plane. We want to check the hypothesis discussed in the last section of the stability against dislocation formation and in particular that the elastic constants remain finite right up to the melting temperature T_m.

As shown by Nelson and Halperin[2] the strain field u_{ij} in the presence of dislocations can be decomposed into a smooth single valued part ϕ_{ij} and a part due to the dislocations. The free energy becomes a sum of an elastic part (eq.1.1) and a part due to the dislocations:

$$F[\phi_{ij},\vec{b}]/kT = \frac{1}{2}\int \frac{d^2x}{a_o^2}\,(2\bar{\mu}\,\phi_{ij}^2 + \bar{\lambda}\,\phi_{kk}^2)$$

$$-\frac{K}{8\pi}\sum_{\vec{r}\neq\vec{r}'} b_i(\vec{r})b_j(\vec{r}')\,\{\delta_{ij}\ln\left|\frac{\vec{r}-\vec{r}'}{a}\right| - \frac{(\vec{r}-\vec{r}')_i(\vec{r}-\vec{r}')_j}{(\vec{r}-\vec{r}')^2}\}$$

$$-\ln y\sum_{\vec{r}} b^2(\vec{r}) \tag{2.8}$$

where K is defined in eq. 2.4 and $y=\exp(-E_c/kT)$ is the fugacity of a dislocation. The free energy of eq. 2.8 describes fluctuations on length scales greater than the dislocation core size a and all processes on a smaller scale are buried in the phenomenological bare elastic constants $\bar{\mu}$ and $\bar{\lambda}$ and the dislocation fugacity y. All interactions which are irrelevant in a renormalisation group sense[30] have been omitted. An implicit assumption in eq. 2.8 is that y is small, which is equivalent to E_c/kT large in the temperature range of interest. This assumption means that we can regard the system as being a linear elastic medium penetrated by a few point like dislocations of Burger's vectors appropriate to the underlying triangular lattice. Note that the only configurations of dislocations with a finite free energy are those with zero total Burger's vector i.e. $\sum\vec{b}(\vec{r}) = 0$.

To analyse the behaviour of the system we ultimately want to evaluate the renormalised elastic constants which are measured by the response of the system to an infinitesmal applied stress. In other words we want to evaluate $\bar{\lambda}$ and $\bar{\mu}$ on very large length scales, and the standard tool for such a calculation is a renormalisation group. There are several equivalent ways of implementing this program[2,31,32,33] but probably the simplest one conceptually is that used in ref. 2, which was first used in the related planar rotor problem by Jose et al[34]. This involves a perturbative expansion in powers of y of an appropriate correlation function but in this expansion only short range fluctuations are averaged.

For the melting problem we perform this analysis for the renormalised compliance tensor S^R_{ijkl} which is the inverse of the tensor of elastic coefficients in the sense

$$S^R_{ijkl}C^R_{klmn} = \frac{1}{2}(\delta_{im}\delta_{jn} + \delta_{in}\delta_{jm}) \tag{2.9}$$

For a system with free boundary conditions, the compliance tensor may be expressed as a correlation function of the macroscopic strain field U_{ij}

$$S^R_{ijkl} = \langle U_{ij} U_{kl} \rangle / a_o^2 \Omega \tag{2.10}$$

where Ω is the area of the system and

$$U_{ij} = \int d^2x\, \phi_{ij}(x) + \frac{1}{2} a_o \sum_{\vec{r}} (b_i(r)\varepsilon_{j\ell} r_\ell + b_j\, \varepsilon_{i\ell} r_\ell) \tag{2.11}$$

The average in eq. 2.10 is calculated with the free energy functional of eq.2.8.

The renormalisation group procedure consists of averaging out dislocations of separation between $a < r < a(1 + \delta\ell)$ and rescaling the core size a to $a(1 + \delta\ell)$. This procedure does not change the compliance because in two dimensions it is dimensionless but yields exactly the same expressions except with y, $\bar{\mu}$ and $\bar{\lambda}$ rescaled from their initial values. The calculations are carried out as a perturbation expansion in the fugacity and rests on the basic assumption of small y. Typical dislocation configurations averaged out are shown in fig. 2.3. A detailed derivation of the rescaling is to be found in refs. 1, 2, 3, 32 but here we merely quote the results.

$$dK^{-1}/d\ell = \frac{3}{4} \pi y^2 \left[2I_o\left(\frac{K}{8\pi}\right) - I_1\left(\frac{K}{8\pi}\right) \right] + O(y^3) \tag{2.12}$$

$$dy/d\ell = \left(2 - \frac{K}{8\pi}\right) y + 2\pi y^2 I_o\left(\frac{K}{8\pi}\right) + O(y^3) \tag{2.13}$$

$$d\bar{\mu}^{-1}/d\ell = 3\pi y^2\, I_o\left(\frac{K}{8\pi}\right) + O(y^3) \tag{2.14}$$

$$d(\bar{\mu}+\bar{\lambda})^{-1}/d\ell = 3\pi y^2 \left[I_o\left(\frac{K}{8\pi}\right) - I_1\left(\frac{K}{8\pi}\right) \right] + O(y^3) \tag{2.15}$$

Here I_o and I_1 are modified Bessel functions which result from the angular terms in the dislocation interaction and since they vary slowly near the melting temperature can be replaced by constants. These angular terms are qualitatively unimportant at the melting transition but become vital above. The presence of the term in y^2 in eq. 2.13 is special to a structure such as the triangular lattice which has elementary dislocations obeying $\vec{b}_1 + \vec{b}_2 + \vec{b}_3 = 0$. Such a term does not appear for melting of a square lattice or the mathematically similar problem of vortices in helium[6,31].

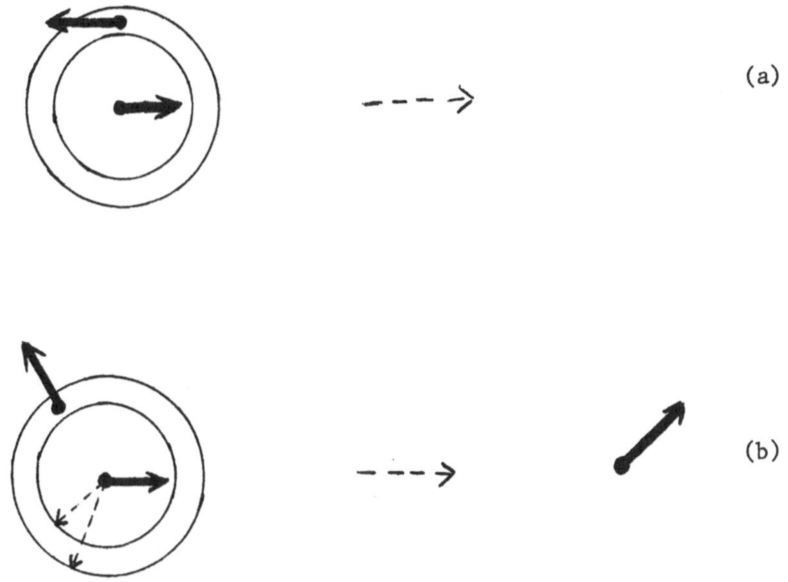

Fig.2.3 Typical dislocation configurations averaged out in the renormalisation procedure. The configuration in (a) is equivalent to no dislocation and causes a rescaling $K \to K + O(y^2)$. The configuration in (b) is equivalent to an elementary dislocation with Burger's vector $\vec{b}_1 + \vec{b}_2$ and causes $y \to y + O(y^2)$.

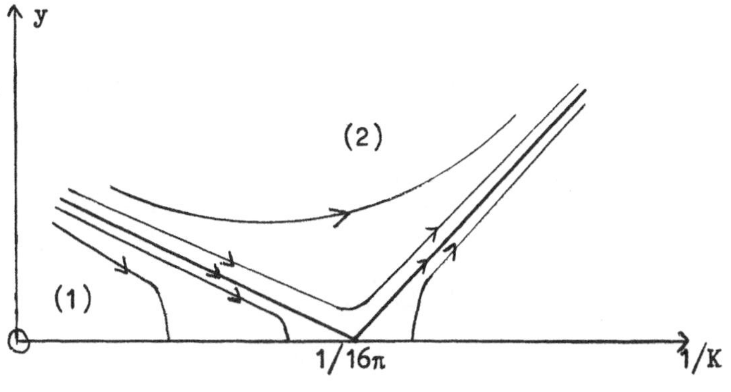

Fig.2.4 The solutions of the recursion relations 2.12 and 2.13 near the melting temperature $K = 16\pi$. Regions 1 and 2 are the low and high temperature phases separated by the line $T = T_m(y)$.

These recursion relations have solutions sketched in fig.2.4. The physical meaning of these recursion relations is that the physical system with fluctuations on length scales greater than a defined by physical parameters K_o and y_o is equivalent to a system with minimum length scale $a\exp(\ell)$ with parameters $K(\ell)$ and $y(\ell)$ as given by the solutions of 2.12 and 2.13. In general, a renormalisation group transformation does not solve anything but merely transforms one problem into another. To extract physical predictions we must perform a real calculation for some values of the parameters $K(\ell)$ and $y(\ell)$.

If the system is at a temperature and vortex fugacity y_o in region 1 of fig. 2.4, at long length scales $y(\ell)$ decreases and eventually goes to zero. This is equivalent to a crystal without dislocations described by the harmonic theory of section I for which we can compute everything. In region 2 on the other hand $y(\ell)$ eventually increases and the system is equivalent to one with many dislocations. At this stage our recursion relations break down but one can use a Debye-Huckel type approximation. This consists of ignoring the discreteness of the Burger's vectors and regarding them as continuous variables. This obviously makes the averaging in eq.2.10 easy since the free energy functional is quadratic in $\vec{b}$. We therefore identify region 1 as the low temperature phase $T < T_m(y)$ in which the renormalised elastic constants are finite and the system will behave as an elastic solid. Region 2 is identified as a fluid phase $T > T_m(y)$. The separatrix leading into the point $K = 16\pi$, $y = 0$ is the critical temperature $T_m(y)$.

In the low temperature phase we use the fact that the renormalised compliance tensor is invariant under the renormalisation group

$$S^R_{ijkl}(\bar{\mu}_o,\bar{\lambda}_o,y_o) = S^R_{ijkl}(\bar{\mu}(\ell)\bar{\lambda}(\ell)y(\ell)) \tag{2.16}$$

to extract information about the elastic constants near the melting temperature. Since $y(\infty) = 0$, there are no dislocations at this length scale and it is easily seen that the particular combination of elastic constants comprising K approaches the universal value of 16π at the melting temperature T_m. Solution of the recursion relations shows that just below T_m

$$\mu(T)/\mu(T_m) - 1 \sim (1 - T/T_m)^{\bar{\nu}} \tag{2.16a}$$

with a similar relation for $\lambda(T)$. The exponent $\bar{\nu} = .37$. Unfortunately, the particular combination $K(T)$ is not easily accessible to experiment and the individual elastic coefficients $\mu(T)$ and $\lambda(T) + \mu(T)$[35] do not themselves approach universal values.

Above the melting temperature, the recursion relations are integrated until $y(\ell)$ is of order unity when one argues that all dislocations are free. This defines a correlation length

$$\xi_+(T) = ae^{\ell^*} \sim \exp b(T/T_m - 1)^{-\bar{\nu}} \tag{2.17}$$

This length has the interpretation of the maximum distance between a pair of dislocations such that they may be considered as bound. These considerations yield exponentially decaying positional correlations

$$C_G(r) \sim \exp -r/\xi_+(T) \tag{2.18}$$

and an unobservable essential singularity in the specific heat

$$C_P \sim \text{analytic part } + \xi_+^{-2} \tag{2.19}$$

A measurement at finite wavelength will yield elastic constants $\mu(q,T)$ vanishing like $q^2\xi_+^2$ so that the discontinuous jumps in $\bar{\lambda}$ and $\bar{\mu}$ at T_m are in the $\vec{q} = 0$ Fourier component only. In the light of this, the unobservable essential singularity in the specific heat is not surprising since a specific heat depends equally on all Fourier components: the discontinuous jump gets translated into something much weaker.

The X-ray structure factor below T_m will show divergences already discussed. At T_m stability requirements puts bounds on $\eta_G(T_m)$ for the lowest reciprocal lattice vector $G_o^2 = 16\pi^2/3a_o^2$ of

$$0 < \eta_{G_o} < 1/3 \tag{2.20}$$

so that for the first few reciprocal lattice vectors, the structure factor is divergent. Above the melting temperature, these peaks are finite with height

$$S(\bar{G}) \sim \xi_+^{2-\eta_{\vec{G}}}$$

This sort of behaviour of the structure factor with the Bragg peaks rising out of a liquid like background as T_m is approached from above is very characteristic of a two-dimensional crystallisation.

Periodic Substrate Potentials

A substrate with a periodic binding potential which is incommensurate with the periodicity of the adsorbate overlayer will not affect this melting transition very much. The only effect is to introduce an extra orientational term into the elastic free energy which tends to align the crystal axes with the axes of the substrate. For small deviations of the two axes by an angle $\theta = \frac{1}{2}\hat{z}.(\vec{\nabla}\times\vec{u})$ the free energy becomes

$$F/kT = \frac{1}{2}\int \frac{d^2\vec{x}}{a_o^2}\left[2\bar{\mu}\phi_{ij}^2 + \bar{\lambda}\phi_{kk}^2 + \bar{\gamma}(\partial_y u_x - \partial_x u_y)^2\right] + F_D/kT \tag{2.22}$$

where the dislocation part

$$F_D/kT = -\frac{1}{8\pi}\sum_{\vec{r}\neq\vec{r}'} b_i(\vec{r})b_j(\vec{r}')\left\{K_1\delta_{ij}\ln\left|\frac{\vec{r}-\vec{r}'}{a}\right| - K_2\frac{(\vec{r}-\vec{r}')_i(\vec{r}-\vec{r}')_j}{(\vec{r}-\vec{r}')^2}\right\}$$

$$- \ln y\sum_{\vec{r}} b^2(\vec{r}) \tag{2.23}$$

The equality between the two coefficients of the logarithmic and angular terms in F_D is broken by the orientational term in (2.22) by the coupling γ

$$K_{1,2} = \frac{4\bar{\mu}(\bar{\mu}+\bar{\lambda})}{2\bar{\mu}+\bar{\lambda}} \pm \frac{4\bar{\mu}\bar{\gamma}}{\bar{\mu}+\bar{\lambda}} \tag{2.24}$$

The recursion relations for this more general problem have been analysed by Young[32] who finds that all the results below T_m go through with the replacements

$$\mu_R \to \mu_R + \gamma_R\ ; \quad \lambda_R \to \lambda_R - 2\gamma_R \tag{2.25}$$

Note that here incommensurate means that the set of adsorbate $\{\vec{G}\}$ and substrate $\{\vec{K}\}$ lattice vectors have no member in common. The X-ray structure factor will have delta function Bragg peaks at $\vec{q} = \vec{K}$ as well as power law peaks at $\vec{q} = \vec{G} + \vec{K}$.

The effect of commensurate substrate potentials is more dramatic. If the binding potential is very weak, the free energy is approximated by

$$F/_{kT} = {}^{F_E}/_{kT} + \sum_{\vec{M}} h_M \sum_{\vec{r}} \exp i\vec{M}.\vec{u}(\vec{r}) \tag{2.26}$$

where F_E is the free energy for the adsorbate in the absence of the substrate in which dislocations are subsumed into a renormalisation of the elastic constants and $\{\vec{M}\}$ is the set of substrate reciprocal lattice vectors common to $\{\vec{G}\}$. As mentioned in section I, the importance of the substrate can be gauged by looking at how the coefficients h_M scale under renormalisation. This is done by studying the correlation function

$$\langle \exp i\vec{M}.[\vec{u}(\vec{r}) - \vec{u}(\vec{o})]\rangle_{h_M=0} \sim r^{-\eta_M(T)} \quad (2.27)$$

and under a length rescaling $h_M \to h_M \exp\lambda_M \ell$ with

$$\lambda_M = 2 - \tfrac{1}{2}\eta_M \quad (2.28)$$

Using eq.1.10 with $\vec{G}$ replaced by $\vec{M}$ we immediately see that, at sufficiently low temperature this is positive for all $\vec{M}$ so that the incommensurate phase is unstable to substrate perturbations. For a sufficiently coarse substrate, like a rare gas on graphite near the $\sqrt{3} \times \sqrt{3}$ structure[7a] λ_M is positive right up to the melting temperature of the adsorbate in the absence of a substrate so that the floating solid[2] is always unstable and a lattice gas description is more appropriate. For a sufficiently fine substrate mesh (large M) there is a temperature window in which both dislocations and the substrate are irrelevant so that the ideal floating solid of section I exists[29]. There is a vast and growing literature on the problem of atoms adsorbed on a periodic substrate, in particular on the transition between a commensurate to incommensurate overlayer.[36,37].

III FLUID PHASES

Although the angular terms in the dislocation interaction are unimportant for the melting of an incommensurate two-dimensional crystal, they have dramatic effects above the melting temperature. On a smooth substrate, they are responsible for a new phase with short range positional order but quasi long-range orientational order. It was first noticed by Mermin[9] that in a harmonic two dimensional crystal ($T<T_m$) there is true long-range order in the orientation of nearest neighbour bonds. Halperin and Nelson[2] introduced an order parameter analogous to the superfluid order parameter

$$\psi(r) = \exp 6i\theta(r) \quad (3.1)$$

where the angle θ has been defined earlier and the factor 6 in eq.3.1 is because of the sixfold symmetry of the underlying lattice. On a smooth substrate θ is an angle relative to an arbitrary axis and for an incommensurate overlayer on a periodic substrate, θ is measured relative to the direction of minimum free energy. This does not necessarily coincide with the substrate axes[38]. It is easily seen that below T_m where the harmonic approximation holds for the elastic free energy

$$\lim_{r\to\infty} \langle \psi^*(\vec{r})\psi(\vec{o})\rangle \neq 0 \quad (3.2)$$

implying long-range orientational order.

Above the melting temperature, the situation is more complicated. If there is residual orientational order, fluctuations in θ on a long length scale should be described by a free energy functional

$$F[\theta]/kT = \frac{1}{2} K_6(T) \int d^2\vec{x}(\vec{\nabla}\theta(\vec{x}))^2 \tag{3.3}$$

The Frank constant $K_6(T)$ which is a measure of the stiffness of the system to local rotations is

$$K_6^{-1}(T) = \lim_{q\to 0} q^2 \langle\theta(\vec{q})\theta(-\vec{q})\rangle/\Omega \tag{3.4}$$

where the average is to be calculated from eq.2.22. Standard isotropic elasticity theory gives the rotation due to a set of dislocations as

$$\theta(\vec{q}) = -ia_o\vec{q}.\vec{b}(\vec{q})/q^2 \tag{3.5}$$

where $\vec{b}(\vec{q})$ is the Fourier transform of the dislocation field $\vec{b}(\vec{r})$. Noting that in the limit $q = 0$ the smooth part of the elastic free energy does not contribute to eq.3.5, the Frank constant is given by

$$K_6^{-1}(T) = \lim_{q\to 0}(a_o^2/\Omega)(q_iq_j/q^2) \langle b_i(q)b_j(-q)\rangle \tag{3.6}$$

This is easily evaluated well above T_m when the vortex core size $a \simeq \xi_+(T)$ by integrating over the $\vec{b}(\vec{q})$ with the result[2]

$$K_6^{-1}(T) = \lim_{q\to 0} \frac{2q^2}{K_1 - K_2 + 4E_c q^2/kT} \tag{3.7}$$

where K_1 and K_2 are defined by eq. (2.24). For an incommensurate solid on a periodic substrate $K_1 \neq K_2$ so that

$$K_6(T) = \infty \tag{3.8}$$

and there is true long-range orientational order which could have been guessed from the physics. More interesting is the situation of a smooth substrate where $K_1 = K_2$ and $K_6(T)$ is finite. Scaling arguments tell us that near the melting temperature

$$K_6(T) \sim \xi_+^2(T) \tag{3.9}$$

In this case there is algebraic decay of orientational correlations

$$\langle\psi^*(r)\psi(0)\rangle \sim r^{-\eta_6(T)} \qquad (3.10)$$

where $\eta_6(T) = 18/\pi K_6(T)$. This algebraic decay is not characteristic of an isotropic fluid in which exponential decay is expected so that above T_m the dislocation theory of melting predicts a new phase of matter called the hexatic phase because of its six-fold symmetry[2].

The final destruction of orientational correlations should take place by a disclination unbinding transition which is completely analogous to the vortex unbinding transition in helium. In the solid phase disclinations have an interaction energy proportional to the square of their separation and can be neglected. They may be regarded as being bound together inside a dislocation core. However, above the melting transition when there are a number of free unbound dislocations, this interaction between disclinations is screened and becomes logarithmic in their separation. At this point one can take over the analysis of a superfluid[6,31] and one finds a universal discontinuous jump in the Frank constant K_6 at the disclination unbinding temperature of $72/\pi$. Above this temperature, there will be free disclinations and the system will behave as an isotropic fluid. Note that this sequence consists of continuous transitions provided no other mechanism intervenes before the stability limit of the two unbinding phenomena is reached.

To conclude this section, a brief discussion of an alternative melting mechanism is in order, namely the grain boundary[39] mechanism. A grain boundary can be thought of as a straight row of dislocations of the same Burger's vector which separates two regions of perfect crystal of differing orientation. This type of organisation of the dislocations is possible because of the angular terms in the interaction but is explicitly excluded by the assumptions implicit in the analysis discussed previously. Recently it was pointed out that the free energy to generate a single small angle grain boundary goes to zero at the same temperature at which dislocations unbind[40] In a recent preprint S.T. Chui[41] has taken this idea further and analysed the free energies required to generate various grain boundary configurations. He concludes that a hexagonal grain boundary network is always generated at a lower temperature than the dislocation unbinding temperature and that this transition is first order if the dislocation core energy is small enough. A two stage melting via an intermediate hexatic phase is possible at a very low dislocation density but if mass density fluctuations are allowed the transition is always first order.

Experiment and Computer Simulations

The evidence from experiment and computer simulations for the existence of a two-dimensional solid is incontrovertible. However, the nature of the melting transition is far from clear and at present is a very contentious subject. Perhaps the most thorough computer simulation to date has been done using Monte Carlo methods by Tobochnik and Chester[27] for point particles interacting by a Lennard-Jones potential. Their simulation suggests that at low densities the melting of the solid is consistent with the dislocation mechanism in that the crucial quantity $K \simeq 16\pi$ at melting. At higher densities where the repulsive cores of the particles are dominating their data is more consistent with a first order transition. Unfortunately, one of the crucial features of the dislocation theory is the intermediate hexatic phase which seems to be extremely difficult to simulate on a computer. The data is consistent with the intermediate temperature region being either a hexatic phase or a two-phase region. There are two basic difficulties here. The first is that very long time scales are involved which could be explained by either hypothesis. The second is that, if a hexatic phase exists, the theory is for fluctuations on a length scale measured in units of a disclination core size which must be several lattice spacings so that the finite size of the system may be important.

Other recent simulations of Lennard-Jones systems using molecular dynamics have been carried out by Abrahams[42] using a constant pressure ensemble. He found strong hysteresis effects at melting which is characteristic of a first order transition. However, the dislocation mechanism involves enormous time scales on the time scale of a computer. In order to guarantee thermal equilibrium after a temperature change, one must allow dislocations to move around and in particular to climb. Since this involves the diffusion of intersitials and vacancies which are present in rather low concentrations near T_m, this must be an extremely slow process. It is possible that the observed hysteresis is due to being out of equilibrium. Much the same can be said about the work of Toxvaerd[43] who interprets his data as indicating a first order transition. Frenkel and McTague[44] interpret their results as being consistent with a hexatic phase but they can be also interpreted as a two phase region.

There have also been a number of simulations by Monte Carlo and molecular dynamics for repulsive soft disc pair potentials of the form

$$V(r) = A_n r^{-n} \tag{3.11}$$

The most extensive simulations have been for $n = 1$[45,46] appropriate to classical electrons, stimulated by the experimental work of Grimes and Adams[17]. For this system, the thermodynamic parameters are a function of a single dimensionless parameter $\Gamma = e^2(\pi\rho)^{\frac{1}{2}}/kT$. Directly from his simulation, Morf[46] inferred a value of Γ at melting of around 130. A theoretical analysis using a linear depression of the "bare" shear modulus $\mu_o(T)$ due to phonon interactions at low temperatures combined with the renormalisation group analysis of dislocations with input parameters $\mu_o(T), \lambda_o = \infty$ and y_o from ref. 40 gave a renormalised shear modulus $\mu_R(T)$ (using eq. 2.14) in excellent agreement with the computer estimates. Moreover, this analysis gave $\Gamma_m = 128$ in good agreement with the experimental value[17] $\Gamma_m \simeq 137 \pm 15$ and an earlier simulation[45] $\Gamma_m \approx 125 \pm 15$. This analysis cleared up an earlier estimate of $\Gamma_m \simeq 79$ by Thouless[47] who used the criterion 2.7 but with the zero temperature value of the shear modulus. However, he found it impossible to guarantee thermal equilibrium near melting so that, despite the good agreement with the theoretical analysis, he was unable to exclude a first order transition. A very recent simulation of the electron system using molecular dynamics reported at this Insitute seems[47a] to give an unambiguous first order transition

Another simulation using soft $1/r^6$ repulsive potentials[48] gives evidence of a two stage melting and a hexatic phase. Again the value of $K = 16\pi$ at melting is satisfied, but in the hexatic phase there is little agreement with theory, so it could just as well be interpreted as a two-phase region. Similar results have recently been reported by J.D. Weeks[49] who comes down in favour of a first order transition.

From this discussion we see that computer simulations of melting are fraught with difficulties and ambiguities. Although the crucial prediction of the dislocation theory $K = 16\pi$ seems to be generally satisfied, it is impossible to eliminate a first order transition. The existence of a hexatic phase is controversial and with present computers the time scales involved are too long to guarantee thermal equilibrium.

The experiment on electrons on helium[17] is in good agreement with dislocation melting but experimental investigation on freely suspended smectic B films show that a solid exists at low temperatures but attempts to measure the shear modulus[22,23] give results inconsistent with theory. This may be due to enormous relaxation times or to technical difficulties. At present we cannot tell if the dislocation theory is just an entertaining model or it really describes melting until much more work is done.

IV DISLOCATION THEORY OF LIQUID CRYSTAL FILMS

Films of smectic B and smectic C can be made as thin as two molecular layers freely suspended in a hole in glass of a few mm. across[19-23]. Such films provide a rich source of interesting systems with many possible phases in two dimensions. The theory of such systems based on the dislocation and disclination theory has been worked out by Halperin, Nelson and coworkers[3,4,50] and in this section we summarise the main ideas and conclusions. Although the films most extensively studied by light[19,20] and X-ray[21] scattering is ferroelectric which can be oriented by a small electric field to obtain a single domain sample we will not discuss the problems associated with this[51]. We shall also ignore any difficulties associated with boundary conditions at the edges of the film.

The extra complication arising in such films is the possibility that the rod-like molecules making up the layers do not lie perpendicular to the plane of the film. This introduces another order parameter measuring the molecular orientation which is coupled to the two already discussed. To describe the orientation of a single molecule we need two angles: γ is the angle the molecule makes with the plane normal and ϕ is the azimuthal angle in the plane of the film. Projecting on to the plane of the film, the orientation of the molecule can be described by a two component vector $\vec{s}$ of length $\sin\gamma$

$$\vec{s} = \sin\gamma\ (\cos\phi,\ \sin\phi) \tag{4.1}$$

If we assume that in the temperature range of interest, $\langle\sin\gamma\rangle \neq 0$ then each molecule can be described by a fixed length xy-like spin variable. We also need two other variables to describe the positions $\vec{r}$ of the centres of gravity of the molecules and the orientation θ of a bond joining the centres of gravity of two neighbouring molecules as in fig. 4.1. Note that the projection of the molecule in the plane of the film is not invariant under a rotation of 180° in contrast to the usual decription of a liquid crystal. In the limiting situation of the molecules lying in the plane of the film this decription is incorrect but can be easily remedied. In what follows we shall assume that the tilt angle γ is not too far from the film normal so that there is no invariance under $\phi \to -\phi$.

A variety of solid and fluid phases follow from the possible couplings between the tilt, orientational, and elastic degree of freedom. To describe the fluid phases, Halperin and Nelson[3] use an effective free energy functional which is valid on very long length scales[50] of the form

$$\frac{F[\theta\ \phi]}{kT} = \frac{1}{2} \int d^2\vec{x}\ \{K_6(\vec{\nabla}\theta)^2 + K_1(\vec{\nabla}\phi)^2 + 2g\vec{\nabla}\theta.\vec{\nabla}\phi - h\cos 6(\theta-\phi)\} \quad (4.2)$$

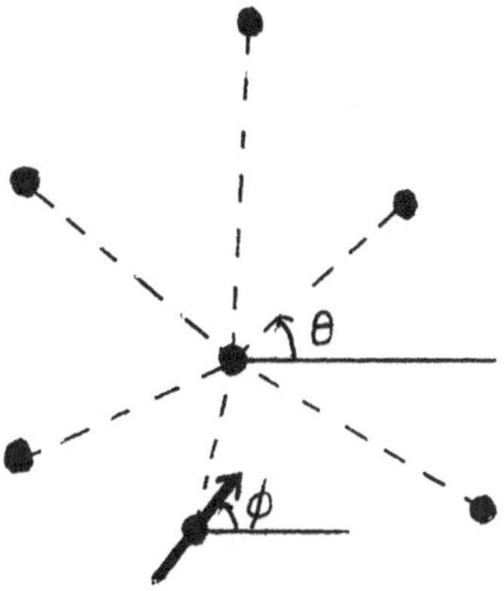

Fig.4.1 A cluster of nearest neighbour molecules in a liquid crystal layer. θ is the bond orientation angle and ϕ the molecular tilt orientation angle. The arrow is the projection of the molecule in the plane of the film.

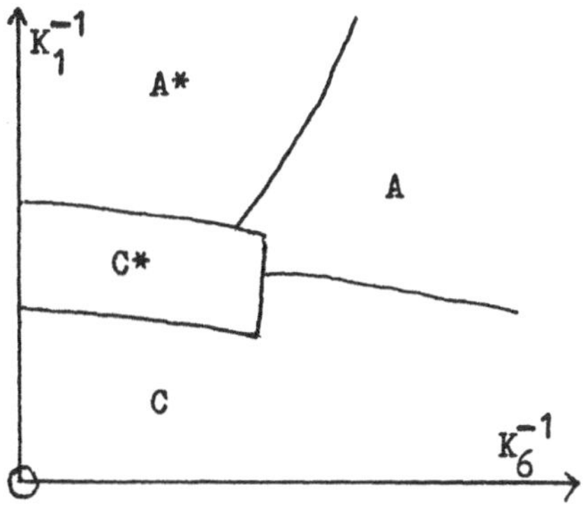

Fig.4.2 A sketch of the possible fluid phases of a layer of tilted molecules. See text for explanation of symbols.

The bond angle $\theta(\vec{r})$ is defined modulo $2\pi/6$ and the phase angle field of the molecules $\phi(\vec{r})$ modulo 2π. The stiffness constant K_6 is a measure of the coupling between different bonds and K_1 is the stiffness constant for fluctuations in the tilt orientations which vanishes as the tilt angle ϕ does. The coupling term g is generated by a length rescaling. There is also a direct coupling between the two angles θ and ϕ represented by the term proportional to h in eq. 4.2. The factor 6 occurs because of the underlying sixfold rotational symmetry. A detailed analysis[3] of the fluid phases gives the phase diagram sketched in fig. 4.2. As the temperature is raised a path will be traced through this diagram from lower left to upper right. Since these lectures are about melting, these phases will not be discussed further although they are very interesting in their own right.

At first sight there should be three solid phases at low temperatures with the bond angle stiffness constant $K_6 = \infty$ corresponding to long-range order in the bond angle θ. There should also be quasi long-range order in the positions of the molecules. If one ignores the coupling of the elastic and molecular orientation ϕ degrees of freedom, the free energy becomes

$$F/kT \;=\; F_E/kT \;+\; \frac{1}{2}\int d^2\vec{x}\{\{K_1(\vec{\nabla}\phi)^2 \;-\; h\,\cos 6(\phi(\vec{x}) \;-\; \theta_o)\} \qquad (4.3)$$

where F_E is the elastic free energy of eq.2.8. The constant bond angle θ_o is because of long-range orientational order. We now have a decoupled problem of an isotropic solid which can melt by the dislocation mechanism and on top of this an xy model with a six-fold symmetry breaking term. This has been analysed in detail by Jose et al[34] who have shown that this leads to three phases. At high temperatures vortices in the ϕ field are relevant for $K_1<2/\pi$ corresponding to disorder in the molecular tilt orientation. At lower temperatures, vortices in ϕ are bound together in neutral pairs and h is an irrelevant variable with renormalisation group eigenvalue

$$\lambda_h \;=\; 2 \;-\; 9/\pi K_1 \qquad (4.4)$$

As the temperature is lowered further, K_1 increases until at $K_1 = 9/2\pi$ the $\cos 6\phi$ perturbation becomes relevant and ϕ is locked to one of the crystal axes. These considerations lead to a phase diagram of the form of fig. 4.3.

All three solid phases have quasi-long-range positional order $C_G \sim 1/r^{\eta_G}$. The three fluid phases A*, C, and C* are distinguished by correlations in the two angle fields θ and ϕ. All have algebraic decay of the bond field θ. The A* phase has exponential decay of ϕ correlations so is identical to the hexatic phase discussed in section 3. The C and C* phases have algebraic decay of ϕ correlations and are distinguished in that long wavelength fluctuations of θ and ϕ are independent in C* whereas in phase C they are tied together.

This picture is very simple and attractive until one realises that in the solid phase the lattice carrying the "spins" is not rigid and one must take magnetoelastic effects into account. The strain field can couple to the tilt orientation by a term

$$w(U_{ij} \;-\; \tfrac{1}{2}\delta_{ij}U_{kk})s_i s_j \qquad (4.5)$$

As shown in ref. 3, this leads to an infinite downwards renormalisation of the shear modulus for all temperatures at which the xy-like phase is stable. Thus, the xy phase on a compressible regular triangular lattice is unstable to shear distortions and a uniaxial phase presumably results leading to a phase diagram shown in fig. 4.4. The phase labelled smectic B is a regular triangular solid with no tilt orientational order equivalent to zero average tilt angle. The phase labelled G is a uniaxial solid with probable long-range order in molecular orientation because of an Ising like anisotropy induced by the anisotropy in the molecular positions.

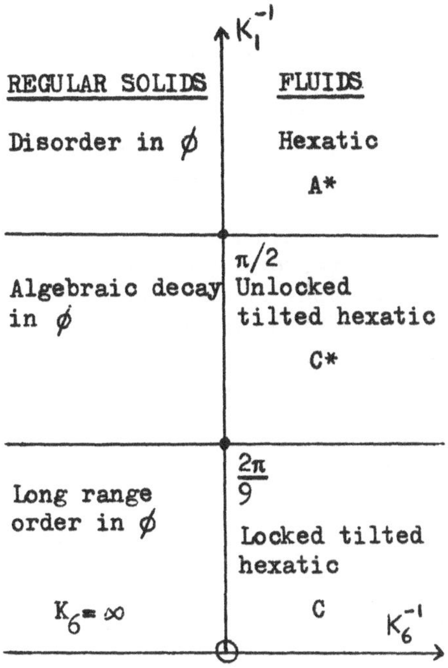

Fig.4.3 The possible phases near the melting temperature ignoring the coupling of strain to molecular orientation. The solid phases have $K_6 = \infty$.

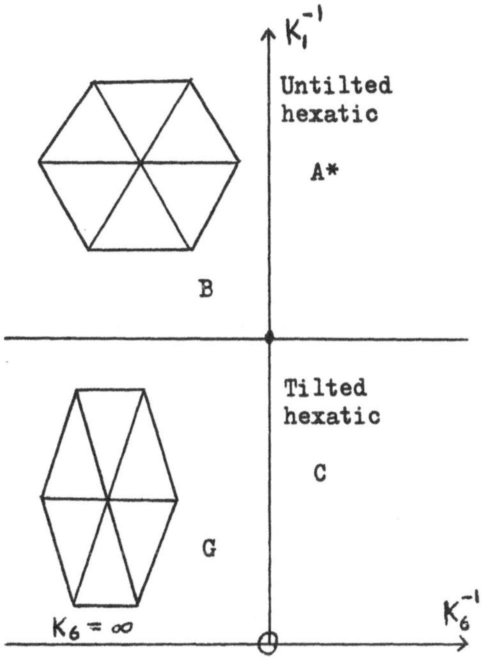

Fig. 4.4 A possible phase diagram including coupling of strain to orientation. The uniaxial solid is identified with smectic G and the regular triangular solid with smectic B.

The melting of the smectic B phase to an untilted hexatic phase will proceed by the dislocation unbinding mechanism already discussed. However, the melting of the smectic G phase is a much more complicated business. Even if the anisotropic solid can be described by linear elasticity theory and we assume that a similar dislocation unbinding mechanism is responsible for melting, the theory requires an understanding of dislocations in an anisotropic elastic medium. The theory of this has been worked out very recently in a very nice formalism by Ostlund and Halperin[4] .

If we restrict ourselves to the most likely situations in a tilted uniaxial solid in which the "spins" point either along one of the six nearest neighbour bond directions or half-way between two of these, there are two perpendicular symmetry axes and the system has rectangular symmetry. The two cases are shown in fig. 4.5. In fig. 4.5a the molecules are assumed to form a distorted triangular lattice in which they line up head to tail parallel to the x-axis. There are three basic Burger's vectors of an elementary dislocation in such a solid, two labelled type II and one of type I with $|\vec{b}_I|>|\vec{b}_{II}|$. In fig. 4.5b the molecules are shown lined up side to side in rows parallel to the x-axis. In this case $|\vec{b}_{II}|>|\vec{b}_I|$. It seems likely that, in analogy to the isotropic triangular lattice, the coupling constants for interactions between the dislocations will be proportional to $b_I^{\,2}$ and $b_{II}^{\,2}$. If this is the case, then one would expect that the dislocations with the shortest Burger's vectors will unbind at a lower temperature. This leads to two types of melting which Ostlund and Halperin[4] have called type I and type II melting.

For both types of melting, they find that the phase above T_m is characterised by algebraic decay of orientational order

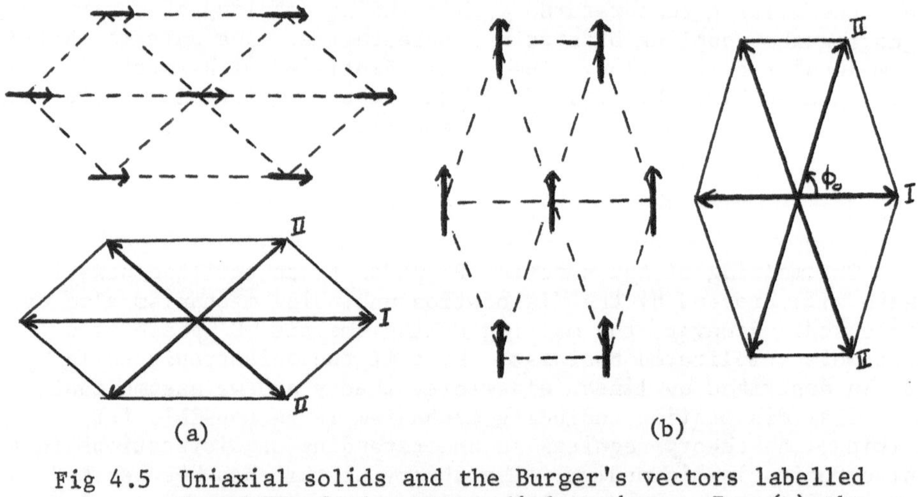

Fig 4.5 Uniaxial solids and the Burger's vectors labelled I and II of elementary dislocations. In (a) the molecules line up head to tail parallel to the x-axis and in (b) side to side in rows parallel to x-axis.

in the limit of very large length scales and this may be loosely termed a two dimensional nematic. In the case of type I melting of the lattice of fig. 4.5b for an experiment on a length scale L with

$$L < \xi_S \sim \exp t^{-\frac{1}{2}} \tag{4.6}$$

one expects that there is a negligible fraction of free dislocations of either type so that a solid-like response is expected. Note that the power of $t = (T/T_m - 1)$ in the correlation length is $\nu = \frac{1}{2}$ in contrast to the isotropic lattice. At the length scale ξ_S, a finite fraction of free type I dislocations are present. However, since two type I dislocations can combine to make a type II, one can argue that both are free. This is true but the concentration of type II dislocations is completely negligible[4] so there are to all intents and purposes no dislocations with Burger's vector in the y-direction. At this length scale, therefore, the molecules are arranged in liquid-like rows parallel to the x-axis with rather well defined spacing between the rows. This structure is a two dimensional version of a smectic liquid crystal which was considered by Toner and Nelson[50] from a different point of view.

The next length scale is ξ_N which may be interpreted as the scale at which dislocations of type II unbind, or rather at which a finite fraction of free type II dislocations are present. These may be regarded as dislocations in the smectic-like rows. Thus for

$$\xi_S < L < \xi_N \propto \xi_S^{p+1} \tag{4.7}$$

the system behaves as a two-dimensional smectic. The power p in eq. (4.7) is a non-universal number[4] depending on the four independent elastic constants, the angle ϕ_o and b_I^2/b_{II}^2.

For $L > \xi_N$, the system behaves like a nematic with two very different Frank constants which become equal due to non-linear phonon interactions[52] at very large length scales ξ_ℓ

$$\xi_\ell \sim \exp(a\xi_S{}^2) \tag{4.8}$$

Note that there are no phase transitions at these various length scales but crossovers from one type of behaviour to another.

For type II melting (fig. 4.5a) a detailed analysis[4] shows there is no smectic-like regime but otherwise things are much the same. The phase diagrams are sketched in fig. 4.6. The critical point T_m is the melting temperature as measured at an infinite length scale.

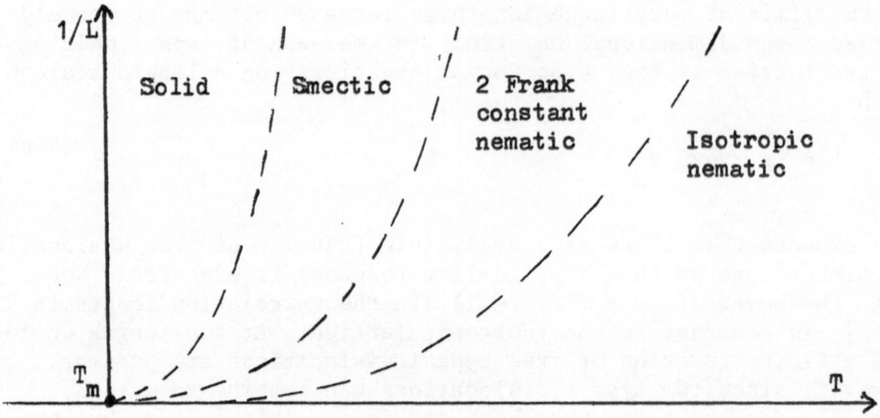

Fig.4.6 Phase diagram of type I melting. L is the length scale of the experiment and the actual melting temperature T_m is at infinite L. The dotted lines are the cross-over regions from one behaviour to another and are the loci of $L = \xi_S$, $L = \xi_N$, $L = \xi_\ell$.

References

1. D.R. Nelson, Phys. Rev. B18 2318 (1978).
2. D.R. Nelson and B.I. Halperin, Phys. Rev. B19 2457 (1979).
3. D.R. Nelson and B.I. Halperin, Phys. Rev. B21 5312 (1980).
4. S. Ostlund and B.I. Halperin, Phys. Rev. B23 335 (1981).
5. A. Zippelius, B.I. Halperin and D.R. Nelson, Phys. Rev. B22 2514 (1980).
6. J.M. Kosterlitz and D.J. Thouless, J. Phys. C6 1181 (1973); Prog. Low Temp. Phys. 7B 371 (1978).
7. a) See for example: Phase Transitions in Surface Films (eds. J.G. Dash and J. Ruvalds) Plenum, New York) 1980.
   b) Ordering in Strongly Fluctuating Condensed Matter Systems (ed. T. Riste) (Plenum, New York) 1980.
   c) Proceedings of the 1980 Summer School on Fundamental Problems in Statistical Mechanics, Enschede, Netherlands - to be published.
   d) Physics of Low Dimensional Systems, Proceedings of the Kyoto Summer Institute (eds. Y. Nagaoka and S. Hikami) Publication Office Progress Theoretical Physics, Kyoto)1979.
8. F. Bloch,Z. Phys. 61 206 (1930); R.E. Peierls, Ann. Inst. Henri Poincare 5 177 (1935): L.D. Landau, Phys. Z. Sowjetunion II 26 (1937).

9. N.D. Mermin and M. Wagner, Phys. Rev. Lett. 17 1133 (1966); N.D. Mermin, J. Math Phys. 8 1061 (1967); N.D. Mermin, Phys. Rev. 176 250 (1968); P.C. Hohenberg, Phys. Rev. 158 383 (1967).
10. H.E. Stanley and T.A. Kaplan, Phys. Rev. Lett. 17 913 (1966); H.E. Stanley, Phys. Rev. 164 709 (1967); Phys. Rev. Lett. 20 150 (1968).
11. V.L. Berezinskii,Sov. Phys. JETP 32 493 (1970); 34 610 (1972)
12. F. Wegner, Z. Phys. 206 465 (1967); B. Jancovici, Phys.Rev. Lett. 19 20 (1967); H.J. Mikeska and H. Schmidt, J. Low Temp. Phys. 2 371 (1970); Y. Imry and L. Gunther, Phys. Rev. B3 3939 (1971).
13. L.D. Landau and E.M. Lifshitz, Theory of Elasticity (Pergamon Press, N.Y.)1970.
14. F.R.N. Nabarro, Theory of Dislocations (Clarendon Press, Oxford) 1967.
15. See for example J.G. Dash, Films on Solid Surfaces (Academic, N.Y.) 1975.
16. For recent developments and further references: Proceedings of International Conference on Ordering in Two Dimensions 1980 (North Holland, N.Y.)
17. C.C. Grimes and G. Adams, Phys. Rev. Lett. 42 795 (1979).
18. D.S. Fisher, B.I. Halperin and P.M. Platzman, Phys. Rev. Lett. 42 798 (1979).
19. C.Y. Young, R. Pindak, N.A. Clark and R.B. Meyer,Phys. Rev. Lett. 40 773 (1978).
20. C. Rosenblatt, R. Pindak, N.A. Clark and R.B. Meyer, Phys. Rev. Lett. 42 1220 (1979).
21. D.E. Moncton and R. Pindak, Phys. Rev. Lett. 43 70 (1979).
22. R. Pindak, D.J. Bishop and W.O. Sprenger, Phys. Rev. Lett. 44 1461 (1980).
23. J.C. Tarczon and K. Miyano, Phys. Rev. Lett. 46 119 (1981).
24. D.J. Bishop and J. Reppy, Phys. Rev. Lett. 40 1727 (1978).
25. P.S. Pershan, J. de Physique Colloque 40 C3 (1979).
26. The sign of the angular term in eq. (2.4) is important and is given wrongly in many textbooks and papers.
27. J. Tobochnik and G.V. Chester, Cornell preprint (1980).
28. A.N. Berker and D.R. Nelson, Phys. Rev. B19 2488 (1979); J.Cardy and D.J. Scalapino Phys.Rev. B19 1428 (1979);
29. S. Ostlund, Harvard preprint 1980.
30. K.G. Wilson and J. Kogut, Phys. Rept. C12 75 (1974).
31. J.M. Kosterlitz, J. Phys. C7 1046 (1974).
32. A.P. Young, Phys. Rev. B19 1855 (1979).
33. D.J. Amit, Y.Y. Goldshmidt and G. Grinstein, J. Phys. A13 1507 (1980).
34. J.V. Jose, L.P. Kadanoff, S. Kirkpatrick and D.R. Nelson, Phys. Rev. B16 1217 (1977).
35. Note that $\mu(T) + \lambda(T)$ is not equal to the bulk modulus B except at $T = 0$.

36. For a recent review see J. Villain in ref. 7B p.221
For recent developments see ref. 16.
37. S.N. Coppersmith, D.S. Fisher, B.I. Halperin, P.A. Lee and W.F. Brinkman, Phys. Rev. Lett. 46 (1981).
38. A.D. Novaco and J.P. McTague, Phys. Rev. Lett. 38 1286 (1977).
39. W.T. Reed and W. Shockley, Phys. Rev. 78 275 (1950).
40. D.S. Fisher, B.I. Halperin and R. Morf, Phys. Rev. B20 4692 (1979).
41. S.T. Chui preprint 1981.
42. F.F. Abraham, Phys. Rev. Lett. 44 463 (1980).
43. S. Toxvaerd, Phys. Rev. Lett. 44 1002 (1980).
44. D. Frenkel and J.P. McTague, Phys. Rev. Lett. 42 1632 (1979).
45. R.G. Gann, S. Chakravarty and G.V. Chester, Phys. Rev. B20 326 (1979).
46. R. Morf, Phys. Rev. Lett. 43 931 (1979).
47. D.J. Thouless, J. Phys. C11 L 189 (1978).
47a. R.K. Kalia, P. Vashista and S.W. de Leeuw, Phys. Rev. B to be published.
48. J.P. McTague, D. Frenkel and M.P. Allen in ref. 16.
49. J.D. Weeks Private Communication
50. J. Toner and D.R. Nelson,Phys. Rev. B23 316 (1981).
51. R.A. Pelcovits and B.I. Halperin, Phys. Rev. B19 4614 (1979).
52. D.R. Nelson and R.A. Pelcovits, Phys. Rev. B16 2191 (1977).

MOLECULAR DYNAMICS STUDY OF 2-D MELTING: LONG RANGE POTENTIALS

R. K. Kalia and P. Vashishta

Argonne National Laboratory

Argonne, Illinois 60439, USA

ABSTRACT

Melting of a two dimensional electron lattice and a two dimensional dipolar solid are studied using molecular dynamics technique. The existence of hysteresis and latent heat of melting are observed, and the melting transitions in the two cases are found to be first order. For an electron lattice the melting occurs between $\Gamma=129\pm3$ whereas in the dipolar solid it is between $\Gamma=62\pm3$, with a transition entropy of 0.3 k_B per particle for both the systems.

INTRODUCTION

Two dimensional solids are known to lack long range translational order at finite temperatures,[1-3] but they have long range orientational order. Kosterlitz and Thouless[4] in 1973 proposed a theory of melting for the 2D solids; the solids in their theory have bound pairs of dislocations and the dissociation of the dislocation pairs is taken to be the cause of melting. Halperin and Nelson[5] in 1978 proposed a two-step melting process for 2D solids; in the first stage a 2D solid melts at a temperature T_M by dissociation of dislocation pairs into a liquid crystal like phase called a "hexatic" phase, which has bound disclination pairs; on further increase in temperature the hexatic phase transforms into an isotropic liquid at a certain temperature T_I, where the disclination pairs become unbound. The two transitions at temperatures T_M and T_I are supposed to be continuous, and according to Halperin and Nelson[5] the melting of 2D solids by a pair of continuous transition is possible provided that it is not preempted by a first order transition.

In this paper we shall describe the nature of the melting transitions in two systems, one where the particles interact via Coulomb interaction $1/r$, and the other is where the pair-wise interaction is the dipole-dipole interaction, $1/r^3$, where $\vec{r}$ is a two dimensional vector. Experimentally, electrons on the surface of liquid helium represent a 2D Coulomb system. The nature of the melting transition in this system has been studied both experimentally[6] and by computer simulation methods.[7-11] Recently, Pieranski[12] has argued that polystyrene spheres ($\sim$3000 Å in diameter) on the water-air interface act as electric dipoles aligned parallel to each other; the interaction in a plane perpendicular to the dipoles is $1/r^3$. The investigation of melting in a 2D dipolar system was recently made through the molecular dynamics method.[13]

In systems where the interaction energy is of the form $V(\vec{r}) = \epsilon(\sigma/|\vec{r}|)^n$, all the properties are fully characterized by a single, dimensionless variable,

$$\Gamma = \beta\epsilon(\sigma/r_o)^n \quad , \tag{1}$$

where $\beta = 1/k_BT$, $\pi r_o^2 = \rho^{-1}$, ρ being the number density. For the Coulomb system $n = 1$ and $\epsilon\sigma = e^2$, where e is the electronic charge.

MOLECULAR DYNAMICS (MD)

The calculations are performed on 100, 256 and 576 particles for the Coulomb system and on 256 particles for the dipolar system. In both cases the particles are confined in a rectangular cell of area A whose sides have a ratio of $\sqrt{3}/2$ so that the cell can accommodate $4M^2$ particles on a triangular lattice. Because of the long range nature of $1/r$ and $1/r^3$ potentials, special care is needed in the calculation of forces and potential energy; they should include not only the interaction of a given particle with all particles in the cell, but also its interaction with all the images in the periodically repeated replicas of the central cell. The forces and the potential energy are calculated by the Ewald method.[14] The Newton's equations of motion were integrated using a predictor-corrector method which involves up to fifth time derivatives of the positions.[15] The time step in the calculations was chosen so that the energy was conserved to at least 1 part in 50,000 even after 15,000 time steps. At each temperature, the system was first thermalized for several thousand time steps; results reported in this paper are based on long MD runs after thermalization. The results are independent of the starting configurations. Near the melting transition calculations extending up to 40,000 time steps were carried out.

RESULTS

In what follows we shall discuss the structural and dynamical properties of the 2D Coulomb and dipolar systems.

a) Coulomb System

MD results for the internal energy E are shown in Fig. 1. The squares correspond to an electron liquid which has been monotonically cooled from a temperature of 1 K ($\Gamma = 36$), and the solid circles represent the results for the electron solid which has been monotonically heated from a low temperature triangular lattice. The presence of hysteresis is evident in the figure, and the hysteresis region is marked by vertical dotted lines. In the temperature range shown in Fig. 1, E seems to vary almost linearly with temperature. The point marked L_2 on the upper line corresponds to the

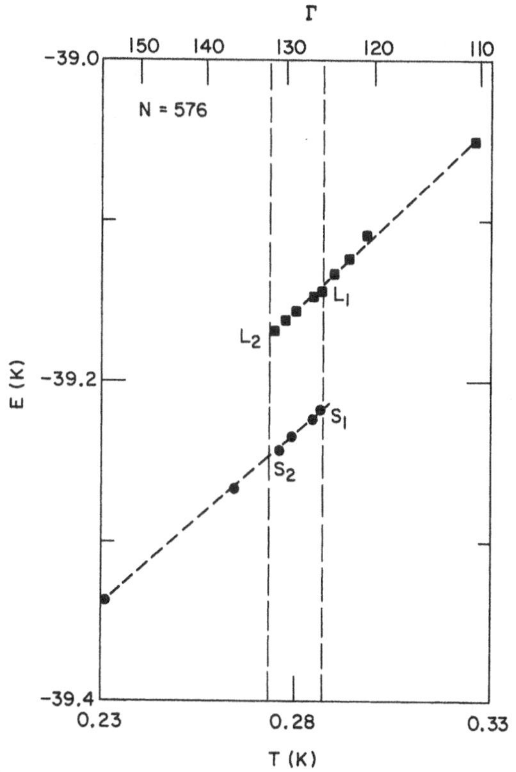

Figure 1. Temperature variation of internal energy for an electron system.

last supercooled liquid state. That the system at L_2 is a liquid is evident from the fact that it has a finite value of self-diffusion constant D = 0.5 cm^2/sec. To further confirm the diffusivity of the supercooled liquid at L_2 we show in Fig. 2 the behavior of mean square displacement $\langle R^2(t)\rangle$, which increases linearly with time t. The point marked S_1 on the lower line in Fig. 1 represents the last superheated solid. It has no diffusion, and its solid-like behavior is evident also from Fig. 2 where the variation of $\langle R^2(t)\rangle$ does not increase with time. The results for the self-diffusion constant show that in the hysteresis region D has two values at each temperature; zero in the solid phase and around 0.6 cm^2/sec in the liquid phase.

The hysteresis in Fig. 1 extends from the superheated solid system at S_1 to the supercooled liquid system at L_2. The hysteresis was found in all three sizes of the system, (N = 100, 256 and 576). In all three cases when the last superheated solid system is heated, it melts abruptly and goes into the liquid phase marked L_1. While at S_1 the diffusion is zero, it acquires a value of $\simeq$ 0.7 cm^2/sec after the system becomes a liquid at L_1. The hysteresis decreases slightly with increase in the size of the system and the transition moves slightly toward a higher value of Γ - 126-132, in good agreement with the emperimental value of

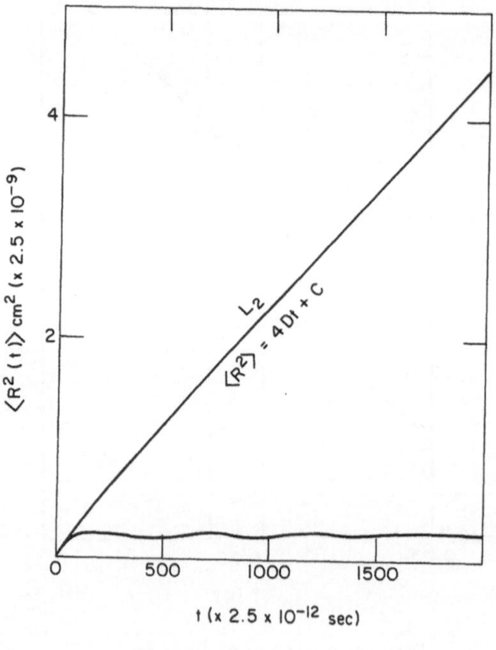

Figure 2. Mean square displacement vs. time.

$\Gamma = 137 \pm 15$. It should be emphasized that the latent heat of melting, $E(L_1) - E(S_1)$, is the same for all three sizes of the system. Taking the middle point in the hysteresis region to be the melting temperature T_M, we find that the change in entropy on melting is $\Delta s \cong 0.3\ k_B$ per particle.

The observed hysteresis in the temperature dependence of E and D, and the attendant latent heat of melting indicate that melting transition in the 2D electron system is first order. To further confirm the nature of the phase transition we have investigated the homogeneous nucleation of an electron solid from a supercooled liquid in a 576 particle system. At time t = 0, the system is cooled by lowering the energy by an amount slightly greater than the latent heat. It is then allowed to run for 20,000 time steps. The time dependence of the instantaneous temperature, T(t), shows that for the first several thousand time steps it fluctuates around a mean value and the system in the E vs T graph lies above the lower line for the solid, but then in a small time interval (~1000 time steps), T(t) increases and settles to a value where the (E, T) point for the system lies on the lower line for the solid, and the self-diffusion constant becomes zero. Afterwards the mean temperature of the system stayed constant for as long as we investigated it, which was several thousand more time steps. The above mentioned variation of T(t) in homogeneous nucleation is unique to systems which undergo first order phase transition.[16,17]

b) Dipolar System

A schematic diagram of an interfacial colloidal crystal system-polystyrene spheres on water-air interface interacting via dipole-dipole interaction -is shown in Fig. 3. In Fig. 4 we show the temperature dependence of total energy $E^* = E/\epsilon$ for 256-particle system at a reduced density $\rho^* = \rho\sigma^2 = \pi^{-1}$. The solid circles represent the MD results for the dipolar liquid which has been monotonically cooled from a reduced temperature $T^* = T/\epsilon = 28.3 \times 10^{-3}$. The solid squares correspond to the dipolar solid phase, and they are obtained by monotonically heating a low temperature triangular lattice. As in the electron system, the points L_2 and S_1 represent the last supercooled liquid and superheated solid systems, respectively. When the system at S_1 is heated, it melts abruptly and goes

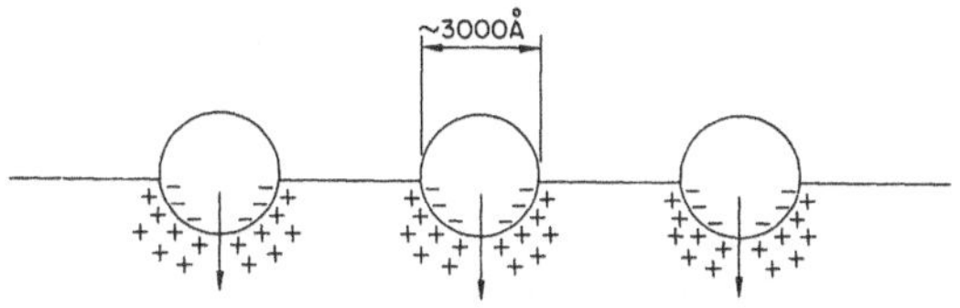

POLYSTYRENE SPHERES ON WATER SURFACE

Figure 3

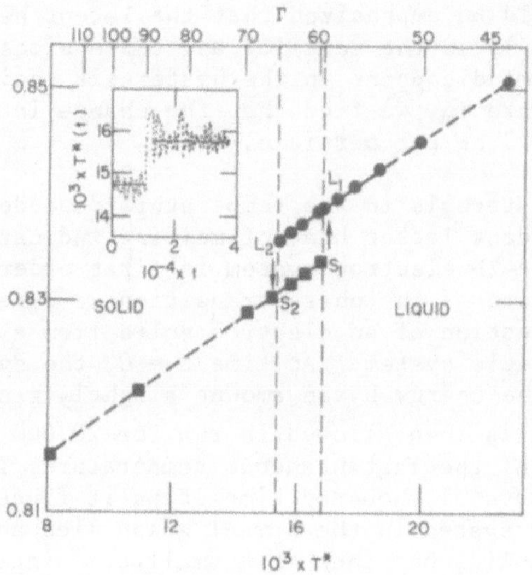

Figure 4 Reduced energy E^* as a function of reduced temperature T^* for a dipolar system in two dimensions. The inset shows the homogeneous nucleation of a dipolar solid from a supercooled liquid state.

into a state marked L_1 where it has a finite mobility. The melting takes place in only about 1000 time steps and after that the system reaches equilibrium and settles into the state L_1.

The homogeneous nucleation of a solid from a supercooled liquid was also investigated along the lines we discussed in the last section. The behavior of $T^*(t)$ is shown in the inset in Fig. 4. During the first 10,000 time steps, $T^*(t)$ fluctuates around a mean value of 14.7×10^{-3} and then in the next 500 time steps, $T^*(t)$ increases and fluctuates around a mean value of 15.8×10^{-3}. The energy E^* and the corresponding mean temperature of 15.8×10^{-3} constitute the point marked S_2 in Fig. 4; at S_2 the system has zero mobility.

The hysteresis in E^* vs T^* for the dipolar system, the presence of latent heat, and the homogeneous nucleation of the dipolar solid from the supercooled liquid, clearly show that the melting transition is first order. The melting transition occurs between $\Gamma = 59\text{-}65$ and the entropy change on melting is again found to be $\cong 0.3\ k_B$ per particle, which is the same as the transition entropy for the electron system.

A remark on the nature of interparticle interaction in interfacial colloidal crystals is in order here. According to Pieranski[12] the experiment shows a long range interaction between the colloidal particles. Since the polystyrene spheres are only partly immersed in water, Pieranski claims that the interaction will vary as $1/r^3$. If the interaction is indeed $1/r^3$, then the properties of the system, as we showed earlier, depend on a single variable $\Gamma = \beta\epsilon(\sigma^2\pi\rho)^{3/2}$ and not on ρ and T separately. A plot of the melting temperature T_M and density $\rho^{3/2}$ will be a straight line in case the interaction is $1/r^3$. Experiments should therefore be carried out to verify if T_M vs $\rho^{3/2}$ is a straight line. It is hoped that the MD results on the melting transition of a dipolar system will stimulate further experimental investigations.

REFERENCES

1. R. E. Peierls, Ann. Inst. Henri Poincare 5, 177 (1935).
2. L. D. Landau, Phys. Z. Sowjetunion II, 26 (1937).
3. N. D. Mermin, Phys. Rev. 176, 250 (1968).
4. J. M. Kosterlitz and D. J. Thouless, J. Phys. C5, L124 (1972).
5. B. I. Halperin and D. R. Nelson, Phys. Rev. Letters 41, 121 (1978).
6. C. C. Grimes and G. Adams, Phys. Rev. Letters 42, 795 (1979).
7. R. C. Gann, S. Chakravarty and G. V. Chester, Phys. Rev. B20, 326 (1979).
8. R. W. Hockney and T. R. Brown, J. Phys. C8, 1813 (1975).
9. R. H. Morf, Phys. Rev. Letters 43, 931 (1979).
10. R. K. Kalia, S. W. deLeeuw and P. Vashishta, Phys. Rev. B (in press).
11. S. K. Sinha, Ordering in 2-Dimensions, North-Holland Publishing Co., 1980; and references therein.
12. P. Pieranski, Phys. Rev. Letters 45, 569 (1980).
13. R. K. Kalia and P. Vashishta, to be published.
14. S. W. deLeeuw, J. W. Perram and E. R. Smith, preprint.
15. C. W. Gear, ANL Report #7126, Argonne National Laboratory (1966); Numerical Initial Value Problems in Ordinary Differential Equations, Prentice Hall, Englewood Cliffs, N. J. (1971).
16. C. S. Hsu and A. Rahman, J. Chem. Phys. 70, 5234 (1979).
17. M. Crentz, L. Jacobs and C. Rebbi, Phys. Rev. Letters 42, 1390 (1979).

ACKNOWLEDGEMENTS

We would like to thank S. W. de Leeuw, A. Rahman, J. M. Kosterlitz, S. K. Sinha, K. Gray, K. Miyano, M. Parrinello, and G. Grest for many stimulating discussions.

EXPERIMENTAL STUDIES OF TWO DIMENSIONAL MELTING

S. K. Sinha*

Argonne National Laboratory
Argonne, IL 60439

ABSTRACT

The nature of the crystalline state in two dimensions (2D) is reviewed, with particular emphasis on the form of the structure factor S(K) peculiar to 2D. Effects due to finite size are discussed, and it is shown that in most cases, this does not severely modify the expected power-law behavior for S(K) at most accessible wavevectors. The experimental determination of 2D solid and fluid phases in adsorbed monolayers is reviewed with particular reference to vapor-pressure isotherms, heat capacity and diffraction measurements. Results of diffraction studies of S(K) in the crystalline phase, and of the melting of incommensurate adsorbed monolayers and liquid crystal films are presented and discussed in the light of the Kosterlitz-Thouless theory of 2D melting. Finally, experiments on the lattice-gas-type melting of commensurate adsorbed monolayers are discussed with reference to theoretical expectations based on q-state Potts model calculations.

1. INTRODUCTION

The melting transition of bulk crystals is known to be strongly first-order in nature, which makes the microscopic processes relating to this transition extremely difficult to study, experimentally or theoretically. However, the situation is different in the case of two-dimensional (2D) crystals. In many cases, there are strong theoretical reasons for believing that the melting process can be continuous, with associated critical behavior, and as a result

---

*Work supported by the U.S. Department of Energy.

there has been an enormous growth of interest in the study of melting transitions in 2D.[1,2] An ideal 2D crystal is truly realized only in the case of a monolayer film, either freely suspended, or on a completely smooth surface. Thus it is also extremely difficult, if not impossible, to study experimentally, except by computer simulation studies. However, many experiments have been carried out on the melting of monolayers physisorbed on the (0001) faces of graphite[1-6] which does constitute a reasonably smooth and homogeneous substrate, particularly if the adsorbate has a periodicity which is incommensurate with the periodicity of the substrate. In this case, the net interaction between the monolayer and the substrate cancels out in the lowest order approximation, although there are still residual effects of the substrate on the melting of the monolayer (which have yet to be treated in detail from the theoretical point of view). In the case where the monolayer atoms are strongly confined to fixed adsorption sites on the substrate, so that we have a so-called "registered phase," the melting is of a different character, being a "lattice solid/lattice liquid" transition. There has also been a great deal of theoretical work done on these types of phase transition.[1,2] These can be second order, which is less surprising, since analogous second order phase transitions also exist in bulk systems. Experiments have also been carried out on the melting of freely suspended thin films of liquid crystals, although even in the case of a 4-layer film (the thinnest case studied so far), three dimensional (interlayer) interactions are observed to be present.[7]

2. CRYSTALLINE ORDER AND MELTING IN 2D

We first discuss 2-D systems where we may, to a first approximation, neglect substrate effects. Before discussing the melting phenomenon, it is worthwhile to discuss what a 2D crystal really is, in view of the fact that it is by now well known that an infinite 2D system cannot show structural long-range order.[8-10] The point is that long-wavelength phonon fluctuations cause the mean square particle displacement to diverge logarithmically with the size of the system. The structure factor $S(\vec{K})$ for the system may be written

$$S(\vec{K}) = \frac{1}{N}\sum_{ij} \exp[-i\vec{K}\cdot(\vec{R}_i-\vec{R}_j)]\ f_K\ (\vec{R}_i-\vec{R}_j) \tag{2.1}$$

where N is total number of atoms, $\vec{R}_i$ denotes the equilibrium position of the ith atom, and

$$f_{\vec{K}}\ (\vec{R}_i-\vec{R}_j) = \langle\exp[-i\vec{K}\cdot(\vec{u}_i-\vec{u}_j)]\rangle \tag{2.2}$$

where $\vec{u}_i$ denotes the displacement of the ith atom.

Even though the absolute mean square displacement becomes very large, the relative displacement between neighboring atoms does not, so that it is still legitimate to invoke harmonic phonon theory. In this case, we may write

$$f_{\vec{K}}(\vec{R}_i-\vec{R}_j) = \exp[-1/2\langle\{\vec{K}\cdot(\vec{u}_i-\vec{u}_j)\}^2\rangle] \tag{2.3}$$

For a bulk crystal, this may be written as a product of $\exp[-2W]$, the Debye-Waller factor, and $\exp[-\langle(\vec{K}\cdot\vec{u}_i)(\vec{K}\cdot\vec{u}_j)\rangle]$, which is usually expanded to yield the Bragg, one-phonon, ... etc., contributions to $S(\vec{K})$. For an infinite crystal in 2D, however, we must proceed differently in the evaluation of Eq. (2.3) because of the long wavelength divergences. We obtain,[10]

$$f_{\vec{K}}(\vec{R}) = \exp\left\{-\eta_{\vec{K}}\int_0^{q_D}\frac{dx}{x}\,[1-J_o(xR)]\right\} \tag{2.4}$$

where q_D is the Debye cut-off wavevector, and

$$\eta_{\vec{K}} = \frac{k_B\,TK^2}{2\pi mc^2 n} = \frac{k_BT\,K^2(3\mu+\lambda)}{4\pi\mu(2\mu+\lambda)} \tag{2.5}$$

where m is the atomic mass, n the number of atoms/unit area, c the sound velocity and μ,λ are the Lame elastic constants. Values for $\eta_{\vec{K}}$ are typically in the range (.01 - 0.2). In deriving Eq. (2.4) we have used the Debye approximation, which is not too bad in view of the fact that it is the low frequency phonons which are dominant. For large R, ($q_DR\gg1$), the integral in Eq. (2.4) can be taken to infinity and we have

$$f_{\vec{K}}(R) = \exp[-\eta_{\vec{K}}\ln(\tfrac{\gamma}{2}q_DR)] = \left(\frac{2}{\gamma q_DR}\right)^{\eta_{\vec{K}}} \tag{2.6}$$

where $\gamma = 1.7810724...$ Substitution in Eq. (2.1) leads to the result (assuming η_K is slowly varying in the vicinity of a reciprocal lattice point),

$$S(\vec{K}) = \sum_{\vec{G}} 4\pi(\gamma/q_D)^{-\eta_G}\frac{\Gamma(1-\eta_G/2)}{\Gamma(\eta_G/2)}\,q^{-2+\eta_G} \tag{2.7}$$

where $\vec{G}$ is a vector of the 2D reciprocal lattice, and $\vec{q}\equiv\vec{K}-\vec{G}$. Thus, there are, for such a system, no delta functions due to the usual Bragg reflections. Rather, the "thermal diffuse scattering" due to the phonon fluctuations has taken over and resulted in the well known "power-law" singularities at the reciprocal lattice points, characteristic of a 2D crystal. For such a system, the crystalline order parameter $\langle\rho(\vec{G})\rangle$ (where $\rho(\vec{K})$ is the Fourier transform of the atomic density) is strictly always zero and the system behaves like a disordered system at the critical point. While true long-range order is absent, neither can the degree of order be termed short-ranged! The question then arises as to the effect on the above result of finite size, since real systems (e.g., monolayers on substrates, or even systems numerically simulated on computers) always have finite linear dimensions characterized by a length L which can vary between (100 - 4000) Å. A finite system has a natural long-wavelength phonon cut-off ($q_L \sim \pi/L$) and thus the divergence of $\langle u^2\rangle$ is prevented and a finite value of the order parameter $\langle\rho(G)\rangle$ and associated Bragg reflections, etc., are restored. However, the effect on $S(\vec{K})$ is more subtle. To see this, let us replace $f_{\vec{K}}(R)$ from Eq. (2.4) by

$$f_{\vec{K}}(\vec{R}) = \exp\left\{-\eta_{\vec{K}} \int_{q_L}^{q_D} \frac{dx}{x} \left[1-J_o(xR)\right]\right\} \tag{2.8}$$

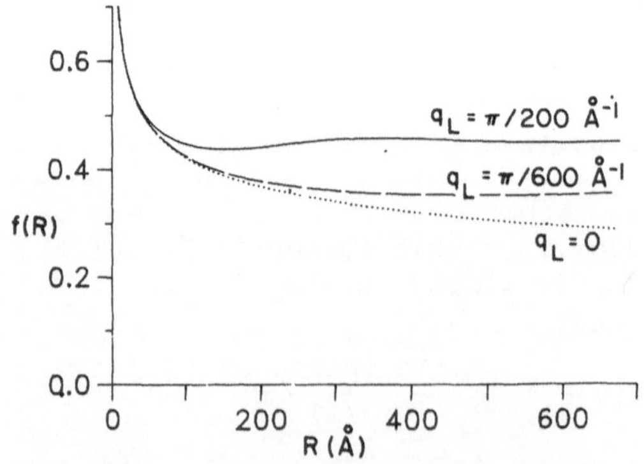

Fig. 1 The positional pair-correlation function for a 2D crystal, f(R) showing the effect of finite sizes for L = 200 Å, 600 Å and ∞. η is chosen as 0.2.

Unlike the expression given by Eq. (2.6), $f_{\vec{K}}(R)$ does not go to zero as $R\to\infty$ but rather $\to \frac{\pi}{q_D L} \eta_K$. Figure 1 shows the relative behavior of the correlation function $f_K(\vec{R})$ for crystallites of different sizes, with a chosen value of $\eta = 0.2$. Because of the finite size of the system, however, the $R\to\infty$ limit of $f_K(\vec{R})$ is not important in determining $S(\vec{K})$. Rather, it is only the behavior for $R \lesssim L$ which matters. For this case, we may write

$$f_{\vec{K}}(\vec{R}) = \left(\frac{2}{\gamma q_D R}\right)^{\eta_{\vec{K}}} \left\{1 + \eta_{\vec{K}} \int_0^{q_L} \frac{dx}{x} [1-J_0(xR)] + \cdots\right\}$$

$$= \left(\frac{2}{\gamma q_D R}\right)^{\eta_{\vec{K}}} \left[1 + \eta_{\vec{K}} \sum_{s=1}^{\infty} \frac{(-\pi^2R^2/4L^2)^s}{2s(s!)^2} + \cdots\right] \qquad (2.9)$$

where it turns out to be an excellent approximation for $R \lesssim L$ to retain only terms of order η.[11] We now return to Eq. (2.1) and remember that the sum i,j is now only over the finite set of atoms in the system. We make the approximation that in the vicinity of $\vec{Q} \simeq \vec{G}$,[12]

$$\frac{1}{N} \sum_{ij} \exp[-i\vec{Q}\,(\vec{R}_i-\vec{R}_j)] \simeq N \exp[-q^2L^2/4\pi] \qquad (2.10)$$

where $\vec{q} = \vec{Q} - \vec{G}$. This is the structure factor of a perfect but finite 2D lattice, and in principle has oscillations due to the finite cut-off. The Gaussian approximation represented by Eq. (2.10) above smears out these oscillations, which is more realistic, considering the statistical randomness of the shapes of the finite crystallites. We may now use the convolution theorem for Fourier transforms to rewrite the sum (2.1) in terms of a real-space integral. The result is that in the vicinity of $\vec{K} \simeq \vec{G}$,

$$S(\vec{K}) = 2\pi\varepsilon/a^2 \int_0^{\infty} dR\, R\, f_{\vec{G}}(R) J_0(qR) \exp(-\pi R^2/L^2) \qquad (2.11)$$

where a is the lattice constant of the 2D crystal and ε is a factor of order unity which is dependent on the geometry of the lattice. Equation (2.11) enables us to write down an analytic expression for $S(\vec{K})$. We obtain

$$S(\vec{K}) = S_o \left[\Phi(1 - \frac{\eta}{2} ; 1 ; - q^2L^2/4\pi) + \eta \sum_{k=1}^{\infty} \frac{\Gamma(k+1-\eta/2)}{2k(k!)^2\Gamma(1-\eta/2)} \left(\frac{-\pi}{4}\right)^k \Phi(k+1-\eta/2; 1 ; - q^2L^2/4\pi) \right] \tag{2.12}$$

where

$$S_o = \varepsilon \left(\frac{L}{a} \right)^{2-\eta} \left(\frac{2}{\gamma\sqrt{\pi}}\right)^{\eta} \Gamma(1 - \frac{\eta}{2}) \tag{2.13}$$

and Φ is the degenerate hypergeometric function. It turns out that only the first term in Eq. (2.12) is important numerically. Figure 2 shows a logarithmic plot of $S(\vec{K})$ as a function of q, where it can be seen that at qL > 10, the power-law behavior for an infinite system [Eq. (2.7)] is obtained, while for small values of qL, the divergence at $\vec{K} = \vec{G}$ is removed and replaced by a function which saturates at a value of $L^{2-\eta}$. Thus, finite size effects will change the observed $S(\vec{K})$ predicted for infinite 2D systems noticeably only if the experimental wave vector resolution is significantly less than ~ (10/L).

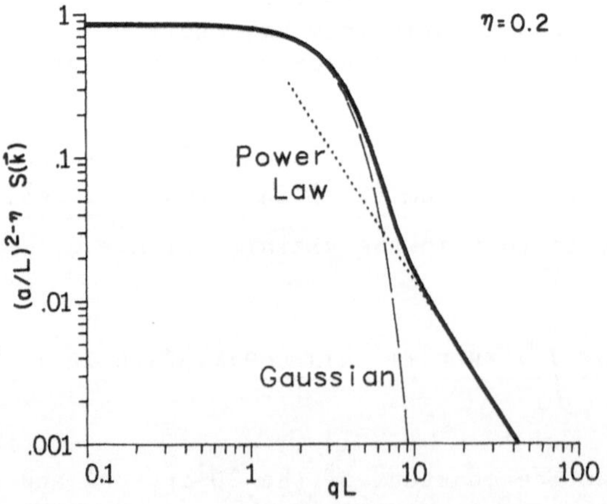

Fig. 2 "Universal curve" showing the behavior of the structure factor for any finite L, for a triangular lattice. (η = 0.2).

We have thus seen that a 2D crystal does indeed possess what might be termed marginal long-range order, with a characteristic signature in the corresponding $S(\vec{K})$. The way in which the transition to short-range order (melting) takes place has been discussed in a well known series of papers[13-15] and is also reviewed in another chapter in this book.[16] For our purposes we simply summarize the salient facts which are relevant to a comparison with experiment. According to the Kosterlitz-Thouless-Halperin-Nelson (KTHN) theory, there exist at finite temperatures a small number of bound dislocation-antidislocation pairs which unbind at a critical temperature T_m, producing free dislocations which can diffuse through the crystal destroying long-range order and also resistance to shear, i.e., one obtains a 2D liquid. This transition proceeds continuously, and is predicted to show no singularity in the heat capacity (the main entropy change taking place above T_m). However, just above T_m the 2D liquid is predicted to show marginal long-range bond-orientational order. For a close packed (triangular) 2D system, this phase has been termed the hexatic phase. In this phase, the spatial density correlation function $f(R)$ is predicted to change from the power-law form given by Eq. (2.6) to an exponential-type decay. We define a correlation function for bond-orientations $f_6(\vec{R}) = \langle \Psi_6(\vec{R})\ \Psi_6(0)\rangle$, where

$$\Psi_6(\vec{R}) = \frac{1}{N}\sum_i \delta(\vec{R}-\vec{R}_i)\left\{\frac{1}{6}\sum_{j=1}^{6} \exp(i6\theta_{ij})\right\} \tag{2.14}$$

where θ_{ij} is the angle the bond between atoms i,j makes with some reference axis. In the 2D crystal phase, $f_6(R)$ is constant [unlike $f(R)$] and in the hexatic phase switches to a power-law decay ($f_6(R) \sim R^{-\eta_6}$). What this means is that if a single 2D crystal were melted into the hexatic phase, and the coherence length of the crystal in a beam of X-rays or neutrons were finite, $S(\vec{K})$ above T_m should not be isotropic, but would still peak around the original crystal reciprocal lattice points. The predicted behavior in the vicinity of a particular $\vec{G}$ is

$$S(\vec{K}) \simeq C\ \xi^{2-\eta_G^*}/[(q^2\xi^2+1)]^{1-\eta_G^*/2} \tag{2.15}$$

where ξ is the correlation length, and η_G^* is the limiting value of η_G at T_m. C is a constant, which must be chosen so that in the limit $\xi\to\infty$, $S(\vec{K})$ goes into the crystalline power-law form given in Eq. (2.7). The form given in Eq. (2.15) is an approximation and rigorously correct only at $q = 0$. In practice, it can also be very well approximated by a Lorentzian in q, since η_G^* is fairly small. The correlation length ξ is predicted to diverge as $T \to T_m^+$ according to[14]

$$\xi = \xi_o \exp\left\{ b(T-T_m)^{-\bar{\nu}} \right\} \tag{2.16}$$

where b is a constant, and the exponent $\bar{\nu}$ generally has the value of 1/2, but for triangular lattices it is expected to be 0.369... .[14] Note that at q = 0, $S(\vec{K})$ diverges as a power of ξ, a characteristic property of scaling behavior for critical phenomena. Equation (2.15) has been calculated for an infinite crystal and it is not known how finite size effects modify $S(\vec{K})$ in this phase. However, on physically intuitive grounds one would expect the result to be valid for $\xi < L$. At still higher temperatures, a second transition to an isotropic liquid is predicted.[14] Halperin and Nelson state that the Kosterlitz-Thouless transition may be pre-empted by a <u>first-order</u> melting transition directly from a 2D crystal to a 2D isotropic liquid. The theory also predicts, at T_m a discontinuous drop from a value of 16π to 0 of the quantity

$$K(T) = \frac{1}{k_B T} a^2 \frac{\mu(\mu+\lambda)}{(2\mu+\lambda)} \tag{2.17}$$

Since there is no predicted singularity in the heat capacity, the experimental signatures of a Kosterlitz-Thouless melting transition are (a) the existence of a hexatic phase (b) the divergence of the correlation length ξ as $T \to T_m^+$ according to Eq. (2.16) and (c) the verification of a 16π discontinuity in K(T) at T_m. A "bump" in the heat capacity <u>above</u> T_m would be additional confirmation of the correctness of the theory.

3. EXPERIMENTAL 2D PHASE DIAGRAMS

We now turn to the experimental situation. The pioneering measurements of Thomy and Duval on rare gases adsorbed on exfoliated graphite[4] showed a progression of "steps" in the coverage vs. vapor pressure isotherms which demonstrated the building up of successive physisorbed adlayers and showed that the adsorption took place on a relatively smooth homogeneous substrate (the basal plane surface of graphite). Figure 3 shows a detailed vaporpressure isotherm for Kr on graphite in the submonolayer region, where more detailed features now become noticeable. A vertical region of the isotherm implies a region of coexistence between two phases (e.g., solid/vapor at the lowest coverages and temperatures or liquid/vapor at higher temperatures and coverages). A point of inflexion or a sudden change in slope without a vertical step would imply a change of phase without a coexistence region. Isotherm measurements such as those in Fig. 3 showed the existence of several phases occurring in the monolayers, many of which have been correctly assigned[4], but a detailed verification of the nature of the phases can only be carried out with a microscopic probe such as diffraction. This has by now been done for a larger number of systems.

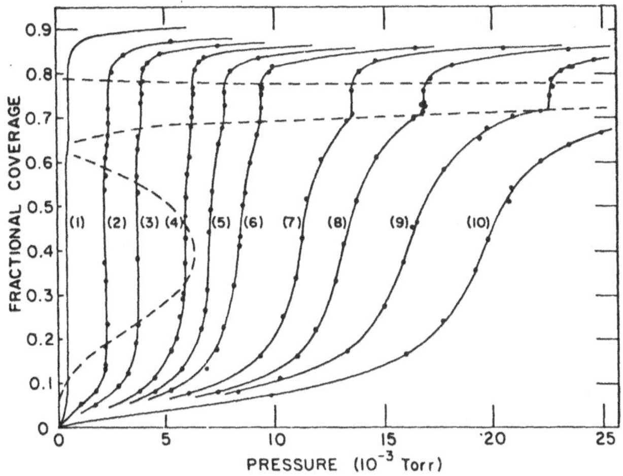

Fig. 3 Vapor pressure isotherms for Kr physisorbed on graphite, from Ref. 4.

Most diffraction (as well as heat capacity) studies have been carried out on a graphite substrate known commercially as grafoil,[3,19] which comes in the form of sheets. The c-axes of the graphite crystallites are partially oriented (with some broad distribution function) about the normal to the sheet while the azimuthal orientation about the c-axis is random. Substrate areas of 20-40 m^2/gm are obtained in this manner. Diffraction measurements of the width of the graphite Bragg reflections indicate crystallite sizes of ~ 400 Å laterally x 100 Å in the c-direction. Of course, the size of the graphite crystallite is simply an upper limit to the allowed size of the adsorbate regions since there may be other imperfections on the substrate surface which limit the latter. More recently, experiments have been carried out on a substrate known as "ZYX graphite" which has a higher degree of orientation and larger crystallite sizes but smaller specific surface area.

In principle, one needs to study the diffraction from the monolayer as a function of both coverage and temperature. For the purposes of mapping out the phase diagram, it is usually not necessary to go into the subtleties of the exact form of $S(\vec{K})$ discussed above, and it suffices to use a simpler form for the diffraction lineshape based on a theory by Warren.[12] This assumes perfect but finite 2D crystallites resulting in the Gaussian form for $S(\vec{K})$ given in Eq. (2.10), centered about the reciprocal lattice points. One must then average suitably over crystallite orientations and fold the resulting lineshape with the experimental resolution function of the diffractometer. (Details are given later in this article). The net result is an asymmetric-looking diffraction peak, as shown in Fig. 4(a). The position of the leading edge yields the magnitude of the reciprocal lattice vector G (and hence the periodicity of the adsorbate layer), and the width of the leading edge is related to the size of the crystallite (and hence a sudden broadening of the line usually indicates the onset of a fluid phase). A low density vapor phase will not show up in a diffraction pattern, so that coexistence of a solid and vapor phase, for example, can be inferred from a rapid and approximately linear decrease in the intensity of the 2D solid diffraction peak and a decrease in crystallite size with increasing temperature. Coexistence of a solid and liquid phase can be inferred from observation of a lineshape corresponding to the superposition of a sharp (solid-like) and a broad (liquid-like) peak [Fig. 4(b)]. Using such data (often in conjunction with vapor-pressure isotherm or heat capacity data, where available) one can obtain a coverage-temperature phase diagram for the adsorbate film. Figures 5(a) and (b) show such phase diagrams obtained for films of Kr and CD_4 adsorbed on graphite. We note the existence of 2D solid phases, both commensurate and incommensurate with the graphite substrate, as well as fluid phases. The differentiation between liquid and high-density vapor phases is a somewhat tricky point, where diffraction measurements are of little help. The existence of a <u>triple point</u> can be inferred from a singularity in heat-capacity measurements at a temperature which is independent of coverage over a certain range of coverages. This

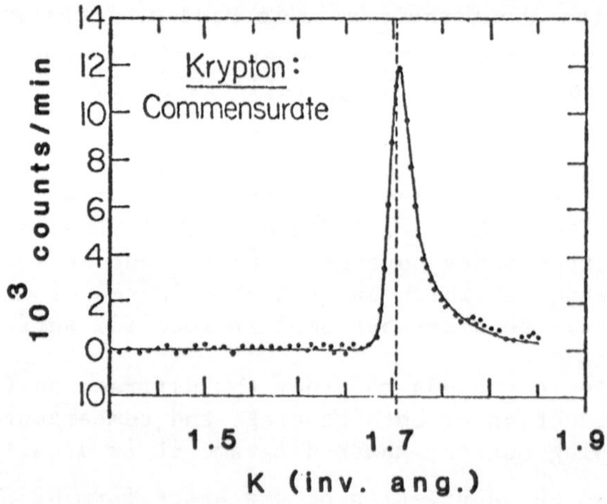

Fig. 4 (a) Diffraction lineshape for (10) peak of Kr on ZYX graphite in the registered phase, from Ref. 21. The smooth curve represents a fit using a Gaussian form for $S(\vec{K})$.

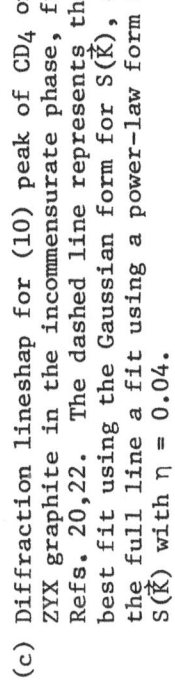

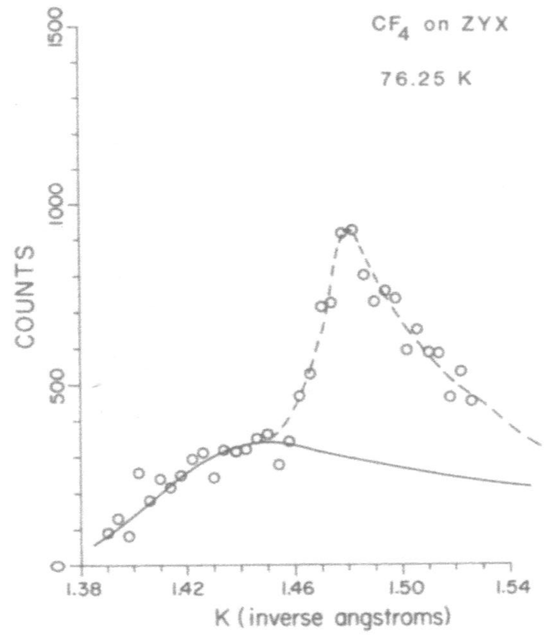

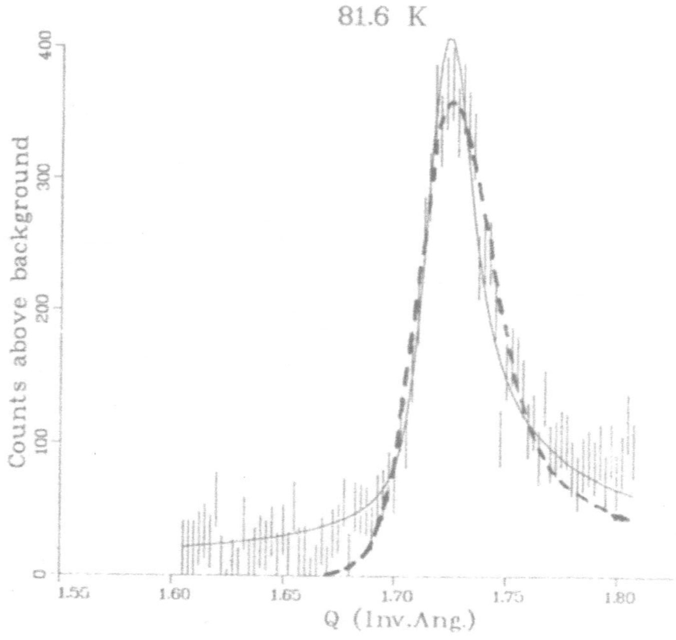

Fig. 4 (b) Diffraction lineshape for (10) peak of CF_4 on grafoil in the (2x2) solid/liquid coexistence region. (P. Dutta, S.K. Sinha and L. Passell, to be published.

(c) Diffraction lineshap for (10) peak of CD_4 on ZYX graphite in the incommensurate phase, from Refs. 20,22. The dashed line represents the best fit using the Gaussian form for $S(\vec{K})$, and the full line a fit using a power-law form for $S(\vec{K})$ with $\eta = 0.04$.

should then be followed by a higher-temperature liquid-vapor critical point at some coverage. However, this is not always observed. In the case of Kr, it is believed[17] that the liquid-vapor transition is preempted so that there is only one fluid phase. In the case of methane, inelastic neutron scattering measurements of the diffusion[18] as well as thermodynamic measurements[4] indicate both a liquid and a vapor phase in the fluid region. We also note that for fixed coverage, coexistence of two phases implies a first-order phase transition (as does the existence of a triple line such as is seen in CD_4/graphite - Fig. 5(b). As a function of temperature, one will traverse the coexistence region with a continuous change of proportion between one phase and the other and thus the phase transition will not look sharp as a function of temperature. On the other hand, the phase boundary between two pure phases may be characterized by a continuous phase transition and will thus paradoxically appear sharper as a function of temperature. Finally, it is worth pointing out that since most such measurements are made by introducing a fixed amount of gas into a sample container of fixed volume and then varying the temperature, a small correction for coverage must be made as the temperature is raised owing to the increase in the vapor pressure and the corresponding small desorption of the gas into the bulk vapor inside the sample container. As a result, the apparently "constant-coverage" scans actually follow the curved paths as indicated in Fig. 5(b).

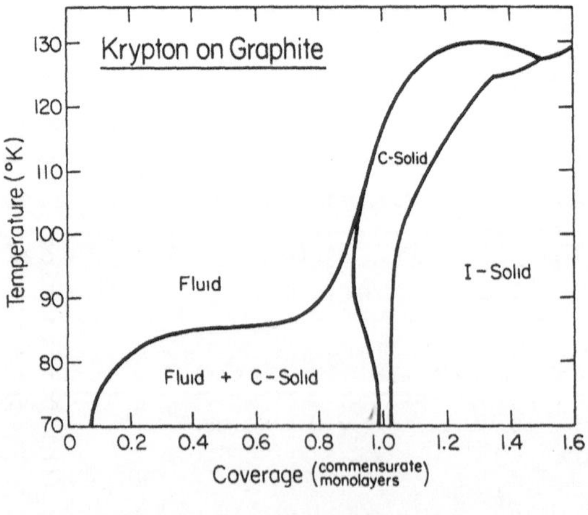

Fig. 5 (a) Phase diagram for Krypton on graphite. The c-solid denotes the $\sqrt{3}x\sqrt{3}$ R 30° structure. The I-solid denotes a compressed incommensurate phase. (Refs. 21, 17).

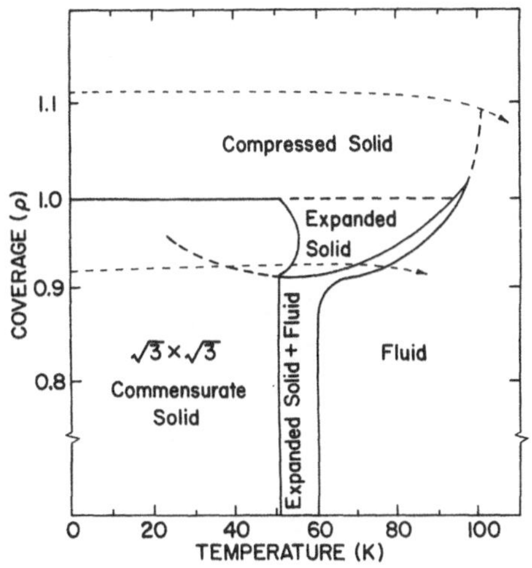

Fig. 5 (b) Phase diagram for CD_4 on graphite (Refs. 26, 20). The arrowed dashed lines represent scans for measurements of the melting process at two different coverages.

4. DIFFRACTION LINESHAPES IN THE 2D SOLID PHASE

We now turn to neutron and X-ray diffraction studies in the 2D solid phase and, in particular, what they reveal about S(K). (LEED studies of adsorbed monolayers have also been carried out. Some of these results will be discussed later). The commensurate or registered 2D solid found in many of these systems is the so-called $\sqrt{3} \times \sqrt{3}$ R 30° structure, where the adsorption sites are at the centers of the hexagonal carbon rings and only second-neighbor sites are occupied, as illustrated in Fig. 6. Increasing coverage causes the $\sqrt{3} \times \sqrt{3}$ structure to pack in to an incommensurate compressed phase, as observed in the case of N_2, H_2, Kr or CD_4, while in the case of CD_4 increasing the temperature causes a thermal expansion to an incommensurate expanded phase. Some systems are always incommensurate on graphite, e.g., Xe (expanded) or Ar (compressed). Most of the analysis has been carried out on the observed (10) diffraction peak of the triangular adsorbate lattice. (We shall not deal here with the special case of the behavior of the lineshape near commensurate-incommensurate transitions).

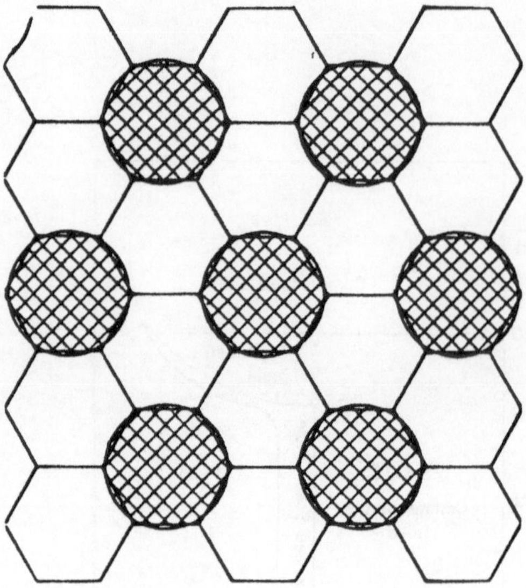

Fig. 6 The $\sqrt{3}\times\sqrt{3}$ R 30° structure on the basal plane of graphite.

A detailed lineshape analysis necessitates a more general treatment than the modified Warren theory conventionally used to analyze 2D diffraction data.[19] Such an analysis has been carried out[20] and the results are as follows: In terms of the true $S(\vec{K})$ we first define an azimuthally averaged structure factor $\bar{S}(K)$ (to take into account the 2D powder nature of the crystallites) given by

$$\bar{S}(K) = \frac{1}{2\pi}\int_0^{2\pi} d\psi \, S(K,\psi) \tag{4.1}$$

where ψ is the angle made by $\vec{K}$ with respect to an arbitrary reference direction in the plane. Averaging now over all orientations of the graphite planes and folding with the resolution function $R(\Delta K)$ of the diffractometer yields the following expression for the intensity $I(K)$ when the diffractometer is set for nominal wave vector transfer K:

$$I(K) = \int_0^{\infty} dK' \, F(K,K') \, K'\bar{S}(K') \tag{4.2}$$

where

$$F(K,K') = \int_0^{K'} dK'' \frac{R(K-K'')}{(K'')^2} (\sin \beta)^{-1} \int_{\beta}^{\pi-\beta} d\sigma P(\sigma) \left[\left(\frac{K'}{K''}\right)^2 - \cos^2\sigma\right]^{-1/2} \tag{4.3}$$

where $\beta = \sin^{-1}\left[1 - \left(\frac{K'}{K''}\right)^2\right]^{1/2}$, and $P(\sigma)$ is the distribution function of the normals of the crystallite planes about their mean directions. $F(K,K')$ is a function which depends only on the substrate and diffractometer used, so that the observed $I(K)$ can be fitted with an assumed functional form of $S(K)$, using Eq. (4.2).

Figure 4(a) shows a (10) diffraction peak from the $\sqrt{3} \times \sqrt{3}$ R 30° phase of Kr adsorbed on ZYX graphite obtained by X-ray diffraction.[21] The fit represents the best fit of an assumed Gaussian form for $S(\vec{K})$ as given by Eq. (2.10). The goodness of the fit reveals that this commensurate phase possesses true long-range order within a given, crystallite size of $\geq$ 600 Å. Figure 4(c) on the other hand shows a (10) diffraction peak from CD_4 adsorbed on ZYX graphite in the incommensurate (compressed) phase, obtained by neutron diffraction.[22] The dashed line represents the best fit possible with a Gaussian $S(\vec{K})$ and as can be seen, it fails to account for the pronounced "tails" seen in the wings of the peak. These tails are believed to represent the fluctuation scattering characteristic of 2D crystals. The full line represents the best fit of a power-law $S(\vec{K})$ [Eq. (2.7)] and accounts quite well for the peak shape. Similar "tails" are seen in the diffraction lineshape of incommensurate Xe on graphite.[21]

Figure 7 shows the values of η vs. temperature for several coverages obtained in this fashion for the ZYX substrate. The data presented here covers only the solid phase. From the discussion of finite size effects in Section 2, it is obvious that finite-size corrections to η will be insignificant in the case of ZYX graphite ($L \geq 600$Å), but may be present (although small) for the grafoil data ($L \sim 200$Å). However, if there are other unknown contributions to the background (e.g., defect-induced diffuse scattering, etc.) which are appreciable, then the values of η could be affected. It is probably best to think of Fig. 7 as representing qualitative trends in the η for the (10) peak of a 2D triangular crystal rather than quantitative values. The rather rapid increase of η at temperatures near the melting transition may be a manifestation of the renormalization of the elastic moduli near melting.[14] Also of interest is the fact that the value of η obtained at the highest

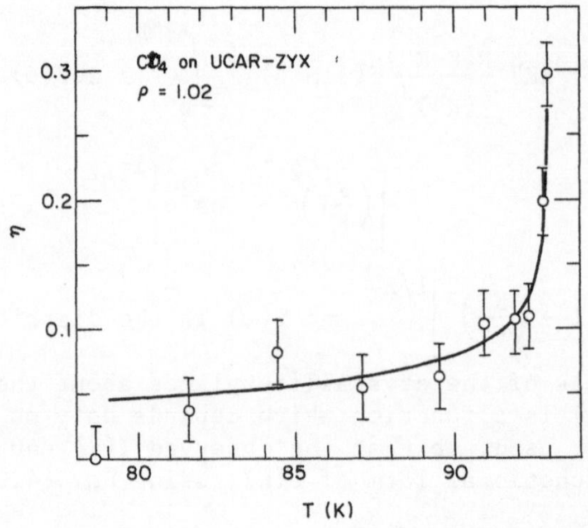

Fig. 7 η versus T for the (10) diffraction line in the incommensurate solid phases of CD_4 on ZYX graphite.

solid temperatures is roughly 0.3. (For the highest coverage, however, it is smaller). In the case where the elastic modulus $\lambda \gg \mu$, so that shear waves provide the main contribution to both η and the value of $K(T_m)$ as given by Eqs. (2.5) and (2.17), it can be easily shown that the limiting value of η for the (10) peak is 1/3, according to the KTHN theory. The effects of a finite λ would be to somewhat reduce from the above value. In this sense the theory is at least consistent with the observations. Also shown in Fig. 7 is a fit to $\eta(T)$ of the form

$$\eta(T) = \frac{T}{3T_m + B(T-T_m)^{0.3696}} \tag{4.4}$$

However, there is a puzzling discrepancy with theory, in the sense that the overall amplitude factor for the power-law form of $S(\vec{K})$ is found to vary with temperature, decreasing as the melting temperature is approached. This may be due to the presence of significant defect concentrations.

Power-law lineshapes have also been observed by Moncton and Pindak[7] in the case of thin 4-layer freely suspended films of the liquid crystal 40.8 in the smectic B phase using very high resolution synchrotron X-ray diffraction data. In this case, one does not have a distribution of orientations of the planes, so that both wings of the diffraction line shape can be clearly distinguished, although domain-reorientation in the plane of the film still requires 2D powder averaging azimuthally. Figure 8 shows an example of the kind of fit obtained, and the resulting values of η are also of the order 0.1-0.2.

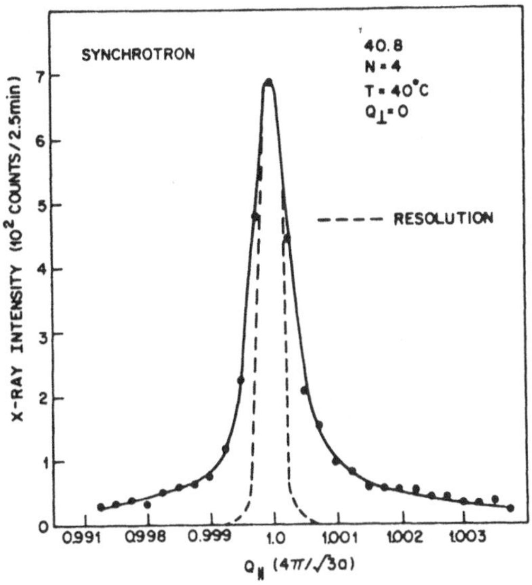

Fig. 8 Synchrotron data obtained on a 4-layer liquid crystal (40.8) in the smectic-B phase. The solid line represents a fit using the power-law form for $S(\vec{K})$. (From Ref. 7).

Thus, in the case of incommensurate adsorbed films or freely suspended films, it appears that the fluctuations do affect the structure factor $S(\vec{K})$ in a way that seems quite different to the corresponding situation in bulk crystals or commensurate monolayers. The effect of the periodic potential of the substrate in the case of adlayers or of interlayer coupling in the case of the suspended film have not been worked out in detail, however, and it is not yet known to what extent they affect the power-law lineshape.

An interesting case arises in the case of <u>commensurate</u> 2D solid phases which are <u>weakly</u> bound to the substrate. In this case while there are no low-frequency acoustic phonons which destroy long-range order, there may still be low enough frequency phonons to be apparent as a significant fluctuation contribution to $S(\vec{K})$. Calculations of this effect on $S(\vec{K})$ for varying strengths of the interaction with the substrate have been carried out by Cleveland et al.[23] We present here a simplified version based on a "gapped" Debye spectrum which has the advantage of being representable analytically. We represent the acoustic spectrum $\omega(q)$ by $(c^2q^2 + \Delta^2)^{1/2}$ instead of cq as before, where Δ is a frequency gap induced by the substrate. Then it may be shown that Eq. (2.4) is replaced by

$$f_K(R) = \exp\left\{-\eta_K \int_0^{q_D} \frac{x\,dx}{x^2+x_0^2}\,[1-J_0(xR)]\right\} \tag{4.5}$$

where $x_0 = \Delta/c$ and η is as before given by Eq. (2.5). The integral may be evaluated to yield

$$f_K(R) = (1+\omega_D^2/\Delta^2)^{-\eta_K/2} \exp[\eta_K\, K_0(x_0R)] \tag{4.6}$$

where ω_D is the Debye cut-off frequency. This produces a correlation function almost identical to that shown in Fig. 1 for the case of the finite incommensurate crystal. For all relevant R values it turns out to be valid to approximate Eq. (4.6) by

$$f_K(R) \simeq [1+\omega_D^2/\Delta^2]^{-\eta_K/2} [1+\eta_K\, K_0(xR)] \tag{4.7}$$

which then yields for $\vec{K} \simeq \vec{G}$,

$$S(\vec{K}) = N\,[1+\omega_D^2/\Delta^2]^{-\eta/2}\,[\exp(-q^2L^2/4\pi) + (2\pi\eta/L^2)(q^2+x_0^2)^{-1}] \tag{4.8}$$

where $\vec{q} = |\vec{K} - \vec{G}|$ as before, and we have assumed the size of the crystal $L >> c/\Delta$. Thus $S(\vec{K})$ is the sum of a Gaussian and a Lorentzian lineshape. The relative amplitudes of the two can be used to determine η in the commensurate phase, as a function of temperature. The width of the Lorentzian can be used to obtain the parameter Δ/c and should be relatively temperature-independent. Finally, the prediction is that the overall amplitude factor scale according to $[1 + \omega_D^2/\Delta^2]^{-\eta/2} \simeq [1 + q_D^2/x_0^2]^{-\eta/2}$ with temperature. CD_4 on graphite at a coverage of $\rho = 0.92$ is in the $\sqrt{3} \times \sqrt{3}$ R 30° commensurate phase for $T < 50K$. However, the fact that it expands out of registry relatively easy implies that the substrate binding is weak in this case, so that the above theory may in fact be applied. In fact, weak "tails" are also observed in this phase around the (10) diffraction peak.[22] The resulting fits using Eq. (4.8) yield a fairly temperature-independent value for (c/Δ) of ∿ 6.2 Å, values of $\eta \sim 0.006T$, and roughly the predicted temperature dependence of the overall scale factor.

5. INCOMMENSURATE MELTING

We now turn to diffraction studies of the melting transition for the incommensurate phases, where one expects the KTHN theory to be applicable. A neutron diffraction study of the melting of argon on grafoil, a system which is always in the incommensurate compressed phase, was carried out by Taub et al.[24] However, since no detailed lineshape analysis was undertaken (the conventional Warren lineshape theory was used for the liquid and solid phases) the results may be regarded as qualitative. A similar study of the melting of incommensurate submonolayer ^{3}He films on grafoil has recently been carried out by Lauter et al.[15] CD_4 on grafoil always melts from an incommensurate (IC) (expanded or compressed) phase [see Fig. 5(b)] and accordingly a neutron diffraction study of the change of the (10) diffraction lineshape was undertaken at two different coverages. In the first ($\rho = 0.92$), the monolayer melts from an expanded IC phase but via a region of solid-liquid coexistence[26] so that the melting is really first-order. Nevertheless, it is possible to look for critical behavior in the fluid phase at temperatures above T_m. In the second coverage ($\rho = 1.09$), the monolayer melts directly from a compressed IC phase to a liquid phase with no (or a very narrow) coexistence region. Thus it might be called a continuous melting process, although this point will be discussed in more detail below. Figure 9 shows the progressive broadening of the lineshape for $\rho = 1.09$, as a function of temperature. The lineshapes in the fluid phase were fitted using the Lorentzian form of $S(\vec{K})$ given by Eq. (2.15). [In these fits the ($\eta_G^*/2$) parameter was ignored in the exponent since its effect is very slight]. This still has to be azimuthally averaged over all crystallite orientations, as in Eq. (4.1). If the hexatic phase really exists, this would be the rigorous way to compare the theoretical Halperin-Nelson $S(\vec{K})$ with experiment. On the other hand, for a truly isotropic fluid, Eq. (2.15) is not valid (since G's do not exist any longer) but ξ can still be taken as some kind of measure of the correlation length in the fluid. It turns out that such Lorentzian structure factors yield a very good fit to the observed peaks in the fluid phase (and even down to temperatures where it has clearly become a solid, since a Lorentzian with a very large ξ cannot be distinguished experimentally from the power law used in the solid phase). In fact, the exact transition temperature from liquid to solid (or to solid/liquid mixtures) is very hard to determine experimentally. Figures 10(a) and (b) show the results for the correlation length ξ vs. T for the two coverages studied. For the high coverage it may be seen that there is a sharp rise as the temperature is decreased, followed by a saturated value of ξ. Of course, ξ will saturate when it reaches a magnitude equal to the crystallite size L. One would place T_m for the corresponding infinite crystal somewhere close to where this occurs. For the lower coverage, the pure Lorentzian structure factor fit is not reliable below 70 K as the lineshapes may be plausibly fitted with a sum of solid and liquid structure factors, with relative weights changing across the transition in such a way as to keep the total amount of material constant (i.e., the fitting procedure is consistent with a solid-liquid coexistence region). Thus again one has only values of ξ in the liquid phase for finite temperatures above the effective T_m.

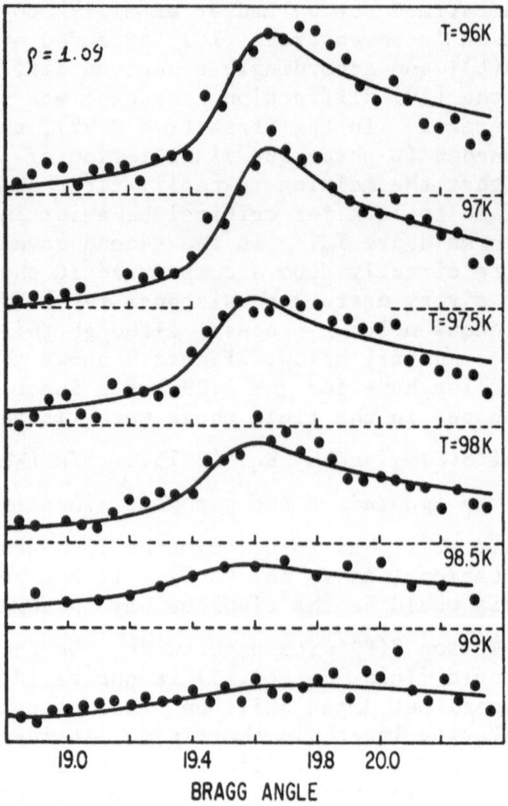

Fig. 9 Diffraction lineshapes for the (10) peak of CD_4 on grafoil at a coverage of ρ = 1.09 as a function of temperature. The full lines represent fits using the Lorentzian form for $S(\vec{K})$. (From Ref. 22).

Excluding the highest values of ξ for the above reasons, fits of the Halperin-Nelson exponential power law for $\xi(T)$ (Eq. 2.16) were done for the above data, using ξ_o, b and T_m as adjustable parameters, and the results are shown in Fig. 10. Also shown are straight power law fits of the form $\xi \sim A(T-T_m)^{-\nu}$. It may be seen that owing to the relative small number of points and the absence of any knowledge of T_m, it is not possible to distinguish between the models and to demonstrate the validity of Eq. (2.16). However, it is clear that ξ does show "precursor" behavior (even for the lower coverage where a first order transition occurs) in the liquid phase that is unknown in bulk liquids. Since the amplitude A of the fitted Lorentzians in the fluid phase was an independent fit-

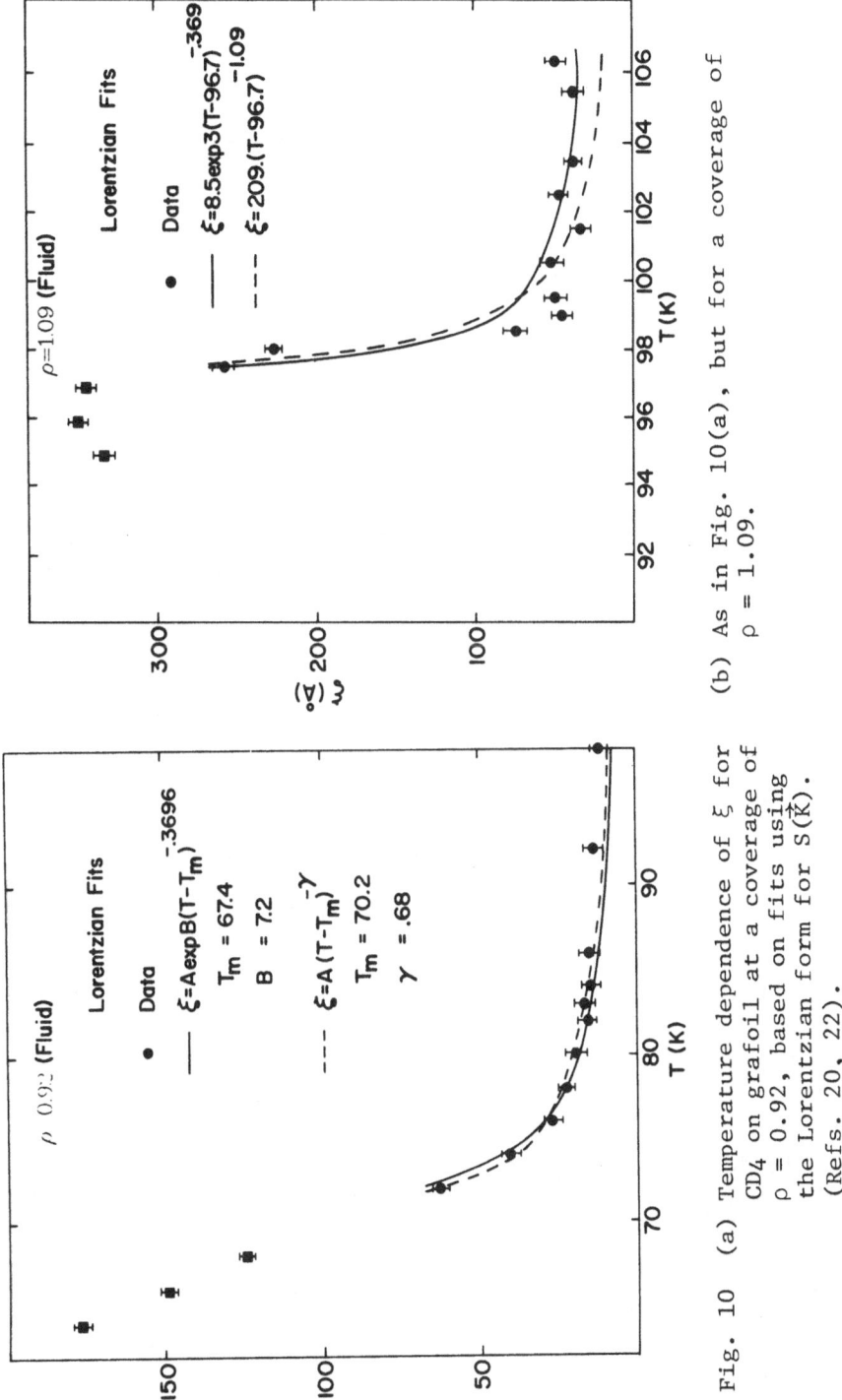

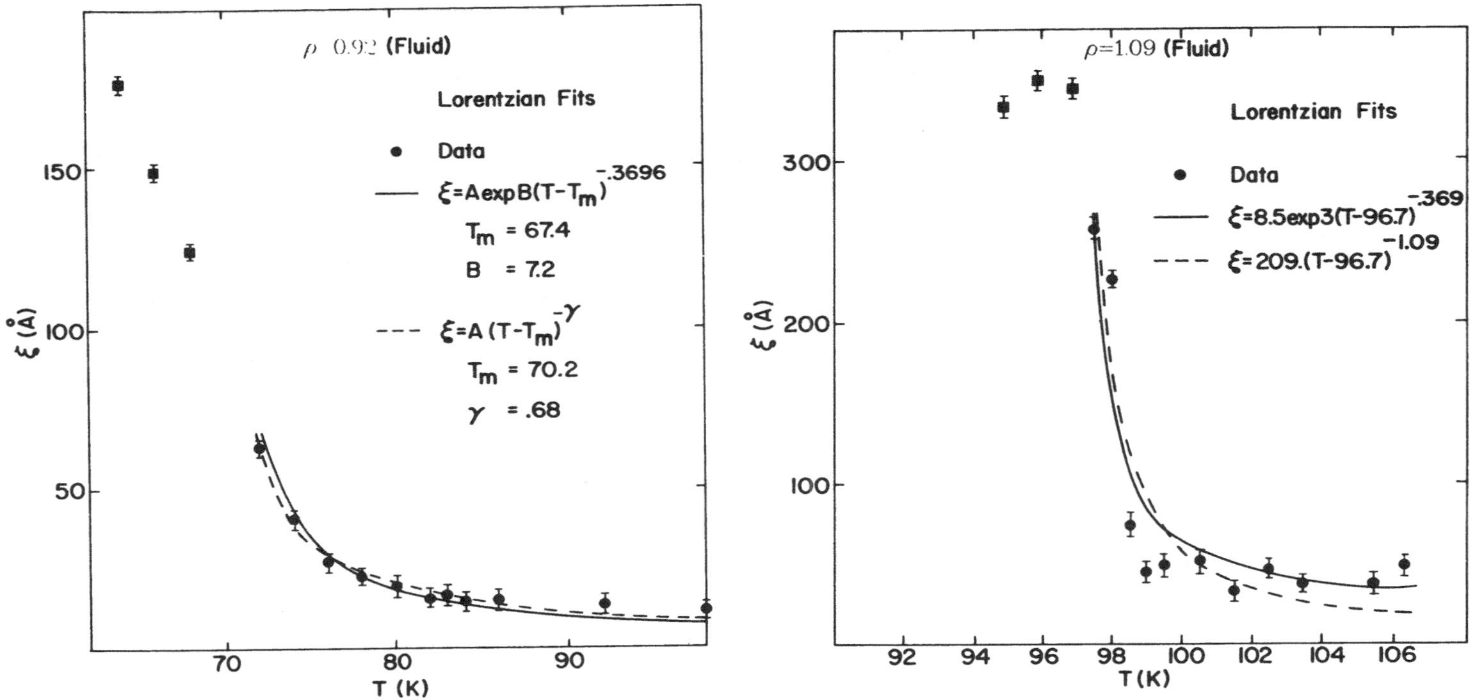

Fig. 10 (a) Temperature dependence of ξ for CD_4 on grafoil at a coverage of $\rho = 0.92$, based on fits using the Lorentzian form for $S(\vec{K})$. (Refs. 20, 22).

(b) As in Fig. 10(a), but for a coverage of $\rho = 1.09$.

ting parameter, it is of interest to test whether A scaled as the $(2-\eta_G^*)$ power of the correlation length ξ. Figure 11 shows a plot of ln A vs. ln ξ where it is seen that this law is indeed obeyed quite well. The exponent η_G^* obtained from these slopes is $\simeq 0$ for $\rho = 1.09$ but appreciable ($\simeq 0.32$) for $\rho = 0.92$, which is consistent qualitatively with the trends apparent from the fits of η to the solid lineshapes at these coverages (Fig. 7.).

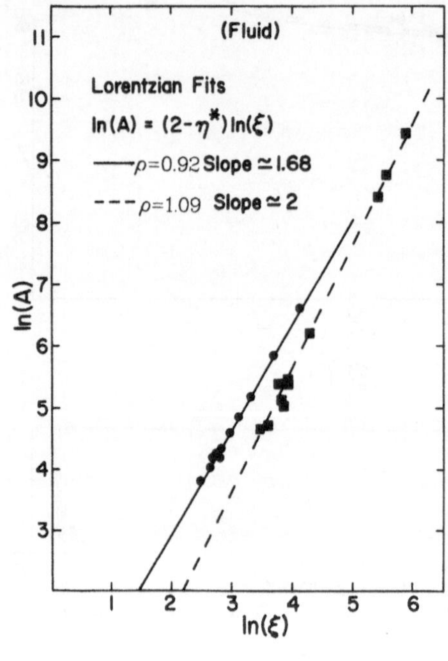

Fig. 11 Dependence of the amplitude factor A of the fitted Lorentzian form of $S(\vec{K})$ on u for CD_4 on grafoil at coverages of $\rho = 0.92, 1.09$. (References 20 and 22).

While for $\rho = 1.09$ no coexistence region for solid and liquid could be determined one cannot unambiguously state that the transition is of higher order, since there are, at this coverage, a significant number of molecules in the second layer.[26] These might adjust themselves so that the effective melting takes place at constant pressure rather than constant coverage, in which case the sharp nature of the transition must be interpreted as being first order.[27] It would be most desirable to have heat-capacity measurements for CD_4 in this region and such measurements are being undertaken.[28]

Measurements similar to those reported here have also been carried out in the case of the incommensurate melting of Xe on graphite using X-ray diffraction.[29] A detailed study of the melting of thin freely suspended liquid crystal films has not been reported, but in an exciting recent development Moncton and Pindak have reported[7] on X-ray diffraction measurements on a <u>thick</u> film of 65 OCB. They find the lineshape well fitted → in the smectic B phase by the Lorentzian form for S(K), i.e., short-range positional correlations, while also observing a sixfold azimuthal anisotropy, as shown in Fig. 12, strongly suggestive of a bulk stacking of hexatic layers in the film. To what extent the interlayer interactions stabilize this hexatic phase is unclear and, clearly, measurements are now required on very thin films of this material. Similar sixfold anisotropy in a positionally disordered liquid phase has been observed by Clarke[30] in CsC_{24}, but here the results have been interpreted in terms of the interaction of the 2D "liquid" of Cs atoms with the hexagonal graphite planes. Thus, at the time of writing, an unambiguous experimental verification of the hexatic phase in a 2D system has yet to be made. However, the KTHN theory predicts critical behavior which is at least qualitatively consistent with that observed around the melting temperature, even though the actual transition may be first order in some, if not most, cases.[31]

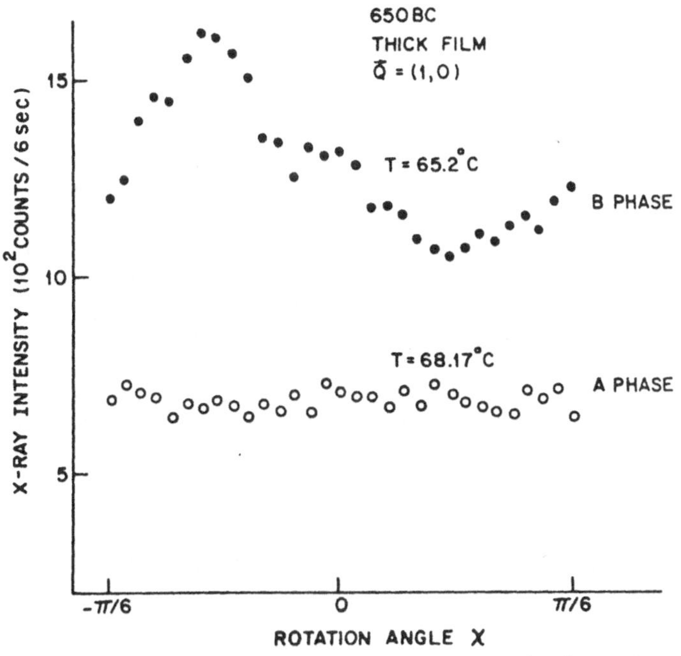

Fig. 12 X-ray diffraction intensity of the (10) peak of a thick film on the liquid crystal 65OBC as a function of azimuthal angle, showing the development of angular modulations in the smectic-B phase. (From Ref. 7).

Attempts have also been made recently to test experimentally the K-T predictions of the jump of the elastic modulus K at melting from 16 to 0 [Eq. (2.17]. Miyano and Tarczon[32] and Pindak et al.[33] performed dynamic torque measurements of the shear modulus of freely suspended liquid crystal films through the smectic B → A transition. No jump of this magnitude was observed, although the results are complicated by the formation of a thin solid layer even in the liquid phase.

Finally, although it is not our purpose to review such work in detail in this article, we discuss briefly the information available to us from computer simulations [Monte Carlo (MC) or Molecular Dynamics (MD)] of the melting of 2D crystals on a smooth substrate. In common with physical experiments, these suffer from the limitation to finite sizes (even though one may have periodically repeated boundary conditions), and in addition from the limitation to time-scales which may be short compared to equilibriation times in "real" systems. McTague et al.[34] have performed MC simulations on a system of 2500 particles interacting with an r^{-6} repulsive potential and find evidence of thermal activation of defects of the type predicted by Kosterlitz and Thouless, an algebraic decay of the positional pair-correlation function below T_m, and an exponential decay above. They also find K(T) (Eq. 2.17) to jump from roughly 16π to zero at T_m, and even evidence for a "hexatic liquid" phase in the sense that they obtain a $g_6(r)$ function which decays algebraically just above T_m. (However, the exponent 1.13, is much too large compared to the value (1/4) allowed by the Halperin-Nelson theory).[14] In the "hot solid," they also found evidence for significant numbers of more complicated defects than dislocation pairs, such as grain-boundary dislocation loops. Just above T_m, free dislocations were observed, as expected on the basis of the KT theory.[13] Tobochnik and Chester[35] in a 1024 particle MC simulation found agreement with the KT predictions for the discontinuity in K(T) at T_m at low densities, but not at high densities. Zollweg[36] in an MC simulation for a system of 1024 hard discs was unable to find evidence for a hexatic liquid phase above T_m. None of these authors was clearly able to establish the nature (first order or continuous) of the transition. On the other hand, Abraham[34] by performing a MC simulation of a 256 particle Lennard-Jones system at constant pressure, was able to observe discontinuities and hysteresis in both the density and the enthalpy across Tm, signalling that the actual transition was first order. MD simulations by Toxvaerd[34] and Kalia and Vashishta[38] also indicate a first-order transition with coexisting solid and liquid phases. The conclusions one may draw from the above studies seem to be that while the actual melting transition may be first order in 2D (at least for finite systems), near T_m at least some of the microscopic behavior predicted by the KT theory appears to be qualitatively correct.

6. COMMENSURATE MELTING

We now turn to studies of the melting of solid films which are commensurate with the substrate. Kr on ZYX graphite melts at coverages $\rho < 1.0$ from the $\sqrt{3}x\sqrt{3}$ R 30° phase into a lattice liquid. It is believed that this order-disorder transition may be well represented by the 3-state Potts model, since there are three sublattices which this phase can choose to occupy. Berker, Ostlund and Putnam[39] have performed an extensive series of real-space renormalization group (RNG) calculations on this system and shown that at low coverages the melting takes place via a solid-fluid coexistence region (i.e., is first-order) while at higher coverages (close to $\rho = 1.0$) one may in fact go from the solid to the fluid via a second-order phase transition. As seen from Fig. 5(a), this is indeed what is observed, and this melting process has been studied by heat capacity[17] techniques, which were in fact used to map out the phase diagram. The heat capacity results show no evidence for a liquid/gas coexistence or a critical point, but indicate instead a single fluid phase of rather rapidly varying density near an <u>incipient</u> critical point. Horn et al.[40] have carried out X-ray diffraction studies of the intensity of the (registered) (10) diffraction peak from Kr on graphite versus temperature for two different coverages, just below and just above the multicritical point separating the line of first-order from that of second-order transitions. In a commensurate 2D solid the crystalline order parameter

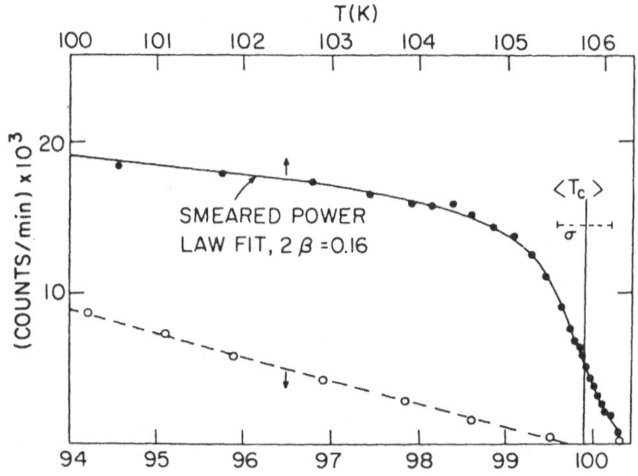

Fig. 13 Peak intensity of (10) diffraction line for Krypton on ZYX graphite at two different coverages, both in the registered $\sqrt{3}x\sqrt{3}$ phase. The dashed line is for a coverage just below the multicritical point, while the solid line is for a coverage just above it. (From Ref. 40).

is non-zero below T_m and so the critical exponent β exists. Figure 13 shows their results for intensity versus T. It is seen that at the lower coverage the behavior is almost linear and suggestive of a solid/liquid coexistence regime. On the other hand, at a slightly higher coverage, the behavior is rather different and more consistent with that expected from the predicted second-order transition. The solid line in Fig. 13 represents the fit of a power law of the form $(1 - T/T_m)^{2\beta}$ "smeared" with a Gaussian distribution of T_m's of width $\sigma = 0.004\ T_m$ to take into account finite size effects. The value of β they obtain from their measurements is $\beta = 0.09 \pm 0.03$. The RNG calculations[39] yield an exponent $\beta = 0.072$ at the multicritical point, and a value of $\beta = 0.105$ along the lambda line. However, the multicritical exponent is expected to dominate in the region where the measurements were performed. Ostlund and Berker have also performed a calculation of the order parameter as a function of temperature which explicitly introduces finite-size effects.[41] Their calculation somewhat justifies the smearing procedure used by Horn et al.,[40] but the agreement with experiment is qualitative.

Of considerable interest is the heat capacity exponent α associated with the melting of the $\sqrt{3}x\sqrt{3}$ structure. This has been measured in the case of ^{3}He and ^{4}He on graphite by Bretz[42] whose results are shown in Fig. 14. Heat capacity results for both grafoil and ZYX graphite substrates are shown and it may be seen that a considerable sharpening of the peak is obtained with the larger crystallite size. Bretz found $\alpha = 0.36 \pm 0.02$. Ostlund and Berker[36] have been able to reproduce this value using RNG theory, but only for a q-state Potts model where q = 20.8! This is probably a consequence of the approximations they had to make in applying their real-space RNG method. It is also possible, however, that quantum effects may change the nature of the transition in the case of He on graphite to that of a different universality class.

As opposed to the $\sqrt{3}x\sqrt{3}$ structure, the 2x2 structure (where only one of four sublattices can be occupied in a given region) is asserted[43] to represent a realization of the 4-state Potts model. This is the highest-state Potts model in 2D which is believed to possess a continuous phase transition and is therefore of some interest. CF_4 on graphite is the only known case of the occurrence of a 2x2 phase in a physisorbed system,[44] but second-order behavior has not been observed at the coverages and temperatures studied. Rather, clear evidence of solid/liquid coexistence has been obtained by neutron diffraction [See Fig. 4(b)]. On the other hand, several chemisorbed systems are known to have 2x2 phases, and a LEED study of the p(2x2) lattice-gas transition in the system O on the (111) surface of Ni has been carried out, in order to compare with the predictions of the 4-state Potts model.[45] A troublesome point in connection with LEED studies is the ever-present possibility of

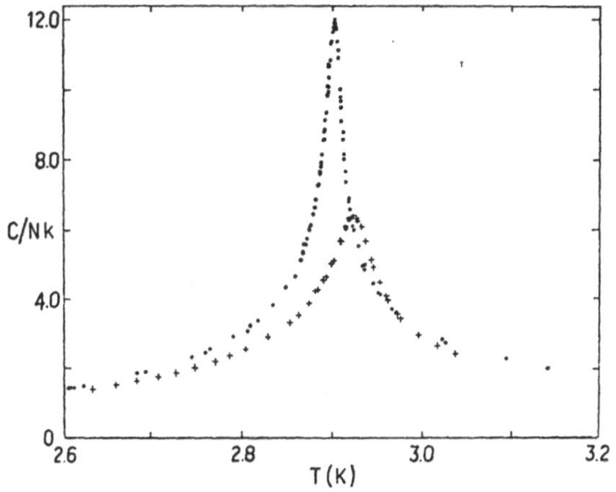

Fig. 14 Heat capacity for ordering peaks of He^4 on ZYX graphite (dots) and on grafoil (+). (From Ref. 42).

multiple-scattering effects complicating the measurements. However, the contention is[45] that these effects are not important within the adlayer but only between the adlayer and the substrate, so that order parameters and peak-widths are not significantly affected. Figure 15 shows the intensity of the (1/2,0) Bragg reflection corresponding to the p(2x2) lattice as a function of temperature at a coverage where there are no coexisting phases at any temperature. The intensities were extracted by fitting the observed peaks with an $S(\vec{K})$ which was the sum of a delta-function (representing the true Bragg reflection) and a Lorentzian (representing the critical scattering), folded with the instrumental resolution. (Since these

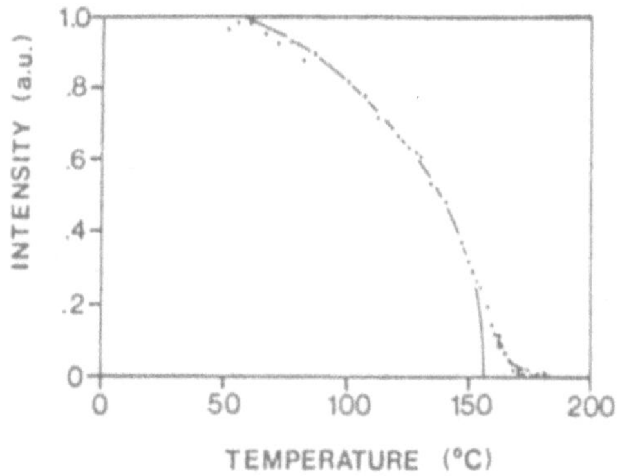

Fig. 15 Intensity of (1/2.0) Bragg reflection versus temperature for (2x2) phase of O/Ni(111). The full line is a fit of the form $(T_m-T)^{2\beta}$ with $T_m = 157°C$ and $\beta = 0.2$. (From Ref. 45).

measurements were done on a single crystal surface, no orientational averaging is necessary). The fitted power law of the form $(T_m-T)^{2\beta}$ yielded $\beta = 0.2 \pm 0.05$, while the 4-state Potts model prediction is 0.0833. A power-law fitted to the temperature variation of the correlation length yielded a critical exponent $\nu = 0.9 \pm 0.1$ while the theoretical prediction is 0.667. Finite size effects may be one source of the discrepancy since they were not allowed for in the fits. Nevertheless, the predictions for the 4-state Potts model have yet to be verified experimentally.

7. CONCLUSION

We have shown how a combination of neutron and X-ray diffraction, LEED, heat capacity and vapor pressure isotherm measurements have led to a greater understanding of the nature of the crystalline and fluid (or disordered) phases in thin films, and of the transitions between them. While the lack of true long-range order in a 2D crystal may be a somewhat academic problem,[31] the associated characteristic power-law behavior in S(K) arising from fluctuation scattering is clearly observed for incommensurate monolayers and freely suspended films. Critical behavior just above the melting transition is also observed and there is evidence for the existence of a hexatic phase in thick liquid crystal films. However, the actual 2D hexatic liquid phase and a continuous Kosterlitz-Thouless melting process cannot yet be said to have been experimentally verified. An interesting question we raise here relates to the question of the dynamics in the hexatic liquid phase. If bond-orientational order is to be preserved over large distances, it is clear that atoms cannot diffuse freely in this phase, rather only a few free dislocations can diffuse. On the other hand, at higher temperatures, true liquid-like diffusion can set in and has in fact been observed for the case of CH_4 on graphite by quasielastic neutron scattering.[18] A study of such diffusion through the melting transition would obviously be of great interest.

For commensurate phases, observation of lattice-gas order-disorder transitions are in reasonable accord with theoretical predictions for the order parameter in the case of the $\sqrt{3}x\sqrt{3}$ structure, but the specific heat exponent cannot be so easily explained. Measurements on melting of a (2x2) structure do not appear to agree with theoretical predictions.

I wish to acknowledge useful discussions and/or collaboration with P. Vora, P. Dutta, L. Passell, M. Nielsen and Y. Imry.

REFERENCES

1. cf. Ordering in Two Dimensions, Proc. Intl. Conf. at Lake Geneva, Wisconsin. S. K. Sinha, ed., North Holland (New York), Amsterdam, (1980), and references therein.
2. cf. Phase Transitions in Surface Films, J. G. Dash and J. Ruvalds, eds., Plenum (New York, London) (1980) and references therein.
3. J. G. Dash, Films on Solid Surfaces, Academic Press (New York) (1978).
4. cf. J. Physique, Colloq. C4, Suppl. No. 10, t. 38 (1977).
5. A. Thomy and X. Duval, J. Chim. Physique 67, 286 (1970).
6. A. Thomy and X. Duval, J. Chim. Physique 67, 1101 (1970).
7. D. E. Moncton and R. Pindak, Phys. Rev. Letters 43, 701 (1979); cf. also, Ref. 1, p. 83 (1980).
8. B. Jancovici, Phys. Rev. Letters 19, 20 (1967).
9. N. D. Mermin, Phys. Rev. 176, 250 (1968).
10. Y. Imry and L. Gunther, Phys. Rev. B3, 3939 (1971).
11. P. Dutta and S. K. Sinha, to be published.
12. B. E. Warren, Phys. Rev. 59, 693 (1941).
13. J. M. Kosterlitz and D. G. Thouless, J. Phys. C 6, 118 (1973).
14. B. I. Halperin and D. R. Nelson, Phys. Rev. B19, 2457 (1979).
15. A. P. Young, Phys. Rev. B19, 1855 (1979).
16. J. M Kosterlitz, these proceedings.
17. D. M. Butler, J. A. Litzinger, G. A. Stewart, and R. B. Griffiths, Phys. Rev. Letters 42, 1289 (1979); D. M. Butler, J. A. Litzinger and G. A. Stewart, Phys. Rev. Letters 44, 466 (1980).
18. J. P. Coulomb, M. Bienfait and P. Thorel, Phys. Rev. Letters 49, 733 (1979).
19. J. K. Kjems, L. Passell, H. Taub, J. G. Dash and A. D. Novaco, Phys. Rev. B13, 1446 (1976).
20. P. Dutta, S. K. Sinha, P. Vora, M. Nielsen, L. Passell and M. Bretz, Ref. 1, p. 169 (1980).
21. R. J. Birgeneau, E. M. Hammons, P. Heiney, P. W. Stephens and P. M. Horn, Ref. 1, p. 29.
22. S. K. Sinha, P. Dutta, P. Vora, M. Nielsen, L. Passell and M. Bretz, to be published.
23. C. L. Cleveland, C. S. Brown and U. Landman, Ref. 1, p. 207 (1980).
24. H. Taub, K. Carneiro, J. K. Kjems, L. Passell and J. P. McTague, Phys. Rev. B16, 4551 (1977).
25. H. J. Lauter, H. Mechert and R. Feile, Ref. 1, p. 291 (1980).
26. P. Vora, S. K. Sinha and R. K. Crawford, Phys. Rev. Letters 43, 704 (1979).
27. I am indebted to M. Nielsen for this suggestion.
28. M. W. Chan, private communication.
29. P. W. Stephens, P. M. Horn, R. J. Birgeneau and P. Heiney, to be published.
30. R. Clarke, N. Caswell, S. A. Solin and P. M. Horn, Phys. Rev. Letters 43, 2018 (1979). Also R. Clarke, Ref. 1, p. 53 (1980).

31. F. F. Abraham, Phys. Rev. Letters 44, 463 (1980); and in Ref. 1, p. 155.
32. K. Miyano and J. C. Tarczon, Ref. 1, p. 175 (1980).
33. R. Pindak, D. J. Bishop, D. D. Osheroff and W. O. Spranger, Ref. 1, p. 465 (1980), cf. also Phys. Rev. Letters 44, 1461 (1980).
34. J. P. McTague, D. Frankel and M. P. Allen, Ref. 1, p. 147 (1980); cf. also D. Frankel and J. P. McTague, Phys. Rev. Letters 42, 1632 (1979).
35. J. Tobochnik and G. V. Chester, Ref. 1, p. 339 (1980).
36. J. A. Zollweg, Ref. 1, p. 331 (1980).
37. S. Toxvaerd, Phys. Rev. Letters 44, 1002 (1980).
38. R. Kalia and P. D. Vashishta, to be published.
39. A. N. Berker, S. Ostlund and F. A. Putnam, Phys. Rev. B17, 3650 (1978).
40. P. M. Horn, R. J. Birgeneau, P. Heihey and E. M. Hammonds, Phys. Rev. Letters 41, 961 (1978).
41. S. Ostlund and A. N. Berker, J. Phys. C 12, 4961 (1979)
42. M. Bretz, Phys. Rev. Letters 38, 501 (1979).
43. E. Domany and M. Schick, Phys. Rev. B20, 3828 (1979).
44. H. J. Lauter, B. Croset, C. Mast and P. Thorel, Ref. 1, p. 211 (1980).
45. R. L. Park, T. L. Einstein, A. R. Kortan and L. D. Roelofs, Ref. 1, p. 19 (1980).

AN UNUSUAL POLYMORPHISM IN THE 2D MELTING :

THE SMECTIC F AND I PHASES

J. J. Benattar

Laboratoire Léon Brillouin
CEN-Saclay, 91191 Gif-sur-Yvette Cedex
France

Among the great variety of liquid crystals one can distinguish two great classes i) the nematic and cholesteric mesophases ii) the smectic phases.

This paper is devoted to the smectic mesophases and particularly to the two dimensional phases.

The smectic phases are all characterized by a periodic packing of molecular layers. In the S_G, S_H and the S_B modification (with untilted molecules) there is a strong coupling between adjacent layers which defines a three dimensional order that we shall examine in the next paragraph. In the S_C and the S_A modification(with untilted molecules),the layers are not correlated and within the layers, there is no translational order ; this constitutes a two dimensional liquid. Meanwhile even if the layer are weakly coupled, an orientational order exists since in the S_C phase, the tilt of the molecules is transmitted over a large number of layers.

For sufficient molecular length the S_F and S_I modifications[1] could appear in the phase diagram of several mesomorphic compounds between the S_G and S_C phases.One can easily understand that molecules with long chains favor the existence of a two-dimensional order since the van der Waals interactions between aromatic cores of adjacent layers are very weak.

The purpose of our work is to study by means of X-ray diffraction the change in order at the $S_G \rightarrow S_F$ and $S_F \rightarrow S_I$ transitions.

The synthesis of homologous compounds of TBBA with the general formula :

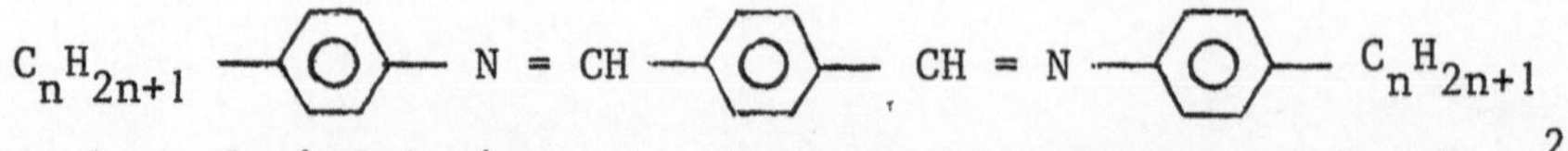
C_nH_{2n+1}
N = CH
CH = N
C_nH_{2n+1}

has been fruitful since the C_5 compound possesses a S_F phase[2,3] and the C_{10} has, in addition, the S_I modification between the S_F and the S_C phases. In this paper we present an investigation of the C_{10} compound which has a particularly rich polymorphism.

Calorimetric and optical measurements[4] show the following sequence :

$$C \xrightarrow{72°C} S_G \xrightarrow{112°C} S_F \xrightarrow{148°C} S_I \xrightarrow{156°C} S_C \xrightarrow{186°C} S_A \xrightarrow{189°C} L_i$$

ΔH : 11.74 Kcal/mole | <<1 cal/mole | ~1 cal/mole | 1340 cal/mole | 115 cal/mole | 1520 cal/mole

One can notice that the $S_G \leftrightarrow S_F$ transition seems to be of second order, the $S_F \leftrightarrow S_I$ transition is weakly of the first order, and the $S_I \leftrightarrow S_C$ is of the first order.

1. STRUCTURE OF THE S_G PHASE

The main characteristic of the S_G phase is that it possesses a three dimensional lattice[5,6] ; indeed the X-ray diffraction patterns of the S_G phase exhibit at small scattering angles the sharp 001 reflections indicative of a layered structure with molecules tilted relative to the layer normal. Further, at large angles six hko Bragg spots form a weakly distortered hexagon (fig.1) corresponding to pseudo-hexagonal packing within the layers. The hko Bragg spots are very sharp and, in addition the patterns also exhibit hkl Bragg spots with $\ell \neq 0$ which clearly establishes the 3d nature of the S_G phase. The presence of twelve diffuse spots located around the hko spots gives evidence of a local "herringbone peaking" of the molecular sections within the layers inside small domains (20 to 30 Å).

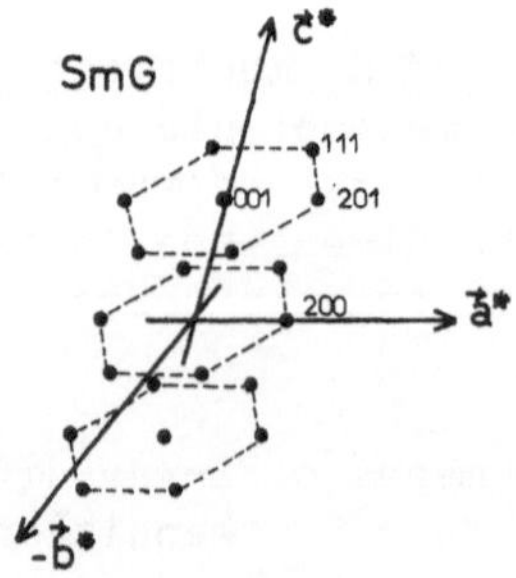

Fig.1. Reciprocal lattice of the S_G phase.

The 3-dimensional structure of the S_G phase can be described by a C-face centered monoclinic cell whose parameters for the C_{10} compound are determined from powder patterns. We find a = 9.69Å, b = 5.14 Å, c = 42.8 Å, β = 114.7°. (c is corresponding to the molecular length)

2. STRUCTURES OF THE 2d S_F AND S_I PHASES

a. S_F *phase*

When one examines the X-ray patterns[3], one notes the presence of the 001 spots at small angles which characterizes a tilted layered structure. At the same time the hkl Bragg spots vanished giving direct evidence of the 2d nature of this phase.

At large angles we find hk0 spots again (as in the S_G phase), forming a pseudohexagon. Those spots are now diffuse and extended along $\vec{c}^*$ which is perpendicular to the layers (fig. 2a). Such a pattern corresponds to a disorder of the smectic layers which is consistent with a 2d structure.

The S_F phase consists of a stacking of weakly coupled layers with a 2d order within the layers. One can define a 3d local cell which is monoclinic with centred C-faces with the following parameters at $T \sim 120°C$, $a = 9.64$ Å, $b = 5.22$ Å, $c = 41.7$ Å and $\beta = 112°$. (Within the layers the 2d cell is rectangular centred since the lattice of the triangular lattice is distorted).

b. *The* S_I *phase*

X-ray patterns for the S_I phase also exhibit 001 spots characteristic of a layered structure and hk0 spots distributed on an hexagon and corresponding to an hexagonal packing within the layers. As in the S_F phase there are no hkl spots ; the structure is also of the 2d kind.

The main and surprising difference from the S_F phase consists in the nature of the order within the layers. On the one hand the 2d reciprocal lattice is rotated by $\pi/6$ in its plane (fig.2a) which means in the real space (fig.2b) that the molecules are now tilted toward the smaller side of the rectangular cell, so the long molecular axis undergoes a reorientation of $\pi/6$. On the other hand, X-ray patterns show very sharp hk0 spots, which means that the order within the layers extends to a longer distance than in the S_F phase although, *the S_I phase appears at an higher temperature than the S_F phase*. If we continue to label b as the two-fold axis, then the cell parameters at $T \sim 152°C$ are $b = 9.00$ Å, $a = b/\sqrt{3}$, c (corresponding to the molecular length) $\simeq 41$ Å and $\beta = 111°$.

To compare the in-plane order in the S_F and S_I phases we recorded powder patterns with a Guinier camera using a focussed monochromatic beam ($\lambda = 1.79$ Å). This method is very well suited for such an analysis since on the one hand, it gives a very high resolution and on the other hand it eliminates the mosaic problem which exists in single domains[7].

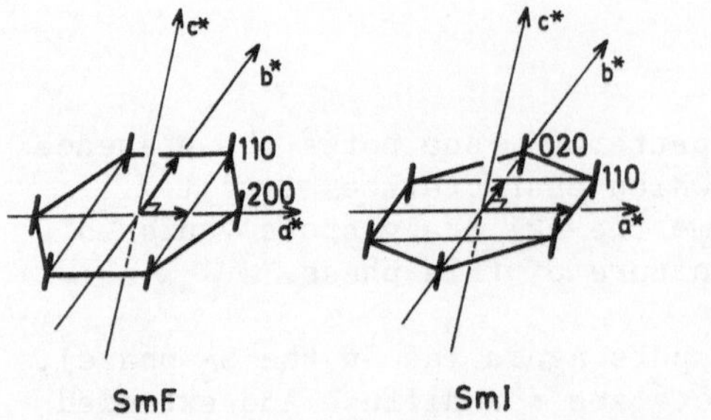

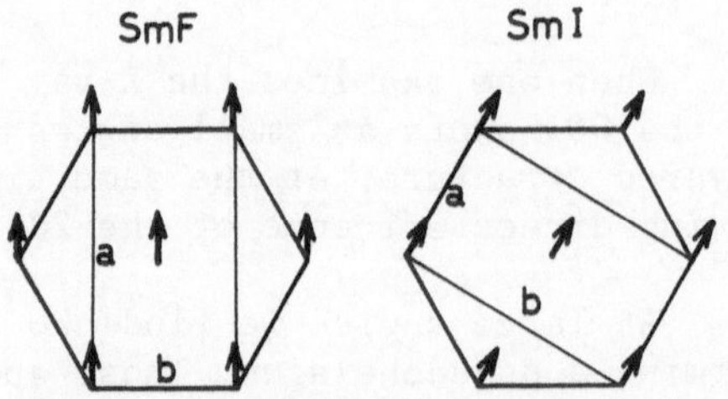

Fig.2a. Reciprocal lattice of the S_F and S_I phases.

Fig.2b. Local cells in the S_F and S_I phases (the arrows represent the molecular projections).

3. ANALYSIS OF THE 2d ORDER

In the S_F phase as well as in the S_I phase, the powder diagrams are composed of 001 rings characteristic of a layered structure and of the superposition of the hk0 reflection forming a diffuse ring rather large in the S_F phase and sharp in the S_I phase.

The nature of the order within the layers will be given by an analysis of the diffuse ring.

As we can see from fig.2a, the 2d reciprocal space consists of rods and the total structure factor $S(\vec{q})$ corresponds to the product of the molecular structure factor $g(q_y)$ along $\vec{c}^*$, by the in-plane $(\vec{a}^*,\vec{b}^*)$ structure factor $f(q_x,q_z)$ given by 2d melting theory.

In a power diagram all the sample orientations are statistically represented so one must integrate the structure factor S over a sphere whose radius is Q (Q being the scattering vector). For a set of reciprocal lattice vectors $\vec{G}$, one finds :

$$I_{\vec{G}}(Q) = \int d^3k\ S(\vec{k})\ \ \delta(|\vec{k}|^2 - |\vec{Q}|^2)$$

where $\vec{k} = \vec{q} + \vec{G}$ and $q_y \parallel c^*$, $q_x \parallel \vec{b}^*$, $q_z \parallel \vec{a}^*$

- The S_F phase : as we have shown for the C_5 compound[7], the diffuse ring of the C_{10} compound (corresponding to the in-plane order) can be analysed by a Lorentzian structure factor characterizing a short range positional order and $f_{\vec{G}}(q_x,q_z) \quad (\xi^2(q_x{}^2 + q_z{}^2) + 1)^{-1}$. At $T \simeq 127°C$ the fit requires a correlation length $\xi = 180 \pm 20$ Å (fig.3a).

- The S_I phase : the line shape of the ring on the S_I phase shows there has been an improvement of the 2d order relative to the S_F phase, although this phase is at a higher temperature than the S_F phase. Taking into account the new orientation of the tilt angle and the resolution function, we have tried to analyse the experimental line shape, as for the S_F phase with a Lorentzian .

This does not give good agreement with the experimental data. On the other hand a de Gennes-Sarma law[8] seems to fit the experimental line shape : $f(q_x,q_z) \propto (q_x^2 + q_z^2)^{(-2+\eta)/2}$ (fig.3b). At $T \sim 152°C$ we find $\eta = 0.24 \pm 0.02$. For the moment, no temperature dependence of η has yet been observed since the temperature range of the S_I phase of the C_{10} compound is only 8°C. The main fact that we can extract from this analysis is that there is a 2d quasi-long range order in the S_I phase very different from the S_F one.

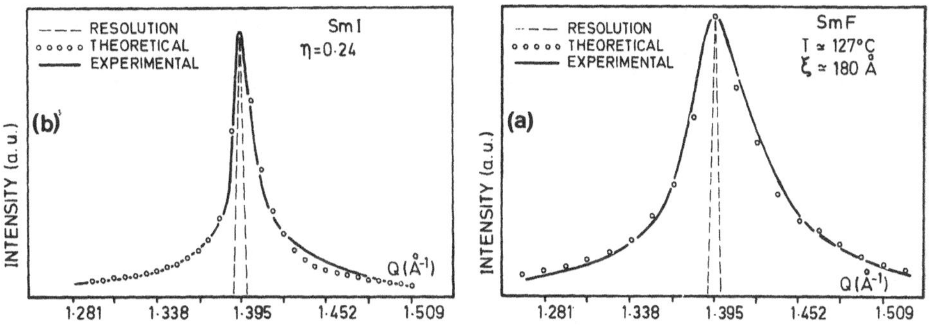

Fig.3. Line shape of the diffuse ring a) in the S_F phase b) in the S_I phase.

4. THE S_I AND S_F PHASES AND THE 2d MELTING THEORY

This is the first observation of an increase in the 2d order with increasing temperature, and no theory predicts such behavior. The theory of Halperin and Nelson[9], based upon ideas of Kosterlitz and Thouless[10], predicts that melting could occur in two steps and they point out that one must consider the successive loss of positional and orientational order :

a) In a 2d lattice true translational order can exist only at zero temperature. In such system, the positional correlation function is given by : $\langle e^{i\vec{G}(\vec{u}(\vec{r}) - \vec{u}(\vec{0}))} \rangle \simeq r^{-\eta_G(T)}$ and then the Bragg δ function of the 3d solid is replaced by $|\vec{Q} - \vec{G}|^{-2+\eta_G}$ for a set of the reciprocal lattice vectors $\vec{G}$. Here, there is a 2d long range orientational order.

b) Above a first critical temperature where there is an unbinding of dislocation pairs, the positional order decreases and the positional correlation function varies as $e^{-r/\xi}$ where ξ is the in-plane correlation length. This leads to a Lorentzian structure factor : $S(Q) \propto (\xi^2|\vec{Q} - \vec{G}|^2 + 1)^{-1}$. The orientational order is modified and follows an algebraic decay. This phase is called "hexatic".

c) In the second step of the 2d melting, there is an unbinding of disclinations pairs which leads to the isotropic phase where both orientational and positional order decay exponentially.

In conclusion, our X-ray study has established the existence of a 2d short range order in the S_F phase which could constitute an example of the hexatic phase (but in a bulk sample).

In the S_I phase, we have shown that a quasi long range order can exist although the S_I phase is at a higher temperature. Further this phase is not special of the C_{10} compound since it has been observed in other materials. The existence of such a transition in compounds[11] with long alphatic chains requires a theory of orientational melting within the planes, for 2d layered systems and an experimental study of the molecular motions.

1. S. Diele, D. Demus, H. Sackmann, Mol. Cryst. Liq. Lett. 56, 217, (1980).
2. A.J. Leadbetter, J.P. Gauchan, B. Kelly, G.W. Gray, J.W. Goodby, J. de Physique, Coll. 40, C3-178 (1979).
3. J.J. Benattar, J. Doucet, M. Lambert, Phys. Rev. A20, 2505 (1979).
4. F. Hardouin, G. Sigaud, Private Communication.
5. J. Doucet, A.M. Levelut, M. Lambert, Phys. Rev. Lett. 32, 301, (1974).
6. J.J. Benattar, A.M. Levelut, L. Liebert, F. Moussa, J. de Physique, Coll. 40, C3-115, (1979).
7. J.J. Benattar, F. Moussa, M. Lambert, J. de Physique 41, 1371 (1980).
8. P.G. de Gennes, G. Sarma, Phys. Lett. A 38, 219 (1972).
9. D.R. Nelson, B.I. Halperin, Phys. Rev. B19, 2457 (1979).
10. J.M. Kosterlitz, D.J. Thouless, J. de Physique Coll. 34, C6-1181 (1973).
11. J. Doucet, P. Keller, A.M. Levelut, P. Porquet, J. de Physique 39, 548 (1978).

PARTICIPANTS

Antoranz, J.C.	Dpto. de Fisica Fundamental - UNED Apartado Correos 50 487 Madrid-3, Spain
Ausloos, M.	Inst. de Physique B5, Université Liege, 4000 Liege Sart Tilman par Liege 1, Belgium
Azouni, M.A.	Lab. d'Aerothermique, 4 ter Route des Gardes, 92190 Meudon, France
Banai, N.	Université Libre de Bruxelles, Campus Plaine, Bd. du Triomphe, C.P. 231, Brussels, Belgium
Benattar, J.J.	Lab. Leon Brillouin, CEN Saclay, 91191 Gif-sur-Yvette - Cedex, France
Berre, B.	Fysisk Institutt, P.O.B. 19, 1432 Ås-NLH, Norway
Bilgram, J.H.	Lab. für Festkörperphysik, ETH, Hönggerberg, 8093 Zürich, Switzerland
Bohr, T.	Physics Lab. I, H.C. Ørsted Institute, Universitetsparken 5, 2100 Copenhagen, Denmark
Brenig, L.	Université Libre de Bruxelles, Campus Plaine, Bd. du Triomphe, C.P. 231, Brussels, Belgium
Bruce, A.D.	University of Edinburgh, Dept. of Physics, Mayfield Road, Edinburgh EH9 EJZ, U.K.

Brey, J.	Dept. of Physics, University of Sevilla, Sevilla, Spain
Büttiker, M.	IBM T.J. Watson Research Center, P.O.B. 218, Yorktown Heights, N.Y. 10598, USA
Büttner, H.	Phys. Institut, Universität Bayreuth, Postfach 3008, 8580 Bayreuth, W-Germany
Chalker, J.T.	Dept. of Theoretical Physics, 1 Keble Road, Oxford, OX1 3NP, U.K.
Croquette, V.	C.E.A. - DPh-G/PSRM - Orme des Merisiers, B.P.No. 2, 91190 Gif-Sur-Yvette, France
Cummins, H.Z.	The City College of New York, Dept. of Physics, 138 Street and Convent Ave., New York, N.Y. 10031, USA
Degiorgio, V.	Ist. di Fisica Applicata, Universita di Pavia, Via Bassi 6, 27100 Pavia, Italy
Dürig, U.	Lab. für Festkörperphysik, ETH Hönggerberg, 8093 Zürich, Switzerland
Feder, J.G.	Dept. pf Physics, University of Oslo, Blindern, Oslo 3, Norway
Fjær, E.	Inst. for almen fysikk, 7034 Trondheim-NTH, Norway
Fossum, J.O.	Inst. for almen fysikk, 7034 Trondheim-NTH, Norway
Giglio, M.	CISE Laboratories, P.O. Box 12081, Milan, Italy
Goldenfeld, N.D.	Theory of Condensed Matter, Cavendisch Laboratory, Madingley Road, Cambridge CB3 OHE, U.K.
Gorman, M.	University of Houston, Dept. of Physics, Huston, TX 77004,USA

Gray, K.E. — Bldg. 223, Argonne National Laboratory, Argonne, Ill. 60439, USA

Guazzelli, E. — L.H.M.P. ESPCI, 10 rue Vauquelin, 75231 Paris, Cedex, France

Hackenbracht, D. — Inst. für Theoretische Physik, Universität Frankfurt, Robert-Mayer Strasse 8, 6000 Frankfurt (Main), W-Germany

Hauge, E.H. — Institutt for teoretisk fysikk, 7034 Trondheim-NTH, Norway

Heldstab, J. — Inst. für Theoretische Physik, Universität Basel, Klingelbergerstrasse 82, 4056 Basel, Switzerland

Hess, W. — Fakultät für Physik, Universität Konstanz, Postfach 5560, 7750 Konstanz, W-Germany

Holden, T.M. — Neutron and Solid State Physics Branch, Atomic Energy of Canada Ltd., Chalk River, Ontario, Canada

Höck, K.H. — Ruhr Universität Bochum, Theoret. Physik III, Universitätsstr. 150, 4630 Bochum, W-Germany

Jensen, M.H. — Physics Laboratory I, University of Copenhagen, Universitetsparken 5, 2100 Copenhagen, Denmark

Jøssang, T. — Institute of Physics, University of Oslo, Blindern, Oslo 3, Norway

Kogon, H.S. — Dept. of Physics, University of Edinburgh, Mayfield Road, Edinburgh EH9 EJZ, U.K.

Kosterlitz, J.M. — Dept. of Mathematical Physics, University of Birmingham, Birmingham B15 2TT, U.K.

Kragler, R. — Fakultät für Physik, Universität Konstanz, Postfach 5560, 7750 Konstanz, W-Germany

Kristensen, J.K.	The Danish Welding Institute, Park Allé 345, 2600 Glostrup, Denmark
Kristensen, W.D.	The Danish Welding Institute, Park allé 345, 2600 Glostrup, Denmark
Landman, U.	School of Physics, Georgia Institute of Technology, Atlandta, Ga. 30332, USA
Landauer, R.W.	IBM, T.J. Watson Reserach Center, P.O.B. 218, Yorktown Heights, N.Y. 10598, USA
Langer, J.S.	Dept. of Physics, Carnegie-Mellon University, Schenley Park, Pittsburgh, Penn. 15213, USA
Leimar, O.	Dept. of Theoretical Physics, Royal Inst. of Technology, S-100 44 Stockholm 70, Sweden
Lekkerkerker, H.	Vrije Universiteit Brussel, Fakulteit Wetenschappen, Pleinlaan 2, 1050 Brussels, Belgium
Lemmens, L.F.	R.U.C.A., Groenenborgerlaan 171, 2020 Antwerpen, Belgium
Libchaber, A.	Groupe de Physique Solides de l'Ecole Normale Supérieure, 24 rue Lhomond, 75231 Paris, Cedex 05, France
Lyons, K.B.	Rm 1A119, Bell Laboratories, 600 Mountain Ave., Murray Hill, N.J. 07974, USA
Marques, M.	Dept. of Physics, Solid State Group, Imperial College, Prince Consort Rd., London SW7, England
Martin, P.C.	Division of Applied Sciences, Harvard University, 217 Pierce Hall, Cambridge, Mass. 02138, USA
Meissner, G.	Universität des Saarlandes, Theoret. Physik, 66 Saarbrücken, W-Germany

Michel, K. — Dept. Natuurkunde, Universiteit Antwerpen, Universiteitsplein 1, 2610 Wilrijk, Belgium

Müller, K.A. — IBM Research Laboratory, Säumerstrasse 4, 8803 Rüschlikon, Switzerland

Møller, H.B. — Forsøgsanlæg Risø, 4000 Roskilde, Denmark

Ono, I. — Dept. of Physics, Tokyo Institute of Technology, Oh-okayama, Meguro-ku, Tokyo 152, Japan

Otnes, K. — Institute for Energy Technology, P.O.B. 40, 2007 Kjeller, Norway

Palm, E. — Inst. of Mathematics, University of Oslo, Blindern, Oslo 3, Norway

Pelce, P. — Fac. des Sciences, Lab. Dynamique et Thermophysique des Fluides, St. Jerome, 13397 Marseille, Cedex 13, France

Pfister, G. — Inst. of Applied Physics, University of Kiel, Olshausenstr. 40-60, Physikzentrum Haus N 61a, 2300 Kiel, W-Germany

Pynn, R. — Institut Laue-Langevin, P.O.B. 156 Centre de Tri, 38042 Grenoble-Cedex, France

Ravndal, F. — Dept. of Physics, University of Oslo, Blindern, Oslo 3, Norway

Riste, T. — Institute for Energy Technology, P.O.B. 40, 2007 Kjeller, Norway

Santos, M. — Lab. de Fisica, Faculdade de Ciencias, 4000 Porto, Portugal

Satija, S.K. — Physics Department, 510B, Brookhaven National Laboratory, Upton, L.I., N.Y. 11973, USA

Sinha, S.K. — Div. Solid State Sciences, Bldg. 233, Argonne Natioanl Laboratory, Argonne, IL 60439, USA

Skjeltorp, A.	Institute for Energy Technology, P.O.B. 40, 2007 Kjeller, Norway
Starobinets, S.	Chemical Physics Dept., Weizmann Institute of Science, Rehovot, Israel 76100
Steiner, M.	Hahn-Meitner-Institut, Glienicker Str. 100, 1000 Berlin 39, W-Germany
Steinsvoll, O.	Institute for Energy Technology, P.O.B. 40, 2007 Kjeller, Norway
Stinchcombe, R.B.	Theoretical Physics Departement, 1 Keble Road, Oxford OX1 3NP, UK
Stokka, S.	Depratment of Physics, 7034 Trondheim-NTH, Norway
Thomas, H.	Dept. of Physics, University of Basel, 4056 Basel, Switzerland
Toledano, P.	Groupe de Physique Teorique, Faculté des Sciences, 33 rue St. Leu 80000 Amiens, France
Valkering, T.	Twente University of Technology, Dept. of Physics, P.O.B. 217, 7500 AE Enschede, The Netherlands
Vaschishta, P.	Argonne National Laboratory, Argonne,Ill. 60439, USA
Velarde, M.G.	Dpto. de Fisica Fundamental-UNED Apartado Correos 50 487 Madrid 3, Spain
Wesfreid, J.E.	Ecole Supérieure de Physique et de Chimie, 10 rue Vauquelin, 75231 Paris, Cedex 05, France
Wunderlin, A.	Inst. für Theoretische Physik, Universität Stuttgart, Pfaffenwaldring 57, 7000 Stuttgart 80, W-Germany

ORGANIZING COMMITTEE

Andersen, E. Jarrett, G. }	Institute for Energy Technology, P.O.B. 40, 2007 Kjeller, Norway

INDEX